Next-Generation Power Systems

Next-Generation Power Systems: Renewable Energy and Electric Vehicle Applications is a comprehensive resource that addresses emerging issues facing the power system industry in its drive toward more sustainable, reliable, and flexible power generation. It discusses technical, operational, and grid integration challenges and solutions for large-scale renewable energy and electric power transmission and generation, including issues related to system stability, reliability, efficiency, and capacity. Integrating foundational theory with realistic tools and applications, along with research-related author insights, this book provides an understanding of the different challenges that exist in operating electricity grids in a sustainable and smart manner. It includes specific content on identifying the overall structure of electricity markets, creating economic models to analyze the behavior of market stakeholders, improving system performance, and managing risk in power systems. It is an invaluable resource for students and researchers focused on energy science, while also providing useful instruction to practicing engineers and professionals in the mechanical or automotive engineering fields.

Features:

- Extensive Understanding: Offers each section of this book as a series of correlated, self-contained tutorials
- Application-Oriented Approach: Mixes basic and advanced analytical material with business logic, case studies, and real-life applications relevant to industry
- Focus on Disruptive Technologies: Discusses hot topics such as vehicle-to-grid (V2G), microgrids, smart charging, digitization, and AI-based energy management
- Power System Challenges Tackled: Addresses grid stability, power quality, protection, reliability, and standards in high renewable and EV penetration scenarios
- Academic and Industry Relevance: Is appropriate for undergraduate and graduate classes and is used as a reference book for researchers, engineers, and policymakers

Next-Generation Power Systems

Renewable Energy and Electric Vehicle Applications

Edited by

Natarajan Balasubramanian Muthu Selvan,
Venkatraman Thiyagarajan, and Cheng Siong Chin

CRC Press
Taylor & Francis Group
Boca Raton London New York

CRC Press is an imprint of the
Taylor & Francis Group, an **informa** business

Designed cover image: Shutterstock

First edition published 2027
by CRC Press
2385 NW Executive Center Drive, Suite 320, Boca Raton FL 33431

and by CRC Press
4 Park Square, Milton Park, Abingdon, Oxon, OX14 4RN

CRC Press is an imprint of Taylor & Francis Group, LLC

ISBN: 9781032950778 (hbk)
ISBN: 9781032955681 (pbk)
ISBN: 9781003585534 (ebk)

DOI: 10.1201/9781003585534

Typeset in Times
by codeMantra

Contents

Preface

The electric power industry has been going through a transformation motivated by the increasing demand for energy, accompanied by environmental issues and new technological development. Conventional centralized power systems are gradually transforming into sophisticated decentralized intelligent networks that integrate renewable generations, advanced power electronic converters, energy storage, and electric vehicles. The energy system is a complex, dynamic system and plays a critical role in all social systems. For the sustainable development of society and economy, the provision of clean energy for people worldwide has become an urgent issue today. In this sense, *Next-Generation Power Systems: Renewable Energy and Electric Vehicle Applications* was born as a comprehensive reference standalone book that presents the fundamental principles and future developments shaping the field of power systems today.

Integration of renewable energy sources (RES) in the form of solar photovoltaic (PV), wind, and hybrid RES on a large scale poses new challenges for operation, planning, and grid protection associated with intermittency, variability, grid stability, protection coordination, and power quality. Meanwhile, the fast-growing number of electric vehicles is reshaping transport and electricity load patterns, challenging generation supply and transmission/distribution grids. EVs are no longer sinking a load onto the grid; they are evolving into an active contributor in supporting and stabilizing the grid through smart charging solutions, vehicle-to-grid marvels, and energy management systems. The investigation of such interchanges is a crucial step to the design of robust and sustainable power systems.

Such aspects of power systems are presented in an innovative content ordering divided into four chapters: Advancements in Hybrid and Electric. Important features of advanced control, along with detailed explanations of how these systems apply to the microgrid environment, are also included in this comprehensive resource. Additionally, this book explores topics that are gaining more popularity with modernization and automation energy system applications, such as digital power systems, artificial intelligence-based energy management systems, demand response (DR) programs for residential and industrial customer-side applications, and cyber-physical security of smart grids.

One key characteristic of this book is its balanced orientation toward theoretical underpinnings and practical applications. Mathematical modeling, system analysis, and control methodology are introduced alongside extensive illustrative examples and case studies derived from practical applications. Such an approach bridges the gap between theoretical principles and their applications to real-world renewable systems and electric vehicle charging infrastructures. It also integrates industry trends with regulation and sustainability to create a complete picture wherever relevant.

This book is addressed to undergraduate and postgraduate students in electrical and electronics engineering and power engineering, as well as researchers, practicing engineers, and energy professionals involved in electric vehicle charging scenarios. It could also serve as a useful resource for policy and planning practitioners engaged in projects related to the energy transition. Our aim is that this book will provide readers with an informed perspective to help address the technical, economic, and environmental issues related to modern power systems, as well as stimulate and drive innovative solutions that contribute to the evolution of clean, efficient, and resilient next-generation energy infrastructures.

Brief Description

Next-Generation Power Systems: Renewable Energy and Electric Vehicle Applications delves into the critical and evolving landscape of modern power systems, driven by the integration of renewable energy sources and the rise of electric vehicles (EVs). This book is an insightful compendium for engineers, researchers, policymakers, and students who are keen on understanding the challenges and advancements in transitioning toward a sustainable energy future.

This book opens with an introduction to the current energy landscape, emphasizing the urgent need to transition from fossil fuels to cleaner, renewable energy sources. It sets the stage by discussing the environmental and economic pressures that drive this transition, along with the technological innovations that make it feasible.

A significant portion of this book is dedicated to renewable energy technologies. The authors provide a comprehensive overview of various technologies, detailing their working principles, advantages, and limitations. The integration of these renewable sources into existing power grids is explored, with a focus on technical challenges such as variability, intermittency, and the need for robust grid infrastructure. This book also discusses advanced grid technologies like smart grids, which enhance the ability to manage and distribute renewable energy efficiently.

The rise of electric vehicles is another focal point of this book. It covers technological advancements in EVs, their impact on power systems, and the symbiotic relationship between EVs and renewable energy. The authors delve into battery technologies, charging infrastructure, and the concept of vehicle-to-grid (V2G) systems, where EVs act as mobile energy storage units that can supply power back to the grid, thereby enhancing grid stability and reliability.

Power electronics play a crucial role in integrating renewable energy sources and managing EV charging. This book provides an in-depth look at the latest advancements in power electronic devices, converters, and control systems. These technologies are essential for efficient energy conversion, transmission, and distribution, ensuring that power systems can handle the dynamic nature of renewable energy and EV loads.

Editors

Natarajan Balasubramanian Muthu Selvan is an associate professor in the Department of Electrical and Electronics Engineering at Sri Sivasubramaniya Nadar College of Engineering (SSN), Chennai, with more than 20 years of teaching experience and over 10 years of dedicated research in the area of power system optimization. Prior to joining SSN, he served at Karunya Institute of Technology and Sciences, Coimbatore, during 2004–2005. He obtained his B.E. (EEE – Distinction) in 2000 and M.B.A. (Finance) in 2002 from Madurai Kamaraj University, followed by an M.E. in 2004 and a Ph.D. in 2013 in power system engineering from Anna University, Chennai. His primary research interests include power system optimization, applications of power electronic controllers in power systems, and renewable energy systems. During his doctoral research, he developed enhanced particle swarm optimization algorithms using Gaussian and Cauchy distributions for solving complex power system optimization problems. His Ph.D. thesis was highly commended by eminent examiners from IISc Bangalore and IIT Madras. He has published more than 30 research papers in refereed international journals and over 40 papers in national and international conference proceedings. He is a recognized Ph.D. supervisor at Anna University and has successfully guided four doctoral scholars in the areas of power electronics and power systems. He has received a funded research grant of ₹5.5 lakhs from SSN Trust for the project on real-time deep learning–based swarm intelligence. He is a senior member of IEEE, life member of ISTE, fellow of the Institution of Engineers (India), and a chartered engineer. He has received prestigious professional awards, organized several conferences and AICTE-sponsored programs, delivered keynote lectures, and actively serves as a reviewer for international journals and technical publications.

Venkatraman Thiyagarajan, associate professor in the Department of Electrical and Electronics Engineering, Shiv Nadar University, Chennai, has about 15 years of teaching experience. He received his B.E. degree in electrical and electronics engineering with distinction and M.E. degree in power systems engineering with distinction from Anna University, Chennai. He received his Ph.D. degree from the Division of Power System Engineering, College of Engineering Guindy (CEG), Anna University, Chennai. He has published 45 research papers in refereed international journals and 75 research papers in the proceedings of various international conferences. He has coordinated and organized various nationallevel workshops and international conferences. He has received several best paper awards for his research papers at various international conferences. He is a recognized supervisor for Anna University. His areas of research include power converters, stability, and control of electric power grids, renewable energy systems, and microgrids. He is the reviewer for many international journals. He is also a senior member of IEEE and a member of the Institution of Engineers, India.

Cheng Siong Chin is a chair professor in intelligent systems modelling and simulation at Newcastle University and an adjunct professor at Chongqing University, School of Automotive Engineering. He received his Ph.D. in applied control engineering at the Research Robotics Centre at Nanyang Technological University (NTU) in 2008 and his M.Sc. (Distinction) in advanced control and systems engineering from The University of Manchester in 2001. Before moving into academia, he worked in the consumer electronics industry for a few years. He obtained six Economic Development Board (EDB) Industrial Postgraduate Programme (IPP) and two Singapore Maritime Institute (SMI) grants from the Singapore government in intelligent systems design, simulation, and predictive analytics as a PI and CoPI. He has made more than 170 publications, including five authored books, and has two Singapore Patent Applications and three US Patents. He also received the DCASE2019 Judges' Award in the IEEE AASP Challenge on DCASE2019. He is an associate editor for the *IEEE Access Journal*, *Applied Artificial Intelligence*, and the IEEE Transportation Electrification Community (TEC) eNewsletter and is an editorial board member of Electronics. He has served as general chair, keynote speaker, session chair, and technical committee member at numerous international conferences.

Contributors

Vadivel Ayyakannu
Mechanical Engineering
Sri Ramakrishna Engineering College
Coimbatore, Tamil Nadu, India

Gowtham Ayyalusamy
Electrical and Electronics Engineering
Sri Ramakrishna Engineering College
Coimbatore, Tamil Nadu, India

Madhu Balasubramanian
Mechatronics Engineering
Velammal Institute of Technology
Chennai, Tamil Nadu, India

Natarajan Balasubramanian Muthu Selvan
Electrical and Electronics Engineering
Sri Sivasubramaniya Nadar College of Engineering
Chennai, Tamil Nadu, India

Manmadha Kumar Boddepalli
Electrical and Electronics Engineering
Aditya Institute of Technology and Management
Tekkali, Andhra Pradesh, India

Selvabharathi Devadoss
Electrical and Electronics Engineering
SRM Institute of Science and Technology
Chennai, Tamil Nadu, India

Somesh Dharmalingam
Electrical and Electronics Engineering
Sri Sairam Institute of Technology
Chennai, Tamil Nadu, India

Anjana Ethirajan
Electrical and Electronics Engineering
Sri Sivasubramaniya Nadar College of Engineering
Chennai, Tamil Nadu, India

Parimalasundar Ezhilvannan
Electrical and Electronics Engineering
Mohan Babu University
Tirupati, Andhra Pradesh, India

Arun Sampaul Thomas Gnanaraj
Artificial Intelligence and Machine Learning
J.B. Institute of Engineering and Technology
Hyderabad, Telangana, India

Muralikrishnan Gopalakrishnan
Electrical and Electronics Engineering
S.A. Engineering College
Chennai, Tamil Nadu, India

Arunnehru Jawaharlalnehru
Department of CSE
SRM Institute of Science and Technology
Chennai, Tamil Nadu, India

Nagendran Jayabalan
Electrical and Electronics Engineering
M.A.M. College of Engineering and Technology
Tiruchirappalli, Tamil Nadu, India

Booma Jayapalan
Electronics and Communication Engineering
PSNA College of Engineering and Technology
Dindigul, Tamil Nadu, India

Ramesh Jayaraman
Electrical and Electronics Engineering
Siddhartha Academy of Higher Education
Tumkur, Karnataka, India

Baskaran Jeevarathinam
Electrical and Electronics Engineering
PSG Institute of Technology and Applied Research
Coimbatore, Tamil Nadu, India

Nandha Gopal Jothiperumal
Electrical and Electronics Engineering
Velammal Institute of Technology
Chennai, Tamil Nadu, India

Sibiraj Kamaraj
Electronics and Communication Engineering
PSNA College of Engineering and Technology
Dindigul, Tamil Nadu, India

Malarvizhi Kandasamy
Electrical and Electronics Engineering
Kumaraguru College of Technology
Coimbatore, Tamil Nadu, India

Rupa Kesavan
Computer Science and Engineering
Sri Venkateswara College of Engineering
Chennai, Tamil Nadu, India

Preetha Krishnasamy
Electrical and Electronics Engineering
Erode Sengunthar Engineering College
Erode, India

Vijayakumar Krishnasamy
Electrical and Electronics Engineering
SRM Institute of Science and Technology
Chennai, Tamil Nadu, India

Darapureddi Vijaya Kumar
Electrical and Electronics Engineering
Aditya Institute of Technology and Management
Tekkali, Andhra Pradesh, India

Selvakumar Kuppusamy
Electrical and Electronics Engineering
SRM Institute of Science and Technology
Chennai, Tamil Nadu, India

Anbarasu Loganathan
Electrical and Electronics Engineering
Erode Sengunthar Engineering College
Erode, Tamil Nadu, India

Vijayaraja Loganathan
Electrical and Electronics Engineering
Sri Sairam Institute of Technology
Chennai, Tamilnadu

Sharan Rabinson Mohanraj
Mechanical Engineering
Sri Ramakrishna Engineering College
Erode, Tamil Nadu, India

Jawahar Marimuthu
Electrical and Electronics Engineering
Mepco Schlenk Engineering College
Sivakasi, Tamil Nadu, India

Yaswanth Murugan
Electronics and Communication Engineering
PSNA College of Engineering and Technology
Dindigul, Tamil Nadu, India

Nirmala Muthusamy
Electrical and Electronics Engineering
Kumaraguru College of Technology
Coimbatore, Tamil Nadu, India

Subramani Nagaraj
Electrical and Electronics Engineering
Sri Ramakrishna Engineering College
Coimbatore, Tamil Nadu, India

Aadithya Nithyanandan
Mechanical Engineering
Sri Ramakrishna Engineering College
Coimbatore, Tamil Nadu, India

Karthikeyan Palaniyappan
Electrical and Electronics Engineering
Erode Sengunthar Engineering College
Erode, Tamil Nadu, India

Sivasubramanian Pandian
Electronics and Communication Engineering
PSNA College of Engineering and Technology
Dindigul, Tamil Nadu, India

Maithili Periyannan
Electrical and Electronics Engineering
Kumaraguru College of Technology
Coimbatore, Tamil Nadu, India

Mahalingam Prabhakar
Centre for Smart Grid Technologies
Vellore Institute of Technology
Chennai, Tamil Nadu, India

Karthikeyan Rajendran
Electrical and Electronics Engineering
Velammal Institute of Technology
Chennai, Tamil Nadu, India

Ramaprabha Ramabadran
Electrical and Electronics Engineering
Sri Sivasubramaniya Nadar College of Engineering
Chennai, Tamil Nadu, India

Bhavani Ramachandran
Electrical and Electronics Engineering
Mepco Schlenk Engineering College
Sivakasi, Tamil Nadu, India

Palanisamy Ramasamy
Electrical and Electronics Engineering
SRM Institute of Science and Technology
Chennai, Tamil Nadu, India

Budi Srinivasa Rao
Electrical and Electronics Engineering
Aditya Institute of Technology and Management
Tekkali, Andhra Pradesh, India

Challa Krishna Rao
Electrical and Electronics Engineering
Aditya Institute of Technology and Management
Tekkali, Andhra Pradesh, India

Dhanasekar Ravikumar
Electrical and Electronics Engineering
Sri Sairam Institute of Technology
Chennai, Tamil Nadu, India

Sarat Kumar Sahoo
Electrical Engineering
Parala Maharaja Engineering College
Berhampur, Odisha, India

Thalapathiraj Sambandham
Department of Mathematics
SRM Institute of Science and Technology
Chennai, Tamil Nadu, India

Thamizharasan Sandirasegarane
Electrical and Electronics Engineering
C.K. College of Engineering & Technology
Cuddalore, Tamil Nadu, India

Sivanandini Saravanakumar
Electronics and Communication Engineering
PSNA College of Engineering and Technology
Dindigul, Tamil Nadu, India

Vetri Ashwath Senthilkumar
School of Electrical Engineering
Vellore Institute of Technology
Chennai, Tamil Nadu, India

Arockiaraj Sesaiya
Electrical and Electronics Engineering
Mepco Schlenk Engineering College
Sivakasi, Tamil Nadu, India

Titus Sigamani
Electrical and Electronics Engineering
K. Ramakrishnan College of Engineering
Tiruchirappalli, Tamil Nadu, India

Muthukaruppasamy Subburaj
Electrical and Electronics Engineering
Velammal Institute of Technology
Chennai, Tamil Nadu, India

Somu Subramani
Department of English
Velammal Institute of Technology
Chennai, Tamil Nadu, India

Esai Vel Murugan Subramanian
Electronics and Communication Engineering
PSNA College of Engineering and Technology
Dindigul, Tamil Nadu, India

Saravanan Subramanian
Electrical and Electronics Engineering
C.K. College of Engineering & Technology
Cuddalore, Tamil Nadu, India

Thirunavukarasu Suryalakshmi
Ivy Mobility Solutions Private Limited
Chennai, Tamil Nadu, India

Shanmuga Prabha Thangaraj
Electronics and Communication Engineering
PSNA College of Engineering and Technology
Dindigul, Tamil Nadu, India

Thambi Ezhilarasan Mangaiyarkarasi Thamizh Thentral
Electrical and Electronics Engineering
SRM Institute of Science and Technology
Chennai, Tamil Nadu, India

Ashish Ragavendra Uthayakumar
Electrical and Electronics Engineering
Sri Sairam Institute of Technology
Chennai, Tamil Nadu, India

Thiyagarajan Venkatraman
Electrical and Electronics Engineering
Shiv Nadar University
Chennai, Tamil Nadu, India

Yuvanshankar Vijayakumar
School of Electrical Engineering
Vellore Institute of Technology
Chennai, Tamil Nadu, India

Mruthunjay Vijayamurali
School of Electrical Engineering
Vellore Institute of Technology
Chennai, Tamil Nadu, India

Gurumoorthy Yohanandhan Rajaa Vikhram
Electronics and Instrumentation Engineering
SRM Institute of Science and Technology
Chennai, Tamil Nadu, India

Shyam Kumar Chandru Yuvakumar
Electronics and Communication Engineering
PSNA College of Engineering and Technology
Dindigul, Tamil Nadu, India

1 Introduction to Electric Vehicles and Renewable Energy Integration in Power Systems

Obstacles, Advancements, and Prospects

Thiyagarajan Venkatraman and
Natarajan Balasubramanian Muthu Selvan

1.1 OVERVIEW

The twin goals of decarbonization and sustainable development are causing a significant change in the global energy landscape. Driven by the need to reduce greenhouse gas emissions and the depletion of fossil fuel supplies, renewable energy sources (RESs), including solar photovoltaic (PV) and wind power, are progressively being integrated into power networks worldwide. Originally seen as auxiliary, these renewable technologies have grown vital for creating next-generation power systems that are more resilient, environmentally friendly, and flexible enough to meet the changing needs of society [1–3].

Still, major technical and practical hurdles prevent the broad integration of renewable energy. Because solar and wind resources are inherently erratic and intermittent, the power supply varies and could jeopardize the reliability and stability of the grid. Advanced storage technologies and innovative grid topologies are required, as traditional energy storage solutions usually cannot adequately reduce these oscillations [4]. Real-time balancing and the effective use of renewable energy are made possible by innovations including metal-ion capacitors, hybrid energy storage systems, and superconducting cables with integrated energy storage.

Particularly with the spread of electric vehicles (EVs), the electrification of the transportation sector is simultaneously changing power consumption patterns and generating new possibilities for grid flexibility [5–7]. Beyond serving only as consumers of electricity, EVs are becoming distributed energy resources capable of providing power back to the grid via technologies like vehicle-to-grid (V2G). This reciprocal flow best illustrates the convergence of the energy and mobility sectors, improves grid stability, and helps integrate renewable energy sources. Combining

DOI: 10.1201/9781003585534-1

electric vehicles with smart grids increases complexity and calls for advanced control, communication, and energy management systems [8–11].

A major component of this change is the development of power electronics. Next-generation devices using materials such as silicon carbide (SiC) and gallium nitride (GaN) deliver higher efficiency, better power density, and improved dependability in applications including electric vehicles and renewable energy conversion. Smart, flexible, and highly automated power systems are being created thanks in large part to technological developments, as well as the Internet of Things (IoT) and artificial intelligence (AI)-driven intelligent energy management systems [12–14].

Furthermore, changing power systems are not only about technical development. Emphasizing microgrids, distributed energy resources (DERs), and cooperative optimization approaches, they call for a review of information flows and grid topologies. In a context marked by distributed generation and changing loads, the consistent supply of power depends on the improvement of grid infrastructure and the use of new market mechanisms.

1.2 INTEGRATION OF RENEWABLE ENERGY: APPLICATIONS OF POWER ELECTRONICS IN SOLAR, WIND, AND HYBRID SYSTEMS

A key component of next-generation electrical networks is the incorporation of renewable energy sources, especially wind and solar (PV, into contemporary power systems. The intermittency and variability inherent in renewable resources can be effectively addressed by hybrid systems that combine solar and wind generation, frequently accompanied by energy storage and sophisticated power electronics. The synergy effect achieved by a hybrid renewable energy system is depicted in Figure 1.1.

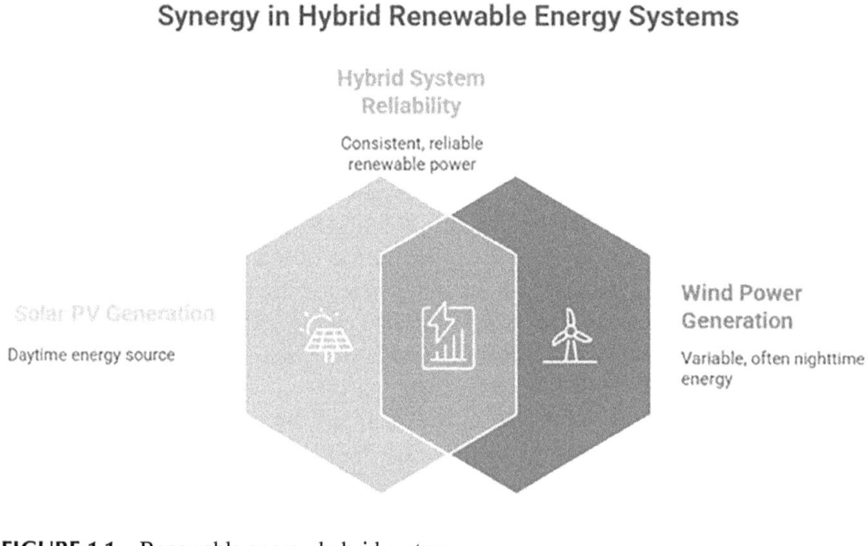

FIGURE 1.1 Renewable energy hybrid system.

1.2.1 Hybrid Systems of Solar and Wind

To mitigate the emission of greenhouse gases while increasing the share of renewables, the adoption of hybrid renewable energy systems (HRES) that incorporate wind energy together with solar photovoltaics is becoming more common. These systems harness solar and wind resources to reduce dependency on backup fossil-fuel generators by smoothing out the total power supply. Power generated by solar technology is most available in the late morning and afternoons (peaking during the day), while wind patterns may provide generation in the evenings, at night, or during cloudy periods.

These systems are particularly useful in standalone, islanded, or microgrid applications lacking access to the grid. Hybrid systems that involve both power sources enable the system to withstand changing environmental conditions, delivering a more consistent and reliable energy supply. The hybrid approach enhances energy management because the combined output can be efficiently allocated to satisfy local demand, stored in batteries, or exported as excess energy to the grid.

1.2.2 Integration Challenges

Despite the advantages, hybrid solar-wind systems face various integration challenges:

Intermittency and Uncertainty: Variability in output due to changes in wind speed and solar radiation impacts power quality and grid stability.
Power Quality Issues: Especially in microgrid and standalone systems, variable loads such as generators can lead to the creation of harmonics, voltage swings, and frequency deviations.
Control Complexity: The interaction among several sources leads to high complexity, as each source and storage device needs to be controlled and optimized for seamless operation and peak performance.

1.2.3 Application of Power Electronics

Power electronics are pivotal in the successful melding of solar and wind resources for hybrid systems. Their functions include, among others:

Energy Conversion and Conditioning: Output from wind generators, which is AC, and the variable DC output from PV panels are converted by power electronic converters (DC-DC boost converters and inverters) to a usable local or grid level. For example, wind turbines with permanent magnet synchronous generators (PMSG) shift to a rectified DC state, which is then converted into AC for grid or load supply. This is done by integrating power conversion from PV outputs and inverting with the aid of buck converters, altering the feed to enhance energy efficiency, termed boost.
Voltage and Frequency Control: Advanced control methods such as PI controllers, fuzzy logic, AI models, among others, are used in power electronics interfaces to control the modulation of voltage and frequency to a steady state, reduce harmonics, and enhance power quality.

In grid-connected scenarios, power electronics interfaces aid in the hybrid system's synchronization to the main grid by governing phase, voltage, and frequency, ensuring reliable and secure operation.

Microgrid Management: As discussed in the preceding chapters in the context of autonomous microgrids, power electronics permit flexible power allocation among generating units, storage units, and loads. It is common practice to use DC distribution lines for better efficiency.

1.2.4 Developments and Prospects

Recent studies emphasize the further development of intelligent power electronics, which incorporate AI and advanced algorithms for adaptive control, fault detection, and optimization in real time. These innovations improve the operational reliability and efficiency of hybrid systems and enable better smart grid and energy market integration with reduced total harmonic distortion (THD) and enhanced system dependability.

1.3 ELECTRIC VEHICLES IN CONTEMPORARY POWER INFRASTRUCTURES

With the latest advances in highways and power systems, the relationship between transportation and electric grids is changing completely due to the addition of EVs. EVs are, now more than ever, becoming active players in energy control, grid equilibrium, and the adoption of renewables. As shown in Figure 1.2, this evolution is

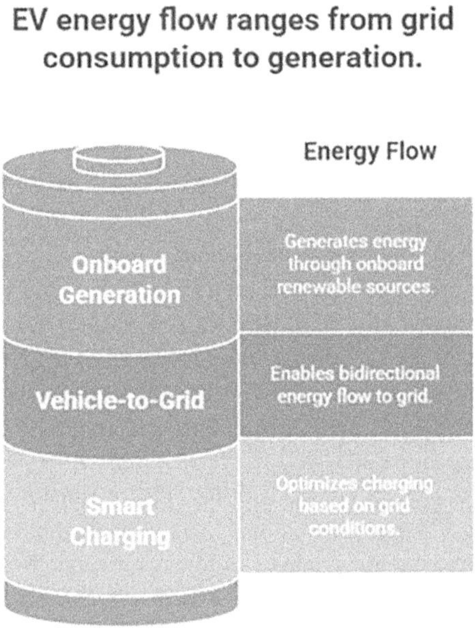

FIGURE 1.2 EV's technological foundations in the contemporary power system.

backed by three fundamental technological pillars: regenerative onboard energy systems, smart charging systems, and V2G technologies. The following sections will elaborate on these technologies.

1.3.1 Vehicle-to-Grid (V2G) Technologies

V2G technologies enable bidirectional energy flow between the power grid and EV batteries. During periods of excessive demand or grid outages, this allows EVs to operate as distributed energy storage devices, providing ancillary services such as voltage support, frequency regulation, peak shaving, and emergency backup. By tapping into the energy storage capacity of thousands of EVs, V2G can significantly enhance grid resilience and flexibility, therefore facilitating the use of intermittent renewable energy sources.

In order to enable grid-to-vehicle (G2V) charging and V2G discharging, bidirectional power electronic converters are vital components of V2G systems.

The characteristics of multiport bidirectional DC-DC converters and Capacitor-Inductor-Inductor-Inductor-Capacitor (CLLLC) resonant converters guarantee seamless energy transfer between EVs, renewable energy, and the grid, including dependability, smooth oscillation into and out of ports, high efficiency, and broad voltage ranges. Moreover, comprehensive real-time control and predictive frameworks are under construction that consider user-specific V2G preferences, vehicle mobility, patterns of movement, and even the nuances of renewable power generation.

Like any other system, V2G has disadvantages that include power system oscillations, cybersecurity concerns, and the potential for harmonics. Additional research is being conducted to address the technical limitations, regulatory structures, and business models required to accelerate acceptance.

1.3.2 Intelligent Charging

Smart charging encompasses the management and planning of EV charging sessions that contribute toward enhancing grid efficiency, lowering costs, and improving user ease. Smart charging adjusts charging rates and schedules based on grid conditions, electricity prices, availability of clean generation, and the level of renewable generation to mitigate peak load impacts and optimize clean energy usage.

Smart charging also streamlines the allocation of charging stations and session planning while lowering operating costs for both EV owners and grid operators through the use of advanced algorithms such as intelligent charging scheduling and optimization techniques. To account for uncertainties in travel patterns, energy demand, and renewable output, these techniques leverage real-time data and predictive analytics to enable robust decision-making.

Moreover, smart charging reveals the architecture for V2G service integration and enables seamless shifts between charging and discharging to grid-support functions. This endeavor is crucial for the evolution of smart grids and widespread electric vehicle adoption.

1.3.3 Energy Generation Onboard

Onboard energy generation refers to the incorporation of renewable energy sources to electrically power vehicles, such as PV panels.

While onboard solar PV increase the driving range, assist in less reliance on outside power sources, and reduce carbon emissions of EVs, its contribution compared to grid charging is modest.

Moreover, by transforming kinetic energy into electrical energy during deceleration and braking, regenerative braking systems capture energy that can be stored in the vehicle's battery for future use. Together with basic electric energy, these systems enhance the sustainability of electric mobility.

Some new studies focus on the integration of onboard solar photovoltaics, advanced battery management, and bidirectional converters to facilitate the use of EVs as mobile energy hubs in decentralized energy networks. These advances support the concept of EVs acting as active grid nodes that can generate and "consume" clean power.

1.4 CASE STUDIES AND USES: ELECTRIC VEHICLES AND RENEWABLE ENERGY INTEGRATION IN CONTEMPORARY POWER SYSTEMS

1.4.1 Integration of Renewable Energy

1.4.1.1 Wide Area Implementation of Solar and Wind Power: Eastern Interconnection in America

Advanced model development of solar penetration in the United States serves as an important as well as a distinctive case. The Eastern Interconnection is among the largest operating power grids globally. In trying to switch from conventional generators to large-scale photovoltaic systems, these researchers built dynamic models. From the analysis, carbon emission mitigation assists, along with high renewable integration in achieving net-zero emissions, but also brings challenges such as diminishing system inertia, enhanced controllability issues, and increased uncertainty in the operational running of the grid. ECRs can expect automation, which devises advanced algorithms for monitoring and controlling energy systems to ensure faultless system execution. The shift toward the adoption of renewable energy sources challenges the existing advanced infrastructure systems, meticulously designed algorithms, and control techniques with refinement, precision, and speed. This competency, the model-solving to reduce costs, needs to state the problems. Renewables intensify interdependencies in decapitated frameworks, accelerating decoupled systems by decomposing systems into components of sustainable systems instead of treating combined systems.

1.4.1.2 RTS-GMLC Test System for Hybrid Wind-Battery Systems

This is one of the examples of hybrid wind-battery system developments. The implementation of battery energy storage systems in wind-powered generation facilities is particularly relevant for the Reliability Test System–Grid Modernization Laboratory

Consortium (RTS-GMLC) test system. This hybrid approach tackles challenges associated with wind power variability. Including batteries at wind farms, according to the study, increases dispatchability, enhances price stability, and supports grid ancillary services such as frequency regulation. Intermittency challenges of renewables create operational challenges because of hybridization. Improved energy storage is one clear answer.

1.4.1.3 Answer to Industrial Needs and Local Sources of Renewable Energy

A case study demonstrating both demand response scheduling at the industrial level and the integration of photovoltaic, wind power, and battery storage showcased the design of a small-scale electricity system around a contiguous industrial process. As a result, this system design sought to minimize greenhouse gas emissions as well as annualized cost. Results proved the integrative potential of modern concepts in electricity management and storage, together with local renewables by demonstrating that batteries not only reduced emissions but also increased market participation flexibility.

1.4.1.4 Frequency Response with High Renewables in the European Grid

The simulations focused on the European interconnected power system included studies on frequency response and inertia for high renewable integration scenarios. Over 700 different cases were run for varying grid inertia and share of renewables. As the results showed, additional measures will have to be taken to control for intermittency and renewables at a large scale, highlighting the importance of frequency control and grid interconnections for stability maintenance.

1.4.2 Electric Vehicles in Modern Power Systems

1.4.2.1 Stability of the Grid and V2G Services

Several field tests have been run to analyze the impact of V2G technology on the stability of the grid and integration of renewables.

In the demonstrations, combined EV fleets provided frequency regulation and peak shaving services, unlocking stored energy during periods of high demand and low renewable output. Their effectiveness was remarkably higher on grids with high renewable energy sources due to the ability to offset volatility and provide frequency stabilization with EVs.

1.4.2.2 Demand Response and Scheduling of Smart EV Charging in Urban Networks

In urban case studies, smart EV charging has been implemented by moving the charging times to when the production of renewable energy is maximized and the grid's demand is minimized. Using real-time data, sophisticated algorithms, and operational costs, these systems attempt to manage renewable energy resources at the lowest possible costs and mitigate peak loads. These advanced recovery programs that combine intelligent charging have demonstrated their ability to significantly reduce system strain and economic burdens linked to the fast uptake of electric vehicles.

1.4.2.3 Integrated Energy Systems Models: Flexibility and Resilience

The integration of EVs, renewable sources of energy, and storage units increases system flexibility and resilience. That is the systematic review of 116 cases of local integrated energy system model studies showed.

These models demonstrated that optimizing storage, renewable energy, and EV charging can significantly improve grid reliability and assist in the achievement of decarbonization targets. The studies also stressed the importance of resolving non-technical barriers and providing efficient flexibility measures for successful integration.

1.4.3 Insights and Outlooks for the Future

Hybridization and Energy Storage: Combining renewables with batteries or other storage technologies will mitigate the unpredictability associated with renewable generation and ensure grid reliability.
Advanced Control and Modeling: The stabilization of systems with high levels of renewable energy integration requires sophisticated dynamic models, frequency response plans, and real-time control.
EVs as Grid Assets: Smart charging and V2G transform EVs from passive loads into active participants of the grid, facilitating the integration of renewables and enhancing the provision of services to the grid.

Successful case studies underscore the importance of integrated planning, which combines advanced control, systems management, and other components, including storage, demand response, and renewable energy sources in power systems engineering to achieve resilient and sustainable power systems.

1.5 ISSUES AND PROSPECTS FOR NEXT-GENERATION POWER SYSTEMS: UTILIZING RENEWABLE ENERGY AND ELECTRIC VEHICLES

1.5.1 Principal Difficulties

1.5.1.1 Flexibility and Grid Stability

Grid stability and operational flexibility are severely hampered by the integration of large percentages of variable RESs, such as wind and solar. Frequency and voltage instability can result from generation fluctuations, necessitating sophisticated control schemes and real-time balancing systems.

The decentralized and erratic nature of RESs and DERs is difficult for current grid architectures to handle because they were created for centralized and predictable power flows.

1.5.1.2 Complexity of Information, Communication, and Control

Robust information and control architectures that can manage large data flows, real-time monitoring, and automated decision-making are essential for next-generation power systems. Communication networks that are safe, scalable, and interoperable are essential, particularly as more devices—including EVs—become connected.

As grids become more connected and digitalized, cybersecurity risks increase and critical infrastructure becomes more vulnerable to potential attacks.

1.5.1.3 Needs for Flexibility and Energy Storage

Large-scale energy storage system deployment is required to mitigate the unpredictability of renewable energy sources and guarantee a steady supply. However, issues with cost, scalability, and lifecycle management continue to plague storage technologies.

As renewable energy sources become more widely used, flexibility demands rise, necessitating new operational procedures, market mechanisms, and investments in both short-term (batteries) and long-term (hydrogen, pumped hydro) storage.

1.5.1.4 Integration of Electric Vehicles

Both opportunities and challenges are brought about by the widespread use of EVs. While smart charging and V2G necessitate complex coordination and infrastructure upgrades, uncontrolled charging can worsen peak demand and put stress on local distribution networks.

It is still technically and legally difficult to guarantee interoperability between EVs, charging stations, and grid management systems.

1.5.1.5 Uncertainty and System Resilience

Extreme weather, supply-demand imbalances, and equipment failures are just a few of the uncertainties that future energy systems must be able to withstand. To verify and enhance system robustness, real-time simulation, hardware-in-the-loop testing, and predictive analytics are required.

Maintaining system performance after 2050 will require a shift from rapid infrastructure expansion to long-term upkeep and renewal.

1.5.2 Prospects for the Future

In order to improve sustainability, efficiency, and dependability, future directions for next-generation power systems place a strong emphasis on incorporating EVs and renewable energy sources into smart grids. Figure 1.3 shows the different developments for building a flexible, intelligent, and resilient power network, facilitating greater penetration of renewable energy sources, lowering emissions, and encouraging the broad use of electric mobility solutions.

1.5.2.1 Advanced Management and Enhancement

To handle the complexity of future grids, decentralized, adaptive, and AI-driven control systems must be developed and implemented. These systems will make it possible to optimize storage, DERs, and flexible loads—including EVs—in real time.

Large-scale, high-dimensional optimization issues for operations and planning may be resolved with the help of quantum computing and sophisticated algorithms.

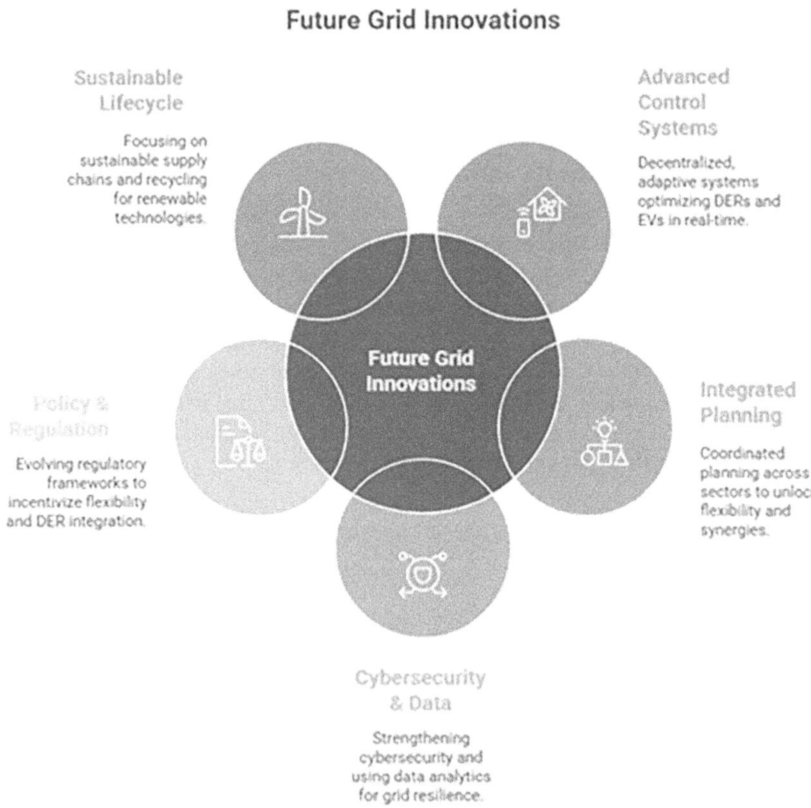

FIGURE 1.3 Future grid innovations.

1.5.2.2 Sector Coupling and Integrated Planning

New sources of flexibility and synergies, like utilizing waste heat from power generation or EV batteries for grid support, will be made possible by coordinated planning across the transportation, heating, and electricity sectors (sector coupling).

Increased resilience and customized solutions for various regions and client needs will be made possible by the growth of microgrids and local energy communities.

1.5.2.3 Improved Data Management and Cybersecurity

As digitalization increases, it will be crucial to adopt secure, resilient communication protocols and strengthen cybersecurity frameworks.

Predictive maintenance, anomaly detection, and enhanced situational awareness will be facilitated by advanced data analytics and machine learning.

1.5.2.4 Market Design, Regulation, and Policy

In order to encourage adaptability, facilitate DER integration, and open up new business models for demand response, storage, and V2G, regulatory frameworks must change.

Market mechanisms that prioritize resilience, flexibility, and ancillary services will become more significant.

1.5.2.5 The Circular Economy and Sustainable Lifecycle

The sustainability of supply chains, recycling, and end-of-life management for batteries, solar, and wind turbines must be prioritized as EV and renewable energy deployments grow.

The long-term viability of next-generation power systems will be aided by research into novel materials, recycling techniques, and circular economy models.

BIBLIOGRAPHY

1. D. Witthaut, K. Aihara, C. Beck, B. Schäfer, and M. Timme, "Modeling the dynamics and control of power systems with high share of renewable energies," arXiv preprint arXiv:2012.05164, 2020. [Online]. The document is accessible at https://arxiv.org/pdf/2012.05164.pdf
2. F. V. Lara-Fanego, J. A. Ruiz-Arias, D. Pozo-Vázquez, J. Santos-Alamillos, and J. Tovar-Pescador, "A review on the complementarity of renewable energy sources: concept, metrics, application and future research directions," arXiv preprint arXiv:1904.01667, 2019. [Online]. The document is accessible at https://arxiv.org/pdf/1904.01667.pdf
3. M. Benedetti, "Microgrid management with weather-based forecasting of energy generation, consumption, and prices," arXiv preprint arXiv:2107.01034, 2021. [Online]. This is accessible at https://arxiv.org/pdf/2107.01034.pdf
4. W.-P. and Zerrahn. Schill, "Integration of renewable energy sources in future power systems: The role of storage," arXiv preprint arXiv:1405.2857, 2014. [Online]. This is accessible at https://arxiv.org/pdf/1405.2857.pdf
5. T. Brown, J. Hörsch, L. Schmedes, and M. Greiner, "How flexible electrification can integrate fluctuating renewables," arXiv preprint arXiv:2301.11096, 2023. [Online]. This is accessible at https://arxiv.org/abs/2301.11096
6. F. Simpson-Porco, J. W. Dörfler, and F. Bullo, "Stability, control, and dynamic modeling of power systems with distributed energy resources," arXiv preprint arXiv:1804.04933v3, 2018. [Online]. This is accessible at https://arxiv.org/abs/1804.04933v3
7. A. Gunkel et al., "Reviewing local and integrated energy system models: Insights into flexibility and robustness challenges," arXiv preprint arXiv:2202.13942, 2022. [Online]. The document is accessible at https://arxiv.org/pdf/2202.13942.pdf
8. H. Lund et al., "Reviewing two decades of energy system analysis with bibliometrics," arXiv preprint arXiv:2103.09917, 2021.
9. Y. Wang et al., "Grid-forming technologies for resilient renewable energy integration into power systems: A review," Semantic Scholar, 2024. [Online]. This is accessible at https://doi.org/10.1109/SST61991.2024.10755449
10. Kumar et al., "Integration of renewable energy in microgrids and smart grids in deregulated power systems: A comparative exploration," Semantic Scholar, 2024. [Online]. The article is accessible at https://doi.org/10.1002/aesr.20240008811
11. X. Zhang et al., "Power systems resilience enhancement through renewable energy integration: Insights and future directions," Semantic Scholar, 2024. [Online]. The article is accessible at https://doi.org/10.1109/ICPEA60617.2024.10498721
12. H. Li et al., "A region-based approach for the operational reliability evaluation of power systems with renewable energy integration," Semantic Scholar, 2024. [Online]. This is accessible at https://doi.org/10.1109/TPWRS.2023.3286088

13. M. Singh et al., "Applications of energy storage systems in power grids with and without renewable energy integration—A comprehensive review," Semantic Scholar, 2023. [Online]. The following is accessible at https://doi.org/10.1016/j.est.2023.107811
14. S. Sharma et al., "Evaluation of bus inertia to enhance dynamic flexibility of hybrid power systems with renewable energy integration," Semantic Scholar, 2023. [Online. The article is accessible at https://doi.org/10.1109/TPWRD.2023.3241721

2 DC-DC Converter Design and Motor Selection for Electric Vehicle Applications

Vijayaraja Loganathan, Dhanasekar Ravikumar, Ashish Ragavendra Uthayakumar, Somesh Dharmalingam, Rupa Kesavan, Arunnehru Jawaharlalnehru, and Thalapathiraj Sambandham

2.1 INTRODUCTION

Global CO_2 emissions reached 36.8 billion tonnes, reflecting a substantial contribution to climate change from various countries [1], which tells us that global CO_2 emissions are rising alarmingly across various industries, driven by high consumer demands and inadequate shifts to low-carbon services. Despite some progress in Europe and North America, industrialisation in Asia exacerbates emissions due to increased material and energy demands [2]. Analyses of G20 countries' emissions from 1971–2021 reveal that the USA, Australia, and Canada consistently generated the highest CO_2 emissions during this period due to energy demand from population growth and industrialisation. Figure 2.1 represents the top emitters; China stands out as the largest contributor, responsible for 31% of global emissions. This is mostly

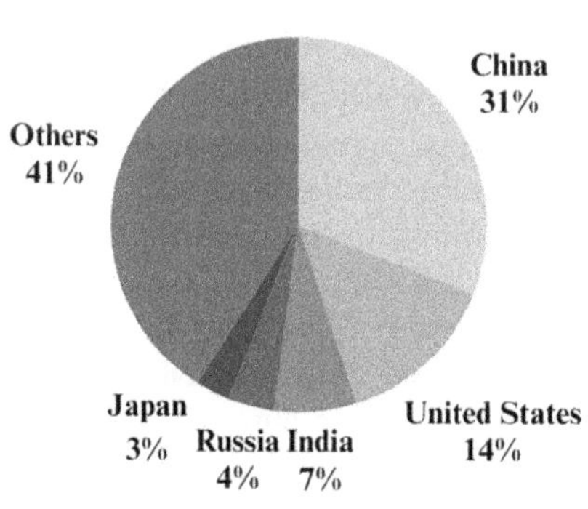

FIGURE 2.1 World's top CO_2 emitters.

DOI: 10.1201/9781003585534-2

because the nation is a global centre for industry, has rapidly industrialised, and depends on coal for energy. The United States follows with 14% of global emissions, largely driven by its high consumption rates, fossil fuel dependency, and widespread use of automobiles. India, another significant emitter, contributes 7% to global CO_2 levels, driven by its large population, industrialisation, and a growing energy demand primarily met by coal. Russia and Japan are next, contributing 4% and 3%, respectively, with Russia's emissions tied to its energy-intensive industries and reliance on fossil fuels, while Japan's emissions come from its industrial and transportation sectors. The other 41% of emissions is produced by other nations, most of which are either developing or have smaller industrial sectors.

CO_2 emissions are influenced by demographic growth, wealth, technology, human capital, urbanisation, and FDI, necessitating country-specific interventions for successful curtailment [3]. Exceeding emission limits is not a phenomenon of developing countries alone; many developed nations exceed their emission targets as well. The United States, Germany, Canada, Australia, and South Korea are all examples of industrialised countries that are above emission limits, with high energy consumption and fossil fuel dependence being central to their emission profiles. These countries have traditionally accounted for the greatest share of world emissions and are still significant emitters because of their standards of living, industrial production, and transportation requirements. In addition, in industrialised nations, CO_2 emissions have fallen with the transition towards renewable energy, changes in energy intensity, and decreases in fossil CO_2 intensity, with prospects for future trends pointing towards sustained declines mainly as a result of the increasing share of renewable energy and reduced fossil CO_2 intensity [4]. In the same manner, developing countries are also contributing major shares of global emissions, with countries like China, India, Russia, Iran, and Saudi Arabia exceeding their emission limits. As these nations industrialise and urbanise, their energy demands increase, usually supplied by coal and oil, which are high-carbon energy sources. As these countries push for economic growth and development, the environmental footprint of their industrialisation is now becoming clearly visible. CO_2 generation from electricity in G7 nations is influenced by the energy sources, with coal and gas increasing, and renewable and hydro decreasing emissions [5].

It is calculated that around 20%–25% of nations globally emit more than their equitable portion of emissions, adding more to global CO_2 levels than their fair share as determined by population, economic progress, and past emissions contributing to climate change. These nations, usually highly industrialised, have large per capita emissions, leading to an unequal apportionment of responsibility for the world's emissions. The imbalance of emissions in these countries is a central challenge in climate talks, as they are perceived to be emitting more than their fair share, yet most developing countries cannot afford to meet their energy needs without using high-carbon-emitting energy resources. Addressing this imbalance is critical for achieving global climate goals and ensuring that both developed and developing nations share responsibility for reducing emissions and transitioning to sustainable energy systems [6] predicts CO_2 emissions for the USA, Asia Pacific, India, and the world from 2020 to 2024, indicating a gradual rise for India and Asia Pacific, while a gradual decrease in emissions is anticipated in the USA. In terms of sector-wise

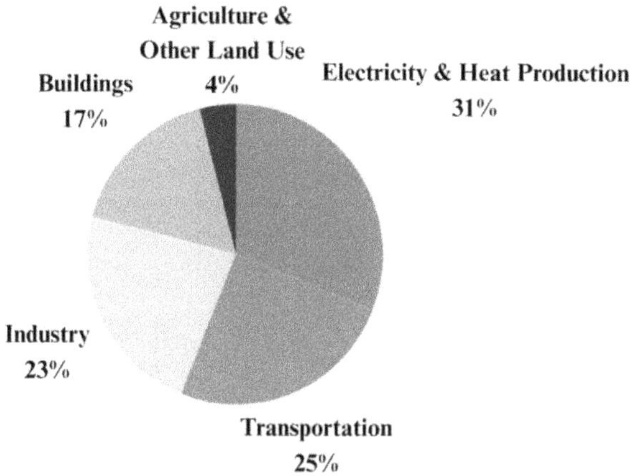

FIGURE 2.2 Sector-wise global CO_2 emissions.

global CO_2 emissions, Figure 2.2 illustrates the major contributors to climate change. Electricity and heat production lead the way, accounting for 31% of global emissions, largely due to the widespread reliance on fossil fuels for the production of electricity, including coal, natural gas, and oil. The transportation sector follows closely with 25%, driven by emissions from road vehicles, aviation, shipping, and rail. Industry is the third-largest emitter at 23%, with manufacturing, cement production, and chemical industries being significant contributors. The buildings sector contributes 17%, primarily from energy use for heating, cooling, and electricity. Finally, agriculture and other land use account for 4%, with emissions coming from livestock, fertiliser use, and land use changes such as deforestation. This distribution highlights the diverse sources of global emissions, underscoring the need for targeted efforts across various sectors to mitigate climate change.

In a recent year, India was the third-largest emitter of CO_2 in the world, with almost 2.7 billion tonnes released. The sector-wise distribution of India's CO_2 emissions is shown in Figure 2.3, which highlights the dominant role of the energy sector, accounting for 68% of total emissions. Within this sector, coal-based power plants are the largest contributor, responsible for 41% of emissions, followed by oil and gas, which together contribute 27%. The industrial sector is next, contributing 22% of India's CO_2 emissions, with iron and steel production accounting for 8%, cement production 7%, and chemicals and petrochemicals contributing 4%. Other industries contribute the remaining 3%. Transportation contributes 8% of emissions, with road transport contributing the most at 6%, whereas aviation, railways, and shipping contribute 1% each. Agriculture contributes 1%, while buildings, together with other sectors, contribute 1% of total emissions. This distribution emphasises the significant role of coal in India's energy sector and the need for a change to cleaner, renewable energy sources to help lower the carbon imprint of the nation.

CO_2 emissions in India are 1.9 tonnes per person, which is relatively low when compared to the global average of 4.7 tonnes. This reflects the country's lower level of individual emissions, driven in part by a large, predominantly rural population and reduced energy use per person relative to developed countries. However, given

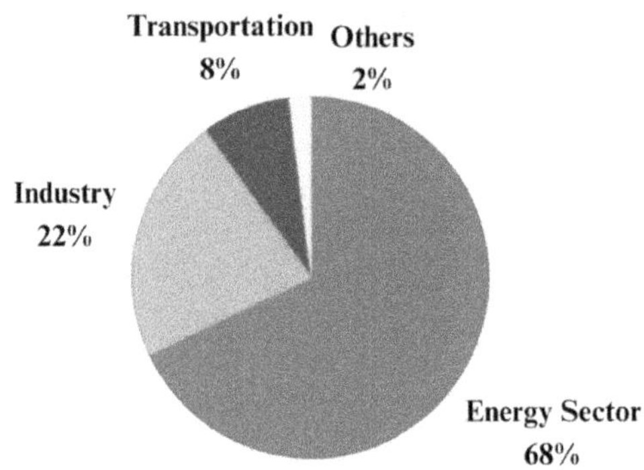

FIGURE 2.3 India CO_2 emissions (sector-wise).

India's vast population and rapid economic growth, its overall CO_2 emissions are significant and increasing. As the country continues to industrialise, urbanise, and meet the growing energy demands of its population, emissions are projected to rise.

CO2 emissions [7] correlate positively with energy consumption and gross domestic product in both immediately and over time, underscoring the need for policies that reduce pollution and promote cleaner energy sources. In response to these challenges, India has committed to reducing its emissions intensity, the number of emissions per unit GDP, by 45% by 2030 compared to 2005 levels. Additionally, India has set the goal of achieving net-zero emissions by 2070, signalling its long-term commitment to addressing climate change, although this target is beyond the global deadline of 2050. India's approach balances the need for economic development with the pressing goal of environmental sustainability, reflecting the complexities of climate action in a developing nation.

It discusses various catalytic conversion methods to mitigate CO_2 by transforming it into hydrocarbons, addressing environmental challenges, and promoting sustainable energy solutions [8]. Electric vehicles, or EVs, are a central part of limiting CO_2 emissions worldwide, and especially in the transportation industry, which is among the largest drivers of climate change. Several factors, such as EV efficiency, battery technology, performance, converter design, motor selection, and overall vehicle design, influence the extent to which EVs minimise the environmental footprint.

Compared to traditional internal combustion engine (ICE) vehicles, EVs are more efficient in that they utilise more of the energy stored in batteries for motion because ICE motors produce larger amounts of energy in the form of heat and friction. EVs simply happen to be powered at much greater efficiencies and thus save energy and emissions, as well as CO_2 emissions. Not only does this efficiency equate to fewer emissions per kilometre, but it also equates to EVs using less power to travel the same distance, an argument for less overall global emissions. Battery technology is also a natural component of the success of EVs.

Improvements in battery chemistry and better lithium-ion technology are rendering batteries more energy-dense and durable, providing longer driving range and quicker recharging time. Apart from lowering the carbon footprint of manufacturing batteries, these technologies are also rendering EVs more convenient to use.

The increased energy density enables EVs to travel longer distances without recharging and without necessarily imparting a heavy load on the car, further enhancing efficiency. Further, technological improvements in manufacturing batteries are rendering the carbon footprint of manufacturing such batteries lower, decreasing the overall environmental cost of EVs. The performance features of EVs are also enhancing their efficiency and attractiveness. EVs are long known to possess a smooth and quiet ride with better acceleration and response. Their capability to provide instant torque enables EVs to possess the potential for high performance, such as excellent acceleration and steep hill climbing without losing power. Further, enhanced regenerative braking systems that capture energy when decelerating in the battery conserve energy and increase the distance the vehicle travels. Such systems are one of the main benefits of EVs, as they offer energy efficiency and lower the frequency at which the vehicle needs to be recharged.

Power electronics and converter design for electric vehicles are critical to increasing efficiency. DC-DC converters control voltage levels between the battery and other parts of the vehicle to transfer power efficiently without massive energy loss. Traction inverters supply the battery's DC electricity as AC power to the motor. Increased efficiency in these devices means increased motor efficiency as well as improved energy consumption control. On-board chargers, which convert grid-supplied AC into DC power to charge batteries, increase the efficiency of a vehicle as a whole. Bidirectional charging allows EVs to supply extra energy to the grid when parked and draw power from it. Such technology in the form of vehicle-to-grid (V2G) systems, can be used to stabilise the grid with the growing use of renewable energy.

Motor choice is another crucial factor that affects EV efficiency and performance. Permanent magnet synchronous motors (PMSMs) are most commonly used due to their high power density, which enables a lighter, more compact motor with excellent torque at low speeds. It is also very efficient and uses less cooling, making it a suitable choice for electric vehicles. Induction motors, which are cheaper and more durable, are also used in some EVs. These motors do not use rare-earth materials, which can reduce environmental cost as well as the cost of production. Brushless DC motors (BLDC) are another suitable choice, with high torque-to-weight ratios, which is of particular interest for hill-climbing and acceleration performance. They also have regenerative braking, which recovers energy when the motor is slowed down. The overall body shape of the vehicle also plays a crucial role in energy efficiency; for example, an aerodynamic shape minimises air resistance, which minimises energy usage, especially at high speeds. Lightweight materials like carbon fibre and aluminium are widely used to minimise the aggregate weight of the vehicle and hence enhance performance and range. Lighter cars need less energy to transport, and that means better utilisation of stored energy in the battery. Thermal management systems also play a very important role in maintaining optimal performance. EVs produce heat when charging the battery and when running the motor, and therefore temperature-management systems need to be used in order to keep the battery and the motor operating at their optimal efficiency, reducing energy loss due to excessive heat.

The effect of EVs on CO_2 emissions will increase even more as the world's electricity grid becomes more and more dominated by renewable energy. While EVs in themselves are very emission-saving, they have their highest environmental impact when driven by electricity from low-carbon electricity generation, i.e., hydro, solar, and wind. As renewable power capacity increases and the grid continues to clean up, carbon emissions per mile travelled by EVs will fall further, as they become an increasingly valuable tool in global CO_2abatement policy. In summary, the combination of increasing energy efficiency, newer motor and battery technologies, more innovative converter design, and overall vehicle design improvements makes EVs a top contender in a clean future, helping to fight climate change by reducing emissions from transport.

2.2 LITERATURE SURVEY

Table 2.1 provides a complete outline of various techniques and methods used in power conversion, motor control, and energy management systems. The studies highlight the significant advancements in efficiency, reliability, and performance across

TABLE 2.1
Survey on Power Conversion and Motor Control

S.No	Methods Used	Advantages	Limitations
[9]	• Reliability predictions using MIL-HDBK-217F and Telcordia SR-332 standards. • PSPICE simulation for temperature and stress factor calculations.	• Comparison of reliability values using two standards. • Useful design data for reliability-focused engineering.	• Older standard lacks mission profile consideration. • Newer standard compensates for outdated component technology.
[10]	• Coupled inductors and transformer for voltage gain. • Active clamping and soft-switching for efficiency.	• Low voltage stresses in both directions. • High efficiency: 95.8% and 96.2%.	• Voltage gains and stresses can be improved through three degrees. • Soft-switching and partially diminished switching losses are implemented.
[11]	• General approach to SC converter modelling and analysis. • Experimental verification using a prototype 2-to-1 SC converter.	• Higher efficiency and power density with reduced output capacitance. • Accurate modelling of output impedance with finite capacitances.	• Finite terminal capacitances affect output impedance and efficiency. • Insufficient input capacitance increases output impedance significantly.
[12]	• A complete review of the design restrictions and operation modes was provided. • Reliability evaluation based on component failure and operational variables investigated.	• High-voltage gain and reduced component count. • Minimum conduction losses and improved reliability.	• Low gain due to high-voltage stress on components. • Low efficiency related to high operating pulse ratio.

(*Continued*)

TABLE 2.1 (*Continued*)
Survey on Power Conversion and Motor Control

S.No	Methods Used	Advantages	Limitations
[13]	• Proposed transfer trip scheme for fault direction identification. • Utilises local voltage and current data for decision-making.	• Reliable protection for converter-dominated power networks. • Superior performance compared to advanced techniques.	• Non-homogeneity affects relay decision reliability during faults. • Communication-assisted schemes depend on unreliable relay decisions.
[14]	• Switched-capacitors (SC) technique for high voltage gain. • Inductor incorporation to address discontinuous input current.	• High-voltage transfer gain using the switched-capacitor technique. • Flexible control options with reduced semiconductor stress.	• Discontinuous input current due to the SC technique. • Duty cycle limitations not imposed by switching methods.
[15]	• Advanced control methods for strong voltage control. • High decoupling and seamless bidirectional power routing.	• Flexible control and high efficiency in power transmission. • Provides galvanic isolation for enhanced safety.	• Traditional controllers can cause system-wide unbalancing issues. • Limited detailed reviews on converter topologies and controls.
[16]	• Analysis of conduction modes for converter design. • Model-based control strategy with current and voltage control loops.	• Higher reduction ratios than conventional step-down converters. • Evaluated performance under steady-state and transient conditions.	• Comparison of efficiency and number of components. • Type of control limitations discussed.
[17]	• Public-accessible component databases for high-temperature characteristics. • Automated communication between finite element simulations and circuit simulators.	• Reduces computational load through exclusion-based workflow. • Automates design for harsh environment performance evaluation.	• Absence of component performance data under harsh conditions. • Time-consuming and computationally heavy design workflow.
[18]	• Coupling inductance technique for voltage conversion. • Proportional–integral control for control of output voltage.	• A high ratio of output voltage conversion was attained. • At full load, the circuit efficiency is 92.48%.	• Low voltage conversion ratio is a drawback of boost converters. • Duty cycle of proposed converter is $\leq$40%.
[19]	• Hybrid PSO-SHO enhanced PI controller. • Large gain Luo converter for PV systems.	• High efficiency of 97.3% achieved. • 2.02% is a low overall harmonic distortion.	• Control difficulties affect system stability and component lifespan. • High cost, maintenance complexity, and resonance issues.

(Continued)

TABLE 2.1 (*Continued*)
Survey on Power Conversion and Motor Control

S.No	Methods Used	Advantages	Limitations
[20]	• Adaline-based estimator for disturbance torque acquisition. • Inertia estimation using an extended sliding mode observer.	• Real-time flux-linkage correction raises estimation accuracy. • Asymptotic error convergence is assured by adaptive feedback gain.	• Rotor flux mismatch in linkage challenges estimation accuracy. • Operating-condition restrictions and zero-derivative constraints have an impact on performance.
[21]	• Use of both active and passive cooling strategies. • Targeted elimination of heat from specific heat sources.	• Effective thermal management enables successful IMD deployment. • Focused heat removal enhances performance significantly.	• Balancing power losses and heat removal is essential for machine design. • IMDs and DMDs use similar thermal control systems for their electrical machinery.
[22]	• Grain-oriented electrical steel application for performance enhancement. • For design optimisation, the Interpolation Multi-Objective Robust Optimisation Algorithm (IMROA) is used.	• Improved torque and reduced iron loss. • Robust optimisation enhances performance and manufacturing tolerance.	• Higher saturation leads to worse pulsation characteristics. • Manufacturing tolerances affect expected performance outcomes.
[23]	• Neural control system. • Regenerative braking for electric vehicle architecture.	• During braking, effective energy recovery is achieved. • Enhanced motor control performance.	• Limited to small-scale electric vehicle applications. • Requires further research for larger vehicle integration.
[24]	• Magnetic equivalent circuit (MEC) for the initial design stage. • Finite element analysis (FEA) for the detailed design stage.	• Rapid analysis with magnetic equivalent circuit (MEC). • Efficient design process reduces computational costs significantly.	• FEA has high analysis time requirements. • Randomness in LHS does not guarantee uniform sampling.
[25]	• Multi-objective and multi-physics design optimisation method. • Sequential subspace optimisation strategy.	• Inexpensive and strong structure. • Excellent fault tolerance and torque density.	• High torque ripples. • High coil temperatures.

(*Continued*)

TABLE 2.1 (*Continued*)
Survey on Power Conversion and Motor Control

S.No	Methods Used	Advantages	Limitations
[26]	• Comparative simulations on PMSM and IPM motors. • A technique for detecting bearing faults based on streak magnetic flux.	• Stray flux detection is immune to noise and consistent. • Enables the precise location of faulty bearings efficiently.	• Signature magnitude of current-based fault detection is small. • Mechanical dampening causes irregular current signatures.
[27]	• Model predictive direct duty-cycle control (MPD2C) approach. • Variable control set (VCS) for duty cycle optimisation.	• Reduces steady-state torque ripples effectively. • Achieves high-precision flux and torque control.	• Torque ripple reduction needs further study. • Parameter robustness requires additional investigation.
[28]	• Finite control set model predictive control (FCS-MPC). • Maximum torque per ampere (MTPA) method.	• Eliminates the weighting factor in the control strategy. • Automatically controls phase angle based on torque.	• Proper selection of the weighting factor is crucial. • Complicated optimisation process for the best phase angle.

different converter designs, control strategies, and fault detection methods. While many approaches show high efficiency and innovative solutions for improved power conversion and system reliability, challenges such as cost, complexity, and component limitations remain. Further research is necessary to optimise these methods for larger-scale applications and to address issues related to system stability, thermal management, and component longevity.

2.3 MODELLING OF ELECTRIC VEHICLE (EV)

2.3.1 Components of EV

EVs depend on a set of components for efficient and consistent performance. A battery pack, usually a lithium-ion pack, stores electric charge and defines the vehicle's performance and range. The electric motor converts energy into mechanical power to move the vehicle with the help of an inverter. A charging system supplies recharging from external power, and an on-board charger converts AC to DC for the battery. A power controller regulates the transmission of energy between components to provide maximum power. A transmission transfers power from the motor to the wheels, with the help of a regenerative braking system to capture kinetic energy at the time of braking. A thermal management system manages battery and motor temperature for safety and performance. The battery management system (BMS) monitors and safeguards the operation of batteries, protecting against overcharging and ensuring maximum efficiency. All these devices form an integrated system that propels EVs using clean and renewable power.

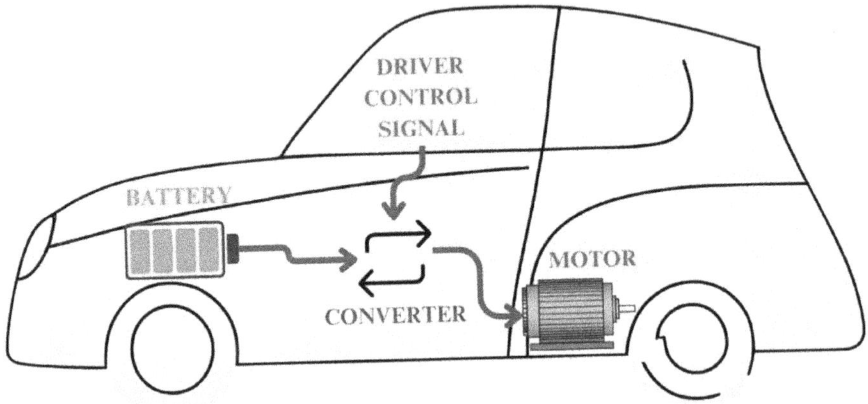

FIGURE 2.4 Block of EV.

2.3.2 Design of EV

Figure 2.4 shows the overall working of an EV. Electrical energy is stored in the battery and is provided as DC power to the converter, which converts and regulates it (to AC if needed) for the motor. The motor converts electrical energy into mechanical energy to run the wheels. The driver control signal, normally from the accelerator, controls the converter output to manage the vehicle's speed and torque. This design provides efficient power transfer and smooth vehicle operation.

2.3.3 Converters

2.3.3.1 Necessity of Converters in EV

EV converters are required devices that enable efficient operation and performance of an EV by controlling and converting electrical power among different powertrain components. Voltage conversion is one of the most significant operations of such converters because EVs utilise battery packs, whereas most of the vehicle systems, such as lighting, infotainment, and HVAC, operate at significantly lower voltages. Converters step down the voltage from the high-voltage battery to an appropriate level at which such auxiliary systems can be utilised without overloading the low-voltage circuits. Besides voltage conversion, converters also play another significant role in power management by controlling the flow of energy between the battery, motor, and other electrical equipment, and the DC-DC converters are classified as illustrated in Figure 2.5.

This regulation optimises energy use, making the EV more efficient and overall better-performing. In addition, charging compatibility requires converters because they enable the EV to be charged from a range of power sources of varying voltage levels and types of current. Charging from a household power supply or a fast-charging station, converters precondition power for charging the battery suitably. Converters also enable energy recovery through regenerative braking by converting kinetic energy to electrical energy for storage in the battery with high efficiency.

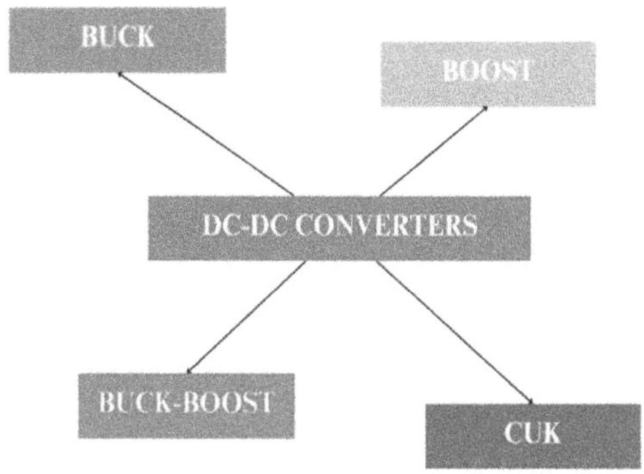

FIGURE 2.5 Types of DC-DC converters.

Proper inductor, capacitor, MOSFET, and diode selection is important to optimise converter efficiency, reduce ripple, and reduce losses. Inductors reduce ripple, capacitors reduce voltage, and low-loss switching devices improve performance. Optimal switching frequencies and soft-switching methods further improve efficiency by reducing switching losses. Finally, some converters are used for motor control for adjusting the power source of the electric motor to change speed and torque, providing exact control over the vehicle's motion. In essence, EV converters are necessary for managing electrical energy, enhancing vehicle efficiency, and enabling advanced features like regenerative braking and motor control.

2.3.3.2 Buck Converter

A buck converter, or step-down converter, is a high-efficiency DC-DC power converter which lowers the input voltage to a lower regulated output voltage, and thus is employed for powering loads operating at lower voltage levels. The fundamental action of a buck converter is to switch the MOSFET on and off extremely rapidly, passing energy through an inductor, diode, and capacitor to produce a constant output voltage. During the conduction of the switch, the current passes through the inductor and accumulates energy in a magnetic field; when the switch is turned off, the inductor delivers this energy to ensure continuity to the output.

The primary elements are the MOSFET switch, powered by a pulse-width modulation (PWM) signal; an inductor to store energy temporarily; a diode for allowing a path for current in the switch-off state; an output capacitor for filtering the voltage ripple; and a controller to control PWM as a function of the desired output voltage as indicated in Figure 2.6. Buck converters employ two wide operation modes: continuous conduction mode (CCM), with inductor current constant, appropriate for heavy loads; and discontinuous conduction mode (DCM), where the inductor current drops to zero in every cycle, being prevalent at light loads. Output voltage is set by the duty cycle of thc switch's on-time, so that the output voltage can be regulated by duty cycle based on the formula $V_{out} = V_{in}*D$. Inductor current ripple is calculated using: $L = (V_{in} - V_{out})*D/(f*\Delta I_L)$ where V_{in} is input voltage, V_{out} is output voltage, D is duty cycle, f is the switching frequency, and ΔI_L is inductor current ripple. Output voltage ripple filtering by a capacitor (C) is: $C = \Delta I_L/(8*f*\Delta V_{out})$, where ΔV_{out} is the

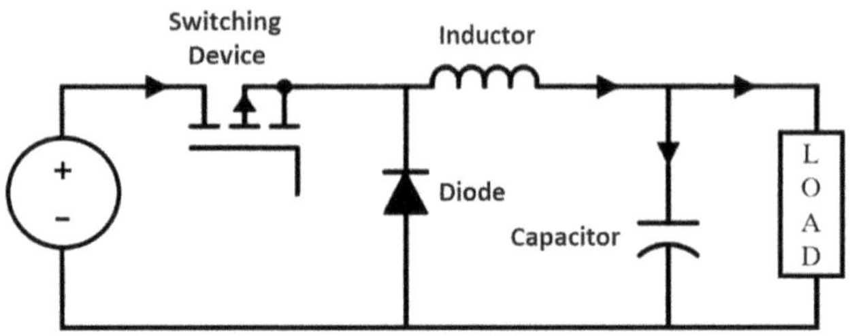

FIGURE 2.6 Circuit of buck converter.

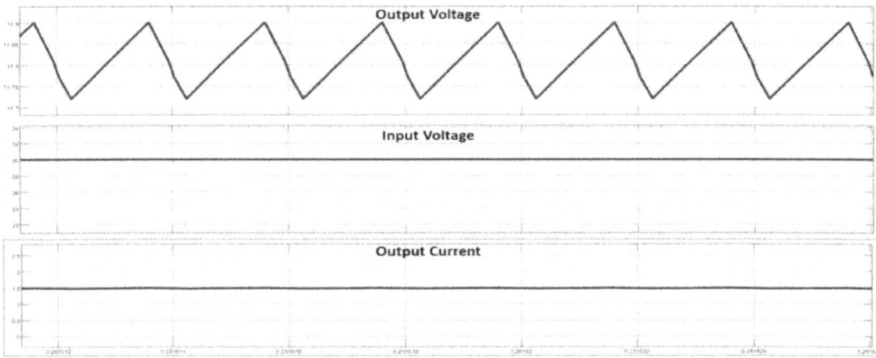

FIGURE 2.7 Waveform of buck converter.

permissible output voltage ripple. Average output current Iout is proportional to input current I_{in} through duty cycle D and can be computed by employing the equation: $I_{out}=I_{in}*D$, where I_{out} is the average output current, and I_{in} is the input current.

In the waveform, switch voltage is rectangular, inductor current is triangular, and output voltage is DC with very low ripple, as indicated in Figure 2.7. Buck converters are efficient, simple to control, provide a wide input voltage range, and have low output voltage ripple when properly designed. They are, however, limited to step-down operation and require special component selection for maximum performance, while their high-speed switching can produce electromagnetic interference (EMI) that may interfere with sensitive electronics nearby.

Control techniques of buck converters include voltage mode, current mode, and hysteretic control, each of which has different benefits regarding stability and response to load variation. Common applications of buck converters comprise power supplies for consumer electronics and battery-powered equipment, LED drivers, and solar charge controllers.

A synchronous type, a derivative topology, replaces a diode with a second MOSFET to reduce losses, improving efficiency, especially in high-current applications. For high-performance applications, multiphase buck converters, utilising several parallel buck stages, can reduce ripple and improve response time. Efficiency is also optimised by the correct selection of components, ideal switching frequencies, synchronous rectification, and adaptive control methods.

To ensure reliable operation, buck converters have protection features such as overcurrent and overvoltage protection and thermal shutdown. It also plays a crucial role in power management in many electronic devices and ensures reliable, efficient power conversion for many types of applications.

2.3.3.3 Boost Converter

A boost converter, or step-up converter, is a DC-to-DC power converter that increases the input voltage to a higher output voltage. The basic mode of operation requires a MOSFET switch, inductor, diode, and capacitor to boost the input voltage to the desired output. When the switch is conductive, energy is stored in the magnetic field of the inductor; when the switch is non-conductive, the energy is released, creating a voltage at the output that is higher than the input.

The key components comprise the MOSFET governed by a PWM signal to regulate energy transfer, an inductor to store energy, a diode to safeguard against reverse flow of current and maintain output stability, a capacitor to filter the output voltage, and a controller to regulate the PWM signal to produce the output voltage as required, as shown in Figure 2.8. The converter works in two operating modes: CCM, in which the inductor current remains greater than zero for the whole cycle, and DCM, in which the inductor current decreases to zero with each cycle, depending on the load. The output voltage is derived as $V_{out} = V_{in}/(1 - D)$, where D represents the duty cycle, which enables the regulation of the output voltage level. The inductor value is derived as $L = V_{in}*(1 - D)/(f*\Delta I_L)$, where V_{in} represents the input voltage, D the duty cycle, f the switching frequency, and ΔI_L the ripple current. A higher value of inductor is necessary to achieve low ripple current or a high value of switching frequency to provide for stability of operation and efficiency. The value of the capacitor is derived as $C = \Delta I_L/(8* f* \Delta V_{out})$, where ΔV_{out} represents the permissible voltage ripple.

At the waveform level, inductor current is triangular, switch voltage is rectangular, and output voltage is a DC level with little ripple, as illustrated in Figure 2.9. The benefits of boost converters are that they have a simple circuit topology, high efficiency, continuous input current, and are able to deliver a higher output voltage than the input. They also have drawbacks such as higher output ripple than buck

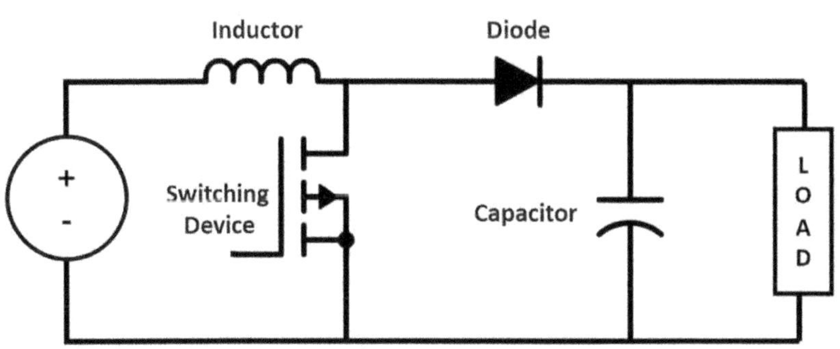

FIGURE 2.8 Circuit of boost converter.

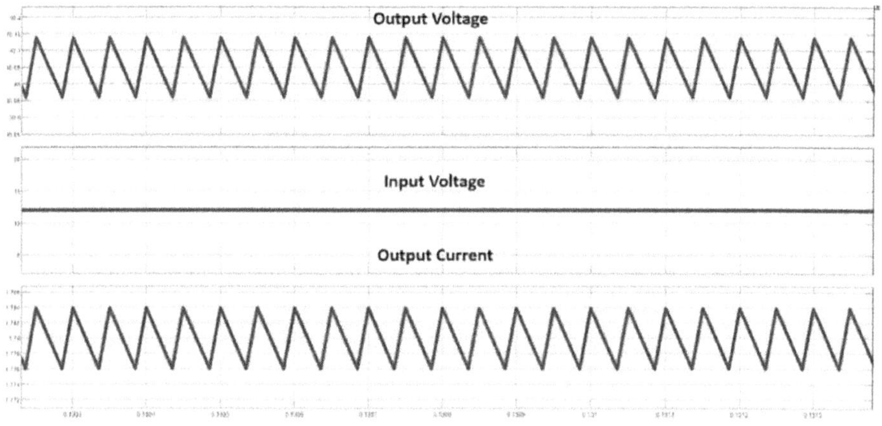

FIGURE 2.9 Waveform of boost converter.

converters, potential instability at high duty cycles, and a right-half-plane zero in the control-to-output transfer function, making control stability challenging.

The control methods employed are voltage mode, current mode, and peak current mode control, each with diverse advantages in terms of stability and load transition response. Common applications for the boost converter include battery supply systems, photovoltaic systems, LED drivers, and power factor correction modules. Design considerations include use of the appropriate inductance for ripple control, choice of the capacitor to facilitate the voltage stability, choice of an efficient switch, and choice of the diodes to reduce the efficiency losses.

A synchronous boost converter implementation substitutes the diode with a second MOSFET, decreasing losses further and improving efficiency, particularly at high currents. Interleaved boost converters, employing several parallel stages with phase-shifted control, can minimise ripple and enhance transient response in high-power applications.

Design issues for boost converters include regulating voltage stress on the devices, avoiding shoot-through in synchronous implementations, and restricting EMI caused by high-frequency switching. Higher performance topologies, such as coupled-inductor, cascaded, and tapped-inductor boost converters, can also be used to improve performance in specialised applications. These issues are significant in designing efficient and robust boost converters for step-up power conversion applications.

2.3.3.4 Buck-Boost Converter

A buck-boost converter is a DC-to-DC power converter that can buck and boost the input voltage, i.e., the input voltage can be higher or lower than the required output voltage. This versatility is achieved by controlling the switch duty cycle such that the converter can adapt to variations in the input voltage.

The fundamental topology consists of a MOSFET switch, inductor, diode, and capacitor, with the inductor in series with the switch, as shown in Figure 2.10. The operation is based on energy storage in the inductor when the switch is closed, followed by the release of energy to the output when the switch is opened leading to

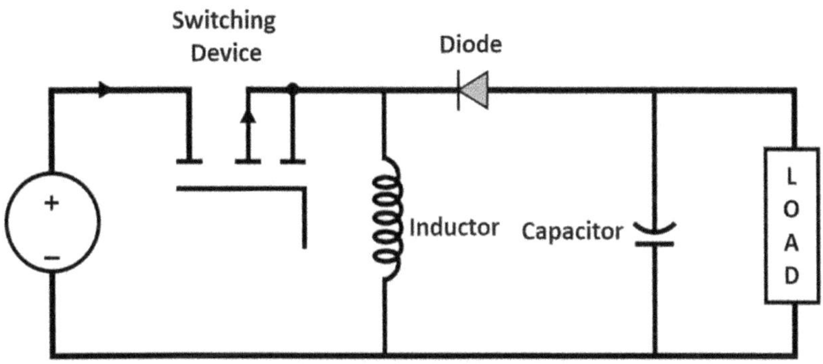

FIGURE 2.10 Circuit of buck-boost converter.

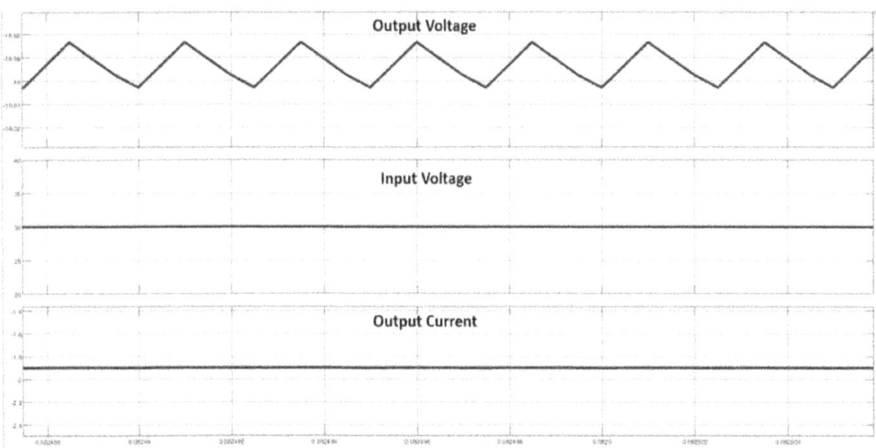

FIGURE 2.11 Waveform of buck-boost converter.

a stable, regulated output voltage. The output voltage may be regulated by the converter using the equation $V_{out} = D*V_{in}/(1-D)$, where D is the duty cycle. Notably, the output voltage in a conventional buck-boost has reversed polarity compared to the input, and requires an additional circuit, such as a non-inverting buck-boost, to correct the polarity. The value of the inductor is calculated by $L = V_{in}*D*(1-\mathrm{D})/(f*\Delta I_L)$, where V_{in} is the input voltage, f is the switching frequency, and ΔI_L is the ripple current. The value of the capacitor is calculated by $C = \Delta I_L/(8*f*\Delta V_{out})$, where ΔV_{out} is the acceptable output voltage ripple.

It operates in two modes of operation: CCM, in which the inductor current is positive for the entire cycle, and DCM, where the inductor current reaches zero in a cycle. The input and output waveforms are shown in Figure 2.11. These modes give the flexibility in the application of the buck-boost converter to various applications, while voltage mode, current mode, and peak current mode control techniques provide stability and dynamic response to load changes.

They are mainly used in solar power systems, battery systems, wide-range LED drivers, and automotive electronics, where varying input voltages necessitate a universal method of power conversion. The converters have some main benefits, such as a broad range of input voltage, universal regulation of output voltage, and a single-stage process for step-down and step-up. The converters possess limitations, including lower efficiency, complex control operation, and increased component stress due to pulsating current.

2.3.3.5 Ćuk Converter

A Ćuk converter is a DC-to-DC power converter with broad applications, utilised to increase and decrease the input voltage while maintaining the output voltage of the same polarity, and hence it is not identical to standard buck-boost converters. It has a unique topological circuit consisting of two inductors, two capacitors, a switching device (in most instances, a MOSFET), and a diode, as shown in Figure 2.12. In contrast to conventional converters, energy transfer occurs only through a coupling capacitor in Ćuk converters, allowing efficient energy transfer at higher frequencies of operation than in inductor-structured systems.

A Ćuk converter stores energy in inductors and transfers it to the capacitor, with constant input and output currents. The output may be stepped down or stepped up relative to the input by the equation $V_{out} = -D*V_{in}/(1-D)$, where D is the duty cycle. The input inductor is given by $L_1 = V_{in}*D/(f*\Delta I_L)$, and the output inductor is $L_2 = V_{out}(1-*D)/(f*\Delta I_L)$. The input capacitor is given by the equation $C_1 = \Delta I_L/(8*f*V_{in})$, and the output capacitor is $C_2 = \Delta I_L/(8*f*V_{out})$, where V_{in} = input voltage, V_{out} = output voltage, ΔI_L = ripple current, f = switching frequency, and ΔV_{in} and ΔV_{out} are voltage ripple values at input and output.

Some of the main benefits of the Ćuk converter are low input and output current ripple, low EMI since currents are continuous, and high efficiency, of the order of 85%–95%. The current continuity aspect can be observed from Figure 2.13. Voltage or current mode control techniques are employed for voltage or current control of the converter, and the converter operates either in CCM or in DCM mode, as per the design specification.

Common applications of the Ćuk converter include power supplies for sensitive electronics, battery charging systems, renewable energy systems such as solar and wind, and power management in electric drives. In spite of these advantages, the converter is more complex in topology compared to buck or boost converters and

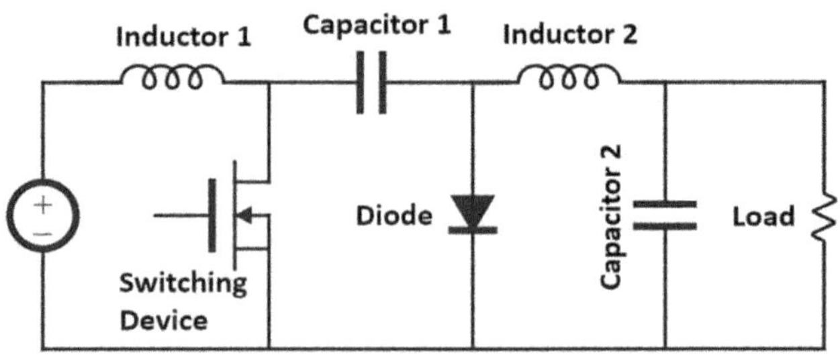

FIGURE 2.12 Circuit of the cuk converter.

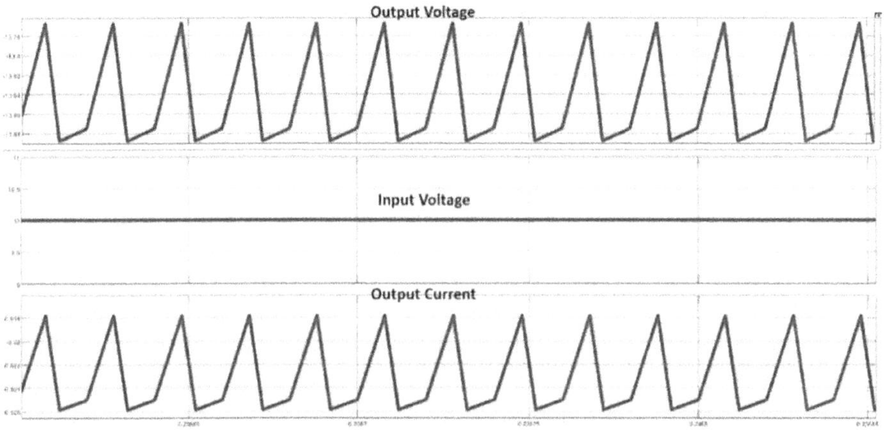

FIGURE 2.13 Waveform of the cuk converter.

requires careful selection of components. For instance, the coupling capacitor, which is necessary for energy transfer, is usually large, and the inductor design has a direct influence on current ripple and efficiency. Furthermore, the greater number of components and the complex control circuit of the converter make it more challenging to design and implement compared to single-function converters.

There are variations such as isolated converters, with a galvanic isolation transformer, and multiphase converters, distributing the load among a number of stages of the converter to enhance power handling and reduce ripple. More sophisticated approaches like synchronous rectification (with a MOSFET replacing the diode to reduce conduction loss) and interleaving (using multiple stages of the converter to achieve higher efficiency and power capacity) are employed to further optimise the operation.

Besides, good printed circuit board (PCB) layout and thermal management must be employed to manage high-frequency switching and minimise parasitic effects and EMI. Unlike other converter topologies, the Ćuk converter has low input and output current ripple and is generally higher in efficiency than average buck-boost converters, but at the expense of being more complicated to implement and design.

2.3.4 Motors

2.3.4.1 Role of Electric Motors in EV

Electric motors are the motive force for EVs, serving as the primary driving force. They take electrical energy from the vehicle battery and convert it into mechanical energy, which is transmitted to the wheels and makes the vehicle move. High efficiency is one of the primary benefits of electric motors, with some motors able to convert as much as 90% of electrical energy into useful mechanical energy. This benefits EV performance and range in a big way. They are capable of producing instant torque and facilitating regenerative braking, enabling quick acceleration, responsive handling, and improved efficiency through the regeneration of kinetic energy into

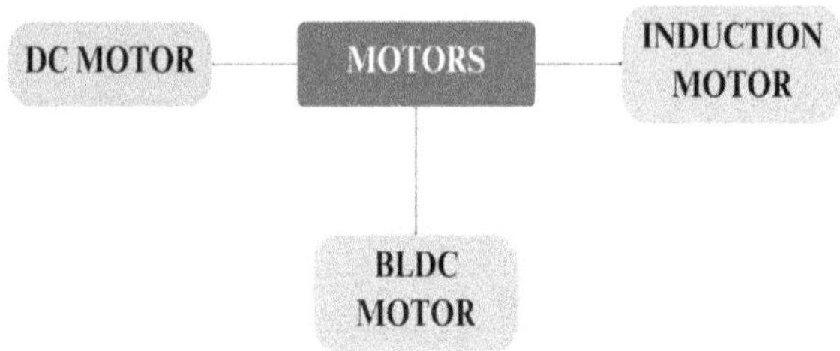

FIGURE 2.14 Types of motors.

electrical energy. The motors are classified with respect to types of current and operating principle, as illustrated in Figure 2.14.

Electric motors are also lightweight and compact in nature relative to ICE, and this makes them more adaptable to vehicle design and overall weight reduction of EVs. With fewer components, their structure is less complex than conventional engines, which leads to lower long-term service costs. Electric motors operate quietly, which helps cut down on noise pollution and makes for a more enjoyable driving experience. Their precise control boosts vehicle dynamics, leading to smoother acceleration, braking, and handling. Plus, some electric vehicles use multiple motors to enhance performance, provide all-wheel drive, and improve traction control. Moreover, electric motors can reverse direction with ease without intricate transmission systems, making design easier, improving reliability, and reducing mechanical failures. These characteristics make electric motors a perfect fit for EVs, providing considerable benefits in performance, efficiency, and driving comfort.

2.3.4.2 Direct Current (DC) Motor

DC motor is an electrical device that directly converts electrical energy into mechanical energy working based on the principle of electromagnetic induction. This motor has some prominent parts, which include the stator, rotor (or armature), commutator, brushes, and shaft, as shown in Figure 2.15.

The stationary part of the motor is called the stator; it usually has permanent magnets or electromagnets which form a fixed magnetic field. The rotor, or armature, is the rotating component and is built of wire coiled around an iron core; it forms a magnetic field when current flows through it. The commutator, a segmented copper cylinder attached to the rotor, reverses polarity at regular intervals so that the armature windings continue rotating. Stationary brushes connect to the commutator, thus providing current to the rotor, and the shaft transmits power from the actuator to the load.

The working principle starts when electric current travels through the armature coil, creating a magnetic field. This interplay between the stator's magnetic field and the rotor's magnetic field makes the rotor turn. As the rotor turns, a commutator continually changes polarity on the armature windings. Brushes provide electrical continuity between the rotating commutator and the power source, performing a very important function.

There are several types of DC motors, each suited to specific applications. Brushed DC motors are the traditional type with brushes and a commutator, while brushless DC motors use electronic commutation to eliminate the need for brushes.

FIGURE 2.15 DC motor.

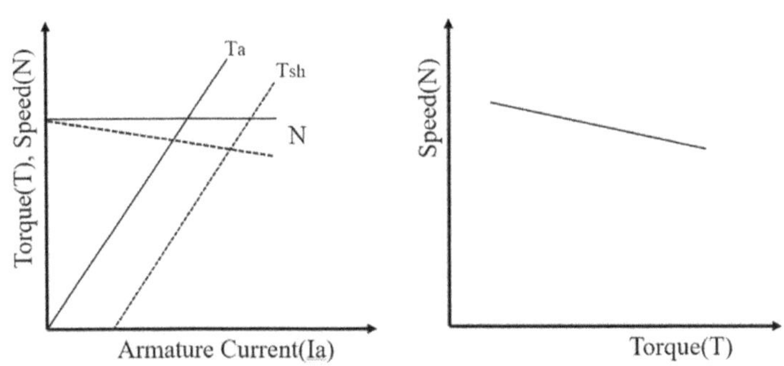

FIGURE 2.16 Characteristics curve of DC shunt motor.

Permanent magnet DC motors feature permanent magnets in the stator, enhancing efficiency for low-power applications. Separately excited DC motors have field windings powered by an independent power source, offering precise control over magnetic field strength. Series DC motors have field windings connected in series with the armature, providing high starting torque but varying speed with load changes. Shunt DC motors, with field windings parallel to the armature, deliver consistent speed under varying loads. The electromagnetic force (EMF) equation of a DC motor is given by $E_b = P\Phi NZ/60A$, where P is the number of poles, Φ is the flux per pole, N is the speed of the motor in rpm, Z is the number of conductors, and A is the number of parallel paths. The back EMF is defined by $E_b = V - I_a R_a$, where V is the supply voltage, I_a is the armature current, and R_a is the armature resistance. Figure 2.16 represents the characteristic curve of a DC shunt motor.

The advantages include simple speed control, high starting torque, and compact size. However, they may require more maintenance due to brush wear and have some limitations in high-power applications compared to AC motors. Despite this, their range and durability make them essential in numerous applications, particularly where speed control and high torque are important. In the automotive sector, the power devices include electric windows, windshield wipers, and cooling fans. Robotics and automation also employ DC motors for controlled movements. They are integral to computer peripherals such as hard drives and printers, industrial machinery, electric vehicles, toys, and small household appliances.

2.3.4.3 Induction Motor

An induction motor, or asynchronous motor, is an AC electric motor that operates based on the principle of electromagnetic induction. Key components and principles that define it include the stator, rotor, and air gap, as illustrated in Figure 2.17. The stator is the stationary outer part of the motor, which consists of a cylindrical core made of stacked laminations with slots to house the three-phase windings. These windings generate a rotating magnetic field when three-phase AC power is applied. The rotor, typically designed as a "squirrel cage" with aluminium or copper bars embedded in a laminated core, is the rotating part. Between the stator and rotor is a small air gap that enables the magnetic field to engage with the rotor.

The working principle is based on electromagnetic induction. When a three-phase AC supply energises the stator windings, a rotating magnetic field is created. This field induces currents in the rotor bars; this interaction between the induced currents and the stator's magnetic field generates torque, causing the rotor to rotate. The rotor's speed is always slightly less than the magnetic field's speed, a phenomenon called slip, which is essential for producing continuous torque.

These are of two general types: the squirrel cage and the wound rotor motors. The squirrel cage motor, extensively applied, has a simple, robust structure with no

FIGURE 2.17 Induction motor.

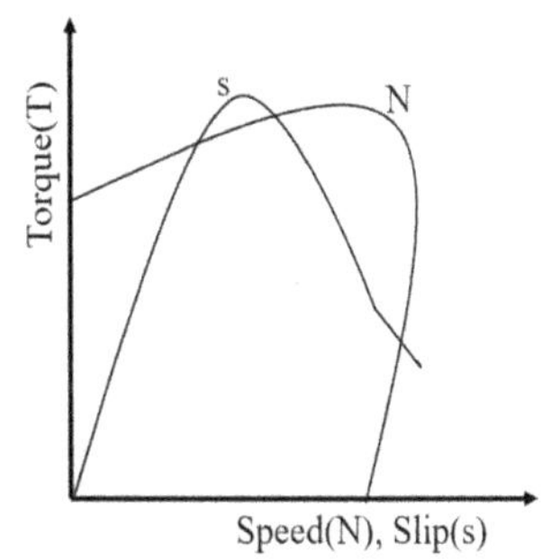

FIGURE 2.18 Characteristics curve of an induction motor.

external contact to the rotor. The wound rotor motor has windings that are connected to slip rings and permit external resistance adjustment, providing improved starting torque as well as controllable speed. Induced EMF is given by the equation $E_{\text{ind}} = vBl$, where v=velocity of the rotor, B=magnetic flux density, and l=length of conductors inside the magnetic field. The slip speed is defined by $N_{\text{slip}} = N_s - N$, where N_s = 120*frequency/poles or synchronous speed and N is the rotor speed of the motor in rpm. The slip is the ratio of the slip speed to the synchronous speed. Figure shows the characteristic curve of an induction motor. Figure 2.18 represents the characteristic curve of a DC shunt motor.

They are valued for their durability, low maintenance, high reliability, and cost-effectiveness. These motors are highly efficient, especially at rated load and speed, making them an excellent choice for applications requiring continuous operation. They are commonly used in industrial machinery, high voltage alternating current (HVAC) systems, elevators, and even electric vehicles, especially those utilising AC induction motor designs. The overall efficiency ofa motor is the ratio of power developed at the shaft, or output power, to the input power.

While speed control for induction motors was traditionally challenging, the development of variable frequency drives (VFDs) has made efficient speed control possible. Despite their high efficiency, induction motors exhibit high inrush currents during startup, requiring starting methods such as star-delta starting to control the initial current surge. Additionally, due to their lagging power factor, power factor correction may be needed in some settings to improve system efficiency.

The durability, reliability, and ease of operation directly from AC power sources make induction motors a good choice in commercial and industrial applications where simplicity and cost-effectiveness are critical.

2.3.4.4 Brushless DC (BLDC) Motor

A brushless DC (BLDC) motor is a type of electric motor that has many advantages over other brushed motors. It consists of a wound electromagnetic coil stator, a permanent magnet rotor supported on a shaft, and no commutator or brushes, as illustrated in Figure 2.19. The structure eliminates brush friction, making the motor efficient and long-lasting. BLDC motors do not utilise mechanical commutation but electronic commutation, whereby an electronic controller supplies the stator

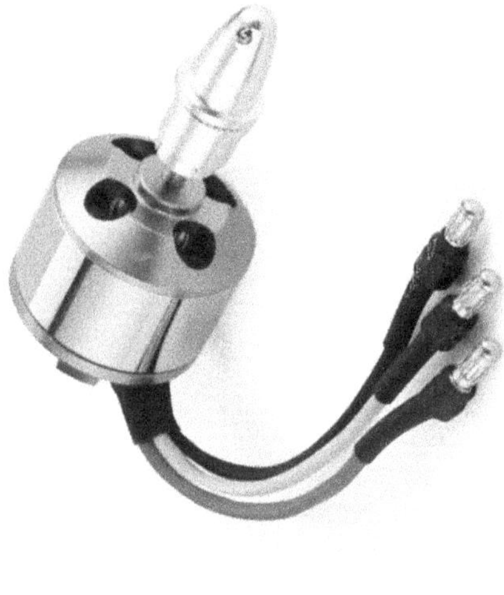

FIGURE 2.19 BLDC motor.

windings sequentially, thereby creating a revolving magnetic field that acts on the permanent magnets of the rotor to induce rotation.

The four main components are stator windings, the permanent magnet rotor, position sensors (such as Hall effect sensors for rotor position detection), and an electronic controller to drive the commutation process. These motors are divided into sensored or sensor less types based on the position sensing implemented by the sensor. Sensored motors use position sensors to enable precise control, whereas sensor less motors determine the rotor position using back-EMF, a voltage developed by the motor itself.

The back EMF E_b is proportional to the angular speed ω and is given by $E_b = K_e * \omega$, where K_e is the back EMF constant. Torque τ is directly related to current as $\tau = K_t * I$, where K_t is the torque constant. The mechanical power is $P_{mech} = \tau * \omega$, and the electrical power input is $P_{elec} = V * I$. Efficiency is the ratio of mechanical to electrical power, $\% \eta = P_{elec} / P_{mech} \times 100\%$. Figure 2.20 shows the characteristic curve of a DC shunt motor.

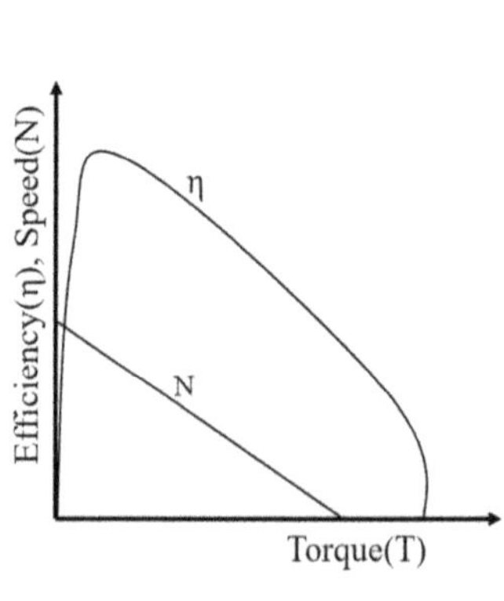

FIGURE 2.20 Characteristics curve of BLDC motor.

BLDC motors are energy-efficient, have a long life, low electromagnetic interference, good thermal dissipation, and accurate speed control. They need a complex control system, and the initial cost is higher due to the electronic control circuitry.

They are used in many applications such as computer fans, hard disk drives, electric vehicles, drones, industrial automation, and medical devices. They also have various control techniques such as trapezoidal (six-step) commutation, sinusoidal control, and field-oriented control (FOC), each being appropriate for various performance demands.

These motors have high starting torque, a linear speed-torque characteristic, a wide speed range, and low vibration and noise. BLDC motors are energy-efficient at variable speeds and loads, and result in less energy consumption than brushed motors. Maintenance is lower since there are no brushes, but occasional bearing and electronic component monitoring is required. Incorporation of sophisticated control algorithms, enhanced power electronics to achieve higher efficiency, reliability, and enhanced performance, and the design of high-performance magnetic materials are the new trends in BLDC motor design.

2.4 DESIGN OF BATTERY AND CONVERTERS

2.4.1 Battery

The battery suitable for EV vehicles is a lithium-ion type. It is a 24 V lithium-ion battery with a 50 Ah capacity, suitable for comprehensive analysis of DC-DC converter and motor suitability. It provides efficient and reliable power for the operation of the motor and also offers good density and long cycle life. Both continuous and peak current demands can be safely handled by the battery, ensuring smooth performance over time. This battery is ideal for applications requiring both power and endurance. Table 2.2 lists the battery specifications utilised in the suggested model.

2.4.2 Design of Boost Converter

A DC-DC boost converter raises the output voltage more than the input voltage in order to stabilise the voltage. It converts energy without affecting the power and modifies the DC voltage and current levels of the output. Figure 2.21 displays the circuit diagram, and Table 2.3 displays the parameters of the boost converter.

TABLE 2.2
Specifications of Battery

Parameters	Value
Type	Lithium-ion
Voltage	24 V
Rated capacity	50 Ah
State of charge (%)	100%

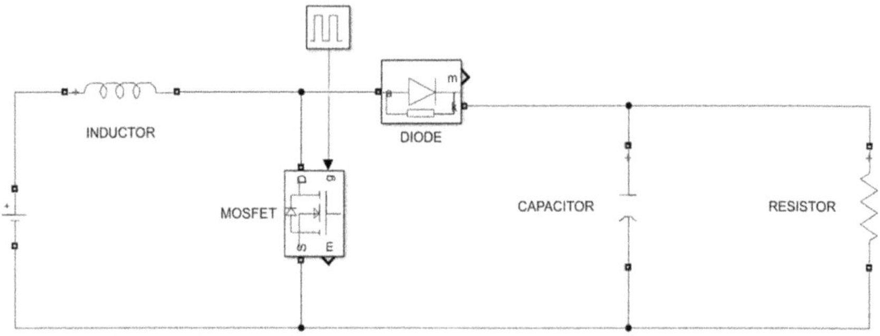

FIGURE 2.21 Equivalent circuit of boost converter.

TABLE 2.3
Parameters of Boost Converter

Parameter	Value
Input voltage	24 V
Output voltage	60 V
Inductor	4.32 μH
Capacitor	416.66 μF
Resistor	3.6 Ω
Duty cycle	60%
Switching frequency	40KHz

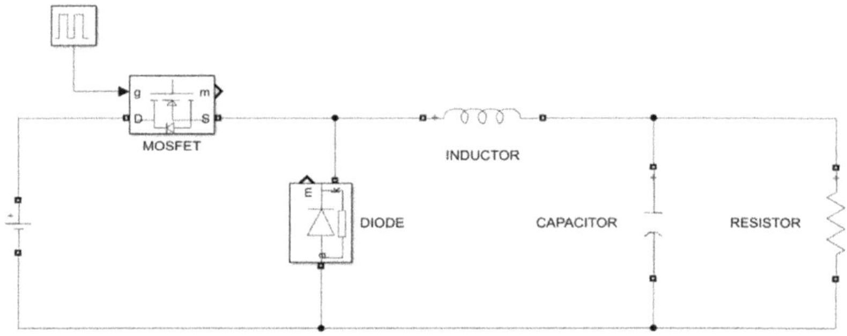

FIGURE 2.22 Equivalent circuit of buck converter.

2.4.3 Design of Buck Converter

A DC-DC buck converter reduces the output voltage below the input value in order to stabilise the voltage. It effectively converts energy while maintaining power stability and adjusting the DC voltage and current levels of the output. The circuit schematic for the buck converter equivalent is displayed in Figure 2.22, and the buck converter parameters are displayed in Table 2.4.

TABLE 2.4
Parameters of Buck Converter

Parameter	Value
Input voltage	72 V
Output voltage	60 V
Inductor	149.7 μH
Capacitor	28.92 mH
Resistor	72 Ω
Duty cycle	40%
Switching frequency	40 KHz

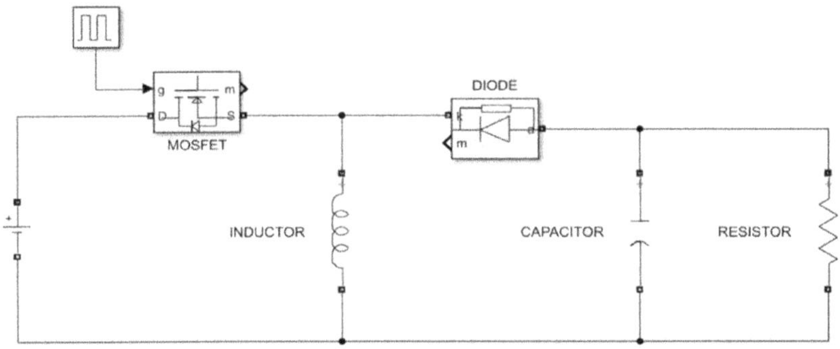

FIGURE 2.23 Equivalent diagram of a buck-boost converter.

TABLE 2.5
Parameters of Buck-Boost Converter

Parameters	Value
Input voltage	24V
Output voltage	60V
Inductor	429 H
Capacitor	14.875 mF
Resistor	120 Ω
Duty cycle	71%
Switching frequency	40KHz

2.4.4 Design of Buck-Boost Converter

A DC-DC buck-boost converter can either increase or lower the output voltage in relation to the input voltage. Providing versatility in power supply management, it efficiently converts energy and modifies the DC voltage and current levels of the output. An analogous circuit schematic is displayed in Figure 2.23, and theconverter specifications are illustrated in Table 2.5.

2.5 SIMULATION OF EV COMPONENTS

2.5.1 Integrating a Battery with Boost Converter System and BLDC Motor

Figure 2.24 shows a simulation of motor drive system, which contains various blocks for the speed control and operation of a BLDC motor in an efficient way.

The EV system is powered by a lithium-ion battery of 36Vnominal voltage and 50Ah capacity. The simulation is conducted in MATLAB 2024a. A 24V output from the lithium-ion battery is fed directly to a DC-DC boost converter. This converter increases the voltage, which is then supplied to a 1,000W BLDC motor. The motor runs at 60V and reaches a speed of 3,000 rpm. The BLDC motor utilised is a standard trapezoidal type.

Figure 2.25 displays the speed vs time graph of the BLDC motor. Initially, the motor speed increases rapidly and reaches its target speed of 3,000 rpm at 0.01 seconds, and reaches a steady state after a short period. This behaviour indicates a fast response time and stable operation of the BLDC motor.

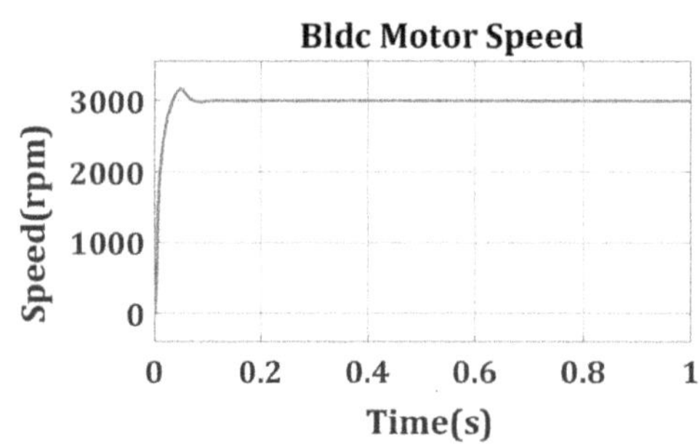

FIGURE 2.24 BLDC motor with boost converter.

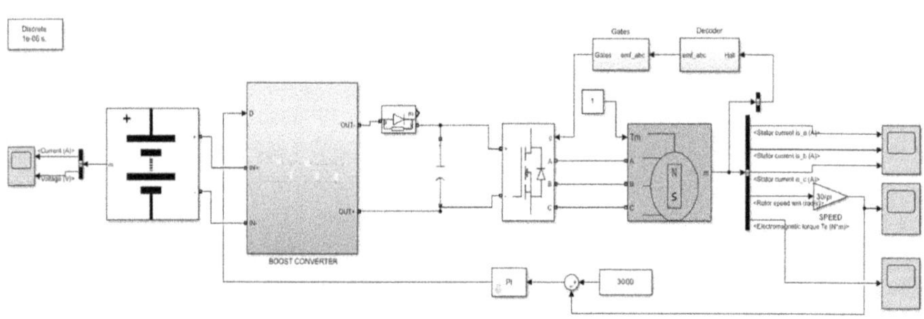

FIGURE 2.25 Output waveform of BLDC motor with boost converter.

2.5.2 Integrating a Battery with Boost Converter System and DC Motor

Figure 2.26 illustrates the simulation of a motor drive system, comprising various blocks designed to efficiently manage the speed and operation of the DC motor.

A lithium-ion battery with a 50Ah rated capacity and 24V nominal voltage powers the system. A DC-DC boost converter, which raises the voltage, is directly attached to the battery's 24V output. Like earlier converter-driven setups, the boosted voltage is fed into a 1,000W DC motor that runs at 60V and reaches a speed of 3,000 rpm.

In Figure 2.27, the speed versus time graph of the DC motor shows that initially the motor accelerates to 2,850 rpm but then drops sharply to around 150 rpm for a brief period of 0.05 seconds. After this dip, the motor gradually recovers and reaches a constant speed of 3,000 rpm by 0.25 seconds, maintaining this steady speed thereafter. This pattern reflects the motor stabilising to its desired operating speed after an initial fluctuation.

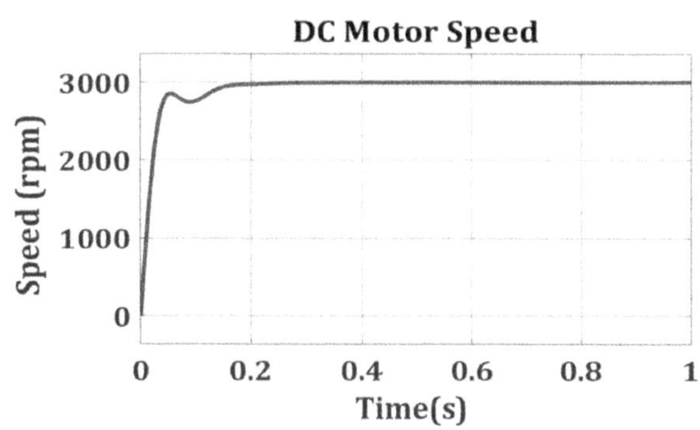

FIGURE 2.26 DC motor with boost converter.

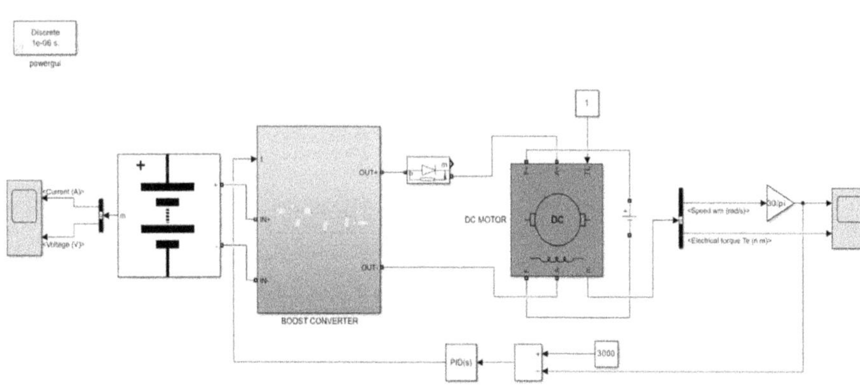

FIGURE 2.27 Output waveform of DC motor with boost converter.

2.5.3 Integrating a Battery with Boost Converter System and Induction Motor

Figure 2.28 presents the simulation of a motor drive system, featuring various blocks designed to efficiently adjust the speed and operation of an induction motor.

A 50Ah rated capacity lithium-ion battery with a nominal voltage of 36V powers the system. A DC-DC boost converter is directly linked to the battery's 24V output, increasing the voltage. After that, a 1,000W induction motor running at 60V and reaching 3,000 rpm is fed the increased voltage.

Figure 2.29 illustrates the speed vs. time of the induction motor. At the beginning, the motor accelerates gradually, reaching 2,200 rpm within the first 0.5 seconds. After a brief fluctuation, it continues to accelerate steadily. The motor eventually reaches its target speed of 3,000 rpm after about 2.5 seconds. Once at the desired speed, the motor maintains a consistent 3,000 rpm with minimal fluctuations, indicating stable operation. Compared to the previous graph, the acceleration is slower, reflecting a more gradual response.

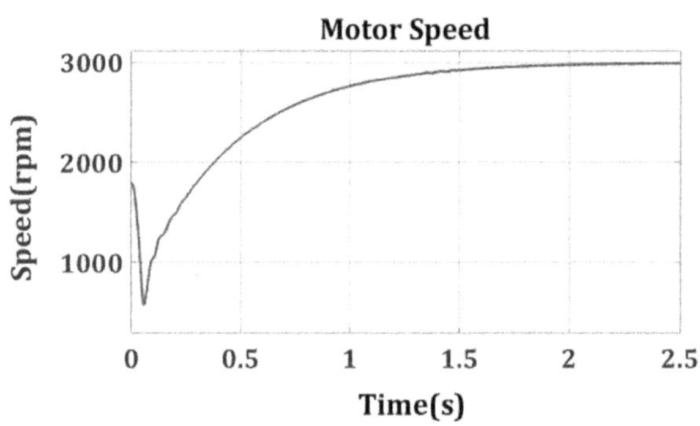

FIGURE 2.28 Induction motor with boost converter.

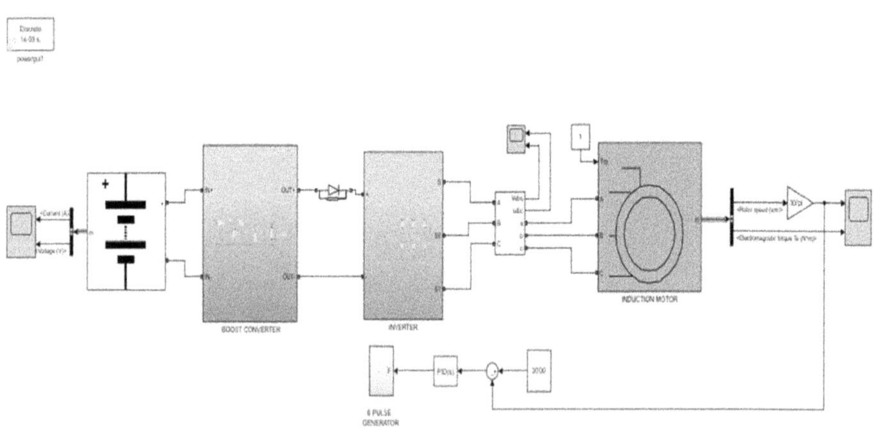

FIGURE 2.29 Output waveform of induction motor with boost converter.

2.5.4 Integrating a Battery with a Buck Converter System and BLDC Motor

Figure 2.30 shows the simulation of a motor drive system, incorporating various blocks designed for effective speed control and operation.

The system is powered by a lithium-ion battery with a nominal voltage of 72V and a rated capacity of 50Ah. The 72V output from the battery is directly connected to a DC-DC buck converter, which steps down the voltage. The reduced voltage is supplied to a 1,000W BLDC motor that operates at 60V and reaches a speed of 3,000rpm. The BLDC motor used is a standard trapezoidal type.

Figure 2.31 shows the speed vs. time graph of the BLDC motor. Initially, the motor speed increases rapidly and reaches its target speed of 3,000rpm at 0.01seconds and reaches a steady state after a short period. This behaviour indicates a fast response time and stable operation of the motor.

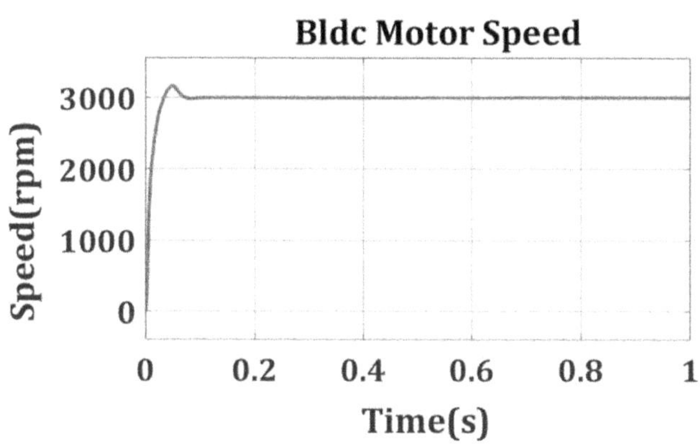

FIGURE 2.30 BLDC motor with buck converter.

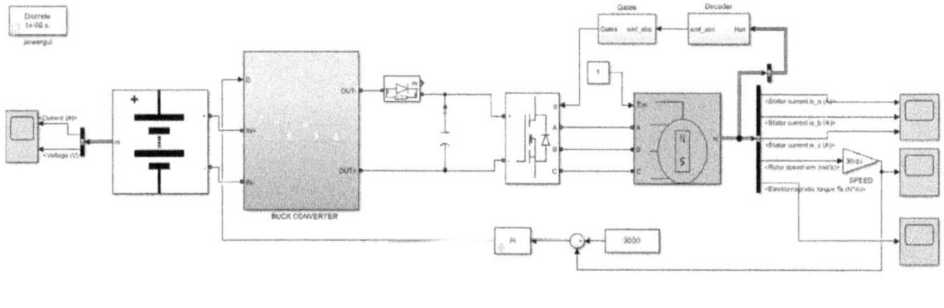

FIGURE 2.31 Output waveform of BLDC motor with buck converter.

2.5.5 Integrating a Battery with Buck Converter System and DC Motor

Figure 2.32 illustrates the simulation of a motor drive system, featuring various blocks designed for optimised speed control and operation of a DC motor.

A lithium-ion battery with a rated capacity of 50Ah and a nominal voltage of 72V powers the system. The battery's 72V output is directly connected to a converter, which reduces the voltage to 60V. Similar to setups that use a boost converter, this 60V output is directly delivered to a 1,000W DC motor that runs at 60V and reaches a speed of 3,000rpm.

Figure 2.33 shows the speed versus time graph of the DC motor. The motor initially accelerates to 2,850rpm but then drops sharply to around 150rpm for a brief period of 0.05 seconds. After this dip, the motor gradually recovers and reaches a constant speed of 3,000rpm by 0.25 seconds, maintaining this steady speed thereafter. This pattern reflects the motor stabilising to its desired operating speed after an initial fluctuation.

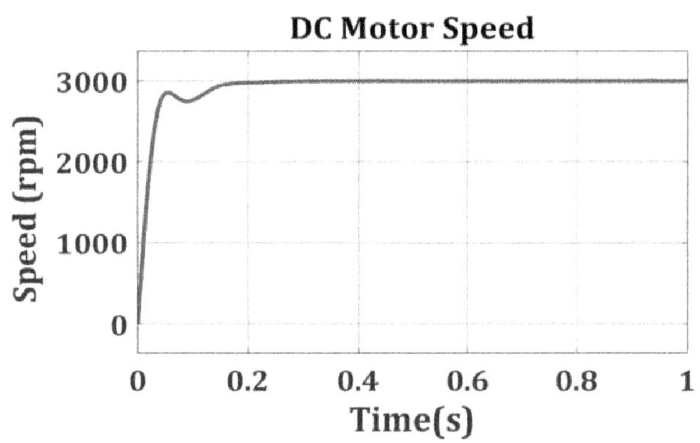

FIGURE 2.32 DC motor with buck converter.

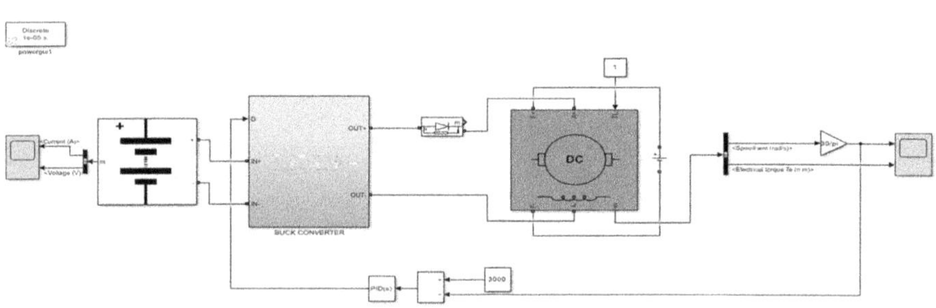

FIGURE 2.33 Output waveform of DC motor with buck converter.

2.5.6 Integrating a Battery with Buck Converter System and Induction Motor

Figure 2.34 showcases the simulation of a motor drive system, consisting of various blocks designed to achieve efficient speed control and operation.

The system is powered by a lithium-ion battery of rated capacity 50Ah and a nominal voltage of 72V. A DC-DC buck converter, which cuts down the voltage to 60V, is immediately connected to the battery's 72V output. After that, the 60V output is fed into a 1,000W induction motor, which runs at 60V and reaches 3,000 rpm.

Figure 2.35 illustrates the speed versus time of the induction motor. At the beginning, the motor accelerates gradually, reaching 2,200 rpm within the first 0.5 seconds. After a brief fluctuation, it continues to accelerate steadily. The motor eventually reaches its target speed of 3,000 rpm after about 2.5 seconds. Once at the desired speed, the motor maintains a consistent 3,000 rpm with minimal fluctuations, indicating stable operation. Compared to the previous graph, the acceleration is slower, reflecting a more gradual response.

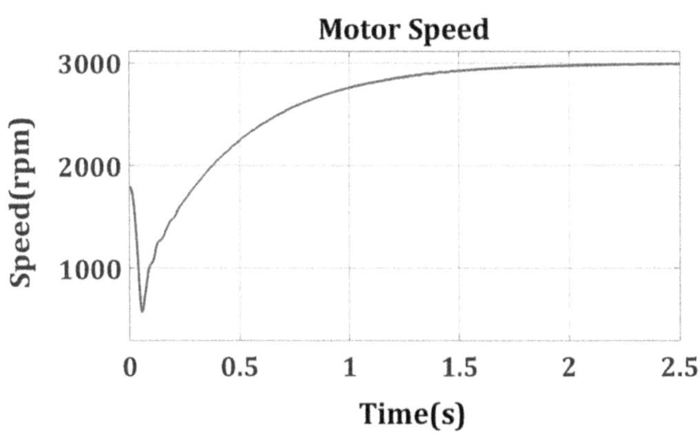

FIGURE 2.34 Induction motor with buck converter.

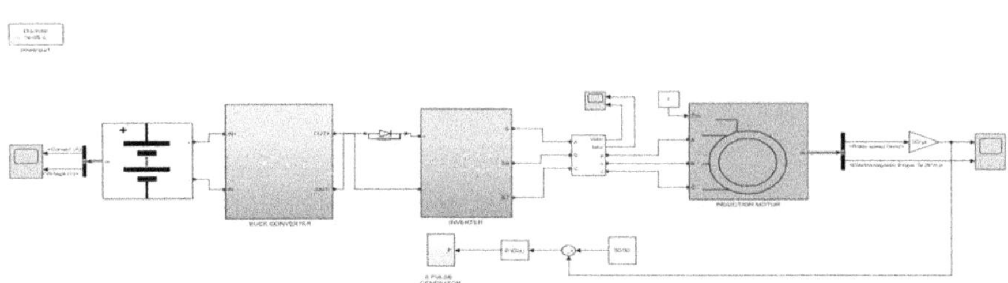

FIGURE 2.35 Output waveform of induction motor with buck converter.

2.5.7 Integrating a Battery with Buck-Boost Converter System and BLDC Motor

Figure 2.36 shows the simulation of a motor drive system, featuring various blocks designed for effective speed control and operation of a BLDC motor.

The system is powered by a Lithium-Ion battery with a nominal voltage of 24V and a rated capacity of 50Ah. The 24V output from the battery is directly connected to boost mode operation, which raises the voltage. The boosted voltage is then supplied to a 1,000W BLDC motor that operates at 60V and achieves a speed of 3,000rpm. The BLDC motor used is a standard trapezoidal type.

Figure 2.37 displays the speed versus time graph of the BLDC motor. Initially, the motor speed increases rapidly and reaches its target speed of 3,000rpm at 0.01seconds, after which it quickly reaches steady-state operation. This behaviour indicates a fast response time and stable operation.

2.5.8 Integrating a Battery with Buck-Boost Converter System and DC Motor

Figure 2.38 illustrates the simulation of a motor drive system, which includes various blocks developed to efficiently adjust the speed and operation of a DC motor.

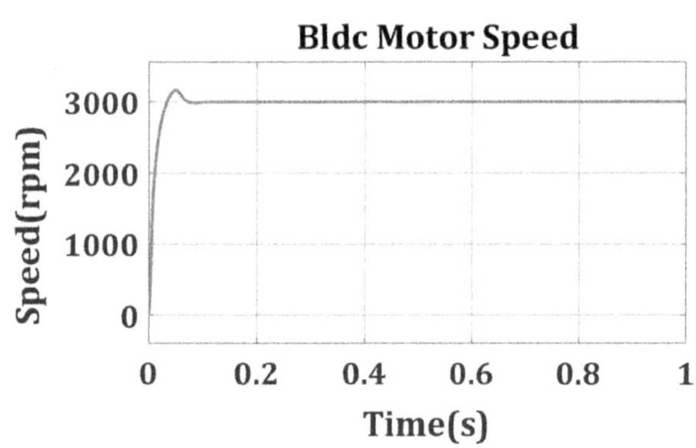

FIGURE 2.36 BLDC motor with buck-boost converter.

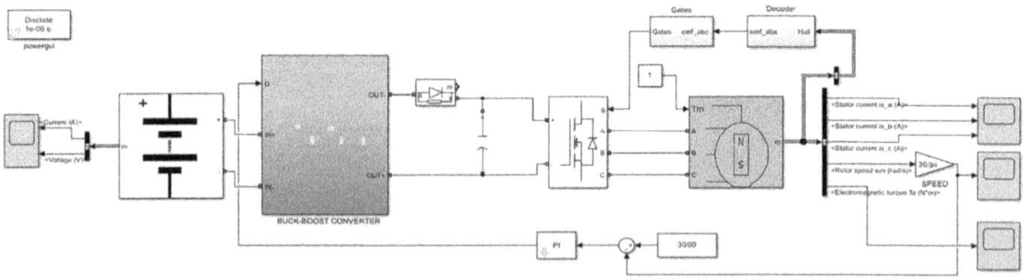

Figure 2.37 Output waveform of BLDC motor with buck-boost converter.

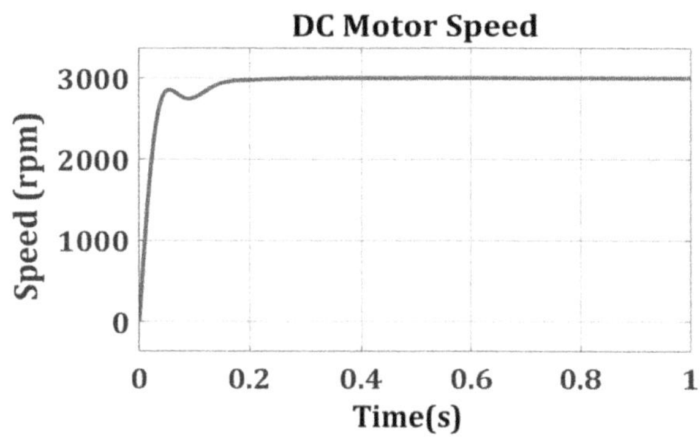

FIGURE 2.38 DC motor with boost converter.

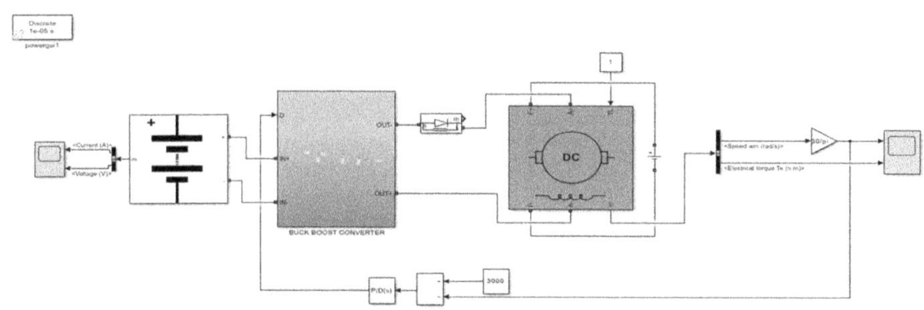

FIGURE 2.39 Output waveform of DC motor with boost converter.

A lithium-ion battery with a rated capacity of 50Ah and a nominal voltage of 24V powers the system. The converter receives the battery's 24V output directly. Like earlier setups utilising voltage converters, this converter raises the voltage to 60V, which is subsequently delivered to a 1,000W DC motor that runs at 60V and reaches a speed of 3,000 rpm.

In Figure 2.39, the speed versus time graph of the DC motor shows that the motor initially accelerates to 2,850 rpm but then drops sharply to around 150 rpm for a brief period of 0.05 seconds. After this dip, the motor gradually recovers and reaches a constant speed of 3,000 rpm by 0.25 seconds, maintaining this steady speed thereafter. This pattern reflects the motor stabilising to its desired operating speed after an initial fluctuation.

2.5.9 Integrating a Battery with Buck-Boost Converter System and Induction Motor

Figure 2.40 shows simulation of a motor drive system, which contains various blocks for the speed control and operation of an induction motor in an efficient way.

A 36V lithium-ion battery with a 50Ah capacity powers the system. The battery's 24V output is fed through a converter. The converter controls the voltage, generating

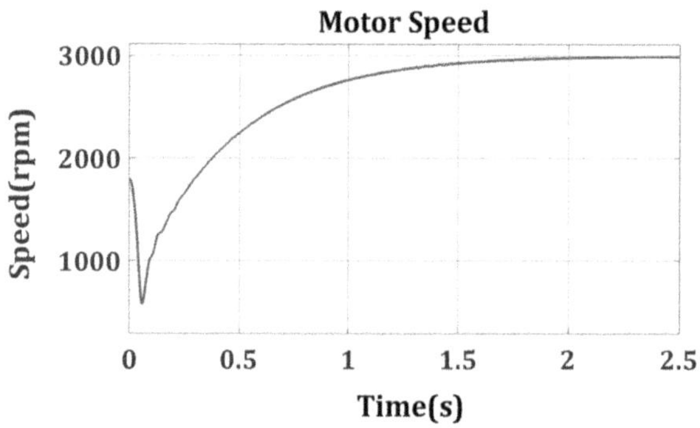

FIGURE 2.40 Induction motor with boost converter.

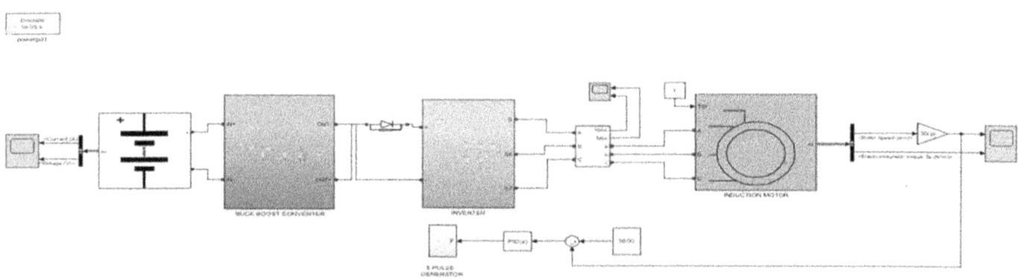

FIGURE 2.41 Output waveform of induction motor with boost converter.

a 60V output that is fed straight into a 1,000W induction motor. This motor has a maximum speed of 3,000rpm and runs on 60V.

Figure 2.41 illustrates the speed versus time of the induction motor. At the beginning, the motor accelerates gradually, reaching 2,200rpm within the first 0.5 seconds. After a brief fluctuation, it continues to accelerate steadily. The motor eventually reaches its target speed of 3,000rpm after about 2.5 seconds. Once at the desired speed, the motor maintains a consistent 3,000rpm with minimal fluctuations, indicating stable operation. Compared to the previous graph, the acceleration is slower, reflecting a more gradual response.

2.6 COST COMPARISON

The cost and scalability of EVs are critical to their widespread adoption, with the power system (battery, motor, converter, and BMS) accounting for 40%–45% of total expenses. Variations in chassis and body design influence material and structural costs, while transmission and wheels remain relatively standardised, offering potential cost optimisations at scale. Electronics and control systems contribute to overall expenses, especially in high-end models featuring advanced connectivity and automation. Additionally, manufacturing and assembly costs depend on production complexity and efficiency. While premium EVs focus on performance and technology, their higher costs limit scalability. Conversely, cost-effective models maximise

TABLE 2.6
Cost Comparison of EV's

Component	MG	Tata	Citroen	Eco-Drive
Power system	\$4,100	\$5,100	\$6,300	\$2,600
Chassis and body	\$2,600	\$3,200	\$4,000	\$2,400
Transmission and wheels	\$1,500	\$1,900	\$2,400	\$1,500
Electronics and controls	\$1,400	\$1,800	\$2,000	\$1,200
Misc. and assembly costs	\$1,500	\$2,700	\$3,100	\$900
Total cost	\$10,100	\$12,700	\$15,800	\$8,600

battery technology, lightweight materials, and effective manufacturing, and thus are more feasible for large-scale adoption. To achieve scalability, future advancements will have to strive to decrease the cost of batteries, adopt modular design, and increase manufacturing efficiency through automation and material technology, thus making EVs more sustainable and affordable worldwide (Table 2.6).

2.7 CONCLUSION

This chapter presents the interaction between DC-DC converters and motor technologies in electric vehicle (EV) applications from an efficiency, stability, and dynamic-response perspective to meet sustainable transport demands. The research emphasises the significance of selecting the right converter topology and motor in order to achieve maximum EV performance. Of the converter topologies under test—buck, boost, and buck-boost—their design parameters, efficiency, and control strategies were comprehensively studied in order to find the most suitable configurations under different situations.

MATLAB/Simulink simulations provided comparative insights into the performance characteristics of brushed DC motors (BDC), induction motors, and brushless DC (BLDC) motors when used with these converters. The results confirm that the BLDC motor is the best of the three motors, with very good speed response and stability. It accelerates very rapidly to 3,000 rpm in 0.1 seconds with very little oscillation, which makes it very effective to use in applications with precise speed control and rapid dynamic responses. While the DC motor also provides good speed control with some initial overshoot, it is less responsive than the BLDC motor. The induction motor, with its inherently higher inductance, however, has slower acceleration and stabilisation and is hence less effective for use in applications requiring instant speed adjustment.

In summary, the BLDC motor, supplemented by a proper DC-DC converter, is the best option for EV applications. Its high response speed, high stability, and high efficiency are entirely consistent with the performance requirements of state-of-the-art electric vehicles, guaranteeing sustainability as well as reliability for transport systems. Such conclusions open the door for further development of EV technology by establishing the important role of power electronics and motor selection in guaranteeing optimal vehicle performance.

REFERENCES

1. N. Bahman, D. Alalaiwat, Z. Abdulmohsen, M. Al Khalifa, S. Al Baharna, M. Al-Mannai, A. Younis. (2022). A critical review on global CO_2 emission: Where do industries stand? *Reviews on Environmental Health*, doi: 10.1515/reveh-2022-0105
2. A. Hussain, F. Khan, O. Albalawi. (2024). Modeling and monitoring CO_2emissions in G20 countries: A comparative analysis of multiple statistical models. *Sustainability*, doi: 10.3390/su16146114
3. C.-C. Lee, Y.-N. Zhao. (2023). Heterogeneity analysis of factors influencing CO_2 emissions: The role of human capital, urbanization, and FDI. *Renewable & Sustainable Energy Reviews*, doi: 10.1016/j.rser.2023.113644
4. R. Wu, J. Wang, S. Wang, K. Feng. (2021). The drivers of declining CO_2 emissions trends in developed nations using an extended STIRPAT model: A historical and prospective analysis. *Renewable & Sustainable Energy Reviews*, doi: 10.1016/J.RSER.2021.111328
5. L.C. Voumik, Md. A. Islam, S. Ray, N.Y.M. Yusop, A.R. Ridzuan. (2023). CO_2emissions from renewable and non-renewable electricity generation sources in the G7 countries: Static and dynamic panel assessment. *Energies*, doi: 10.3390/en16031044
6. H. Wang, Y. Wang. (2022). Estimating CO_2 emissions using a fractional grey Bernoulli model with time power term. *Environmental Science and Pollution Research*, doi: 10.1007/s11356-022-18803-0
7. T. Choudhry, U. Nawaz, K. Azeem, G. Syed, A. Haider, S. Ahmad. (2023). Carbon emissions, environmental distortions, and impact on growth. *Energy Economics*, doi: 10.1016/j.eneco.2023.107040
8. M.D. Garba, M. Usman, S. Khan, F. Shehzad, A. Galadima, M.F. Ehsan, A.S. Ghanem, M. Humayun. (2021). CO_2 towards fuels: A review of catalytic conversion of carbon dioxide to hydrocarbons. *Journal of Environmental Chemical Engineering*, doi: 10.1016/J.JECE.2020.104756
9. D. Butnicu. (2024). Comparative study for DC-DC converter output bank's reliability evaluation using prediction standards MIL-HDBK-217F vs. telcordia SR-332. *Energies*, doi: 10.3390/en17163957
10. A.H. Khanbabaie, M.N. Azari, T. Nouri, M. Shaneh. (2024). A bidirectional isolated DC-DC converter with low voltage stresses for energy storage systems integration. *IET Power Electronics*, doi: 10.1049/pel2.12725
11. Y. Zhu, Z. Ye, R.C.N. Pilawa-Podgurski. (2024). Modeling and analysis of switched-capacitor converters with finite terminal capacitances. *IEEE Transactions on Power Electronics*, doi: 10.1109/tpel.2023.3313562
12. S. Harini, C. Nallaperumal, A. Hosseinpour. (2024). Analysis, design, and reliability evaluation of a modified multi-port quasi-resonant converter with high gain. *IET Power Electronics*, doi: 10.1049/pel2.12691
13. S. Paladhi, Q. Hong, C. Booth. (2024). A reliable accelerated protection scheme for converter-dominated power networks. *IEEE Transactions on Industry Applications*, doi: 10.1109/tia.2024.3397640
14. F. Mohammadi, A. Khorsandi. (2024). Dual-input single-output high step-up DC–DC converter for renewable energy applications. *IET Power Electronics*, doi: 10.1049/pel2.12646
15. Z. Qi, Md. B. Hossain, Md. R. Islam, Md. A. Rahman. (2024). Advancement in converter topology, control, and power management: Magnetic linked power converter for emerging applications. *IEEE Access*, doi: 10.1109/access.2024.3424872
16. G. Vazquez-Guzman, C.J. Rodríguez-Cortés, P.R. Martínez-Rodríguez, J.M. Sosa, D. Langarica-Córdoba, M. Malinowski, E. Stevens-Navarro. (2024). Design of a step-down non-isolated DC-DC n-cell converter with a high reduction ratio. *IEEE Access*, doi: 10.1109/access.2024.3438090

17. Y. Xiao, Z. Zhang, T.-G. Zsurzsan, M.A.E. Andersen, M. MacFadyen, Y. Liao, R. Peña-Alzola, W. Yuan, M. Zhang. (2024). Design automation using exclusion-based hierarchical computation for power electronics converters in harsh environments. *IEEE Open Journal of Power Electronics*, doi: 10.1109/ojpel.2024.3439593
18. A. Bilsalam, N. Ketprapajun, P. Insri, H. Sawachan, D. Guilbert, P. Thounthong. (2023). Analysis and small-signal modelling technique for support bus DC-link of front-end coupling inductance high step-up single switch boost converter in low voltage renewable source. *IET Power Electronics*, doi: 10.1049/pel2.12638
19. S. Prakash, K. Boopathy. (2024). High speed BLDC motor for grid tied PV based EV system using hybrid PSO-spotted hyena optimized PI controller. *International Journal of Applied Power Engineering*, pp. 768–782, doi: 10.11591/ijape.v13.i3
20. C. Yang, B. Song, Y. Xie, S. Lu, X. Tang. (2022). Stable simultaneous inertia and disturbance torque identification for SPMSM drive systems subject to mismatched rotor flux linkage. *IEEE Journal of Emerging and Selected Topics in Power Electronics*, vol. 10, no. 2, pp. 2445–2462, doi: 10.1109/JESTPE.2021.3111122
21. R. Wrobel. (2022). A technology overview of thermal management of integrated motor drives – Electrical machines. *Thermal Science and Engineering Progress*, doi: 10.1016/j.tsep.2022.101222
22. J.-C. Son, J.-Y. Kim, J.-W. Choi, D.-K. Lim, H.-K. Yeo. (2022). Performance enhancement of the IPMSM for HEV applications using grain-oriented electrical steel and design optimization. *IEEE Access*, vol. 10, pp. 46599–46607, doi: 10.1109/ACCESS.2022.3170896
23. E. Quintero-Manriquez, E.N. Sanchez, M.E. Antonio-Toledo, F. Munoz. (2021). Neural control of an induction motor with regenerative braking as electric vehicle architecture. *Engineering Applications of Artificial Intelligence*, doi: 10.1016/J.ENGAPPAI.2021.104275
24. J.-G. Lee, D.-K. Lim. (2021). A stepwise optimal design applied to an interior permanent magnet synchronous motor for electric vehicle traction applications. *IEEE Access*, doi: 10.1109/ACCESS.2021.3105119
25. X. Sun, B. Wan, G. Lei, T. Xiang, Y. Guo, J. Zhu. (2021). Multi-objective and multi-physics design optimization of a switched reluctance motor for electric vehicle applications. *IEEE Transactions on Energy Conversion*, doi: 10.1109/TEC.2021.3078547
26. V. Gurusamy, K.H. Baruti, M. Zafarani, W. Lee, B. Akin. (2021). Effect of magnets asymmetry on stray magnetic flux based bearing damage detection in PMSM. *IEEE Access*, doi: 10.1109/ACCESS.2021.3076779
27. Z. Zhou, C. Xia, T. Shi, Q. Geng. (2021). Model predictive direct duty-cycle control for PMSM drive systems with variable control set. *IEEE Transactions on Industrial Electronics*, doi: 10.1109/TIE.2020.2978721
28. A. Fereidooni, S.A. Davari, C. Garcia, J. Rodriguez. (2021). Simplified predictive stator current phase angle control of induction motor with a reference manipulation technique. *IEEE Access*, doi: 10.1109/ACCESS.2021.3070790

3 A Cascaded Quadratic Boost DC-DC Converter for Renewable Energy Integration

Yuvanshankar Vijayakumar,
Vetri Ashwath Senthilkumar,
Mruthunjay Vijayamurali,
and Mahalingam Prabhakar

3.1 INTRODUCTION

Renewable energy is crucial in addressing the issue of climate change and reducing dependence on fossil fuels. The use of renewable energy sources (RES) such as solar, wind, and hydropower not only reduces greenhouse gas emissions but also significantly aids in meeting electrical energy demand. The energy converted from various RES is integrated to form a microgrid. However, the output voltage obtained from photovoltaic (PV) panels is low [1]. A suitable power electronic converter is essential to fruitfully utilize the energy converted from RES. Figure 3.1 depicts the schematic diagram of a solar PV-fed DC microgrid. The output obtained from the DC bus can be utilized in AC form by connecting an inverter. Due to modern-day DC native loads, a DC bus is gaining popularity [2]. High-gain DC–DC boost converters play a vital role in the effective utilization of renewable energy.

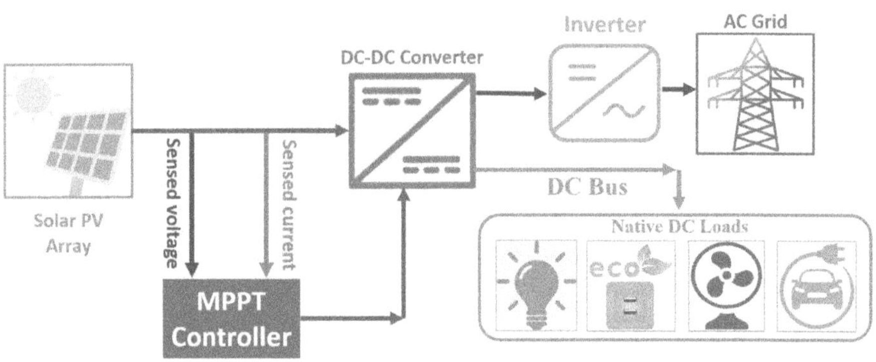

FIGURE 3.1 Schematic diagram of a solar PV fed DC bus.

DOI: 10.1201/9781003585534-3

They increase the efficiency of the overall system by significantly stepping up the input voltage obtained from the PV panels and reducing power loss during energy transfer.

The voltage obtained from a classical boost converter (CBC) is insufficient to meet the load voltage levels, despite operating the switch at duty ratio values greater than 0.8. Moreover, due to the simultaneous conduction of the semiconductor devices, additional losses occur in the CBC at extreme duty ratio values [2]. Hence, gain extension cells such as coupled inductors (CIs) [3], voltage multiplier cells (VMCs), diode-capacitor multipliers (DCMs), switched inductors (SIs) and switched capacitors (SCs) are generally adopted in a CBC structure [4].

Generally, single-switch boost-derived converters employ one or more of the above-mentioned gain extension techniques to achieve high voltage conversion ratios [4]. By adopting hybrid combinations of gain extension techniques, quadratic boost converters (QBCs) and their variants are synthesized and discussed in [5–10]. However, due to the higher voltage and current stress on the switch, the achievable voltage gain is limited.

The voltage amplification that can be achieved in CI-based converters depends on the turns ratio values [11–13]. When the turns ratio is extremely high (>4), voltage spikes appear across the switch due to the leakage inductance of the CI. These voltage spikes are reduced by using clamp circuits, which increase the component count.

In VMC-based converters, the voltage amplification depends on the number of additional cells added to the CBC [14–16]. Therefore, more components are needed, which constrains the voltage gain of VMC-based circuits. Additionally, the power switches and diodes in the converters are susceptible to impulsive charging, further lowering the efficiency of VMC-based solutions.

An alternative to VMC is the DCM cell, which uses two fewer components than the VMC [17–19]. However, the voltage gain yielded by the additional cell remains unaltered. Although DCMs are slightly more efficient compared to VMCs, they suffer from the same drawback as VMC-based boost-derived converters. By adopting suitable gain extension techniques and embedding them within boost-derived configurations, higher-order voltage gain can be obtained. The converters described in [20,21] belong to QBC variants. The converters described in [22–31] use the QBC as a basic building block to achieve voltage gain values that are cubic or quartic (power of 4) functions. In all these converters, there are multiple switches. Further, due to the use of multiple switches, the voltage and current stress on the devices is reduced. Moreover, using an interleaved stage closer to the input port aids in nullifying the input current ripple [21–24,31,32].

The converters described in [25–28] adopt switched capacitor (SC) cells in conjunction with CIs to achieve a quadratic voltage gain function. Due to the use of SC cells, the component count increases. However, due to the high voltage gain capability, the converters utilize the components more effectively.

In [29,30], the interleaving concept is employed to reduce input current ripple. Hence, the converters draw ripple-free current from the input port while yielding high voltage gain values. Despite using more components, these converters possess good voltage gain capabilities.

In this chapter, two QBC structures are cascaded and a converter that yields a voltage gain function that is the fourth power of that of the CBC structure is described. The subsequent sections describe the power circuit, operating principle, design details, and experimental results.

3.2 POWER CIRCUIT OF THE PROPOSED CONVERTER AND ITS DESCRIPTION

Figure 3.2 shows the power circuit diagram of the proposed cascaded quadratic boost converter (CQBC). In the circuit diagram, the elements L_1, D_1, D_2, and S_1 combine to operate as a CBC. Capacitor C_1 acts as the boost-stage output capacitor, besides serving as the source to the subsequent CBC structure. The second CBC structure is formed by the elements L_2, S_1, and D_3. The output from the second CBC stage is stored in capacitor C_2. Since two CBC stages are cascaded in series, the net output voltage across C_2 will be the square of that of the CBC. Thus, the first quadratic boost converter (QBC-1) is synthesized. The second QBC (QBC-2) is synthesized exactly like QBC-1. The elements comprising QBC-2 are L_3, D_4, D_5, S_2, C_3, L_4, and D_6. Since the output from QBC-1 acts as the input to QBC-2 (through C_2), the proposed converter is named a cascaded quadratic boost converter (CQBC).

The operating principle of the proposed CQBC is explained using two modes with the following assumptions:

i. All the semiconductor devices are ideal.
ii. All the passive components are ideal.
iii. The inductors and capacitors are pre-charged, i.e., the circuit is analyzed under periodic steady-state conditions.

3.2.1 MODE 1 ($0 < t < t_1$)

The switches are operated in a synchronous manner at a 50 kHz switching frequency to reduce the size of the magnetic elements. During mode 1, at time $t = 0$, S_1 and S_2 are turned ON. Therefore, the current starts to flow from the voltage source and charges the inductor L_1 through D_2. Inductor L_2 begins to store energy from C_1 through S_1. Likewise, in QBC-2, capacitor C_2 acts as a voltage source and charges L_3 through D_5 and S_2. Similarly, capacitor C_3 charges L_4 through S_2. Due to the energy storage process occurring in L_1, L_2, L_3, and L_4, the diodes D_1, D_3, D_4, and D_6 are reverse-biased. Mode 1 ends at time $t = t_1$ when the inductor currents reach their respective maximum values. The equivalent circuit during mode 1 is depicted in Figure 3.3(a). The governing equations during mode 1 are expressed by (3.1)–(3.4).

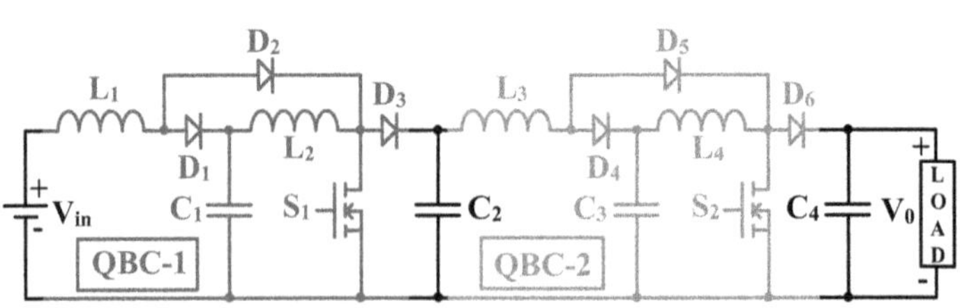

FIGURE 3.2 Power circuit diagram of the proposed cascaded quadratic boost converter (CQBC).

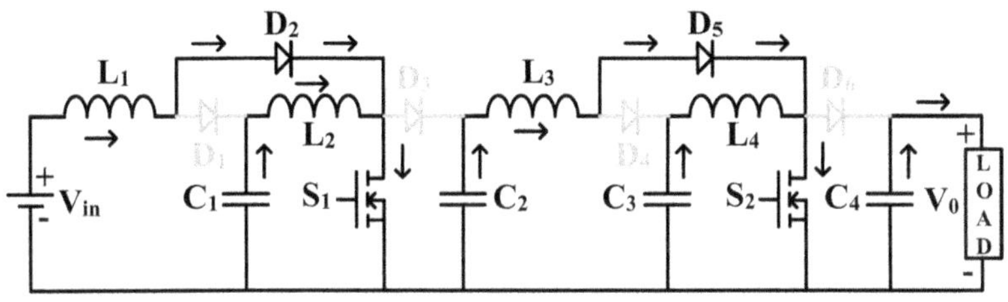

FIGURE 3.3a Equivalent circuit of the CQBC during Mode 1.

$$V_{L_1} = V_{\text{in}} \tag{3.1}$$

$$V_{L_2} = V_{C1} \tag{3.2}$$

$$V_{L_3} = V_{C2} \tag{3.3}$$

$$V_{L_4} = V_{C3} \tag{3.4}$$

3.2.2 Mode 2 ($T_1 < T < T_2$)

Mode 2 commences at time $t = t_1$ when both switches S_1 and S_2 are turned OFF. Since the switches are turned OFF, the inductors release their stored energy and transfer it to the capacitors. For instance, due to the electrical inertia of L_1, diode D_1 is forward-biased while D_2 is reverse-biased. Thus, L_1 transfers its stored energy to C_1 and charges it. Similarly, inductor L_2 charges C_2 through D_3. In QBC-2, L_3 transfers its stored energy to C_3 and charges it, while L_4 charges C_4 through D_6. Due to the energy transfer process occurring in all the inductors, the diodes D_1, D_3, D_4, and D_6 conduct, while D_2 and D_5 are reverse-biased. Mode 2 ends at time $t = t_2$ when the inductor currents reach their respective minimum values. The next switching cycle commences when gate pulses are applied to S_1 and S_2. The equivalent circuit during mode 2 is depicted in Figure 3.3(b). The governing equations during mode 2 are expressed using (3.5)–(3.8). The characteristic waveforms are portrayed in Figure 3.4.

$$V_{L_1} = V_{\text{in}} - V_{C1} \tag{3.5}$$

$$V_{L_2} = V_{C1} - V_{C2} \tag{3.6}$$

$$V_{L_3} = V_{C2} - V_{C3} \tag{3.7}$$

$$V_{L_4} = V_{C3} - V_{CO} \tag{3.8}$$

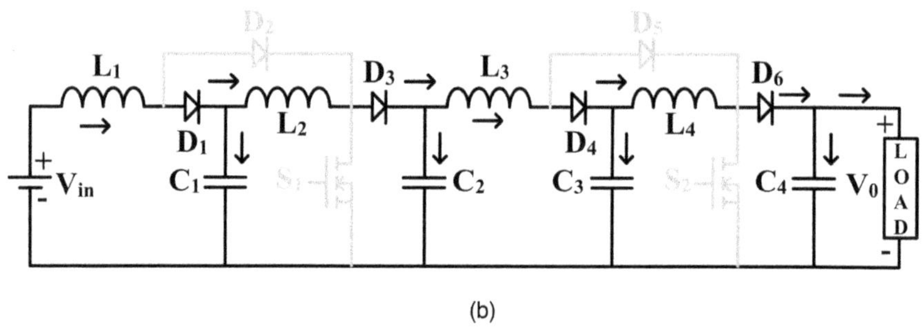

FIGURE 3.3b Equivalent circuit of the CQBC during Mode 2.

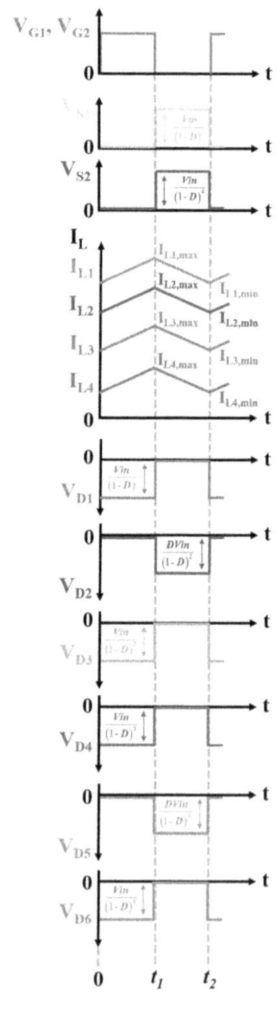

FIGURE 3.4 Characteristic waveforms of the circuit variables used in the proposed CQBC.

3.3 CQBC ANALYSIS UNDER STEADY-STATE

The design equations are derived in this section. First, the voltage gain of the CQBC is derived using the volt–second balance principle.

3.3.1 Voltage Gain of the Proposed CQBC

By applying Kirchhoff's voltage law (KVL) during mode 1 and mode 2, the voltages induced across the inductors L_1–L_4 are expressed using (3.9)–(3.16).

$$V_{L1,\text{ Mode }1} = V_{\text{in}} \tag{3.9}$$

$$V_{L1,\text{ Mode }2} = V_{\text{in}} - V_{C1} \tag{3.10}$$

$$V_{L2,\text{ Mode }1} = V_{C1} \tag{3.11}$$

$$V_{L2,\text{ Mode }2} = V_{C2} - V_{C1} \tag{3.12}$$

$$V_{L3,\text{ Mode }1} = V_{C2} \tag{3.13}$$

$$V_{L3,\text{ Mode }2} = V_{C2} - V_{C3} \tag{3.14}$$

$$V_{L4,\text{ Mode }1} = V_{C3} \tag{3.15}$$

$$V_{L4,\text{ Mode }2} = V_{C3} - V_{CO} \tag{3.16}$$

The proposed converter's voltage gain is derived using the volt–second balance concept. Upon simplification, the overall voltage gain (M) is given by (3.17).

$$M = \frac{V_O}{V_{\text{in}}} = \frac{1}{(1-D)^4} \tag{3.17}$$

where D is the duty ratio of the switches. In the proposed CQBC, both switches are operated at the same duty ratio value.

3.3.2 Voltage Rating of the Switches

The switches are located at the output end of the individual QBC structure. Since the proposed CQBC is developed from a classical QBC, the voltage impressed across the switches is derived and expressed using (3.18) and (3.19).

$$V_{S_1} = \frac{V_{\text{in}}}{(1-D)^2} \tag{3.18}$$

$$V_{S_2} = \frac{V_{\text{in}}}{(1-D)^4} \tag{3.19}$$

3.3.3 Voltage Rating of the Diodes

When S_1 and S_2 are ON, D_2 and D_5 are forward-biased, while D_1, D_4, D_3, and D_6 are reverse-biased. The diodes D_2 and D_5 turn OFF when the switches are turned OFF.

Their voltage stress expressions are derived and presented in (3.20)–(3.25).

$$V_{D_1} = \frac{V_{in}}{(1-D)} \tag{3.20}$$

$$V_{D_2} = \frac{D \cdot V_{in}}{(1-D)^2} \tag{3.21}$$

$$V_{D_3} = \frac{V_{in}}{(1-D)^2} \tag{3.22}$$

$$V_{D_4} = \frac{V_{in}}{(1-D)^3} \tag{3.23}$$

$$V_{D_5} = \frac{D \cdot V_{in}}{(1-D)^4} \tag{3.24}$$

$$V_{D_6} = \frac{V_{in}}{(1-D)^4} \tag{3.25}$$

3.3.4 Design of Passive Components

In this subsection, the design equations for obtaining the values of the inductors and capacitors employed in the proposed CQBC are derived and presented. From basic principles, the design equations for the energy storage elements are derived and expressed in (3.26)–(3.33).

$$L_1 = \frac{D \cdot V_{L1}}{f \cdot I_{rip}} \tag{3.26}$$

$$L_2 = \frac{D \cdot V_{L2}}{f \cdot I_{rip}} \tag{3.27}$$

$$L_3 = \frac{D \cdot V_{L3}}{f \cdot I_{rip}} \tag{3.28}$$

$$L_4 = \frac{D \cdot V_{L4}}{f \cdot I_{rip}} \tag{3.29}$$

$$C_1 = I_{in} \frac{(1-D) \cdot D}{f \cdot V_{rip}} \tag{3.30}$$

$$C_2 = I_{\text{in}} \frac{(1-D)^2 \cdot D}{f \cdot V_{\text{rip}}} \tag{3.31}$$

$$C_3 = I_{\text{in}} \frac{(1-D)^3 \cdot D}{f \cdot V_{\text{rip}}} \tag{3.32}$$

$$C_4 = I_0 \frac{D}{f \cdot V_{\text{rip}}} \tag{3.33}$$

where f is the switching frequency, I_{rip} is the ripple magnitude of the individual inductor current, and V_{rip} is the voltage ripple magnitude across the individual capacitors.

3.4 EXPERIMENTAL RESULTS OBTAINED FROM PROTOTYPE CQBC

To validate the proper operation of the proposed CQBC, a laboratory prototype model was designed and fabricated with the specifications listed in Table 3.1. The components shown in Table 3.2 were used to construct the prototype converter. The STM32F411RE Nucleo-64 microcontroller was used to generate the gating

TABLE 3.1
Specifications of the Proposed CQBC

Parameters (Symbol)	Values
Input voltage (V_{in})	16 V
Output voltage (V_0)	400 V
Output power (P_0)	100 W
Switching frequency (f)	50 kHz
Duty ratio (D)	0.55

TABLE 3.2
Components Used in the Prototype CQBC

Circuit Element	Device Type	Part Number (Specifications)
Switch (S_1, S_2)	MOSFET	IRFB4410Z (100 V, 88 A, 8 mΩ)
Diodes (D_1, D_2)	Fast recovery diode	DSS 16-01A (100 V, 16 A, 0.64 V)
Diodes (D_3–D_5)	Fast recovery diode	BYV29X-600 (600 V, 9 A, 1.26V)
Inductor L_1	Ferrite core	PCV2–184–05L (180 µH, 5 A)
Inductor L_2	Ferrite core	PCV2–394–05L (390 µH, 5 A)
Inductor L_3	Ferrite core	PCV2–105-02L (1000 µH, 2 A)
Inductor L_4	ETD 49/25/16 Ferrite core	2 mH, 1 A, 50 kHz
Capacitor C_1	Electrolytic	100 µF, 100 V
Capacitor C_2	Electrolytic	47 µF, 250 V
Capacitors C_3, C_4	Electrolytic	B43644J65 66M067 (56 µF, 500 V)

signals and implement the closed-loop control algorithm. A MIC4424 dual low-side MOSFET driver was employed to amplify the gate pulses and apply them to the gate terminals of the power switches. The key waveforms were captured using a mixed-domain oscilloscope (MDO4014C) along with differential high-voltage and current probes. Figures 3.5(a) and 3.5(b) depict the photograph of the prototype CQBC and the experimental setup.

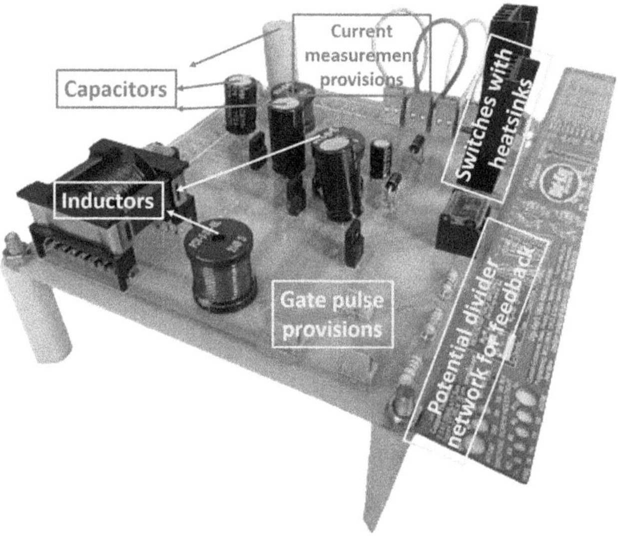

(a)

FIGURE 3.5A Photograph of the prototype CQBC.

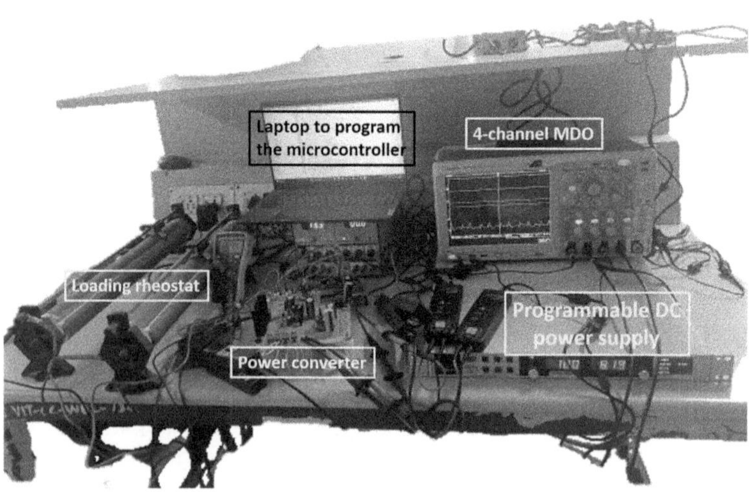

(b)

FIGURE 3.5B Photograph of the experimental setup.

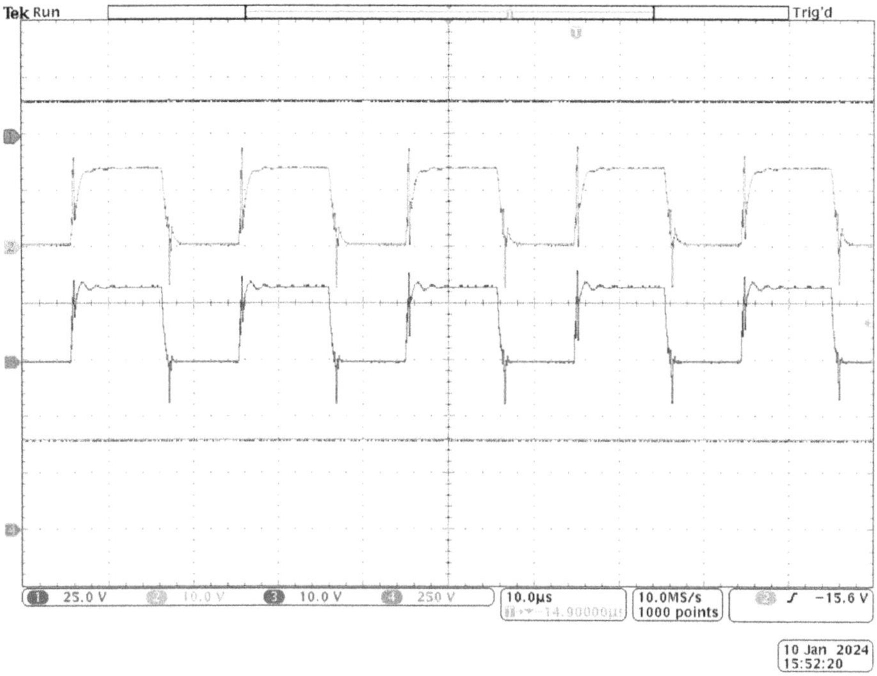

FIGURE 3.6 Experimental waveforms of input voltage (CH1), gate pulses to S_1, S_2 (CH2, CH3, respectively), and the output voltage (CH4).

Figure 3.6 depicts the voltage gain obtained from the prototype CQBC. With an input voltage of 16 V and the switches operated at a duty ratio of 0.55, the proposed CQBC practically delivers an output voltage of 400 V. From the waveforms presented in Figure 3.6, the proper gating of the switches is verified. Based on the input–output voltage levels, the proposed voltage gain technique is validated under experimental conditions.

Figure 3.7 depicts the voltage stress of the switches used in the prototype CQBC. Switch S_1 is located within QBC-1, while S_2 is in QBC-2. Consequently, their voltage stress levels are different. The voltage stress across S_1 (CH2) is around 80 V, which is in accordance with the analytically predicted value. The voltage stress of S_2 (CH4) is observed to be close to 400 V. Since it is located close to the output port, its stress equals the output voltage. Moreover, the synchronized operation of the two switches at a 50 kHz frequency is also verified.

The voltage stress values experienced by the diodes are validated through the experimental results captured and presented in Figure 3.8. The operation of D_4 (CH1), D_5 (CH2), and D_6 (CH3) is compared with S_2 (CH4). From the waveforms, the voltage stress on the diodes closely matches the analytical values predicted using (3.23)–(3.25). Depending on their location, the diodes are subjected to different voltage stress values. Since D_5 and D_6 are located closer to the output port, their voltage stress equals the output voltage. Moreover, diodes D_5 and D_6 operate

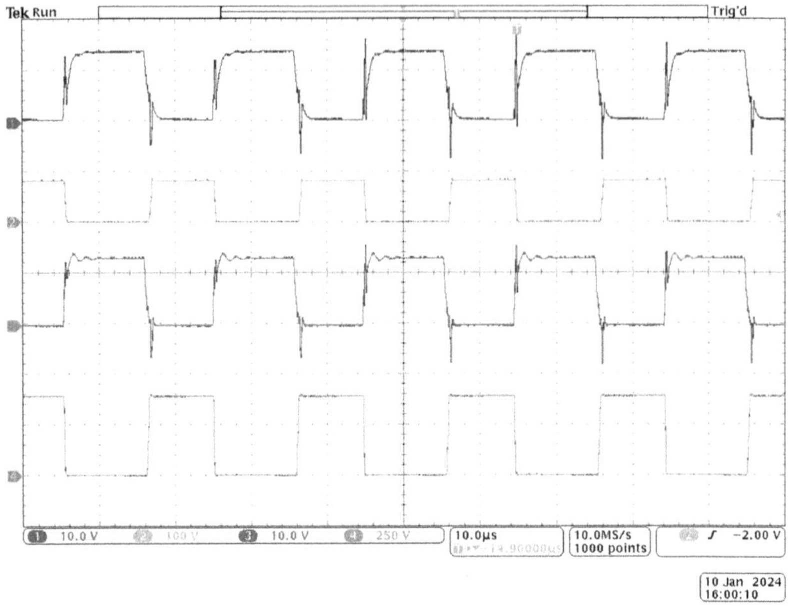

FIGURE 3.7 Experimental waveforms of gate pulses (CH1, CH3), and voltage stresses of S_1, S_2 (CH2, CH4, respectively).

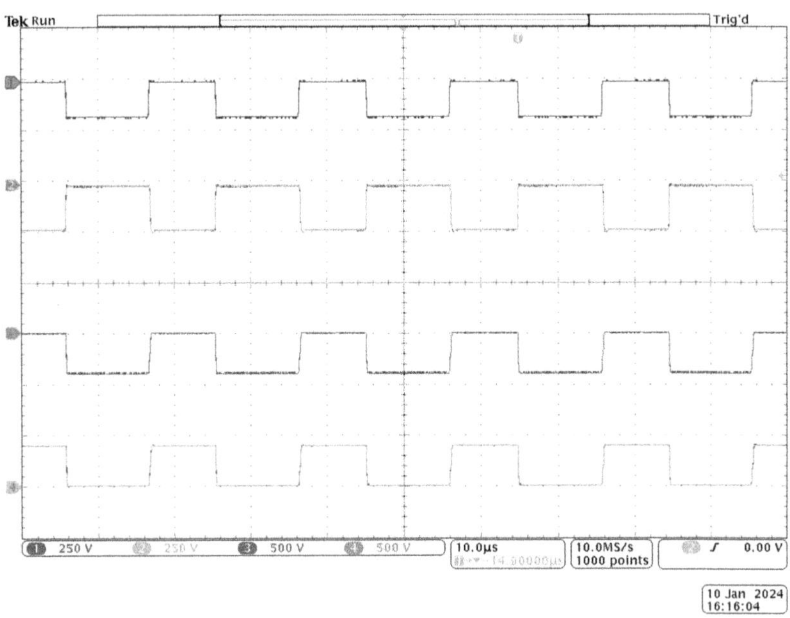

FIGURE 3.8 Experimental waveforms of voltage across the diodes D_4 (CH1), D_5 (CH2), and D_6 (CH3) correlated with the operation of S_2 (CH4).

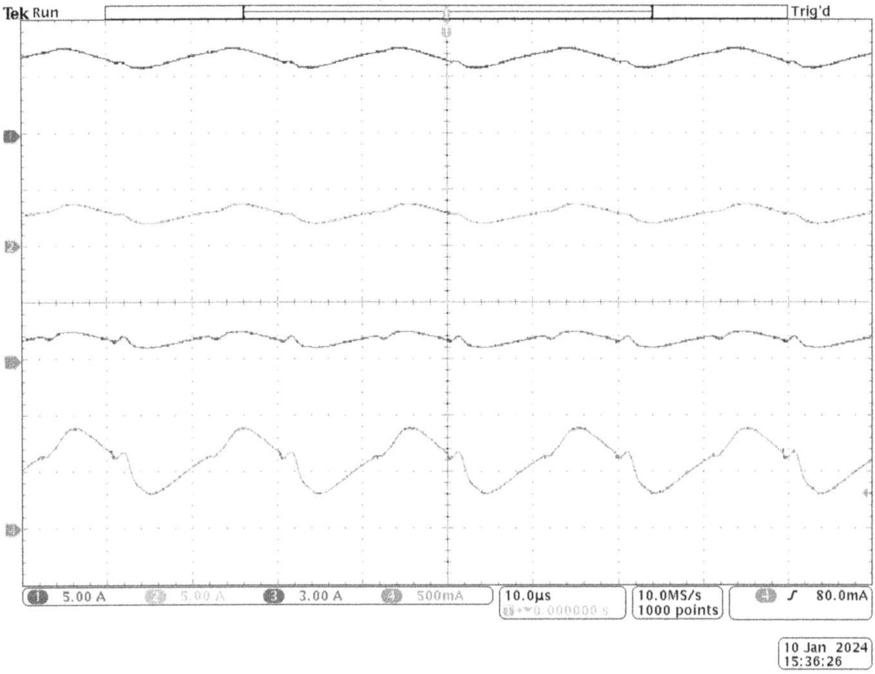

FIGURE 3.9 Experimental waveforms of current through L_1 (CH1), L_2 (CH2), L_3 (CH3) and the load (CH4).

complementarily to each other. In relation to S_2, D_4 and D_6 complement the operation of S_2, while D_5 operates in the same manner as S_2.

The experimental waveforms of current through the inductors L_1, L_2, and L_3 are captured and presented in Figure 3.9 along with the load current waveform. The ripple observed in the inductor current waveforms is in accordance with the design value. Moreover, as observed from the waveforms, the three inductors store energy and release it simultaneously, as predicted in the operating principle. The magnitude of the load current waveform confirms that the proposed CQBC delivers 100 W of power to the load.

The waveforms verifying the operating efficiency of the proposed CQBC are presented in Figures 3.10a and 3.10b. When the load on the prototype converter is 100 W, it draws an input current of 6.8 A at a 16 V input supply. The output voltage is 400 V, and the proposed CQBC operates at an efficiency of 91.11%. At a 120 W load power rating, the proposed CQBC operates at an efficiency of 89.26%. The practical efficiency values closely match the simulated values, as shown in Figure 3.11.

Across the passive elements, power loss occurs due to parasitic effects. Likewise, the active components also experience losses. These losses are computed using the expressions provided in [32]. The equations are presented in (3.34)–(3.36) for clarity.

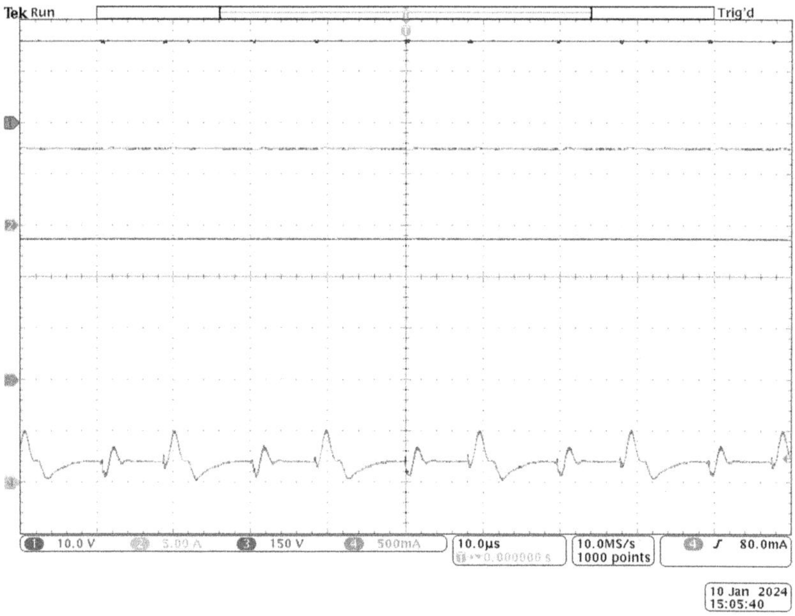

(a)

FIGURE 3.10a Experimental waveforms to compute efficiency at 100 W. V_{in} (CH1), I_{in} (CH2), V_0 (CH3), and I_0 (CH4).

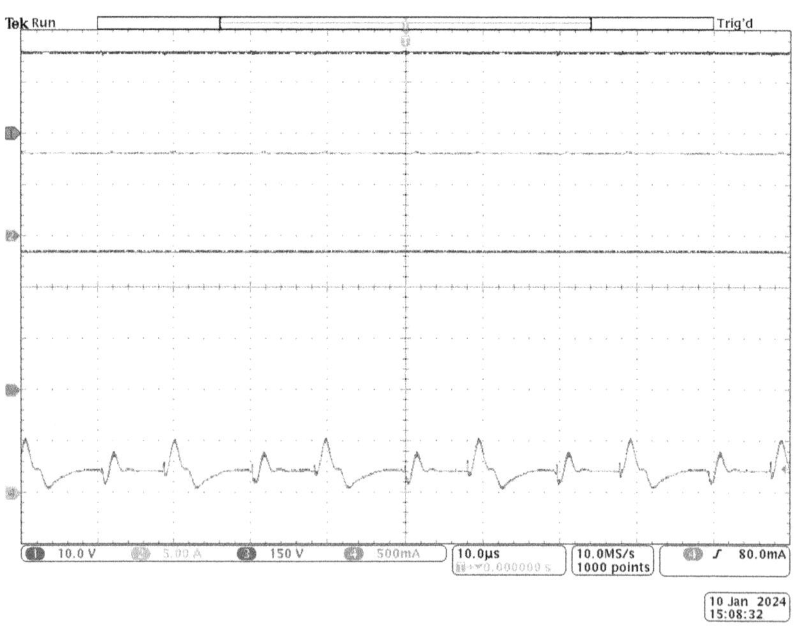

(b)

FIGURE 3.10b Experimental waveforms to compute efficiency at 120 W. V_{in} (CH1), I_{in} (CH2), V_0 (CH3), and I_0 (CH4).

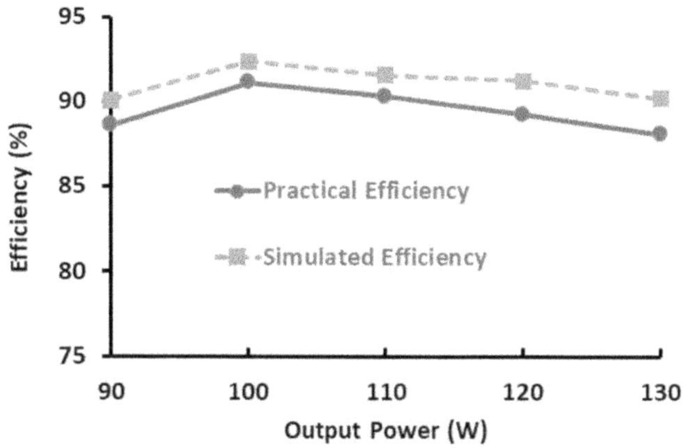

FIGURE 3.11 Efficiency curve of the proposed CQBC at various output power ratings.

$$P_{\text{loss\_switch}} = \underbrace{\left(I_{\text{switch,RMS}}\right)^2 \times R_{\text{DS\_ON}}}_{\text{Conduction loss in the switch}} + \underbrace{V_{\text{body-diode\_ON}} \times I_{\text{body-diode}}}_{\text{Loss across the body-diode}} + \underbrace{P_{\text{switch\_ON}} + P_{\text{switch\_OFF}}}_{\text{Switching loss}} \tag{3.34}$$

$$P_{\text{loss\_diode}} = V_{\text{diode\_ON}} \times I_{\text{diode}} + \frac{V_{\text{diode\_OFF}} \times I_{\text{rr}}}{2} \tag{3.35}$$

$$P_{\text{loss\_CI}} = P_{\text{copper}} + P_{\text{iron}} = (I_{L_1P})^2 \times R_{L_1P} + (I_{L_2P})^2 \times R_{L_2P} + (I_{L_1S})^2 \times R_{L_1S} + (I_{L_2S})^2 \times R_{L_2S} + P_{\text{iron}} \tag{3.36}$$

The switching losses of one MOSFET are represented by $P_{\text{switch\_ON}}$ and $P_{\text{switch\_OFF}}$ during the turn-ON and turn-OFF states, respectively. Across a diode, its ON-state forward voltage drop is indicated by $V_{\text{diode\_ON}}$, while during the OFF state, the voltage across a single diode is denoted by $V_{\text{diode\_OFF}}$. The term I_{rr} represents the diode reverse-recovery current. The copper and iron losses across the magnetic elements are indicated by P_{copper} and P_{iron}, respectively. Figure 3.12 shows the loss distribution profile of the proposed converter.

The proposed converter is intended to be connected to a common DC bus. Hence, its output voltage is regulated by implementing a closed-loop control algorithm. Figure 3.13 depicts the schematic block diagram implemented through the microcontroller. The analog output of the proposed CQBC is sensed using a potential-divider network and applied to the in-built analog-to-digital converter (ADC) pins of the microcontroller unit (MCU). The digital output obtained from the ADC pins is compared with a preset desired value. Based on this comparison, an error signal is obtained. The error acts as the input to the proportional–integral–derivative (PID) controller. The PID controller is implemented in digital form by suitably programming the MCU. The PID controller provides input to the pulse-width modulation

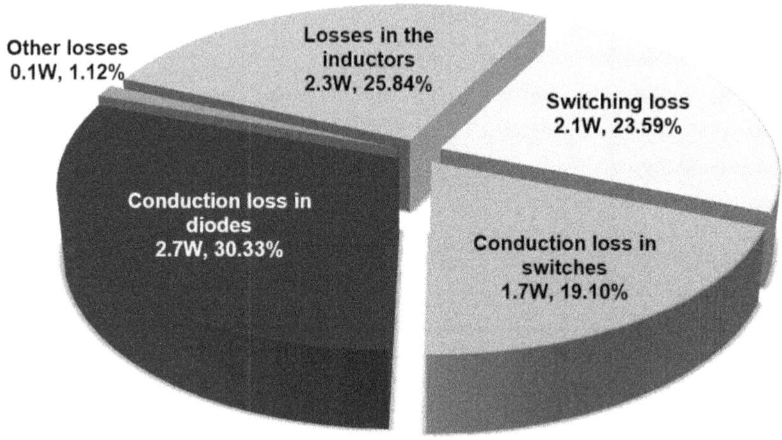

FIGURE 3.12 Loss distribution profile of the proposed CQBC.

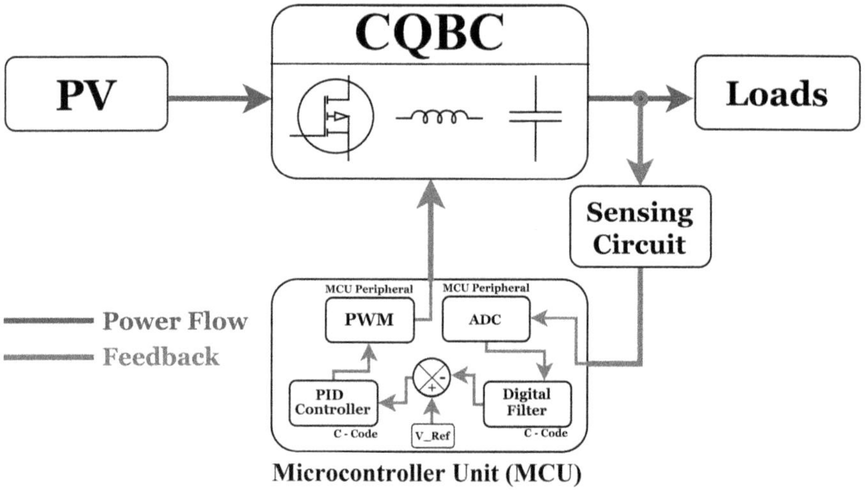

FIGURE 3.13 Schematic block diagram representing the closed-loop control implementation.

(PWM) block. Based on the input to the PWM block, the modified duty ratio for the switches is obtained. Thus, the output voltage is regulated to the desired level.

Figures 3.14a and 3.14b depict the dynamic performance of the proposed CQBC under closed-loop conditions. The input voltage is varied in a stepped manner from 16.3 to 20.1 V. Due to the control algorithm implemented in the microcontroller, the converter's output voltage is restored to 400 V quickly. The overshoot and undershoot in the output voltage magnitude during the input voltage variation are minimal and within safe limits. Moreover, the input current magnitude is adjusted appropriately to meet the required load power demand.

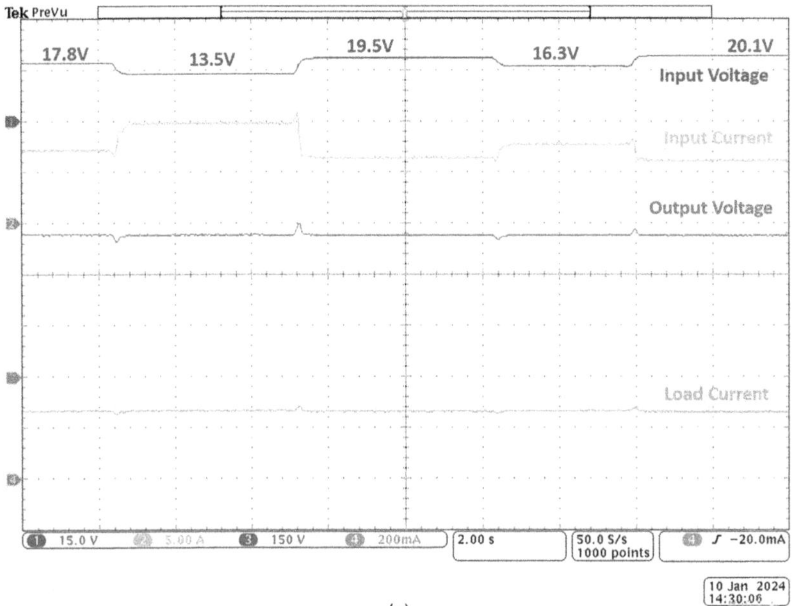

(a)

FIGURE 3.14a Line voltage regulation profile of the proposed CQBC.

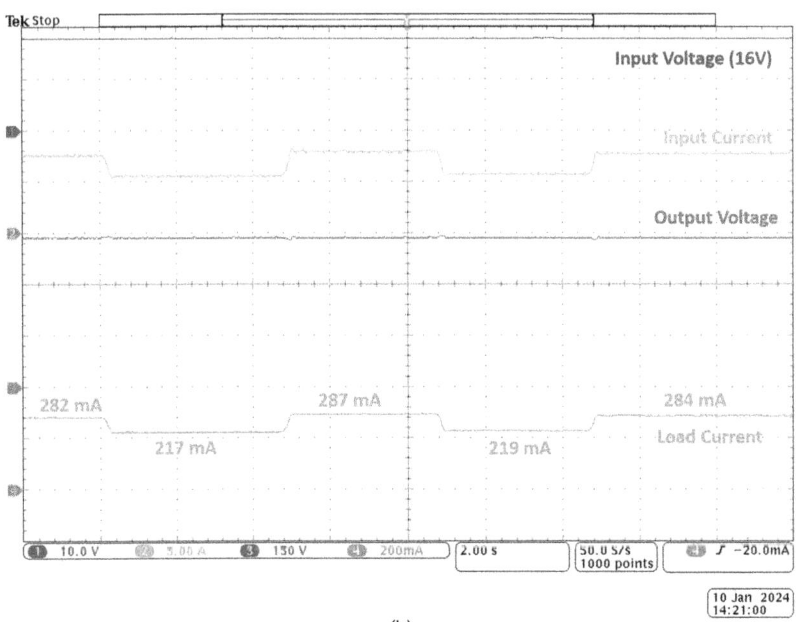

(b)

FIGURE 3.14b Voltage regulation profile of the proposed CQBC when load current is varied.

TABLE 3.3
Comparison of the Proposed CQBC

Attribute	[21]	[22]	[26]	[31]	Proposed CQBC
Input voltage	24	18	27	18	16
Output voltage	380	380	400	400	400
Voltage gain	15.833	21.11	14.81	22.22	25
Duty ratio	0.65	0.5	0.5	0.5	0.55
Voltage gain method	Voltage lift	IBC + lift capacitor + QBC	QBC with CI and DCM	IBC + lift capacitor + cubic cell	Cascaded QBC

In Figure 3.14b, the output voltage regulation performance of the proposed CQBC is shown when the load current undergoes stepped variations. The load current magnitude is varied in the range of 217–287 mA, while the line voltage is maintained constant at 16 V. Based on the demanded load current magnitude, the closed-loop control restores the output voltage to 400 V. During this dynamic condition, the input current is adjusted simultaneously according to the load demand. Thus, the performance of the CQBC under dynamic conditions is verified and observed to be satisfactory.

3.5 COMPARATIVE ANALYSIS

In this section, the proposed converter is compared with several converters presented in the literature. Table 3.3 provides the comparison details. The proposed converter is compared with converters that possess quadratic, cubic, and quartic voltage gain profiles.

Figure 3.15 depicts the voltage gain profiles of all the compared converters. All the converters listed in Table 3.3 exhibit excellent voltage gain capability. They employ various gain extension mechanisms, such as the voltage lift technique and combinations of CI, QBC, and DCM, to achieve high voltage conversion ratios. The superior voltage gain capability of the converter presented in [31] is observed. However, the voltage gain of the proposed CQBC is also close to that of the converter in [31]. Due to the cascaded QBC structure, the proposed converter yields the high voltage gain values at safe duty ratios. Thus, the proposed CQBC is expected to be an excellent choice for integrating a PV source with a common DC bus.

3.6 CONCLUSION

In this chapter, a non-isolated high-gain DC–DC converter with quadratic voltage gain capability was presented. The proposed DC–DC converter was developed by cascading two QBC structures. The presented CQBC exhibits a voltage gain profile that is quartic (power of 4) compared to the voltage gain obtained from a classical boost converter (CBC). Due to the cascaded QBC structure, a very high voltage gain of 25 is practically achieved when the switches are operated at a safe duty ratio of $D = 0.55$. Owing to the adopted gain extension technique, the voltage stress of the

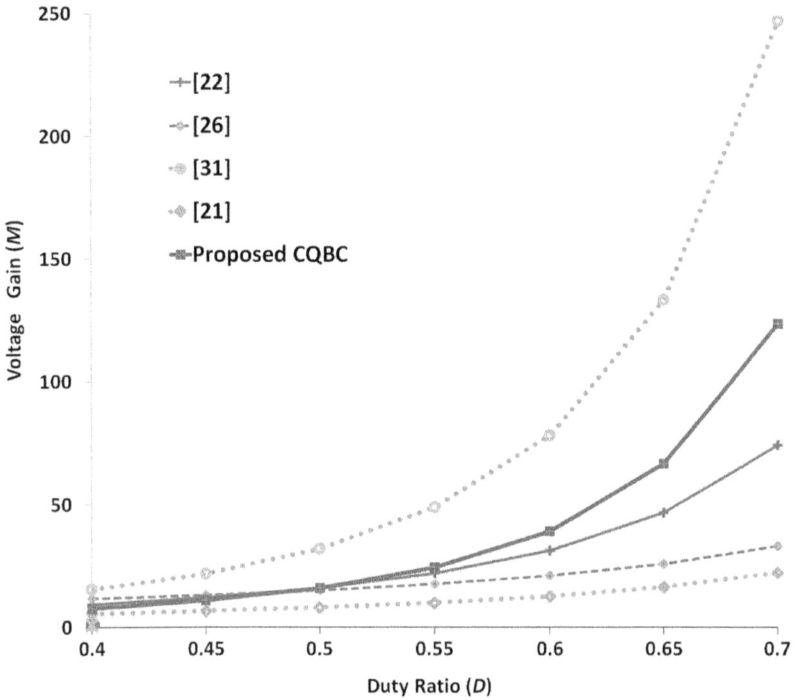

FIGURE 3.15 Voltage gain plots of all the converters that are compared with the CQBC.

switch employed in QBC-1 is just 20% of the rated output voltage. The voltage stress on the diodes employed in the QBC-1 structure is also only a fraction of the output voltage. However, due to progressive gain extension, the voltage stress on the diodes increases progressively. Consequently, the switch employed in QBC-2 is subjected to a voltage stress level equal to the output voltage. The proposed gain extension concept was validated through experiments on a 16–400 V, 100 W prototype converter. The prototype CQBC yielded the desired voltage gain of 25 when the switches were operated at 50 kHz with a nominal duty ratio of 0.55. Additionally, by implementing a simple closed-loop control algorithm, the CQBC's output voltage was regulated against line voltage and load current variations. Under experimental conditions, the proposed converter exhibited a swift dynamic response when subjected to stepped variations in input voltage and load currents. Considering these beneficial features, the proposed converter is expected to be a strong candidate topology for interfacing a low-voltage photovoltaic (PV) input to a high-voltage DC bus.

REFERENCES

1. Tarzamni, H., Gohari, H. S., Sabahi, M., Kyyrä, J., Nonisolated high step-up DC–DC converters: Comparative review and metrics applicability. *IEEE Transactions on Power Electronics*, 2024, 39, 1, 582–625.

2. Valarmathy, A.S., Prabhakar, M., High gain interleaved boost-derived DC-DC converters – A review on structural variations, gain extension mechanisms and applications. *e-Prime – Advances in Electrical Engineering, Electronics and Energy*, 2024, 8, 100618.
3. Samuel Vijay, J., Gna, K., Prabhakar, M., High gain interleaved quadratic boost DC-DC converter. In *2019 2nd International Conference on Power and Embedded Drive Control (ICPEDC)*, Chennai, India, 2019, 390–395.
4. Valarmathy, A.S., Prabhakar, M., Coupled inductor based high gain DC-DC converter for renewable energy application. In *2019 2nd International Conference on Power and Embedded Drive Control (ICPEDC)*, Chennai, India, 2019, 516–521.
5. Mirzaee, A., Arab Ansari, S., Moghani, J., Single switch quadratic boost converter with continuous input current for high voltage applications. *International Journal of Circuit Theory and Applications*, 2020, 48, 4, 587–602.
6. Thirukkuralkani, K.N., Srinivasan, K., Karthikeyan, V., A single switch hybrid quadratic converter for grid-integrated renewable energy system. *International Journal of Circuit Theory and Applications*, 2023, 51, 9, 4276–4293.
7. Ahmad, J., et al., Performance analysis and hardware-in-the-loop (HIL) validation of single switch high voltage gain DC-DC converters for MPP tracking in solar PV system. *IEEE Access*, 2020, 9, 48811–48830.
8. Mahmood, A., et al., A non-inverting high gain DC-DC converter with continuous input current. *IEEE Access*, 2021, 9, 54710–54721.
9. Nilanjan, T., Nilanjan, P., Meenakshi, J., Prabhakar, M., Reconfigurable high step-up DC to DC converter for microgrid applications. *IET Power Electronics*, 2024, 17, 9, 1023–1034.
10. Babaei, E., Mashinchi Maheri, H., Sabahi, M., Hosseini, S.H., Extendable nonisolated high gain DC–DC converter based on active–passive inductor cells. *IEEE Transactions on Industrial Electronics*, 2018, 65, 12, 9478–9487.
11. Singh, K. A., Prajapati, A, Chaudhary, K. High-gain compact interleaved boost converter with reduced voltage stress for PV application. *IEEE Journal of Emerging and Selected Topics in Power Electronics*, 2022, 10, 4, 4763–4770.
12. Chakraborty, U., Kashyap, P., Kapoor, H., Mahalingam, P., Interleaved high step-up DC-DC converter for photovoltaic applications. In *2020 IEEE First International Conference on Smart Technologies for Power, Energy and Control (STPEC)*, Nagpur, India, 2020, 1–6.
13. Valarmathy, A.S., Prabhakar, M., High gain interleaved DC-DC converter with ripple-free input current and low device stress. *International Journal of Electronics*, 2024, 4, 722–747.
14. Valarmathy, A.S., Prabhakar, M., Non-isolated high gain DC-DC converter with ripple-free source current. *Scientific Reports*, 2024, 14, 973.
15. Valarmathy, A.S., Prabhakar, M., Comparison of coupled-inductor based interleaved high gain DC-DC converter topologies. *2021 7th International Conference on Electrical Energy Systems (ICEES)*, Chennai, India, 2021, 364–369.
16. Kashyap, P., Prabhakar, M., Non-isolated high step-up DC-DC converter with low input current ripple. In *2021 IEEE International Power and Renewable Energy Conference (IPRECON)*, IEEE, Kollam, India, 2021, 1–6.
17. Amaleswari, R., Prabhakar, M., High gain multi-input converter with inherent current sharing mechanism for photovoltaic application. *Solar Energy*, 2022, 233, 118–133.
18. Amaleswari, R., Prabhakar, M., Non-isolated multi input DC-DC converter with current sharing mechanism. *International Journal of Electronics*, 2021, 108, 2, 237–264.
19. Reddy, A. B., Mahato, S. N., Tewari, N., High-voltage lift DC–DC converter with reduced switch stress. *IEEE Journal of Emerging and Selected Topics in Power Electronics*, 2024, 12, 2, 1730–1741.

20. Li, G., Amirabadi, M., Chen, X., Lehman, B., The methodology of constructing the quadratic converters. *IEEE Journal of Emerging and Selected Topics in Power Electronics*, 2022, 10, 6, 6586–6606.
21. Samuel Vijay, J., Gna, K., Prabhakar, M., Interleaved quadratic boost DC–DC converter with high voltage gain capability. *Electrical Engineering*, 2020, 102, 651–662.
22. Samuel Vijay, J., Gna, K., Prabhakar, M., Non-isolated DC–DC converter with cubic voltage gain and ripple-free input current. *IET Power Electronics*, 2020, 13, 16, 3675–3685.
23. Samuel Vijay, J., Gna, K., Prabhakar, M., Coupled inductor-based DC–DC converter with high voltage conversion ratio and smooth input current. *IET Power Electronics*, 2020, 13, 4, 733–743.
24. Samuel Vijay, J., Gna, K., Prabhakar, M., Ultra-high gain DC-DC converter based on interleaved quadratic boost converter with ripple-free input current. *International Transactions on Electrical Energy Systems*, 2020, 30, 11, e12622.
25. Rezaie, M., Abbasi, V., Kerekes, T., High step-up DC–DC converter composed of quadratic boost converter and switched capacitor. *IET Power Electronics*, 2020, 13, 17, 4008–4018.
26. Hajiabadi, M.K., Mosallanejad, A., Salemnia, A., A high step-up DC-DC converter based on coupled inductor and diode-capacitor with improved voltage gain and efficiency, and reduced voltage stress for photovoltaic systems. *International Journal of Circuit Theory and Applications*, 2023, 51, 5, 787–806.
27. Subhani, N., May, Z., Alam, M.K., Khan, I., Hossain, M.A., Mamun, S., An improved non-isolated quadratic DC-DC boost converter with ultra-high gain ability. *IEEE Access*, 2023, 11, 11350–11363.
28. Bharathsimha, R., Mahato, S.N., Nilanjan, T., Dual switch ultra-high gain DC–DC converter with low voltage stress. *AEU – International Journal of Electronics and Communications*, 2024, 173, 154995.
29. Ashok Bhupathi Kumar, M., Krishnasamy, V., Quadratic boost converter with less input current ripple and rear-end capacitor voltage stress for renewable energy applications. *IEEE Journal of Emerging and Selected Topics in Power Electronics*, 2022, 10, 2, 2265–2275.
30. Rahimi, R., Habibi, S., Ferdowsi, M., Shamsi, P., An interleaved quadratic high step-up DC-DC converter with coupled inductor. *IEEE Open Journal of Power Electronics*, 2021, 2, 647–658.
31. Atul Thiyagarajan, R., Adhvaidh Maharaajan, A., Aditya Basawaraj, S., Sankar Ram, C., Prabhakar, M., High step-up DC-DC converter with quartic voltage gain. Intelligent solutions for smart grids and smart cities. *Lecture Notes in Electrical Engineering (LNEE)*, 2023, 1022, 63–75.
32. Sakthiram, T., Rahul, S., Yogesh, L., Prabhakar, M., Real-time monitoring and control of a PV-fed enhanced cubic voltage gain converter for DC microgrid. *Computers and Electrical Engineering*, 2024, 120, B, 109761.

4 Low Switch Count Multilevel Inverter Topologies for Electric Vehicle Traction Inverters

Thamizharasan Sandirasegarane, Saravanan Subramanian, Thiyagarajan Venkatraman, Baskaran Jeevarathinam, and Ramesh Jayaraman

4.1 INTRODUCTION

In the last few decades, gasoline vehicles have been considered a feasible mode of transportation for domestic and cargo applications. Internal Combustion (IC) Engine Vehicles (ICEVs) have several drawbacks, including high carbon emissions contributing to global warming and climate change, as well as economic and maintenance concerns. This paved the way for the evolution of Electric Vehicles (EVs) as a zero-emission and feasible solution for the drawbacks associated with ICEVs. Present-day EVs, with the adoption of new technologies, are completely different from the EVs available before the start of the twentieth century. The modern EV powertrain is composed of a battery pack, power converter, traction machine, and gear transmission system. There are four types of EV architectures, namely hybrid, battery, fuel cell, and PV-based EVs.

In the EV powertrain shown in Figure 4.1, the traction inverter plays a key role in energy transmission by converting DC (Direct Current) power from the battery pack into AC (Alternating Current) to drive the traction motor. It plays a decisive

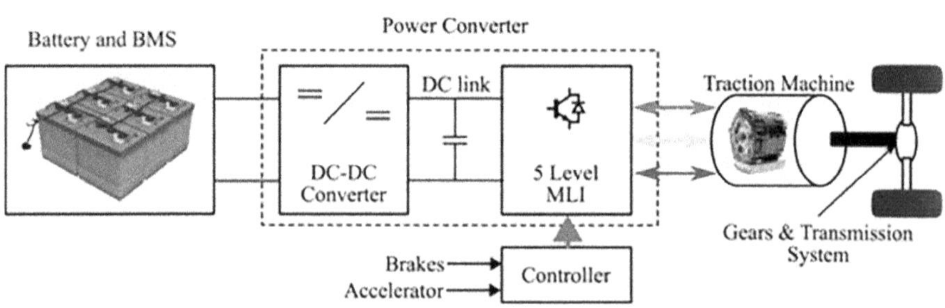

FIGURE 4.1 Basic block diagram of EV power train.

DOI: 10.1201/9781003585534-4

role in adjusting voltage and frequency to regulate the speed, torque, and efficiency of the traction motor. It also performs battery charging during regenerative braking by converting kinetic energy into electrical energy and ensures that the system operates within safe regions. EVs typically employ a traditional two-level inverter as the traction inverter to translate DC power into AC power suitable for driving the traction motor. It is modular in structure and easy to control. However, it has several limitations: high harmonic distortion due to lower-order harmonics in the pole voltage, which leads to increased motor losses and heating; high switching losses and voltage stress on the switching devices; and lower efficiency at high-power applications such as large trucks and grid-connected charging systems.

Multilevel inverters (MLIs) offer better performance at high voltages due to their inherent property of generating pole voltage in small voltage steps with lesser harmonic distortion, eliminating the need for filters to smooth the voltage waveform close to a sinusoidal shape, and employing low-voltage switches. There are three basic topologies at the time of inspection: diode-clamped, flying capacitor, and cascaded H-Bridge inverters. The first two variants involve a larger number of diodes and clamping capacitors to produce stepped voltage, and complex control is required to eliminate common-mode voltages and capacitor voltage balancing issues. The cascaded H-Bridge inverter is predominantly used in high-voltage and medium-power applications due to its modular structure and usage of lower-voltage switches. On the other hand, increasing voltage levels requires scaling up several H-Bridges in its structure, which increases the switch count and, consequently, inverter cost. This chapter highlights several reduced-count topologies, their structure, and operation suitable for traction applications. These reduced-count topologies are very peculiar in their structure and mode of operation compared with traditional CHB-MLIs. This chapter is framed to explore the design equations for achieving symmetrical/asymmetrical operations and the degree of freedom in producing voltage levels by combining other reduced-count topologies. Section two of this article narrates the basic operating principles of two-level inverters and their performance characteristics, followed by a comprehensive analysis of the reduced-count topologies presented in the literature and suitable for EV applications in the next section. The last section summarizes the inferences drawn from the presented topologies.

4.2 TWO-LEVEL INVERTER

Inverters play a crucial role in traction applications due to their ability to operate at high voltages and power levels. Passenger Electric Vehicles (EVs) utilize various inverter topologies and semiconductor technologies. Most commercially available EVs currently employ a conventional two-level inverter topology. For example, models like the Chevrolet Volt, Toyota Prius, Nissan Leaf, and Audi e-Tron incorporate a two-level topology with double-sided cooling insulated gate bipolar transistor (IGBT) modules (Lee et al., 2013). Tesla, on the other hand, integrates different technologies: the Model S utilizes a two-level inverter with Silicon Carbide Metal-Oxide-Semiconductor Field-Effect Transistors (SiC MOSFETs), while the Model 3 employs IGBT modules.

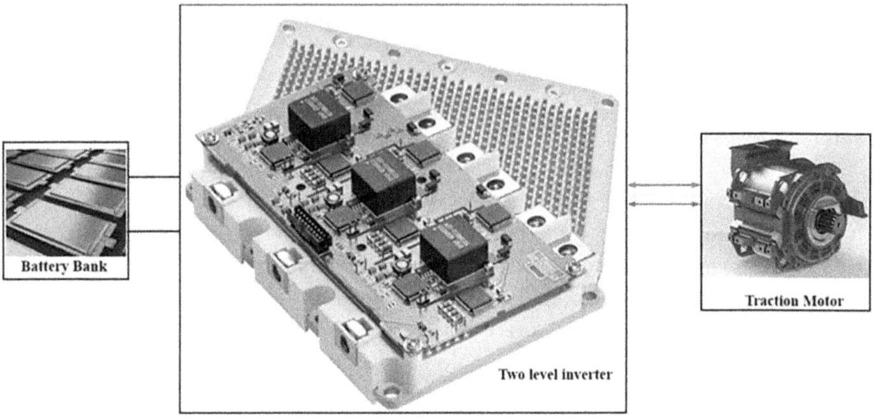

FIGURE 4.2 Two-level inverter.

The two-level inverter portrayed in Figure 4.2 is constituted using fast-switching semiconductor devices, and the input DC is obtained from batteries or renewable energy sources. The power circuit has the ability to produce two or three levels according to the mode of operation of the inverter. Normally, the two-level output is obtained from the circuit as (+Vs or −Vs), and it creates harmonic distortion, high Electromagnetic Interference (EMI), and a large *dv/dt*. This drawback restricts its usage in high-voltage EV applications. These drawbacks can be mitigated by increasing the voltage levels to reduce harmonic distortion, lower EMI, and decrease voltage stress across the devices. Sinusoidal PWM (SPWM) is commonly used to extract variable voltage and frequency to achieve the desired torque, speed, and frequency from the traction motor for vehicle propulsion.

4.3 REDUCED COUNT TOPOLOGIES

The key philosophy behind the development of MLI is to synthesize pole voltage closer to a sinusoidal shape to achieve reduced harmonic distortion, lower voltage stress, reduced EMI, and decreased bearing noise. Traditional topologies such as Neutral Point Clamped, Flying Capacitor, and Cascaded H-Bridge inverters require a large number of switches, which increases complexity and cost. However, the practicability of MLI is restricted in industrial applications due to the increased switching rate of components and the challenge of voltage balancing in capacitors as the number of voltage levels increases. To overcome these challenges, many studies have introduced optimized MLI topologies that achieve multilevel output with fewer switches. These topologies have been used in place of two-level inverters for driving traction motors with the desired speed-torque characteristics. The various EV inverter structures, their voltage levels, switches used, applications, and EV models that employed them are tabulated in Table 4.1.

Cascaded Multilevel Inverters (MLIs) are increasingly utilized in Electric Vehicle (EV) applications due to their ability to cascade several H-bridges from multiple DC sources, typically batteries, to generate high-quality voltage waveforms, enhance

TABLE 4.1
Inverter Structures, Voltage Levels, and Switches Used in EV Model

Inverter Structure	Voltage Levels	Switches Used	Application	EV Models
Two-Level Inverter	300–800V	IGBT, SiC MOSFET	Standard passenger EVs	Nissan Leaf, Chevrolet Volt, Audi e-Tron (Lee et al. 2013)
Three-Level NPC (Neutral Point Clamped)	600V–1.5kV	IGBT, SiC MOSFET	High-performance EVs, mild hybrids	Some experimental EVs, e-trucks (Ronanki and Williamson, 2018)
Cascaded H-Bridge MLI	1.5–15kV	IGBT, SiC MOSFET, GaN FETs	High-power applications, electric ships	Not commonly used in EVs (Ronanki and Williamson, 2018)
Flying Capacitor MLI	1–3kV	IGBT, MOSFET	Research and specialized EVs	Limited research applications (Shivakumar et al., 2020)
Hybrid Multilevel Inverter	800V–3kV	Combination of IGBT, MOSFET, GaN	Ultra-high efficiency EVs, trucks, buses	Experimental EVs, electric buses (Shivakumar et al., 2020)

motor performance, and improve overall system efficiency. Cascaded H-Bridge Multilevel Inverters (CHB-MLIs) are widely used in EV applications due to their high efficiency, modularity, reduced harmonic distortion, and improved power quality. CHB-MLIs use lower-frequency switching, which results in reduced switching losses and improves inverter efficiency. This translates to longer battery life and an extended driving range for EVs (Rodriguez et al., 2002). Their use of multiple lower-voltage sources (battery modules) improves system reliability and reduces insulation stress on the inverter (Escalante et al., 2002). Additional voltage levels can be added by stacking more H-bridge cells, allowing EV designers to customize the powertrain voltage and performance based on vehicle requirements. The modular structure reduces maintenance needs and simplifies fault diagnosis (Maharjan et al., 2008). Due to bidirectional power flow capability, CHB-MLIs efficiently handle regenerative braking by converting braking energy into a usable form to charge the battery, thereby improving energy efficiency and battery life (Du et al., 2009). Their modular and scalable nature makes them ideal for next-generation EVs, especially with advancements in wide bandgap semiconductors like Silicon Carbide (SiC) or Gallium Nitride (GaN) devices (Huang et al., 2011).

Other MLI variants for achieving reduced component count include: (1) Hybrid MLI, which combines different inverter topologies to achieve a balance between performance and switch reduction; (2) Asymmetrical MLI, which uses different DC voltage sources to achieve more output levels with fewer components; and (3) Switched Capacitor MLI, which utilizes capacitors to generate additional voltage levels, reducing the need for extra switches. These topologies help improve efficiency, reduce size, and lower overall system costs, making them suitable for applications requiring high power quality with minimal hardware complexity.

A multilevel cascaded inverter (Tolbert et al., 1998) is presented for EVs, highlighting its potential to make EV systems more accessible and safer by isolating each low-voltage battery through switching devices. A transformer-less multilevel converter (Verma et al., 2011) is designed for high-voltage and high-current hybrid electric vehicles, emphasizing the benefits of the cascaded topology in such applications. A 12-switch inverter is used to develop a 31-level multilevel inverter (Saravanan et al., 2024), minimizing the number of switching devices required for EV applications and thereby enhancing system efficiency and reliability. A cascaded multilevel inverter topology (Zhang and Gao, 2018) is developed that utilizes a single battery with a high-frequency link and low-voltage rating MOSFETs, aiming to improve motor drive performance in EVs. Reduced Switch Count Multilevel Inverter topologies with isolated DC sources have gained more attention in EV applications due to their flexibility in configuring their structure and operating states to optimize switching requirements. MLIs with isolated DC sources offer reliability and scalability without constraints while maintaining high-quality output waveforms with enhanced efficiency. The forthcoming discussion portrays the structure, operation, and design equations for any voltage level of recent reduced-count MLIs for both symmetrical and asymmetrical operation.

4.3.1 Series Parallel Switched Multilevel DC-Link Inverter

The Series Parallel Switched Multilevel DC-Link Inverter (SPSMLDCLI) topology for single-phase supply (Ramkumar et al., 2012), shown in Figure 4.3, is composed of three DC sources and six controlled switches to produce 7 levels in symmetrical and 15 levels in asymmetrical configurations. It is noted that the topology uses both

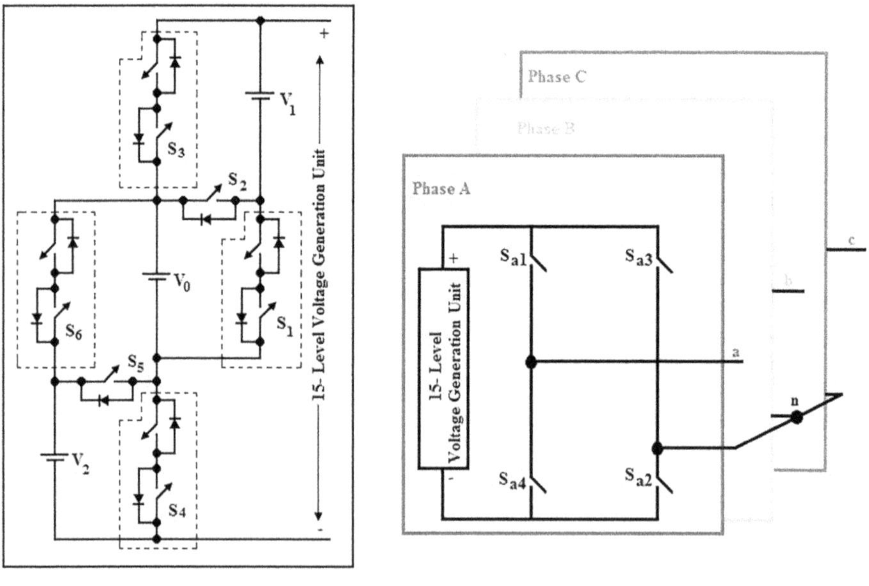

FIGURE 4.3 Series parallel switched multilevel DC-link inverter.

unidirectional and bidirectional switches in its structure to achieve voltage levels. The switches (S_1, S_3, S_4, and S_6) are bidirectional, while the switches (S_2 and S_5) are unidirectional. The bidirectional switches help overcome the problem of inter-looping across the DC sources. The level-synthesizing unit is operated at high-frequency switching, which uses MOSFETs, while the H-bridge inverter for polarity reversal operates at line frequency and employs IGBTs. The topology can be scaled up by adding the switches (S_7, S_8, and S_9) and a DC source (V_3) to the tail end of the voltage source (V_2). The SPSMLDCLI has the option of connecting the DC sources in series or parallel with the voltage source (V_o) in asymmetrical modes of operation. Switches S_1 and S_6 are conducted to operate the voltage sources (V_1 and V_2) in parallel, and switch S_2 is turned ON to connect them in series to obtain the voltage level ($V_1 + V_2$) in the output voltage. The voltage source (V_o) occupies one step voltage value to allow addition/subtraction in association with the remaining voltage sources connected in the SPSMLDCLI. The three-phase topology is constituted using three single-phase SPSMLDCLI units tied in Y or Δ connection.

The SPSMLDCLI topology is well suited for induction motor drive applications due to its inherent advantages of free voltage unbalancing and modular structure, allowing easy scaling for any number of voltage levels. The SPSMLDCLI also allows for cell balancing in its structure through proper switching sequences when using lithium-ion cells. Figure 4.4 depicts a comparison of the power components required between SPSMLDCLI and Cascaded H-bridge MLI (CHBMLI). The design equation for a generalized SPSMLDCLI is presented in Table 4.2.

The Series Parallel Switched MLI (SPSMLI) uses a reduced number of switches and passive elements, making it more compact and ideal for space-constrained EV applications. The series-parallel switched approach allows for dynamic voltage-level control, optimizing power extraction from the battery. This helps maintain voltage balance across multiple battery modules, improving reliability and extending battery life. SPSMLI enables bidirectional power flow, supporting effective regenerative braking by switching between series and parallel configurations to optimize energy recovery. The multilevel output of SPSMLI improves torque control, reduces harmonics, and enhances motor performance, leading to better driving range and smoother acceleration.

4.3.2 Level Dependent Source Concoction Multilevel Inverter Topology

The Level Dependent Source Concoction (LDSC) MLI (Jose and Titus, 2016) is a modified version of SPSMLDCLI designed to elevate the output voltage levels, as shown in Figure 4.4a. The structure uses half-bridge cells connected in place of switch S_2 in SPSMLDCLI (Ramkumar et al., 2012) to enable symmetrical operation. The SPSMLDCLI requires three switches and one DC source, whereas the LDSCMLI needs two DC sources and four switches for each cell added to the preceding DC sources. The SPSMLDCLI generates 3 levels, while LDSCMLI produces 5 levels for each cell addition. It is observed that the Level Dependent Source Concoction MLI (LDSCMLDCLI) has two degrees of freedom in its structure regarding cell addition in cascade connection—either in the vertical or horizontal direction, as shown in Figure 4.4b. The operation in symmetrical mode is studied for a 7-level inverter,

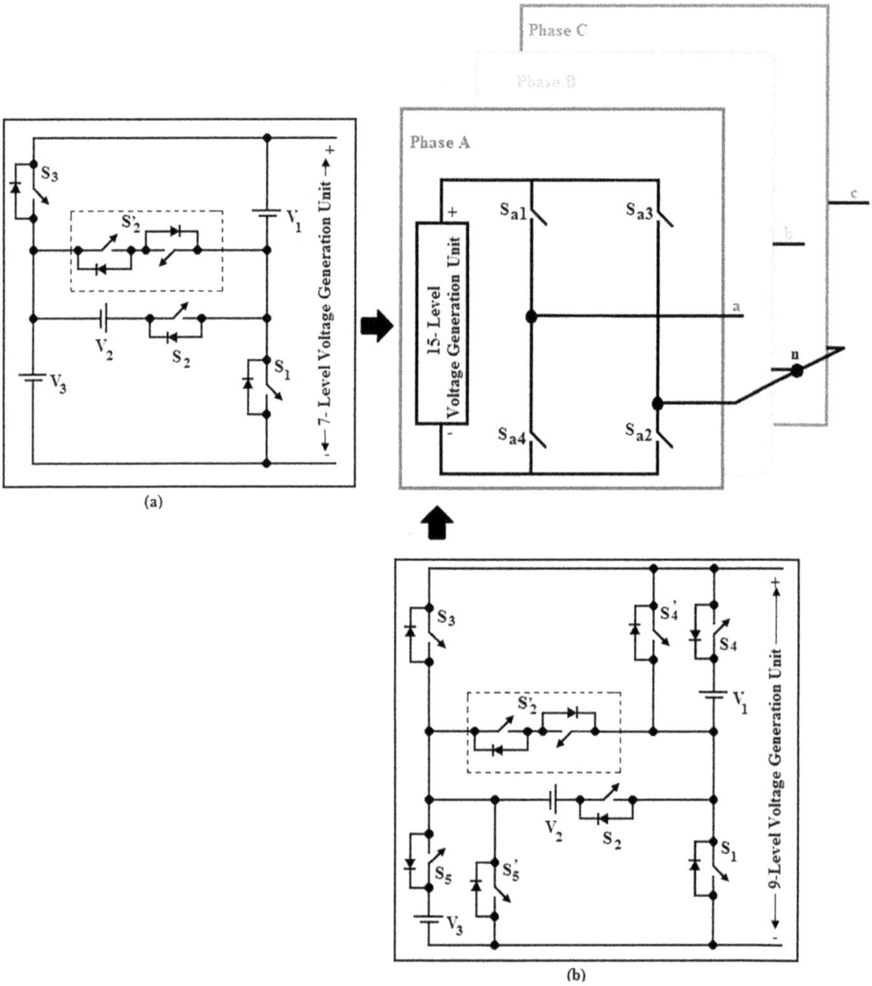

FIGURE 4.4 Level-dependent source concoction (LDSC) MLI.

TABLE 4.2
Design Equations for '*l*' Level

Parameters	Switches	Gate Drivers	DC Sources	Diodes
SPSMLI	$[(5\times l)-7]/2$	$[(3\times l)-1]/2$	$(l-1)/2$	1

as portrayed in Figure 4.4a. Switches S_1 and S_3 are turned on to connect the voltage sources V_1 and V_3 in parallel, and switch S_2 is turned on to connect the voltage source V_2 with V_1 and V_3. The bidirectional switch S_2' is used to bypass the voltage source V_2 while V_1 and V_3 are in addition. Switches S_1, S_2, and S_3 are unidirectional IGBTs, which require the same number of gate drivers compared with SPSMLDCLI. The LDSCMLI (Jose and Titus, 2016) is a single-phase topology, and it can be easily extended to a three-phase MLI suitable for three-phase induction motor drive applications. The design equation for LDSCMLI is tabulated in Table 4.3 to explore the possibility of expanding to higher voltage levels.

Due to lower power losses in this topology, it offers reduced motor heating, improved battery-to-motor efficiency, and extended battery life. It optimizes energy utilization, increasing EV driving range. The MLI has the specific feature of series-parallel battery switching, which ensures balanced power distribution and reduced voltage stress on components. The MLI produces high-quality output, enhancing torque control and smooth operation, resulting in better acceleration.

4.3.3 Dual Bridge DC-Link Multilevel Inverter Topology (DBMLI)

A compact MLI topology shown in Figure 4.5, named the Dual Bridge Multilevel Inverter (DBMLI) topology (Thamizharasan et al., 2013), is developed to offer reduced power components compared with CHBMLI. The topology uses two

TABLE 4.3
Design Equations for 'c' Cells

Parameters	Switches	Gate Drivers	DC Sources
LDSCMLI	$(5\times c)+4$	$4\times(c+1)$	$(2\times c)+1$

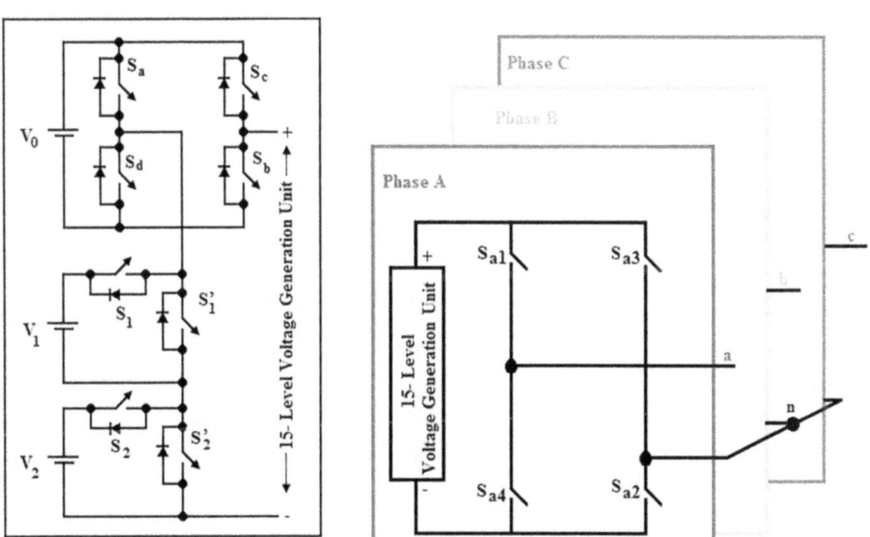

FIGURE 4.5 Dual Bridge MLI (DBMLI).

TABLE 4.4
Design Equations for '*l*' Levels

Parameters	Switches	Gate Drivers	DC Sources
LDSCMLI	$(l+5)$	$(l+5)$	$(l-1)/2$

H-bridges in its structure to justify the name "Dual Bridge," and out of the two H-bridges, one bridge is used for the level generation module, while the other one is for polarity reversal. The DBMLDCLI incorporates half-bridge cells cascaded with the level-synthesizing module to increase the voltage levels.

The switched diode cell can also be employed in place of half-bridge cells to attain the same number of levels. The main drawback of this topology is that the polarity generation bridge has to block high DC-link voltage, which limits its application and make it complicated in fault-tolerant operation. The three-phase topology can be obtained in two ways: connecting three single-phase topologies in Y or Δ, or cascading the voltage generation unit with a three-phase six-switch inverter to drive a BLDC motor drive. The DBMLI is configured in symmetrical or asymmetrical configurations in the order of 2 or 3 (Thamizharasan et al., 2013) to increase the output voltage to higher levels. The operation for symmetrical configuration begins with the voltage level (V_o) by switching on the switches (S_a, S_b, S_1', and S_2'), and the voltage level ($V_{1+}V_2$) is obtained by connecting the switches (S_1, S_2, S_b, and S_d). The voltage level ($V_{o+}V_{1+}V_2$) is produced by triggering the switches (S_1, S_2, S_a, and S_b), respectively. This topology is well-suited for BLDC motor EV drives. Table 4.4 shows the design equation for DBMLI for any voltage levels.

The Dual Bridge Multilevel DC-link Inverter (DBMLDCLI) is a promising solution for EV traction systems, offering a high-efficiency, compact, and low-harmonic alternative to conventional inverters. Its ability to provide higher voltage levels with fewer components makes it a strong candidate for next-generation EV powertrains, leading to improved driving range, smoother motor control, and effective regenerative braking.

4.3.4 Cross-Switched DC Sources Multilevel Inverter Topology (CSMLI)

The Cross-Switched DC Sources MLI (Thamizharasan et al., 2014), shown in Figure 4.6, uses an H-bridge inverter and a module with a string of DC sources connected in a cross fashion in a ladder structure. The cross-switched DC sources module has the flexibility of extending any number of voltage sources. This module has fewer switches than CHBMLI for the same number of DC sources. To produce 7 levels, the CSMLI requires 10 switches, whereas CHBMLI requires 12 switches. This count is drastically reduced in CSMLI compared with CHBMLI. The topology is switched using hybrid modulation, which combines PWM and fundamental switching to reduce switching loss. The H-bridge inverter is operated in PWM mode, while the cross-switched DC source module works in square wave operation mode. The level addition is performed only in the cross-switched module by adding DC sources in the reverse direction and operating at fundamental switching to reduce switching loss.

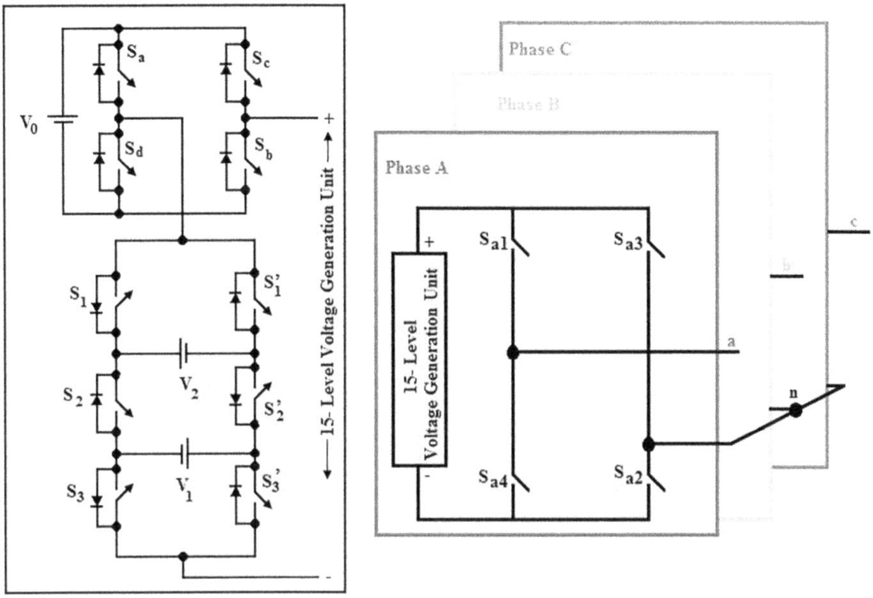

FIGURE 4.6 Cross-switched MLI (DBMLI).

TABLE 4.5
Design Equations for '*l*' Levels

Parameters	Switches	Gate Drivers	DC Sources
CSMLI	$(l+1)$	$(l+1)$	$(l-1)/2$

The cross-switched module is configured with a voltage ratio of 2 or 3 to offer an asymmetric scale of operation. The cross-switched structure can be extended to a three-phase form by tying three single-phase inverters. The topology is modular and can be cascaded to any form of MLI to operate as an inverter drive for EV applications. The design equations for the cross-switched MLI are presented in Table 4.5.

The CSMLI is an efficient, compact, and high-performance topology ideal for EV applications. By reducing switching losses, improving power quality, and enhancing system reliability, CSMLI contributes to better EV performance, extended battery life, and optimized regenerative braking. This topology is a promising next-generation inverter solution for electric mobility.

4.3.5 MLI with DC Sources Arranged in Matrix Structure

The topology shown in Figure 4.7 (Thamizharasan et al., 2015) is composed of an array of DC sources connected in reverse direction in the form of a matrix ($r \times c$), where 'r' and 'c' are the rows and columns of the matrix structure, respectively.

The structure arranges the DC sources in a matrix form and can be scaled up for any number of levels. The main advantage of the MLI structure is that it requires a fewer number of switches in the current conduction path by restricting the rows to a maximum of 2. By this arrangement, the MLI structure follows the cross-switched MLI, extending in the horizontal direction using bidirectional switches as intermediate switches and unidirectional switches at the outer columns. The topology uses fewer gate drivers while increasing the number of voltage levels. The topology is shown for a 9-level inverter, which uses four DC sources, six unidirectional IGBTs, and three bidirectional IGBTs. The structure operates in both symmetrical and asymmetrical voltage ratios, with a ratio of 2. The voltage source (V_1) appears in the output voltage by connecting the switches (S_2, S_5, and S_7) and (S_1, S_4, and S_8) during the positive and negative cycles, respectively.

It is inferred that the topology with the structure shown in Figure 4.7 can be cascaded to operate at different voltage combinations to attain higher voltage levels. The topology can also be obtained in hybrid form by combining it with basic topologies. The input DC sources can also be obtained from PV sources and capacitors. This provides insight into a topology employing capacitors as input sources, which

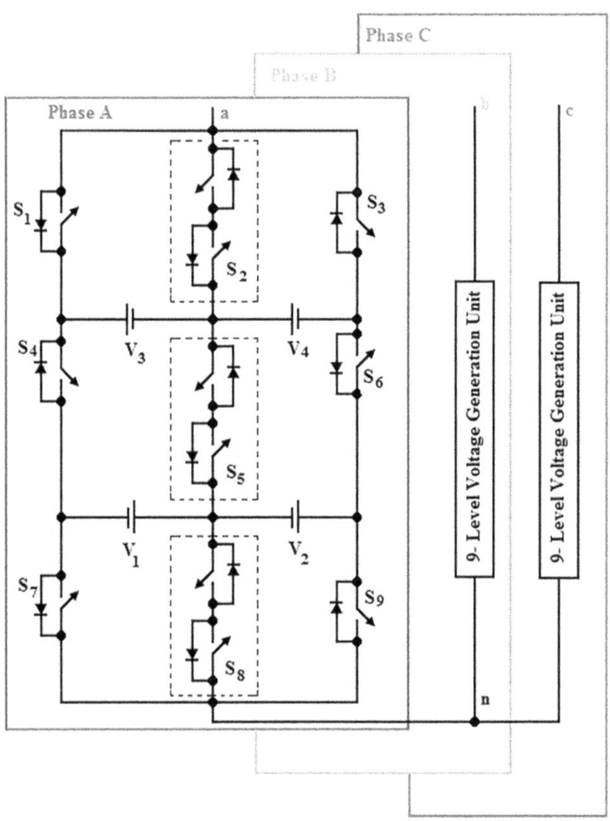

FIGURE 4.7 MLI with DC sources arranged in matrix structure.

TABLE 4.6
Design Equations for Topology for MLI (Thamizharasan et al., 2015)

Topology	IGBTs	Driver Circuits	Input DC Sources	Current Conducting IGBTs
Single-phase MLI	$(2\times c)\times(r+1)$	$(r+1)\times(c+1)$	$(r\times c)$	$(r+1)$
Three-phase MLI	$(6\times c)\times(r+1)$	$3\times(r+1)\times(c+1)$	$3\times(r\times c)$	$3\times(r+1)$

requires more care in balancing the capacitor voltages. The topology can be configured as a switched capacitor topology with a single DC source in one row and a capacitor in the other row for voltage boosting capability. The design equation for the MLI structure is shown in Table 4.6.

The Matrix-Structured DC Source Multilevel Inverter (MSDC-MLI) is a highly efficient, scalable, and compact solution for EV powertrains. By utilizing a dynamic matrix configuration, this topology optimizes battery integration, improves energy efficiency, and enhances motor performance. With its low switch count, improved power quality, and effective regenerative braking support, MSDC-MLI stands out as a promising next-generation inverter technology for electric mobility.

4.3.6 New Multilevel Inverter Topology with Reduced Current Conducting Devices

A new H-6 inverter depicted in Figure 4.8 (Maalmarugan et al., 2017), with an array of DC sources arranged on either side, is proposed to offer a reduced switch count. The structure uses one bidirectional IGBT with a DC source as a single unit, thereby reducing gate driver circuits. The bidirectional IGBTs can be realized with two IGBTs in a common emitter configuration or two MOSFETs with a common source configuration; however, a single bidirectional IGBT is available in the electronic market. The topology uses symmetrical DC sources with different values to operate in an asymmetrical configuration. The topology is configured with minimal DC sources and arranged in a cascaded connection to offer asymmetrical operation with a voltage ratio in the order of 2 or 3. The topology has the provision of utilizing DC capacitors in place of isolated DC sources with a single DC source to reduce the usage of a higher number of DC sources in its structure. The structure for a 9-level inverter is shown in Figure 4.10, composed of four DC sources, four bidirectional IGBTs, and six unidirectional IGBTs. The topology is well suited for EC traction inverter applications due to its inherent feature of generating a quality output voltage with lower harmonic distortion. The operation for the voltage level ($V_1 + V_2$) follows the switching of the devices (S_1, S_2, S_6 and S_8) in the ON-state condition. The topology is operated using hybrid PWM, with PWM switching at the level generation part and fundamental switching at the H-6 inverter part. Table 4.7 portrays the design equations for the topology shown in Figure 4.8.

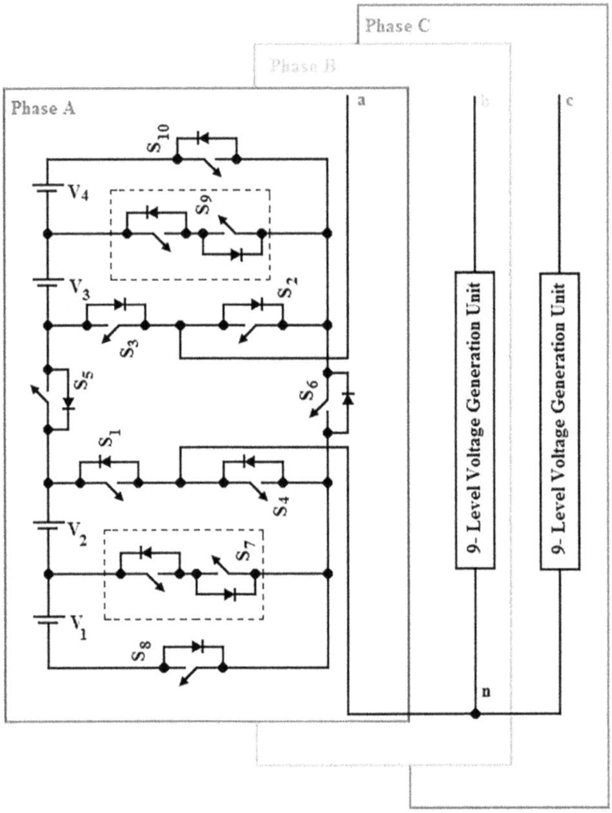

FIGURE 4.8 MLI with reduced current conducting devices.

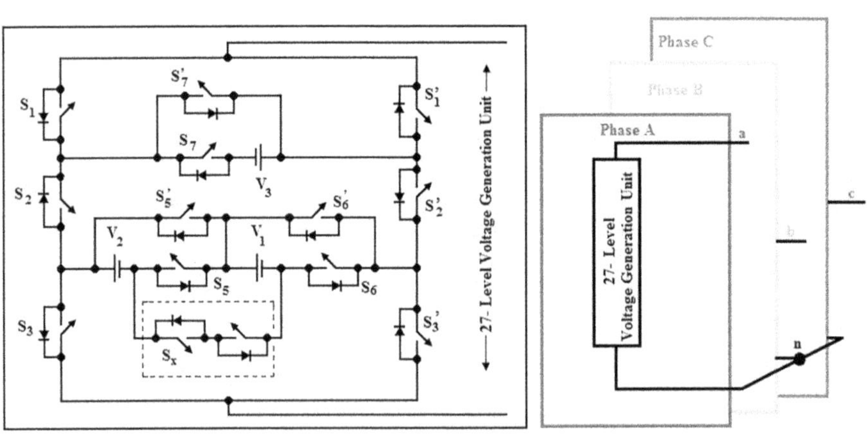

FIGURE 4.9 Flexible rung ladder structured MLI.

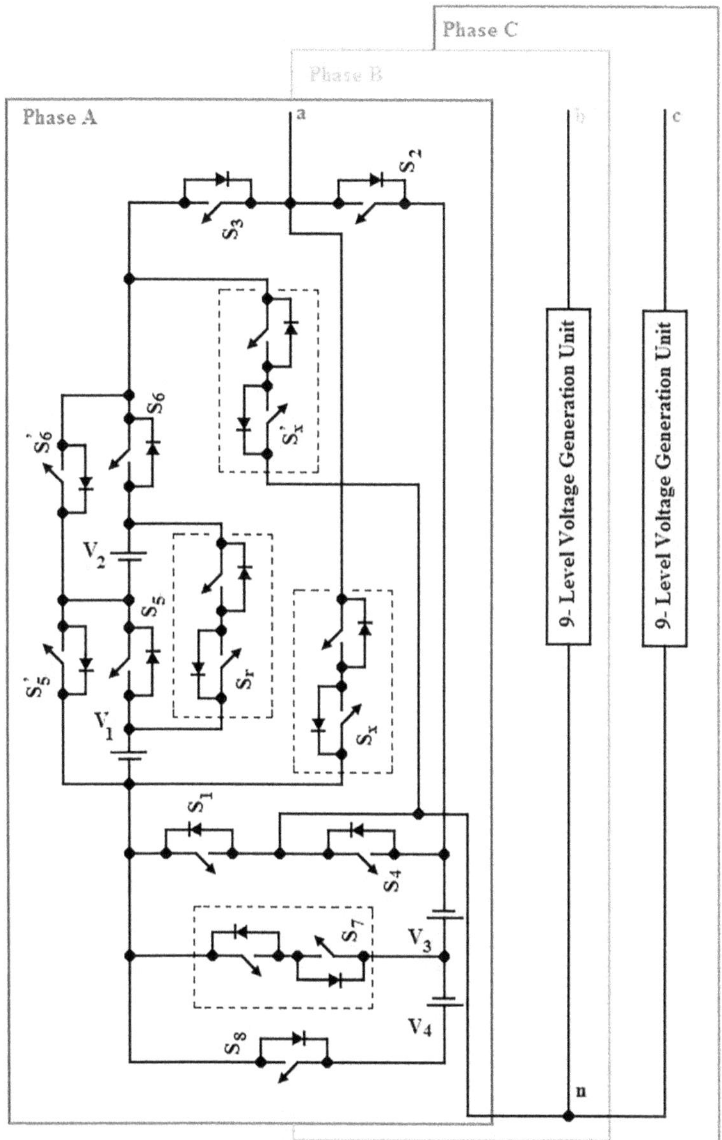

FIGURE 4.10 New tapped sources stack succored modified HX Bridge MLI.

4.3.7 Flexible Rung Ladder Structured MLI

The MLI topology (Chandrasekaran and Mohanty, 2020), represented in Figure 4.9, comprises voltage synthesizing modules formed by [voltage sources V_1 and V_2, switches (S_5, S_5', S_6, S_6', and S_x)] as module (d), and the voltage source (V_3) with switches (S_7 and S_7') as another module (g), which are arranged in reverse direction

TABLE 4.7
Design Equations for the Topology Shown in Figure 4.10 (Maalmarugan et al., 2017)

Parameters	Symmetrical	Asymmetrical
No. of DC sources	n_ds	n_ds
No. of switches	n_ds + 6	n_ds + 6
No. of output levels	(2 × n_ds) + 1	(4 × n_ds) − 1
Maximum voltage	n_ds × V_{dc}	((2 × n_ds) − 1) × V_{dc}
Standing voltages on all the switches	n_ds × V_{dc}	((2 × n_ds) − 1) × V_{dc}
$(S_1 - S_n)$ and $(S_1' - S_n')$		
H-bridge switches	2 × n_ds × V_{dc}	((4 × n_ds) − 2) × V_{dc}

in a ladder structure connected across the H-6 inverter. The dual voltage source cell (*d*) is used to perform asymmetrical operation for the voltage ratio of 2 or 3. For level addition, the switch pairs (S_5, S_5' and S_6, S_6') are switched to add the voltage sources, while for subtraction, the switch (S_x) plays a key role in reversing the polarity of the voltage sources for level generation. This particular arrangement makes this topology well suited for asymmetrical operation. A few voltage modules (*d* and *g*) can be connected individually or in combination to attain the desired number of voltage levels (*l*). For the MLI topology using (*d* and *g*) modules, the number of voltage levels (*l*) will be $(3 \times ((4 \times d) - 1)$, $((4 \times g) - 1)$, and $((6 \times ((2 \times d) + g) - 3))$ in the output voltage.

4.3.8 New Tapped Sources Stack Succored Modified HX Bridge MLI

A multilevel inverter is an alternative to two-level inverters, which provides a high-quality output waveform through small voltage steps from isolated DC sources, DC-link capacitors, or other renewable energy sources. On the other hand, it suffers from a higher number of power components required for synthesizing higher voltage levels. A new variant of the multilevel inverter is devised with a view to minimizing the total power components compared to traditional multilevel inverters. A new topology (Chandrasekaran et al., 2021), depicted in Figure 4.10, is composed of an array of DC sources (V_3–V_4) and dual DC sources (V_1–V_2) with an H-bridge inverter. The dual source module (V_1–V_2) performs arithmetic functions with the voltage sources (V_3–V_4) when the magnitudes of the voltage sources are not equal. The switches (S_5 and S_6) are complementary to the switches (S_5' and S_6'), which connect/bypass the voltage sources (V_1–V_2) to the load, while the switch S_r helps the voltage source (V_3) to be subtracted from V_1. The switches (S_7–S_8) connect the voltage sources (V_3–V_4) to the load, and the switches (S_x and S_x'), (S_1, S_2, S_3, and S_4) render polarity reversal. The DC sources can be derived from DC-link capacitors or other renewable energy sources. The developed topology requires $(n+9)$ switches, while the cascaded multilevel inverter needs $(4n)$, where '*n*' is the number of DC sources used to produce '*l*'

TABLE 4.8
Design Equation for the Topology Shown in Figure 4.10

MLI Structure Components	Cascaded H-Bridge	Diode Clamped	Flying Capacitor	Proposed
Main switches	$2(l-1)$	$2(l-1)$	$2(l-1)$	$(l+17)/2$
Clamping diodes	—	$2(l-3)$	—	—
DC split capacitors	—	$(l-1)/2$	$(l-1)/2$	—
Clamping capacitors	—	—	$(2l-6)/2$	—
DC sources	$(l-1)/2$	1	1	$(l-1)/2$

levels, and the design equations are tabulated in Table 4.8. The MLI topology utilizes a minimum number of switches in the conduction path to acquire various levels of output voltage. The topology can be configured to produce different varieties of voltage levels with possible values of V_1.

4.3.9 New Asymmetric H-6 Structured Multilevel Inverter

The MLI structure (Annadurai et al., 2024) shapes the output voltage closely to a sine wave, which inherently attenuates low-frequency harmonics without wave-shaping passive components. Traditional MLIs restrict their applications in variable drive applications, which drives the development of new MLI topologies with reduced power switches and DC sources.

An H-6 inverter presented in Figure 4.11 is used in series with an array of isolated DC input sources (V_1–V_2) and (V_3–V_4) on either arm to produce all possible values of V_1. The input DC sources (V_1–V_2 and V_3–V_4) use the switches (S_8–S_9 and S_{11}–S_{12}) for synthesizing voltage steps, and the switches (S_1–S_6) perform polarity reversal of the output voltage. The voltage source (V_1) provides the smallest step in the output voltage, while the other sources (V_2 and V_3–V_4) create voltage steps in multiples of (V_1) in the output voltage. The switching devices (S_7, S_{10}, and S_{13}) perform arithmetic operations between the voltage sources (V_1–V_2) from (V_3–V_4). PV (photovoltaic) panels, FC (fuel cells), or other storage elements such as capacitors or batteries can be used as input voltage sources. It is observed that four switches in the level-synthesizing part are switched at PWM, and three switches in the H-bridge operating at fundamental switching facilitate lower switching loss. The design equations are presented in Table 4.9.

4.3.10 Tri-Source Symmetric Cascaded Multilevel Inverter Topology

A new MLI (Jeyarama et al., 2023), represented in Figure 4.12, uses three DC sources as a module with a minimum number of semiconductor switches to achieve lower device count and power loss. The topology uses three identical voltage sources to work as a symmetrical module with three bidirectional IGBTs and two unidirectional IGBTs.

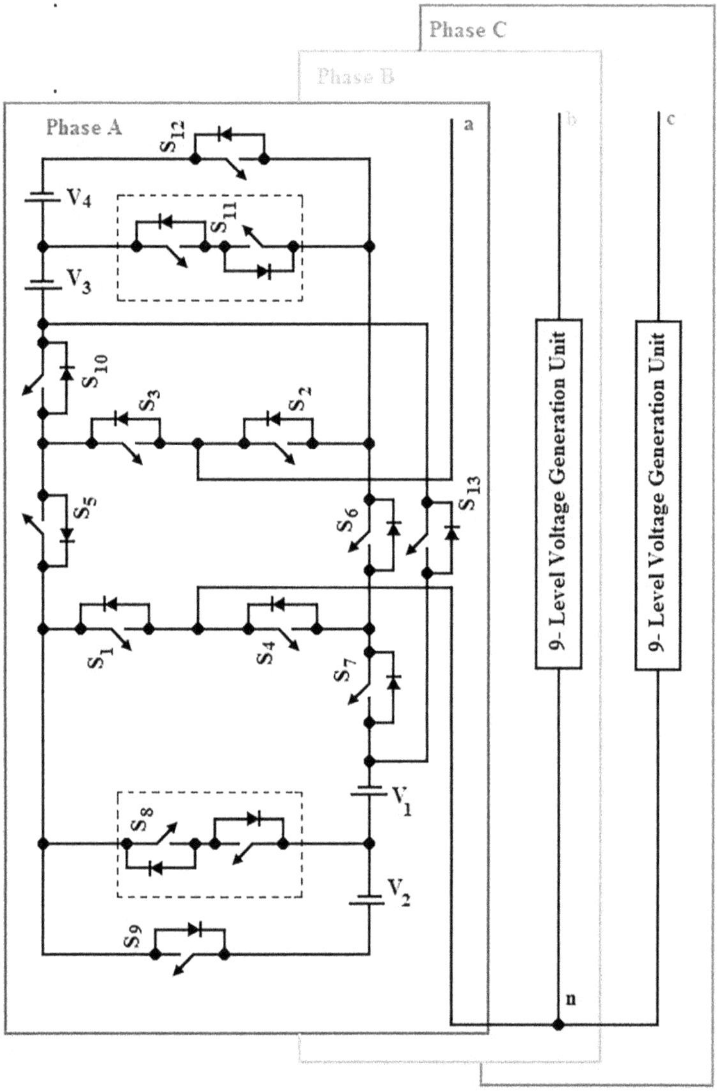

FIGURE 4.11 New asymmetric H-6 structured multilevel inverter.

TABLE 4.9
Design Equations to Obtain '*l*' Levels

Parameters	Switches	Gate Drivers	DC Sources
Cross-switched MLI	$13(l-1)/2$	$13(l-1)/2$	$4(l-1)/2$

The topology appears to offer fewer switches in the current path, which helps enhance higher voltage levels with fewer cascaded cells. The other module (Jeyarama et al., 2023) uses seven switches to operate as an asymmetrical module to achieve more voltage levels with a variety of voltage ratios. An H-bridge inverter is included to alternate the polarity of the output voltage.

The switches (S_2 and S_3) are turned on to add the voltage sources (V_1, V_2, and V_3) across the load. The switches (S_1, S_3, and S_4) are turned on to connect the voltage sources (V_1 and V_2) in parallel. The MLI shows more variations with different values of source voltages, and five mathematical relations have been introduced to configure the symmetrical module in an asymmetrical configuration to achieve higher voltage levels with fewer switches. Table 4.10 presents the mathematical relations for the topology shown in Figure 4.12.

TABLE 4.10
Design Parameters for CHBMLI and Tri-source MLI for *'u'* Units

Multilevel Inverter Structure	CHB Topology	Proposed
Main switches	$(4\times u)$	$(7\times u)+4$
Max. device count in the conduction path	$(2\times u)$	$(2\times u)+2$
SDCs	(u)	$(3\times u)$
Total standing voltage on all switches	$(4\times u)\times V_{dc}$	$(21\times u)\times V_{dc}$

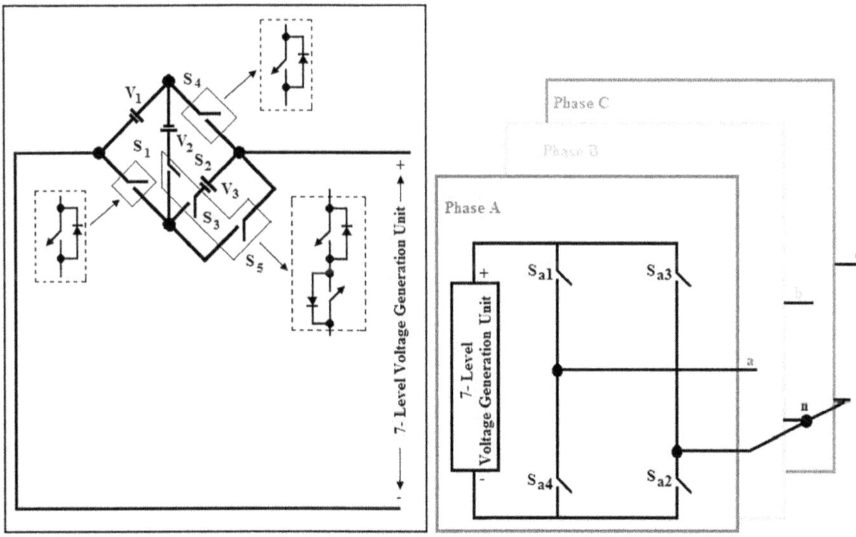

FIGURE 4.12 Tri-source symmetric cascaded MLI.

4.3.11 Switched Configuration-Based Novel Multilevel Inverter Topology

The switched configuration-based novel MLI (Krishnakumar et al., 2022), shown in Figure 4.13, uses voltage sources (V_1–V_4) and the switches (S_1–S_{10}) to synthesize multilevel DC-link voltage, and an H-bridge inverter with switches (S_{a1}–S_{a4}) for polarity reversal in the suggested topology. The diode (D_1) facilitates the function of avoiding inter-looping problems. The voltage sources (V_1–V_3) act as a permanent module, and the voltage source (V_4) is used as an additional module to elevate the voltage levels to a higher number. Using the voltage sources (V_1–V_3) alone in the MLI, it generates a 7-level stepped voltage. The switches (S_1, S_3, S_4, S_5, S_{a1}, and S_{a2}) and (S_1, S_3, S_4, S_5, S_{a3}, and S_{a4}) are switched in sequence to produce level-1 (V_{dc}) voltage. Different varieties of sources with asymmetrical voltage values have been chosen to configure the suggested topology in asymmetrical operation.

4.3.12 Reduced Switch Count MLI with Modified Pulse Width Modulation

The structure of the new inverter topology (Thiyagarajan et al., 2018) with a polarity-changing unit is shown in Figure 4.14. This topology consists of a basic unit and an H-bridge unit. The basic unit of the proposed structure is made of two DC voltage sources and four main switches. The voltage source (V_1) and the switches (S_1 and S_3) form a half-bridge cell, and the voltage source (V_2) along with switches (S_2 and S_4) is coined as a T-shape. These two arrangements are connected in parallel in such a way that the two voltage sources are connected in series. This arrangement produces a 5-level stepped voltage with the help of an H-bridge inverter, and the design equations are presented in Table 4.11. The switch (S_2) is a bidirectional IGBT

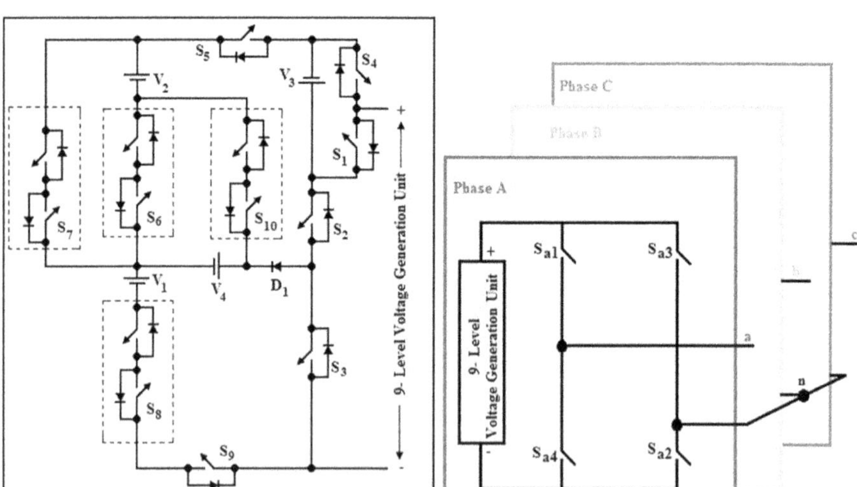

FIGURE 4.13 Switched configuration-based novel multilevel inverter topology.

TABLE 4.11
Design Equations to Obtain 'u' Units

Parameters	Switches	Gate Drivers	DC Sources
MLI	$5(u)+4$	$4(u)+4$	$2(u)$

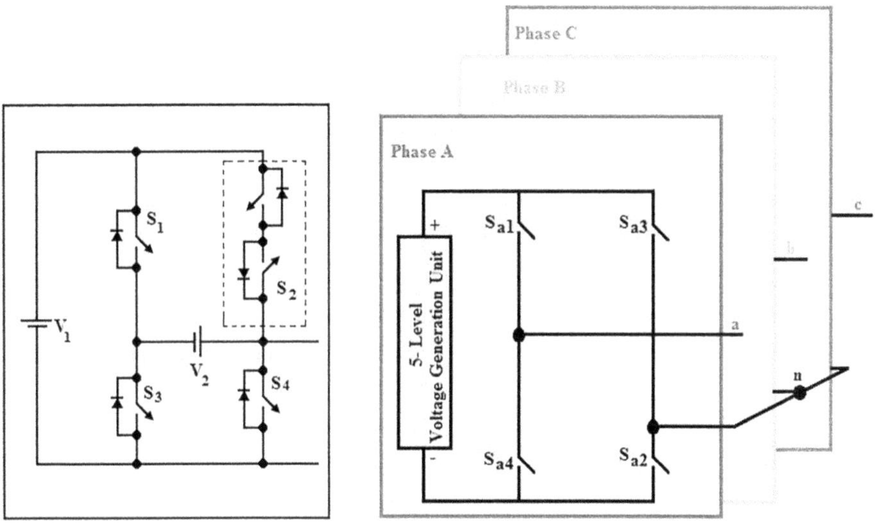

FIGURE 4.14 Reduced switch count MLI with modified pulse width modulation.

to allow bidirectional current flow and unidirectional voltage blocking in the circuit. The switch (S_1) is operated to connect the voltage sources in a cascaded connection. A modified PWM is employed to produce stepped voltage. The presented topology produces 7 levels with a voltage ratio of 2, and trinary operation is not possible with this arrangement. Six algorithms have been suggested to obtain a variety of voltage sources to operate in asymmetrical operation. These algorithms are employed with several units in cascaded connection. This topology offers reduced blocking voltage at the level generation unit; however, the full DC-link voltage is required to be blocked by the H-bridge inverter switches. Fault tolerant is possible only in the level generation unit, and any malfunction occurring in the H- bridge unit results in failure to synthesize stepped voltage. This topology is well suited for battery EV applications.

4.3.13 Modified Symmetrical Inverter Topology

A new structure (Thiyagarajan, 2020) of the MLI inverter topology and H-6 bridge inverter is shown in Figure 4.15. The structure comprises five DC sources and two bidirectional IGBTs. The structure is very peculiar in the MLI family, as it has two T-type double DC sources arranged across a single DC source (V_3) on each side

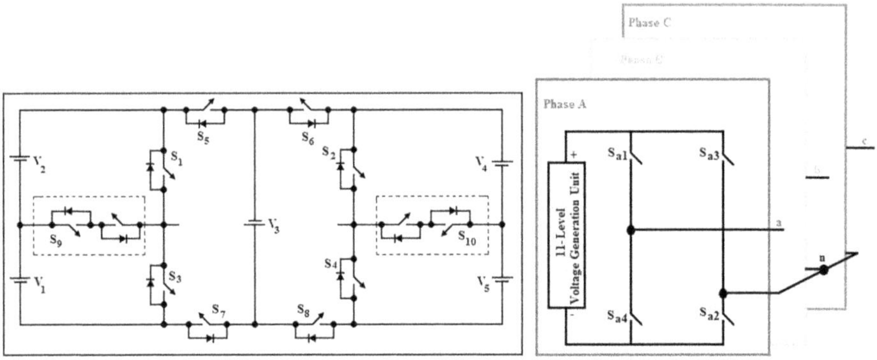

FIGURE 4.15 Modified symmetrical inverter topology.

through the H-6 inverter. The operation is very interesting, with switching of the switches in possible ways to connect each DC source to the load. The voltage source (V_1) and the switches (S_9, S_4, S_8, and S_7) are switched to connect the load. Similarly, the switches (S_5, S_8, S_2, and S_3) connect the voltage sources (V_1, V_2, V_3, V_4, and V_5) to the load. This configuration generates an 11-level stepped voltage across the load. The switches (S_9 and S_{10}) are bidirectional IGBTs used to clamp the voltage sources (V_1, V_2, V_4, and V_5) to the load. Fault tolerant is possible only in this topology at reduced voltage levels. This topology is well suited for battery EV applications.

4.4 SUMMARY

As multilevel inverters (MLIs) become increasingly vital for both high-power and low-power applications, researchers have explored innovative topological solutions tailored to specific needs. New MLI architectures have been introduced, aiming to achieve high output resolution while minimizing the number of components. However, in the effort to reduce the device count, certain trade-offs arise, impacting key characteristics of the topology. These include challenges in maintaining charge balance for equal load distribution among input sources, optimizing the switching of power devices with different ratings, and enabling configurations with asymmetric input sources.

REFERENCES

Chandrasekaran, K.; Mohanty, N.K., A flexible rung ladder structured multilevel inverter. *Tehnički vjesnik* 2020, 27, 1044–1049.

Chandrasekaran, K.; Mohanty, N.K.; Ranganathan, S., A new tapped sources stack succored modified HX bridge MLI. *Circuit World* 2021, 47, 2, 164–172.

Du, Z.; Tolbert, L.; Ozpineci, B.; Chiasson, J., Fundamental frequency switching strategy for multilevel inverters. *IEEE Transactions on Industrial Electronics,* 24, 1, 25-33, 2009.

Escalante , M.F.; Vannier, J.C.; Arzande, A., Flying capacitor multilevel inverters and DTC motor drive applications. *IEEE Transactions on Industrial Electronics*, 49, 4, 809-815, 2002.

Huang, A.Q.; Crow, M.L.; Heydt, G.T; Zheng, J.P.; Dale, S.J., The future renewable electric energy delivery and management (FREEDM) system. *IEEE Transactions on Industrial Electronic,* 49, 1, 133–148, 2011.

Maalmarugan Jayabalan, M.; Jeevarathinam, B.; Sandirasegarane, T., Reduced switch count pulse width modulated multilevel inverter. *IET Power Electronics* 2017, 10, 10–17.

Jose, S.; Titus, S.; A level dependent source concoction multilevel inverter topology with a reduced number of power switches. *Journal of Power Electronics* 2016, 16, 1316–1323.

Krishnakumar, V.; Anbarasan, P.; Ramash Kumar, K.; Venmathi, M., Novel switched configuration-based multilevel inverter topology for industrial applications. *Journal of Electrical and Computer Engineering*, 8743334, 12 2022.

Lee, S.; Lee, K.; Sul, S., A novel two-level inverter topology for electric vehicle applications. *IEEE Transactions on Power Electronics* 2013, 28, 12, 5568–5576.

Maharjan, L.; Inoue, S.; Akagi, H., A transformer-less energy storage system based on a cascade multilevel PWM converter. *IEEE Transactions on Industrial Electronics*, 44, 5, 1621–1630, 2008.

Jeyarama Pradeep; Vengadakrishnan, K.; Palani, A.; Sandirasegarane, T., A new tri-source symmetric cascaded multilevel inverter topology with reduced power components. *Circuit World* 2023, 49, 4, 431–444.

AnnaduraiRadhakrishnan, Arasan, E.S.; Ramalingam, B.C.; Chandrasekaran, K., A new asymmetric H-6 structured multilevel inverter with reduced power components. *Symmetry* 2024, 16, 72.

Ramkumar, S.; Kamaraj, V.; Thamizharasan, S.; Jeevananthan, S., A new series parallel switched multilevel dc-link inverter topology. *International Journal of Electrical Power & Energy Systems* 2012, 36, 1, 93–99.

Rodriguez, J., Multilevel inverters: A survey of topologies, controls, and applications. *IEEE Transactions on Industrial Electronics*, 49, 4, 724–738, 2002.

Ronanki, D.; Williamson, S.S., A comprehensive review of multilevel inverter topologies for electric vehicle applications. *IEEE Transactions on Vehicular Technology* 2018, 67, 2, 1102–1115.

Saravanan, K.; Sivasubramanian, M.; Gopinath, N.P., A 31 L multilevel inverter topology with less switching devices for hybrid electric vehicle applications. *Science Reports* 2024, 14, 27459.

Shivakumar, S.; Das, A.; Agarwal, P., Analysis of multilevel inverters for high-power electric vehicle traction applications. *IEEE Access* 2020, 8, 21245–21257.

Thamizharasan, S.; Baskaran, J.; Ramkumar, S., A new cascaded multilevel inverter topology with voltage sources arranged in matrix structure. *Journal of Electrical Engineering and Technology* 2015, 10, 4, 1552–1557.

Thamizharasan, S.; Baskaran, J.; Ramkumar, S.; Jeevananthan, S., A new dual bridge multilevel dc-link inverter topology. *International Journal of Electrical Power & Energy Systems* 2013, 45, 1, 376–383.

Thamizharasan, S.; Baskaran, J.; Ramkumar, S.; Jeevananthan, S., Cross-switched multilevel inverter using auxiliary reverse-connected voltage sources. *IET Power Electronics* 2014, 7, 1519–1526.

Thiyagarajan, V., Modified symmetrical inverter topology and switching pulse generation using logic gates. *Materials Today: Proceedings* 2020, 33, 3864–3869.

Tolbert, L.M.; Peng, F.Z.; Habetler, T.G., Multilevel inverters for electric vehicle applications. *Power Electronics in Transportation (Cat. No.98TH8349)*, Dearborn, MI, 1998, 79–84.

Thiyagarajan Venkatraman; Periasamy, S., Multilevel inverter topology with modified pulse width modulation and reduced switch count. *Acta Polytechnica Hungarica* 2018, 15, 2, 141–167.

Verma, A.K.; Thakura, P.R.; Jana, K.C.; Buja, G., Cascaded multilevel inverter for hybrid electric vehicles. In *India International Conference on Power Electronics*, New Delhi, India, 2011, 1–6.

Zhang, C.; Gao, Z., A cascaded multilevel inverter using only one battery with high-frequency link and low-rating-voltage MOSFETs for motor drives in electric vehicles. *Energies* 2018, 11, 1778.

5 Advanced Power Electronic Converters for Power System Applications

Palanisamy Ramasamy, Thambi Ezhilarasan Mangaiyarkarasi Thamizh Thentral, Vijayakumar Krishnasamy, Selvakumar Kuppusamy, and Selvabharathi Devadoss

5.1 INTRODUCTION

The importance of power electronics converters in various power system applications is crucial and continuously increasing day by day, which is mostly driven by the growing adoption of renewable energy sources, electric vehicle transportation, and the evolution of smart grids [1]. The advanced power electronics converters are used to convert energy from one form to another, like inverters, rectifiers, choppers, and cycloconverters [2–4]. These converters enable the precise control and modulation of power, allowing for high efficiency and adaptability across real-time and industrial applications. As the energy sector transitions towards sustainability, advanced power electronics converters need an indispensable role in synchronizing renewable sources, stabilizing power quality, and improving energy flows to the grid [5–7].

In modern power systems, the power converters and their topologies play a major role in regulating power, controlling, and optimizing the flow of electrical energy across different systems with minimized harmonic contents [8,9]. Here, the power converters include various advanced power switches, which have less conduction and switching losses; this leads to their use in various power system applications. Power converters are critical for incorporating renewable energy sources like solar, fuel cells, and wind into the power grid [10,11]. Since these sources generate variable DC or AC outputs that do not match the grid's requirements, these advanced power converters are primarily used to convert and regulate the output to associate with grid standards [12]. This ability is crucial as renewable energy sources are inherently variable and dependent on environmental conditions. Power converters play a significant role in maintaining grid stability, particularly in grids with high renewable penetration. This system manages various factors like reactive power, voltage magnitude, phase angle, and frequency, which are most needed for preventing disturbances and ensuring efficient and consistent power transfer [13].

 DOI: 10.1201/9781003585534-5

Converters empower frequency regulation by balancing supply and demand, responding quickly to fluctuations, and supporting frequency adjustments, and voltage support through reactive power control, which helps maintain voltage within safe limits, especially in areas far from central production sources [14]. The harmonic level is minimized to improve power quality and reduce power quality issues like voltage sag, swell, and fluctuations caused by nonlinear loads [15].

Pulse Width Modulation (PWM) methods are very important in modern power electronics and electrical engineering due to their role in efficiently controlling power delivered to electrical devices, especially in converters and motor drives [16]. The various PWM controllers include some key points to provide efficient power transfer. PWM allows for precise control of power by varying the magnitude and width of voltage pulses, which can control the system to adjust the average power delivered without dissipating energy as heat, making it a highly efficient control method for any power converter [17,18]. PWM controllers help diminish harmonic distortion in all output waveforms of the converter, leading to improved overall power quality of the system [19]. This is especially valuable in applications like power supplies and renewable energy systems, where harmonics can affect system performance and efficiency [20]. Different PWM methods enable smooth and precise control of motor speed and torque in AC and DC motor drives, contributing to dynamic response and higher efficiency, which plays a major role in industrial motor control [21].

The switching losses of the system are minimized by implementing suitable PWM techniques; by switching at specific intervals, PWM reduces the time that transistors operate in the linear region, minimizing switching losses and improving the thermal management of power switches [22]. PWM methods are versatile and can be applied across a wide range of applications, from medium voltage to high power applications, making them adaptable to various power levels and control schemes, which are more flexible and scalable [23]. The performance of the system is improved, like in inverters, by using PWM techniques that help in converting DC to AC with a smooth sinusoidal output, essential for applications that necessitate steady and clean power, such as grid-tied systems, Photovoltaic (PV) inverters, Flexible AC Transmission Systems (FACTS), High Voltage Direct Current (HVDC) systems, and UPS [24,25].

Advanced power electronics converters are very important for real-time applications in power systems for enhancing the efficiency and stability of the network. Converters like DC-AC inverters and MPPT controllers help integrate solar and wind power into the grid by stabilizing and regulating variable energy output, which is a key application in renewable energy systems [26]. The converters used in applications like UPQC, IPFC, STATCOM, and UPFC control reactive power and improve voltage stability in power transmission, optimizing grid system performance. HVDC converters enable long-distance, low power loss transmission, supporting efficient energy transfer from generating stations and grid interconnections. AC-DC converters in EV chargers ensure efficient power conversion, enabling fast and reliable charging infrastructure, which is much more suitable for EV charging technology [27].

The efficiency of advanced power electronics converters in modern power system applications is very high due to flexibility and the ability to reduce losses during the transmission of power. The equipment used in these applications, like inverters and MPPT controllers, has efficiency of nearly 95%–98% [28]. These high efficiencies help reduce energy and power losses when converting and feeding renewable energy sources into the grid system, ensuring maximum power utilization from variable sources like solar and wind. FACTS converters, such as IPFCs, STATCOMs, and UPFCs, are premeditated for real-time reactive power control and voltage stabilization [29]. Their efficiency is naturally 98% or higher, enabling minimal energy loss while enhancing the stability and performance of the power grid [30].

HVDC systems are highly effective for long-distance power transmission, with efficiency rates ranging from 95% to 98%. The use of modern voltage source converters and other HVDC technologies has further enhanced efficiency, particularly in reducing losses in the conversion between AC and DC [31]. EV chargers with advanced converters can achieve efficiencies of around 90% or higher, especially in fast-charging EV applications. The efficiency depends on the quality of the converter design and PWM methods, with modern strategies focusing on reducing losses during both AC to DC conversion and DC to battery charging for EV applications [32,33].

This chapter explores the vital role of advanced power electronics converters in modern power systems, focused on their application in renewable energy integration, grid management, and electric vehicle power. The field of power electronics has undergone significant evolution, driven by the increasing demand for efficient, reliable, and sustainable energy conversion and control in modern power systems. Over the years, the development of new materials, advanced control techniques, and more well-organized converter topologies has transformed the way power is generated, transmitted, distributed, and used.

5.2 TYPES OF POWER ELECTRONICS CONVERTERS

The advanced power converters are used to familiarize and convert power from one form to another mainly from AC to DC and vice versa, or to adjust the voltage or frequency of the power being transported [34]. The converters are used in various applications such as industrial systems, electrical drives, FACTS devices, and variable speed drives, which play a critical role in modern power systems by enabling well-organized energy conversion, control, and conditioning in a wide collection of applications, from renewable energy systems to variable industrial motor drives and power supplies [35].

5.2.1 Rectifiers

The rectifiers, also called AC-DC converters, are converters that change AC to DC. This adaptation is most important for powering electronic devices, charging batteries, and supplying DC power to various systems that require stable voltage conversion [36]. Rectifiers are mostly used in power supplies, motor drives, renewable energy sources, and industrial drive applications. There are several types of rectifiers, each with specific characteristics suited for different applications. The various types of rectifiers include half-wave and full-wave rectifiers. AC-DC converters are essential components in modern power systems, providing the necessary DC voltage for a wide range of applications. The choice of rectifier type half-wave, full-wave, or bridge depends on the application requirements in terms of efficiency, ripple reduction, and power levels. With ongoing advances in semiconductor technology, rectifiers are becoming more efficient and reliable, playing a critical role in the power supply of electronic devices and systems [37].

AC-DC converter systems are used in numerous applications, including power supplies to convert AC from the electrical grid into DC to power electronic devices such as computers, televisions, and telecommunications equipment; battery charging, where AC-DC converters are used in battery chargers for devices like smartphones, electric vehicles, and renewable energy storage systems; industrial equipment for powering motors and control systems in industrial machinery; and renewable energy, where in solar and wind power systems, AC-DC converters help change the DC from wind turbines into AC for grid compatibility or DC for storage.

The Interleaved Vienna Rectifier is an advanced power electronics converter used for AC-DC conversion, specifically designed to improve efficiency, reduce harmonic distortion, and enhance the overall performance of rectification systems, especially in high-power applications such as renewable energy sources, electric vehicle charging stations, and industrial power supplies. It combines the Vienna Rectifier topology with interleaving, a technique where multiple converters operate in parallel with a phase shift between them to segment the load. The Vienna Rectifier is a three-phase AC-DC converter that uses a three-level diode-clamped structure to achieve high PF and low total THD during AC to DC conversion. It typically uses three switches and three diodes to rectify AC to DC and can provide a high-quality DC output while minimizing input current harmonics.

The Interleaved Vienna Rectifier works by incorporating multiple Vienna Rectifier stages, typically two or three stages, operating in parallel with a phase shift between the stages. Each stage operates at a different time, which reduces the overall ripple current and improves the efficiency of the system. Each of the rectifier stages operates with a specific phase shift in voltage magnitude and phase angle. This phase shift ensures that the current drawn from the AC supply is more evenly distributed across the various stages of the rectifier. Each interleaved Vienna rectifier stage works in the same way as a classical Vienna rectifier, which is used to convert AC to DC. Each stage uses a three-level diode-clamped converter with switches to rectify the AC voltage. The power structure of the interleaved Vienna rectifier is shown in Figure 5.1.

The interleaved design splits the load current between various stages, reducing the ripple current and the stress on each converter stage. This also results in a higher current-handling capability without requiring larger components. Its capability to handle higher power while minimizing component sizes and harmonic distortion makes it an attractive choice for modern power electronics applications. The DC output from the interleaved stages is then combined, providing a high-quality DC voltage with reduced ripple content and lower THD. The Interleaved Vienna Rectifier is a highly efficient AC-DC converter topology that combines the advantages of the Vienna rectifier and interleaving. It offers enhanced efficiency, reduced ripple current, and better power quality, making it ideal for medium-voltage and high-power applications.

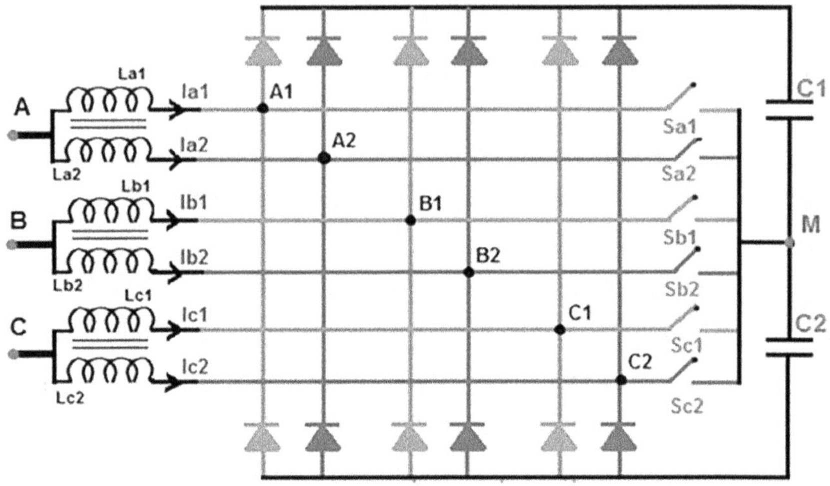

FIGURE 5.1 Power structure of the interleaved Vienna rectifier.

5.2.2 Inverters

Inverter construction is a crucial part of modern power systems, as it enables the conversion of DC to AC, which is important for the operation of most electrical grids, renewable energy systems, and numerous industrial, commercial, and residential applications [38]. Inverters are essential components in modern power system applications, ensuring the conversion, synchronization, and efficient management of energy. Their ability to convert DC to AC, synchronize renewable energy sources with the grid system, optimize motor control, improve power quality, and enable backup power makes them a basis for energy-efficient control, reliable systems, and sustainable energy systems. Whether in renewable energy, electric vehicles, industrial systems, or microgrids, inverters enable seamless power transformation and utilization across various applications.

The 3-Level NPC Inverter is an advanced inverter topology used for DC–AC conversion, providing an AC output with three voltage levels instead of the traditional two (positive and negative) seen in conventional inverters. This topology helps reduce harmonic distortion, improve efficiency, and enable higher voltage handling in power systems [39]. The three-phase three-level Neutral Point Clamped Inverter (NPC-MLI) is shown in Figure 5.2. It consists of three legs; each leg has four IGBT switches, 4 clamp diodes, and two DC-link capacitors. The applied DC input voltage is balanced using the DC-link capacitors, wherein under ideal conditions the voltage across the capacitors is VC1 = VC2. Thus, each capacitor splits the voltage to $V_{dc}/2$, and the voltage stress is limited to one capacitor level through the clamping diodes. Each leg (phase) can operate in three switching states: $+V_{dc}/2$, 0, and $-V_{dc}/2$.

In a 3-level NPC inverter, each leg has three switching modes ($+V_{dc}/2$, 0, $-V_{dc/}2$), resulting in a total of 27 switching states. Among these, there are three null (or zero)

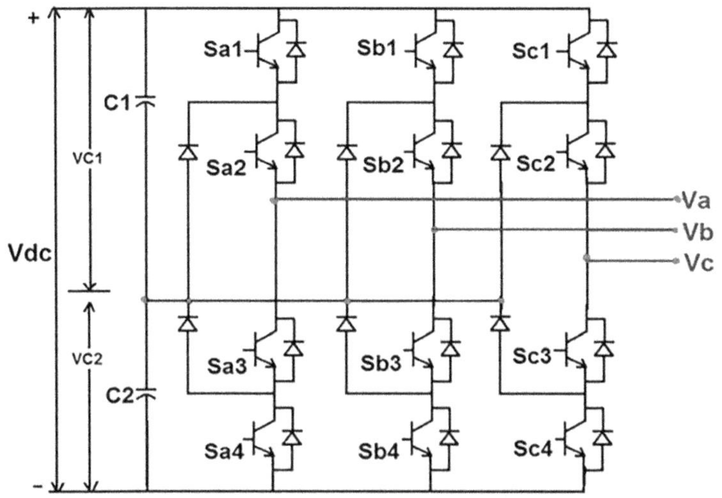

FIGURE 5.2 Circuit diagram of 3-Phase 3-Wire NPC inverter.

vectors, 12 small vectors, and six medium and large vectors. The NPC 3-Level Inverter is an advanced, high-performance topology that enhances the efficiency and quality of power conversion. By utilizing three voltage levels, it reduces harmonics, improves efficiency, and provides better voltage control, making it highly suitable for high-power applications, renewable energy systems, and motor drives. Although more complex in design compared to traditional inverters, its benefits in power quality and system performance make it a popular choice in modern power electronics [40].

5.2.3 Matrix Converters

A 3-Phase Matrix Converter is a direct AC–AC converter that converts a 3-phase AC input into a 3-phase AC output without the need for a transitional DC link. Unlike traditional converters that use a DC bus (such as AC–DC–AC converters), the matrix converter operates using direct switching, providing high efficiency and compactness, which makes it suitable for applications requiring smooth and efficient AC power conversion [41]. A 3-phase matrix converter consists of nine bidirectional switches arranged in a 3×3 matrix configuration, as shown in Figure 5.3. Each switch connects an input phase to an output phase. These switches are typically IGBTs or MOSFETs in modern applications, with anti-parallel diodes used for freewheeling. The matrix converter does not require a DC intermediate stage; instead, it directly connects the 3-phase AC input to the 3-phase AC output. This results in no DC-link capacitors or inductors, making the system more compact and reducing losses compared to traditional AC–DC–AC converters.

The matrix converter operates by switching the 3-phase input supply (A, B, C) directly to the 3-phase output supply (X, Y, Z) with controlled output voltage magnitude and frequency. The output voltage of the matrix converter is obtained

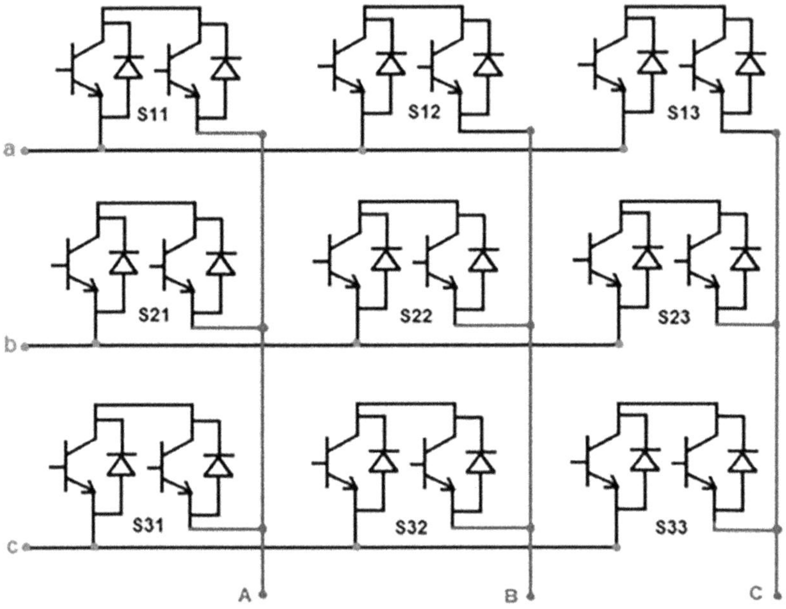

FIGURE 5.3 Circuit diagram of 3-to-3 Phase Matrix converter.

by selecting a suitable input supply at the right time using appropriate switching pulses. PWM methods such as SVM or DTC are generally used to generate the switching pulses and states, which control the timing and sequence of the switches to produce a high-quality 3-phase output voltage waveform. The switching of the matrix converter is usually controlled by advanced modulation techniques such as SVM, modified SVM, and other advanced SVM methods, which enhance the switching sequences to reduce harmonic distortion and achieve a 3-phase sinusoidal output voltage. The output frequency and voltage can be controlled by adjusting the timing of the switching states, allowing for variable-frequency drive and industrial applications.

The converter uses nine bidirectional switches to connect each input phase source to an output phase of the system. The switches are turned ON and OFF in a precise sequence based on the modulation technique used for control. The result of the switching action is an AC output on phases X, Y, and Z, where the voltage and frequency can be controlled by adjusting the switching sequence with the help of an advanced PWM controller. Unlike traditional converters, the matrix converter does not require a DC-link capacitor or inductor to store energy, which simplifies the converter structure and increases efficiency.

The 3-Phase Matrix Converter is a versatile and efficient AC-–AC converter that eliminates the need for an intermediate DC stage, offering compactness, high efficiency, and precise control. Its bidirectional power flow capability, low harmonic distortion, and suitability for various applications make it a promising solution for modern power conversion systems.

5.3 ADVANCED CONTROL TECHNIQUES FOR CONVERTERS

To overcome the mentioned issues, an alternative type of inverter called the multilevel inverter is developed. The main features of a multilevel inverter are as follows: the topology allows it to process high voltage and high power while generating low harmonics without the use of transformers or series-connected power switching devices [42]. The efficiency is very high, about 95%. The efficiency of the multilevel inverter is greater than or equal to 95% for electrical loads greater than 10% of rated power, whereas the conventional 2-level inverter or VSI does not achieve 90% efficiency for electrical loads greater than 30% of rated power. The multilevel inverter system has more than 98% efficiency for load conditions greater than 30% of rated power. It produces an m-level or step staircase output voltage waveform without using any transformer for line-frequency switching; the output voltage almost reaches a pure sinusoidal waveform by increasing the number of inverter levels. The m-level inverter has m−1 steps or levels.

The EMI of the system is very low because its switching frequency is much higher than that of a conventional two-level inverter. It increases the voltage handling capability if the devices operate within their safe operating limits. Generally, a multilevel inverter topology with a proper PWM control strategy can provide better performance, lower cost, lower EMI, and higher efficiency than a conventional 2-level inverter. In practical applications, PWM inverters are mostly used in power electronic circuits. Compared to square-wave inverters, PWM inverters provide higher quality output power, which is more suitable for variable-speed drives. PWM inverters are used to control the output voltage and current and to minimize the harmonic content in the output voltage. The widths of the pulses are modified to change the output voltage. Generally, PWM methods are classified as discussed in the following sections.

5.3.1 Single Pulse Width Modulation

In single PWM, every half cycle has only one pulse. By adjusting the width of the pulse, the output voltage of the inverter can be controlled. The switching pulses are generated by comparing the rectangular reference signal (A_r) with the triangular carrier signal (A_c). Figure 5.4 shows the generation of switching pulses using single pulse width modulation. The fundamental frequency of the inverter output voltage (V_0) can be obtained from the frequency of the input signal [43].

The amplitude modulation index of this technique is defined as,

$$M = \frac{A_c}{A_r} \tag{5.1}$$

The main drawback of single pulse width modulation is the presence of higher harmonic content in the output voltage.

5.3.2 Sinusoidal Pulse Width Modulation (SPWM)

The SPWM technique is a widely used method for controlling power semiconductor switches, where gating pulses are generated by comparing a reference signal with a high-frequency carrier signal. In this technique, a sinusoidal waveform serves as the

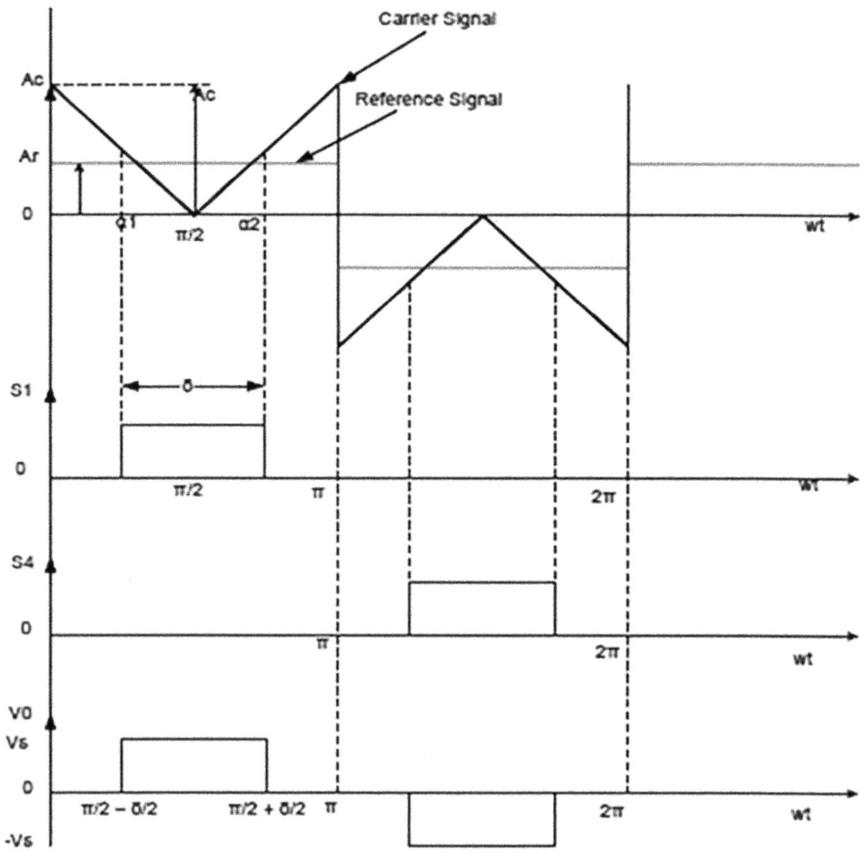

FIGURE 5.4 Generation of a switching pulse using a single PWM.

reference signal, while a triangular waveform functions as the carrier. The intersection of these two signals determines the switching instants, resulting in gating pulses whose widths are proportional to the instantaneous amplitude of the sine wave, as illustrated in Figure 5.5.

The output frequency (f_o) of the NPC inverter is determined by the frequency of the reference signal. The root mean square output voltage (V_o) is governed by the modulation index, which in turn is influenced by the peak amplitude of the reference signal. The inverter output voltage is regulated through the gating pulses generated by the modulation controller. Furthermore, the number of pulses produced per half cycle depends on the frequency of the carrier signal [44].

$$M_a = \frac{V_c}{V_{car}} \tag{5.2}$$

V_c – Peak magnitude of the reference control signal, V_{car} – Peak magnitude of the carrier signal.

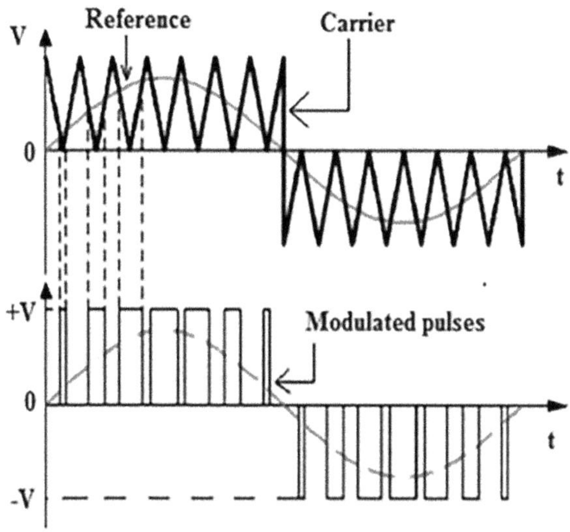

FIGURE 5.5 Sinusoidal pulse width modulation technique.

5.3.3 Space Vector Pulse Width Modulation (SVPWM)

Space vector pulse width modulation (SVPWM) is an algorithm to control pulse width modulation and is used to generate alternating current signals, in addition to reducing total harmonic distortion (THD), as shown in Figure 5.6. It directly uses the control variable states, which are given by the power electronic system, and discriminates each switching vector as a vertex point in the hexagon space plane [45]. SVPWM is a digital PWM method, so it is very simple to control the duty cycle of every power switch. SVPWM has additional advantages due to the utilization of the full DC-link voltage and operation in the hexagon (entire modulation region). The SVPWM concept can be observed as a 'weighted average use of the output voltage to attain the maximum fundamental voltage value with reduced harmonic content level'. The output sinusoidal voltage is a stable amplitude state vector rotating at a constant frequency. The flowchart for sector and triangle determination is shown in Figure 5.7.

5.3.4 Sliding Mode Control (SMC)

It is a robust nonlinear control technique used in dynamic systems to achieve desired performance by forcing the system to "slide" along a predefined surface in the state space. It is widely applied to control systems that are subject to uncertainty, disturbances, or nonlinearities, such as power electronics, robotics, and motor control. Sliding mode control refers to the behavior of the modified system once it reaches and stays on the sliding surface of the application. In this mode, the system's dynamics are governed by the dynamics of the sliding surface, which are usually simpler and can be designed for stability and robustness. The control law forces the system

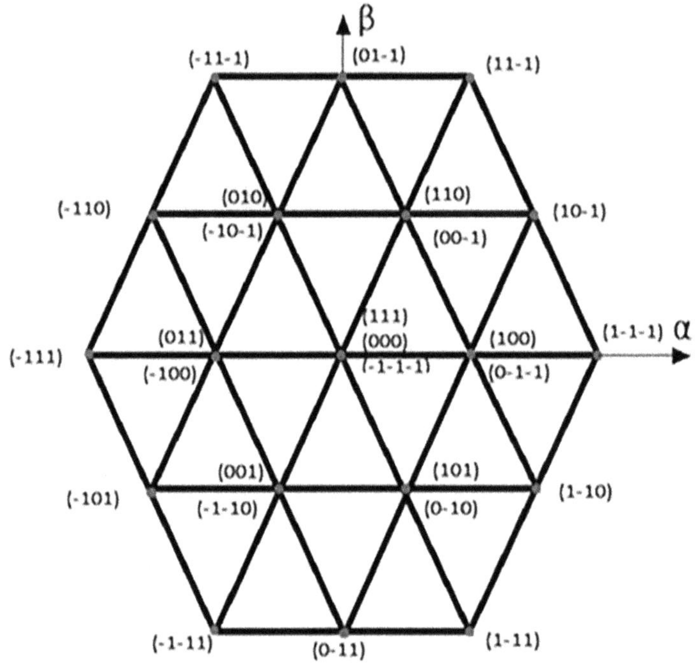

FIGURE 5.6 Hexagonal 2D-SVPWM for 3-level inverter.

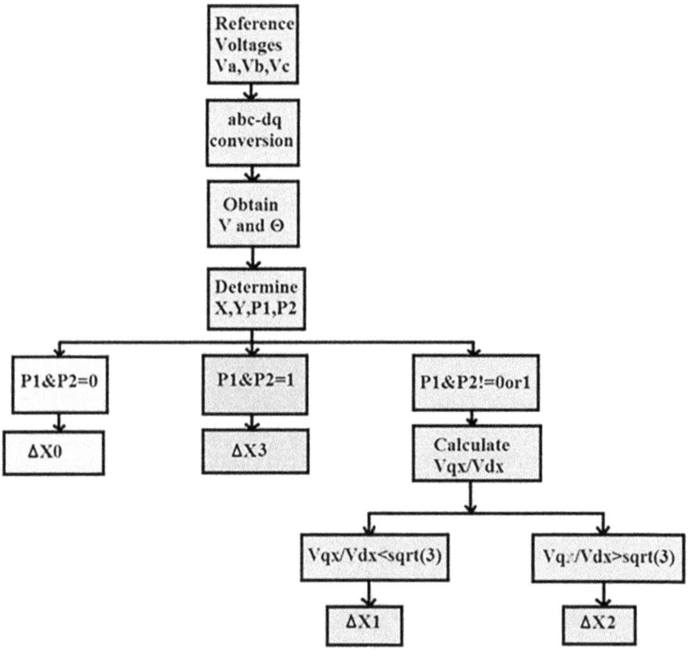

FIGURE 5.7 Flowchart for sector and triangle determination.

to slide along this surface, ensuring that the state of the system stays on the surface with highly efficient performances, even in the presence of disturbances or model uncertainties.

The control input for the required system is designed so that when the system state is off the sliding surface, the control law drives it toward the surface. Once the state reaches the surface of the particular application requirement, the system is forced to slide along it. The control law is typically discontinuous, meaning it switches between two distinct values to push the state toward the sliding surface. This discontinuity is what gives SMC its robustness, but it can lead to chattering near the surface. The discontinuous nature of the control law leads to chattering near the sliding surface, which can be undesirable in practice. Techniques like boundary layers or higher-order sliding mode control can help mitigate this issue. Implementing sliding mode control requires fast and accurate switching of the control input, which may be challenging in systems with limited computational power or in hardware with limited switching capabilities.

SMC is an advanced and robust control method most suited for systems with uncertainties and nonlinear conditions. By forcing the system state to follow a sliding surface, SMC ensures that the system's output performance remains stable and predictable despite external disturbances or model inaccuracies. While challenges like chattering exist, advanced techniques and applications in various fields such as power electronics, robotics, and automotive systems demonstrate its practical utility.

5.3.5 Model Predictive Control (MPC)

Model Predictive Control is a progressive control system that uses a precise model of the system to predict imminent behavior and make control decisions based on this calculation of the system. The basic idea behind MPC is to improve control inputs over a finite prediction horizon, ensuring that the system behaves optimally while also satisfying limitations on the system and control variables within a particular range. MPC needs an extrapolative model of the system, which designates how the system state evolves over time. This model can be a linear or nonlinear system, and it is typically characterized in state-space form or as a difference equation. The model is used to forecast future system states over a specified prediction horizon, given the current state, previous state, and the control inputs. At each time step, MPC solves an optimization problem that determines the best arrangement of control inputs that minimize a cost function. The objective is to achieve the desired system performance while satisfying constraints.

MPC operates on a receding horizon principle. This means that at each time step, the controller uses the current state to solve an optimization problem for the next several time steps (the prediction horizon), but only the first control input in the optimal sequence is applied to the system. After applying the first control input, the horizon is shifted forward, and the optimization problem is solved again with the updated state. This process repeats at each time step. The control action applied to the system is the first control input from the optimal sequence, and only this input is implemented at each step. This approach makes MPC receding, as the optimization moves forward with new measurements and updated system states.

An optimization algorithm is employed to find the arrangement of control inputs over the prediction horizon that minimizes a cost function. The cost function typically includes tracking error, which is the difference between the predicted system state and the anticipated state. Control effort: The magnitude of the control inputs, penalizing excessive actuation. The optimization problem may also incorporate constraints, such as actuator limits, safety margins, and operational boundaries. It is a flexible control strategy that optimizes system performance while respecting constraints and handling uncertainties. It is particularly useful in complex systems where future behavior can be predicted and where constraints must be respected. While MPC provides optimal control, its computational demand and reliance on accurate models can present challenges. Nonetheless, its widespread applications in industries such as power systems, automotive, process control, and robotics demonstrate its effectiveness and versatility in modern control systems.

5.4 CONVERTERS IN RENEWABLE ENERGY APPLICATIONS

Renewable energy such as solar, wind, and hydro have seen exponential growth in use due to their environmental benefits and sustainability. However, renewable energy sources are intermittent and variable, necessitating efficient conversion and integration with the grid [46]. Power converters play a critical role in the incorporation of renewable energy into power systems. These advanced converters manage energy conversion, control, and power flow between renewable energy sources and the grid or any battery-based storage systems. Here, converters are much useful in providing smooth synchronization with the grid system, enhancing system stability and reliability, and ensuring efficient consumption of renewable energy sources. This discussion explores the different types of converters used in renewable energy applications, their working principles, and their significance in improving the performance of renewable energy-integrated systems [47].

Renewable energy sources, due to their intermittent nature, require power electronic converters to accomplish energy flow effectively and guarantee constant and stable grid power integration. The role of power converters in renewable energy applications includes the following functions:

a. **Voltage and Frequency Regulation***:* Power converters help in adapting voltage and frequency to ensure control so that renewable energy sources can be combined with the grid system. For instance, in wind and solar power systems, inverters adjust the output to match the grid's frequency and voltage, ensuring stable operation.
b. **Grid Connection and Power Flow Control***:* In renewable energy system applications, converters enable the system to transmit power from the renewable energy source to the grid system. Inverters enable the necessary synchronization between the DC output of PV panels or wind turbines and the AC grid system without the use of any transformer or coupled inductor unit. Converters also ensure that the power delivered to the grid is stable, meets power quality standards, and is properly synchronized with the output unit.

c. **Efficiency and Loss Minimization**: Advanced power converters are designed and implemented with high efficiency to reduce switching and conduction losses, which is essential in renewable energy applications where energy conservation is a primary requirement. Advanced topologies and semiconductor technologies have led to noteworthy developments in converter efficiency and reliable performance.

Power converters are important for the effective operation of renewable energy systems, ensuring that energy produced from sources like solar, wind, and hydro can be converted, stored, and transferred with negligible losses. The various converters include AC-DC, DC-AC, DC-DC, and AC-AC converters, which are crucial in managing energy conversion in applications such as voltage regulation, monitoring, and grid synchronization. As renewable energy technologies continue to progress, the role of power converters in enhancing energy efficiency, ensuring stability, and improving system reliability will further increase system performance, driving advancements in sustainable power generation.

5.4.1 Energy Storage Systems (ESS)

Energy storage systems (ESS) are crucial for enabling the efficient incorporation of renewable energy sources like solar and wind into the power grid, as well as enhancing the reliability and stability of energy supply. Bidirectional converters are pivotal in energy storage systems, especially for applications involving batteries and other storage technologies, such as supercapacitors or flywheels. These converters manage the flow of energy in both directions, enabling the storage of excess energy and the retrieval of stored energy when required. This concept is essential for harmonizing supply and demand in renewable energy systems.

- Large-scale storage systems are often used to stabilize the grid system by absorbing excess energy during periods of high renewable generation and discharging when energy demand exceeds power generation.
- Bidirectional converters help in frequency regulation by storing energy when the grid frequency is high and discharging it when the grid frequency is low, ensuring stable grid system operation.
- In solar applications, various bidirectional converters are used in solar battery storage systems to charge batteries during the day and discharge them when solar energy is unavailable, such as during the night or cloudy days.
- Bidirectional converters enable households or businesses with solar panels to store excess energy in batteries for later use, reducing dependence on the grid.
- Bidirectional converters are used in electric vehicle systems that lead to Vehicle to Grid (V2G) and Grid to Vehicle (G2V) technology. The EV battery can discharge energy back to the grid, providing supplementary storage and control for the grid management system, allowing the vehicle to charge during periods of low energy demand.

- In microgrid systems, various bidirectional converters enable energy exchange between the renewable energy system, storage, and local load or grid system. The converter supports both the storage of surplus energy and the discharge of energy to supply the load or export power to the grid unit.

Bidirectional converters are essential to energy storage systems, providing important functionality for the effective charging and discharging of energy in EV applications. This system enables the efficient use of energy in various renewable energy applications, ranging from grid-connected systems to hybrid electric vehicles and microgrid setups. With their ability to standardize energy flow, preserve system efficiency, and improve the reliability and longevity of storage devices, bidirectional converters are vital in maximizing the benefits of renewable energy and advancing the transition to more sustainable energy systems. As renewable energy adoption grows, the role of various bidirectional converters will continue to expand, ensuring that energy storage remains a consistent and effective solution for handling recurrent power generation systems.

5.4.2 Challenges in Renewable Energy Integration and Solutions

Due to increasing challenges with respect to power demand and synchronization with grid-connected systems using more sustainable energy sources and renewable energy integration, these challenges arise primarily from the inherent inconsistency and intermittency of renewable energy sources such as biomass, solar, wind, and hydropower. Even so, various innovative ideas and solutions using advanced control technologies, improved grid system management, and enhanced policy frameworks are being established to address these issues and provide efficient solutions for the seamless incorporation of renewable energy into the power grid. Renewable energy sources like solar and wind are sporadic; their output can vary based on different environmental conditions. Solar power generation depends on sunlight, radiation, and temperature, which can be affected by the time of day, weather conditions, or seasons, while wind power generation depends on wind speed and direction. These factors can lead to fluctuations in power generation, making it problematic to match supply with demand.

5.5 POWER CONVERTERS IN VARIOUS APPLICATIONS

In renewable energy, it enables the efficient conversion of power from sources like solar panels and wind turbines, ensuring proper intergration with the grid system while maintaining power quality. In various electric vehicles, converters accomplish charging and discharging processes using different methodologies, optimizing battery performance and enhancing energy efficiency control. In industrial settings, converters control the operation of motors, drives, and robotics, providing precise speed and torque control. Additionally, in power grids, converters help stabilize voltage, control frequency, and enable efficient power distribution across diverse sources and loads, contributing to the overall consistency and effectiveness of modern electrical power systems.

5.5.1 HVDC and FACTS

Power converters are critical components in HVDC systems and FACTS controllers. Their significance can be understood by analyzing their roles and impact in these systems. HVDC systems use power converters to convert AC to DC at the sending end and DC back to AC at the receiving end. These converters enable well-organized long-distance power transmission and interconnectedness between asynchronous grids.

Key functions of power converters in HVDC include enabling the transmission of power in DC form, reducing losses over long distances. Converters can produce or absorb reactive power, helping stabilize the AC grid. HVDC systems connect grids operating at different frequencies or standards. Fast control of power flow improves system stability during disturbances. Benefits enabled by power converters in HVDC include lower line losses compared to AC, especially over long distances. Smaller conductors and reduced right-of-way requirements lower costs and environmental impact. Converters provide precise control over the magnitude and direction of power flow.

FACTS devices use power converters to regulate voltage, control power flow, and improve the stability and reliability of AC transmission systems. Key applications of power converters in FACTS devices include maintaining voltage stability by injecting or absorbing reactive power, controling power flow by injecting a voltage in series with the line, combining shunt and series compensation to provide comprehensive control of power flow and voltage, and dynamically adjusting line impedance. Benefits of power converters in FACTS include enhancing the system's ability to maintain voltage within acceptable limits during disturbances, enabling better utilization of existing transmission infrastructure, optimizing power flow to minimize line losses, and providing rapid response to transient and dynamic instabilities in the grid.

Modern power converters use semiconductor devices such as IGBTs and GTOs to achieve high efficiency, reliability, and controllability. These advantages enable participation of renewable energy sources and enhance system flexibility through higher efficiency, smaller physical footprints, and modular designs for future expansion. Power converters are indispensable in HVDC and FACTS technologies for their ability to enhance the efficiency, reliability, and flexibility of power transmission and distribution systems. They are foundational to the modern electrical grid, supporting sustainable energy integration and improving overall grid performance.

5.5.2 Electric Vehicle and Charging Infrastructure

Power converters are critical in EV charging infrastructure as they enable efficient and reliable power transfer between the grid and the EV battery. They perform crucial functions such as AC to DC conversion in Level 1 (120 V) and Level 2 (230 V) chargers, and DC-DC conversion in ultra-fast chargers, guaranteeing compatibility with various grid standards and battery specifications. Furthermore, power converters can control voltage, current, and power flow, enhancing charging speed and efficiency while protecting the battery from overcharging or thermal stress. Advanced converters also provide for bidirectional energy flow, enabling V2G functionality, which enhances grid stability and energy management by allowing EVs to act as distributed energy storage systems [48].

EV chargers are classified based on their power delivery and charging speed. Level 1 chargers use standard household AC outlets and provide slow charging, suitable for overnight use. Level 2 chargers operate on 240 V and offer faster charging, making them ideal for residential, commercial, and public settings. DC Fast chargers directly supply DC power to the battery, bypassing the onboard converter and delivering significantly higher power, enabling rapid charging in minutes, making them crucial for long-distance travel and public charging stations.

Power electronics converters are the backbone of EV chargers, performing essential functions for AC-DC or DC-DC conversion and enabling accurate control of voltage, current, and power flow. Rectifiers are used in Level 1 and Level 2 chargers to convert AC to DC, while DC Fast Chargers employ advanced power converters with high-frequency switching for efficient DC power conversion and transmission. These converters guarantee compatibility between grid power and EV battery charging levels and storage methods, improve energy efficiency, and integrate features such as power factor correction and reduced harmonic levels to minimize grid impact.

Bidirectional chargers enable two-way energy flow between EVs and the grid, supporting V2G technology. This innovation allows EVs to act as distributed energy resources by feeding stored energy to the grid during peak demand or outages, enhancing grid stability and energy management. Bidirectional power converters are crucial for this functionality, providing seamless transition between charging and discharging modes while maintaining battery health through intelligent algorithms and control systems. Efficiency and optimization of EV charging systems focus on minimizing energy losses and maximizing charging speed while preserving battery life. Advanced power converters with high-efficiency switching devices reduce thermal losses and improve power density for system applications. Various optimization techniques, such as dynamic load management and real-time energy monitoring, ensure efficient use of available grid power. Smart chargers further enhance efficiency by integrating renewable energy sources and adapting charging patterns based on grid condtions and user preferences.

5.5.3 Power Quality Improvement and Harmonic Mitigation

The power converters have an animated role in power quality development and harmonic extenuation by ensuring the efficient and stable operation of electrical power systems. They actively filter out harmonic levels, which are distortions in the electrical outputs caused by non-linear loads such as variable frequency drives and electronic devices, preventing adverse effects like overheating, equipment damage, and inefficiencies. Advanced converters, like active power filter units, inject compensatory signals to cancel harmonic distortions, thereby reinstating waveform quality. Additionally, power converters help normalize voltage levels, correct power factor, stabilize frequency, and improve overall power quality, system reliability, and performance. Their dynamic and precise control capabilities make them indispensable for maintaining compliance with stringent power quality standards in various modern electrical networks.

Power electronic converters play a crucial role in improving power quality by delivering accurate control over voltage, current, and frequency in electrical network systems. Power converter devices, such as inverters, choppers, matrix converters, and compensators, actively address issues like voltage sags, swells, fluctuations, and harmonics. They allow fast and adaptive responses to disturbances, ensuring steady and reliable power delivery to sensitive equipment. By integrating advanced switching technologies and various control algorithms, power electronic converters enhance efficiency, maintain power factor, and minimize disruptions, making them essential for modern power systems.

Harmonic filters are essential for reducing output waveform distortions caused by non-linear loads. Passive filters, comprising inductors, capacitors, and resistors, are cost-effective system solutions designed for specific harmonic frequencies but may lack flexibility and control. Active filters use power electronics to dynamically inject compensating currents or voltages to cancelf harmonics in real-time. While passive filters are suitable for fixed loads, active filters provide superior performance, adaptability, and scalability for varying load conditions, ensuring compliance with harmonic standards.

APFs are advanced power electronic devices used for real-time harmonic mitigation, power factor correction, and reactive power compensation for various applications. Shunt APFs are connected in parallel to the load and inject currents to neutralize harmonics and preserve sinusoidal waveforms in the system. They are highly effective in dynamic and non-linear load environments, improving overall power quality and reducing stress on electrical infrastructure. APFs also contribute to energy efficiency by minimizing losses and enhancing grid reliability and system performance.

DVRs in power electronics are designed to protect sensitive loads against voltage sags, swells, and temporary interruptions. Through the injection or absorption of voltage using series compensation FACTs devices, DVRs provide a constant supply to critical equipment. They are used extensively in industries like manufacturing, data centers, and hospitals, where even slight voltage disturbances can lead to extensive disruptions or damage. With their high efficiency and fast response times, DVRs are crucial for achieving uninterrupted and high-quality power supply in contemporary power systems.

5.5.4 Smart Grid and Microgrid Applications

Power converters play critical roles in smart grid and microgrid systems, facilitating efficient energy organization, the integration of renewable energy sources, and improved grid stability. In smart grids, converters provide bidirectional energy transfer, allowing real-time power distribution, demand response, and V2G interactions. In microgrids, they transform and condition power from various sources like solar panels, wind turbines, and energy storage systems to provide smooth integration and stable operation in both grid-connected and islanded modes. High-frequency switching and smart controls in advanced converters improve power quality, handle load variations, and provide voltage and frequency stability, which are essential for the reliability and efficiency of today's energy systems.

Converters are crucial in smart grids as they provide seamless integration of renewable energy sources, energy storage, and consumer loads into the grid. They also provide bidirectional power flow to support energy trading, V2G services, and demand-side management. Converters also improve grid reliability and flexibility by offering real-time voltage, current, and frequency control. Sophisticated power electronics in converters also enhance energy efficiency and cater to the dynamic requirements of smart grids, with maximally optimized energy distribution and minimal losses.

Power converters are an essential part of microgrid operation, facilitating the integration of distributed energy sources such as solar panels, wind power, and energy storage systems. They perform essential functions like converting renewable-based DC power into AC for grid compatibility and vice versa for storage. Converters provide seamless transitions from grid-connected to islanded mode and offer stability during such fluctuations. Their ability to control power flow and maintain voltage and frequency standards makes them critical for the efficient and fault-tolerant operation of microgrids. Microgrid power management systems depend significantly on converters to provide efficient energy distribution and stability. Converters adjust power flow dynamically between generation, storage, and loads, maximizing energy utilization according to demand and supply. They provide load balancing, avoid overloading, and ensure constant power supply even in cases of grid failure. With the inclusion of sophisticated control algorithms, converters also support renewable energy prediction, peak shaving, and economic dispatch, maximizing the efficiency and sustainability of the microgrid.

Converter-based systems play a key role in energy routing and fault detection in contemporary grids and microgrids. They allow power from several sources to be routed to loads with high accuracy, maximizing the use of available energy. Advanced monitoring and control capabilities in converters identify and isolate faults in real time, limiting interruptions and preserving vital equipment. With quick response to grid instability and preservation of power quality, such systems contribute to higher operational reliability and protection against cascading failures.

5.5.5 High-Efficiency Power Conversion

Power converters play a key role in high-efficiency power conversion since they allow for accurate control over voltage, current, and frequency in electrical systems. With the use of innovative semiconductor technologies like Silicon Carbide and Gallium Nitride, contemporary converters reduce switching and conduction losses to a large extent, greatly enhancing energy efficiency [49]. They play a crucial role in application areas such as the integration of renewable energy, electric vehicles, and industrial drives, where effective energy transfer minimizes operational costs and environmental degradation. Power converters also ensure maximum utilization of energy using methods such as soft switching, PWM, and real-time monitoring to minimize energy loss and improve system performance.

Performance parameters and efficiency measurements for converters include power conversion efficiency, THD, power factor, and load regulation. Power conversion efficiency, expressed as a percentage, is the ratio of output power to input power, which represents energy loss during conversion. Other essential parameters, including THD and power factor, measure the quality of the output waveform and the effect on the grid. High-performance converters have optimal efficiency under all load conditions to ensure reliable operation and prevent energy dissipation. High-efficiency converter design consists of choosing sophisticated semiconductor devices like Silicon Carbide or Gallium Nitride, which have reduced switching and conduction losses. Methods like optimized circuit topologies, pulse-width modulation, and adaptive control algorithms yield greater efficiency by reducing energy loss. Electromagnetic compatibility design minimizes interference and allows stable operation. Compact module designs also increase power density, which supports efficient energy conversion in space-limited applications.

Loss-minimizing techniques such as soft switching and resonant converters play a central role in maximizing converter efficiency. Soft switching methods, including Zero Voltage Switching and Zero Current Switching, minimize switching losses through transitions at zero voltage or current. Resonant converters utilize the natural resonance of inductors and capacitors to ensure efficient energy transfer with minimal switching and conduction losses. These methods are especially useful when high-frequency operation is employed, as regular hard switching will cause sinificant energy loss. Efficient thermal management is important in ensuring the reliability and efficiency of power converters. Thermal techniques, e.g., heat sinks, forced air cooling, and liquid cooling, remove heat produced during operation. Novel thermal interface materials and heat spreaders maximize heat conduction, and on-die or on-silicon thermal sensors ensure real-time temperature feedback. Advanced topologies, e.g., planar transformers and chip-scale cooling, enhance thermal performance, ensuring converters stay within safe temperature levels and achieve high efficiency over extended durations.

5.5.6 Future Trends and Emerging Technologies

Power converters are at the forefront of future directions and new technologies because they facilitate innovation in energy systems, transportation, and industrial automation. With increasing renewable energy resources, smart grids, and energy storage systems, converters are required to integrate and control various power flows effectively. New technologies like wide bandgap semiconductors are setting new standards of converter efficiency, power density, and thermal performance. Power converters also play a critical role in electric mobility, wireless power transmission, and bidirectional energy systems such as V2G. As industries shift toward electrification, automation, and distributed energy systems, converters will remain crucial to making sustainability, reliability, and sophisticated energy management a reality.

Solid-state transformers are now seen as a revolutionary technology that substitutes conventional electromagnetic transformers with power electronic-based transformers. With the use of high-frequency switching and semiconductor technology, SSTs have various benefits such as reduced size, weight, and improved control over voltage, current, and power flow. SSTs facilitate more effective grid connection of renewable power, aid in smart grid development, and exhibit sophisticated functionalities such as voltage regulation, power quality improvement, and interfacing with distributed energy resources. Their ability to perform flexible, high-efficiency, and compact power conversion is opening the door to more robust and versatile energy systems. New-generation converters are being engineered with innovative materials and new topologies to maximize power conversion efficiency and minimize system size and complexity.

New technologies in the form of high-temperature superconductors, high-capacity capacitors, and low-loss magnetic materials are being integrated to enhance power density, thermal management, and overall reliability of converters. Advanced topologies such as multilevel converters, matrix converters, and resonant converters provide greater efficiency, lower harmonic distortion, and improved scalability for high-power applications. These technologies play an important role in addressing the changing requirements of electrification, renewable energy, and automation. Artificial intelligence and machine learning are being increasingly incorporated in power electronics to improve performance, reliability, and real-time control.

Machine learning algorithms can anticipate and adjust to varying load conditions, allowing converters to operate more efficiently by dynamically changing parameters. Machine learning algorithms may be used to analyze massive amounts of operational data to detect trends, diagnose faults, and enable predictive maintenance, enhancing the overall lifespan and reliability of power systems. The integration of AI and ML technologies in power electronics makes it possible to develop more intelligent and autonomous systems that can enhance energy efficiency, minimize downtime, and support more sophisticated applications like smart grids and electric vehicles [50].

5.6 CONCLUSION

Advanced power electronics converters are key to the future of power system applications, with efficiency, flexibility, and sustainability being enhanced. Advanced converters using wide bandgap semiconductors, solid-state transformers, and novel topologies provide seamless power flow management for complicated systems such as renewable energy grids, electric vehicles, and smart grids. Their capacity to reduce energy losses, optimize power quality, and improve system reliability puts them at the heart of future and modern energy infrastructure. Continuous technological advances in materials, control methods, and artificial intelligence hold the potential for even greater efficiency and functionality, making modern power electronics converters a necessity for the shift to a more sustainable, intelligent, and resilient global energy system.

REFERENCES

1. Martínez-García, M.S.; de Castro, A.; Sanchez, A.; Garrido, J. Word length selection method for HIL power converter models. *International Journal of Electrical Power & Energy Systems* 2021, 129, 106721.
2. Zhang, G.; Chen, H.; Samson, S.Y.; Chi, K.T. Impedance strengthening and weakening networks for power converter analysis and design. *IEEE Transactions on Power Electronics* 2021, 36, 9717–9721.

3. Sarker, M.R.; Saad, M.H.M.; Olazagoitia, J.L.; Vinolas, J. Review of power converter impact of electromagnetic energy harvesting circuits and devices for autonomous sensor applications. *Electronics* 2021, 10, 1108.
4. Tan, B.; Zhao, J.; Netto, M.; Krishnan, V.; Terzija, V.; Zhang, Y. Power system inertia estimation: Review of methods and the impacts of converter-interfaced generations. *International Journal of Electrical Power & Energy Systems* 2022, 134, 107362.
5. Meraj, M.; Bhaskar, M.S.; Reddy, B.P.; Iqbal, A. Non isolated DC–DC power converter with high gain and inverting capability. *IEEE Access* 2021, 9, 62084–62092.
6. Almutairi, A.; Sayed, K.; Albagami, N.; Abo-Khalil, A.G.; Saleeb, H. Multi-port PWM DC-DC power converter for renewable energy applications. *Energies* 2021, 14(12), 3490.
7. Hassan, M.A.; Su, C.L.; Chen, F.Z.; Lo, K.Y. Adaptive passivity-based control of dc-dc boost power converter supplying constant power and constant voltage loads. *IEEE Transactions on Industrial Electronics* 2021, 6, 1093.
8. Huang, L.; Zhen, J.; Lygeros, J.; Dörfler, F. Quadratic regularization of data-enabled predictive control: Theory and application to power converter experiments. *IFAC Papers on Line* 2021, 54, 192–197.
9. Łakomy, K.; Madonski, R.; Dai, B.; Yang, J.; Kicki, P.; Ansari, M.; Li, S. Active disturbance rejection control design with suppression of sensor noise effects in application to DC–DC buck power converter. *IEEE Transactions on Industrial Electronics* 2021, 69, 816–824.
10. Ye, S.; Zhou, D.; Yao, X.; Blaabjerg, F. Component - level reliability assessment of a direct-drive PMSG wind power converter considering two terms of thermal cycles and the parameter sensitivity analysis. *IEEE Transactions on Power Electronics* 2021, 36(9), 10037–10050.
11. Pape, M.; Kazerani, M.A. Generic power converter sizing framework for series-connected dc offshore wind farms. *IEEE Transactions on Power Electronics* 2021, 37, 2307–2320.
12. Meral, M.E.; Çelik, D. Mitigation of DC-link voltage oscillations to reduce size of DC-side capacitor and improve lifetime of power converter. *Electric Power Systems Research* 2021, 194, 107048.
13. Komuro, Y.; Honda, S.; Kurooka, K.; Warigaya, R.; Tanaka, F.; Uchida, S. A 43.0% efficient GaInP photonic power converter with a distributed Bragg reflector under high-power 638 nm laser irradiation of 17 W cm^{-2}. *Applied Physics Express* 2021, 14, 052002.
14. Palanisamy, R.; Vijayakumar, K. Minimization of common mode voltage and capacitor voltage unbalance of 3-phase 4-wire 3-level NPC inverter using 4D-SVM. *IEEE Transactions on Power Electronics* 2022, 10, 202–212.
15. Liu, C.; Bai, H.; Ma, R.; Wang, Y.; Gao, F. Suppression of chattering in the real-time simulation of the power converter. *IEEE Transactions on Power Electronics* 2021, 36, 11944–11952.
16. Yao, J.; Wang, S.; Luo, Z. Modeling, analysis, and reduction of radiated EMI due to the voltage across input and output cables in an automotive non-isolated power converter. *IEEE Transactions on Power Electronics* 2021, 37, 5455–5465.
17. Faraji, R.; Ding, L.; Esteki, M.; Mazloum, N.; Khajehoddin, S.A. Soft-switched single inductor single stage multiport bidirectional power converter for hybrid energy systems. *IEEE Transactions on Power Electronics* 2021, 36, 11298–11315.
18. Sikorski, A.; Falkowski, P.; Korzeniewski, M. Comparison of two power converter topologies in wind turbine system. *Energies* 2021, 14, 6574.
19. Parvez, M.; Pereira, A.T.; Ertugrul, N.; Weste, N.H.; Abbott, D.; Al-Sarawi, S.F. Wide bandgap DC–DC converter topologies for power applications. *Proceedings of the IEEE* 2021, 109, 1253–1275.

20. Liu, J.; Dong, D.; Zhang, D. A hybrid modular multilevel converter family with higher power density and efficiency. *IEEE Transactions on Power Electronics* 2021, 36, 9001–9014.
21. Ahmad, S.S.; Urabinahatti, C.; Prasad, K.N.; Narayanan, G. High-switching-frequency SiC power converter for high speed switched reluctance machine. *IEEE Transactions on Industry Applications* 2021, 57, 6069–6082.
22. Palanisamy, R.; Vijayakumar, K. Minimization of common-mode voltage of three-phase five-level NPC inverter using 3D space vector modulation. *Journal of Circuits Systems and Computers* 2020, 29, 2050229.
23. Cheng, H. C.; Chen, P. H.; Su, Y. T.; Chen, P. H. A reconfigurable capacitive power converter with capacitance redistribution for indoor light-powered battery less internet-of things devices. *IEEE Journal of Solid-State Circuits* 2021, 56, 2934–2942.
24. Wang, J.; Sun, K.; Xue, C.; Liu, T.; Li, Y. Multi-port DC-AC converter with differential power processing DC-DC converter and flexible power control for battery ESS integrated PV systems. *IEEE Transactions on Industrial Electronics* 2021, 69, 4879–4889.
25. Song, W.N.; Wu, M. Virtual direct power control scheme of dual active bridge DC–DC converters for fast dynamic response. *IEEE Transactions on Power Electronics* 2018, 33, 1750–1759.
26. Rahman, M.A.; Islam, M.R. Design of a multiloop control structure for load-disturbance attenuation and power mismatch mitigation in isolated multiport power converters. *IEEE Transactions on Industrial Electronics* 2022, 69, 8984–8996.
27. Dutta, S.; Hazra, S. A digital predictive current mode controller for a single-phase high-frequency transformer-isolated dual-active bridge DC-to-DC converter. *IEEE Transactions on Industrial Electronics* 2016, 63, 5943–5952.
28. Yu, Z.; Zeng, J.; Liu, J. Terminal sliding mode control for dual active bridge DC–DC converter with structure of voltage and current double closed loop. In *Proc. Austral. New Zealand Control Conf. (ANZCC)*, Melbourne, Australia, 2018, 11–15.
29. Chen, W.H.; Yang, J.; Guo, L. Disturbance-observer-based control and related methods—An overview. *IEEE Transactions on Industrial Electronics* 2016, 63, 1083–1095.
30. Palanisamy, R.; Thamizh Thentral, T.M.; Ramesh, M. Implementation of four dimensional space vector modulation for five phase voltage source inverter. *Ain Shams Engineering Journal* 2021, 12, 2891–2898.
31. Nguyen, D.D.; Fujita, G.; Bui-Dang, Q. Reduced order observer-based control system for dual-active-bridge DC/DC converter. *IEEE Transactions on Industrial Applications* 2018, 54, 3426–3439.
32. Liao, M.; Li, H.; Wang, P.; Chen, Y. Machine learning methods for power flow control of multi-active-bridge converters. In *Proc. IEEE 22nd Workshop Control Model. Power Electron. (COMPEL)*, Cartagena, Colombia, 2021, 1–7.
33. Lee, Y.J.; Bak, Y.; Lee, L. Control method for phase-shift full-bridge center-tapped converters using a hybrid fuzzy sliding mode controller. *Electronics* 2019, 8, 705–712.
34. Saeed, J.; Wang, L.; Fernando, N. Model predictive control of phase shift full-bridge DC–DC converter using Laguerre functions. *IEEE Transactions of Controlled Systems and Technology* 2022, 30, 819–826.
35. Al-Bayati, A.K.M.; Altürk, E.; Al-Araji, A.S. Development of a fuel cell energy controller design for an electric vehicle engine via a PIDPSO robust control algorithm. In *Proc. IEEE Conf. Energy Convers. (CENCON)*, China, 2023, 126–131.
36. Kim, C.E. Optimal dead-time control scheme for extended ZVS range and burst-mode operation of phase-shift full-bridge (PSFB) converter at very light load. *IEEE Transactions on Power Electronics* 2019, 34, 10823–10832.
37. Zheng, X.; Yang, X. Command filter and universal approximator based backstepping control design for strict-feedback nonlinear systems with uncertainty. *IEEE Transactions of Automatic Control* 2020, 65, 1310–1317.

38. Ravindran, R.; Sathiasamuel, C.R.; Palanisamy, R. MSVM-based hybrid energy-fed quasi-Z-source cascaded H-bridge inverter for grid-connected system. *International Transactions on Electrical Energy Systems* 2021, 12, e13139.
39. Ge, S.S.; Wang, C. Direct adaptive NN control of a class of nonlinear systems. *IEEE Transactions on Neural Networks* 2020, 13, 214–221.
40. Zheng, X.; Yu, X.; Jiang, J.; Yang, X. Practical finite-time command filtered backstepping with its application to DC motor control systems. *IEEE Transactions on Industrial Electronics* 2024, 73, 2955–2964.
41. Padmarajan, B.V.; McGordon, A. Blended rule-based energy management for PHEV: System structure and strategy. *IEEE Transactions on Vehicle Technology* 2016, 65(10), 8757–8762.
42. Li, J.; Zhou, Q.; He, Y.; Williams, H. Driver-identified supervisory control system of hybrid electric vehicles based on spectrum guided fuzzy feature extraction. *IEEE Transactions on Fuzzy Systems* 2020, 28, 2691–2701.
43. Ezzat, A.A.; Liu, L.; Hochbaum, D.S. A graph-theoretic approach for spatial filtering and its impact on mixed-type spatial pattern recognition in wafer bin maps. *IEEE Transactions on Semiconductor Manufacturing* 2021, 34, 194–206.
44. Gupta, S.; Bhaskaran, B.; Sarangi, S.; Abdollahian, A. A novel graph coloring based solution for low-power scan shift. In *Proc. IEEE 37th VLSI Test Symp (VTS)*, Monterey, California, 2019, 1–6.
45. Wolaver, D. Basic constraints from graph theory for DC-to-DC conversion networks. *IEEE Transactions on Circuit Theory* 1972, 19, 640–648.
46. Zhang, B.; Qiu, D. *Sneak Circuits of Power Electronic Converters* (1st ed.). Hoboken, NJ: Wiley, 2014.
47. Palanisamy, R.; Vijayakumar, K. Three-phase 3-level Z-source NPC inverter using modified 3D-space vector modulation. *International Journal of Ambient Energy* 2020, 12, 234–242.
48. Cuk, S. General topological properties of switching structures. In *Proc. IEEE Power Electron. Specialists Conf.*, San Diego, California, 1979, 109–130.
49. Raveendran, V.; Andresen, M.; Liserre, M. Graph theory-based power routing in modular power converters considering efficiency and reliability. In *Proc. IECON-44th Annu. Conf. IEEE Ind. Electron. Soc.* Washington, DC, 2018, 1237–1242.
50. Liserre, M.; Raveendran, V.; Andresen, M.; Graph-theory-based modeling and control for system-level optimization of smart transformers. *IEEE Transactions on Industrial Electronics* 2020, 67, 8910–8920.

6 Energy Storage Systems in Renewable Energy Grids

Anbarasu Loganathan,
Muralikrishnan Gopalakrishnan,
Preetha Krishnasamy,
and Karthikeyan Palaniyappan

6.1 INTRODUCTION TO RENEWABLE POWER GRIDS AND ENERGY STORAGE

This introduction explores the importance of renewable power grids, the challenges they face due to intermittency, and the critical role of energy storage in addressing these issues, ensuring grid stability, and enabling a domestic, more sustainable energy future.

6.1.1 Renewable Power Grids and Their Importance

Renewable power grids, often referred to as green grids or clean energy grids, are designed to primarily or entirely run on energy generated from renewable sources. Unlike traditional fossil-fuel-based grids, which rely on coal, oil, and natural gas, renewable grids harness natural resources that are sustainable and low-emission. The transition to renewable grids is vital for several reasons:

- **Reducing Greenhouse Gas Emissions**: Fossil fuel combustion is one of the primary sources of GHGs, which trap warmth in the Earth's atmosphere and contribute to climate change. Renewable grids allow countries to significantly reduce their carbon footprint, helping to meet international climate goals such as those established by the Paris Agreement.
- **Improving Community Strength**: Fossil fuel power plants emit pollutants, including sulfur dioxide and nitrogen oxides, which are linked to breathing and circulatory illnesses. Renewable energy production produces minimal air contaminants, leading to domestic air and better community fitness outcomes.
- **Enhancing Energy Security and Independence**: Renewable grids allow countries to produce energy domestically, plummeting need on introduced fuels and shielding them from flucuations in global energy prices. This enhances energy security and supports a more stable, self-reliant energy infrastructure [1] (Figure 6.1).

 DOI: 10.1201/9781003585534-6

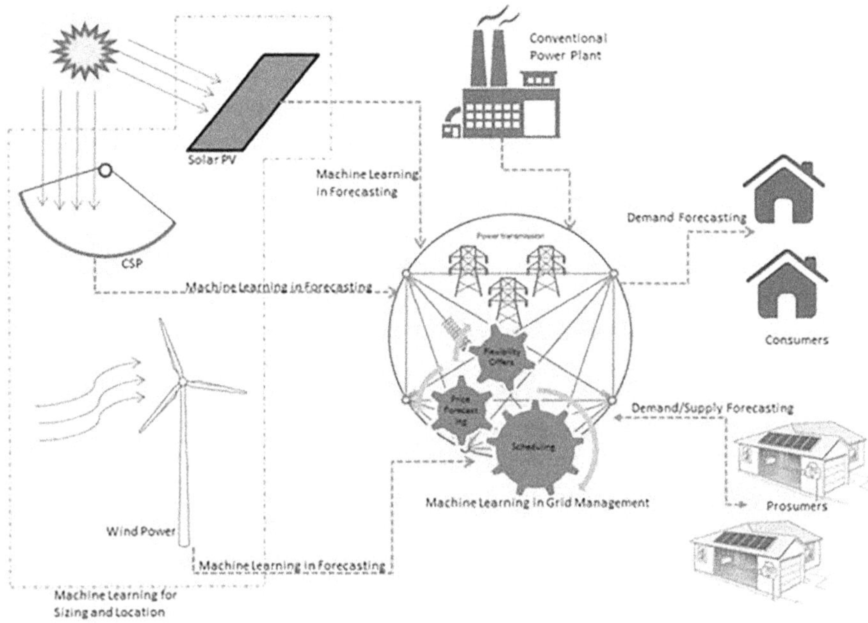

FIGURE 6.1 Renewable power grids.

- **Financial Development and Profession Formation**: The renewable energy sector has become an important driver of job creation and technological advancement. Investment in renewable power grids fosters innovation and supports economic growth, creating jobs in industrial, connection, upkeep, and investigation and growth.

However, while renewable power grids offer these considerable benefits, they also introduce new challenges, particularly in ensuring a reliable, uninterrupted supply of electricity. Renewable bases such as astral and breeze are integrally variable, producing power only when conditions are favorable—during sunlight hours for solar and when the wind is blowing for wind turbines. These intermittency issues highlight the essential need for vigorous energy storage solutions that can store excess vigor when generation is tall and then dispatch it when renewable output is low.

6.1.2 The Role of Sustainable Power Storage in Power Grids

Energy storage is essential for incorporating renewable energy into the power grid effectively. Storage systems capture extra energy once renewable generation exceeds demand and release it during periods when generation falls short, such as in the the evening or during overcast, windless conditions. This capability is vital for upholding grid stability, balancing supply and demand, and safeguarding a reliable, reliable power source [2]. Key roles of energy storage in renewable grids include:

- **Balancing Intermittency**: Energy storage helps mitigate the variability of renewables. Solar and wind output fluctuates with weather and time, which can lead to inconsistencies between generation and demand. Storage systems smooth these fluctuations, storing extra energy produced during high generation periods and discharging it when generation drops. This balance is critical for avoiding disruptions and ensuring that the grid can meet demand at all times.
- **Peak Load Shifting and Demand Management**: Energy storage allows for load shifting, where electricity is stored during times of low demand or high renewable output and discharged during peak demand. This assists in decreasing the need for fossil-fuel-powered peaker plants, which are typically activated during high-demand periods. By reducing peak demand, storage systems can lower grid costs, stabilize electricity prices, and reduce emissions from peak load generation.
- **Frequency and Voltage Regulation**: The grid's frequency and voltage must remain within certain bounds to ensure stable and safe operation. Frequency fluctuations can occur when there is an imbalance between supply and demand, while voltage can vary based on the amount of power moving through the grid. Energy storage systems can deliver fast-response frequency and voltage regulation, helping to keep these metrics within safe ranges and preventing service disruptions or equipment damage.
- **Reducing Renewable Curtailment**: When renewable generation is high but demand is low, the extra energy may be curtailed or wasted because there is no immediate demand for it on the grid. Energy storage captures this surplus energy, reducing the need for curtailment and making renewable resources more efficient. This maximizes the utilization of renewable energy infrastructure and ensures that as much clean energy as possible is used.
- **Emergency Backup Power and Reliability**: Energy storage also enhances grid pliability by providing backup power during emergencies, such as extreme weather events or outages. In areas prone to natural disasters, storage systems can confirm an incessant power supply for dangerous substructure, such as infirmaries and emergengy facilities. Some storage systems offer "black start" capability, meaning they can help resume the network autonomously in the event of a blackout, which is essential for recovery and resilience.

6.1.3 Types of Vigor Storing Solutions

Here remain numerous kinds of energy storage technologies, each with its own advantages and applications:

- **Lithium-Ion Batteries**: Widely used in residential, commercial, and grid-scale applications, lithium-ion batteries are ideal for temporary storage and can respond quickly to demand fluctuations. They are compact, efficient, and have become more affordable, though their storage duration is typically limited to hours.

- **Pumped Hydro Storage**: One of the oldest and most recognized storage methods, pumped hydro uses energy to pump water to an elevated tank. When electricity is needed, the water flows back down, turning turbines and generating power. Pumped hydro is ideal for large, long-duration storage; nonetheless, it requires specific geographic conditions.
- **Flow Batteries**: Flow batteries store energy in liquid electrolytes contained in external tanks. They offer long-duration storage and can be scaled up by increasing the electrolyte volume. Flow batteries are promising for large applications, although they are still in the early stages of widespread adoption.
- **Thermal Energy Storage**: This method involves storing energy as warmth, which can be advancedly rehabilitated back into electricity or used straight for boilers and refrigeration. Focused astral control plants, for example, store thermal energy in molten salts, which can provide power even when sunlight is unavailable.
- **Hydrogen and Fuel Cells**: Excess renewable electricity can be used to produce hydrogen through electrolysis. The hydrogen can be stored and later converted back to electricity in fuel cells, providing a flexible, emission-free storage option. Hydrogen storage is appropriate for established energy storage but is still being scaled up for widespread use.

6.1.4 The Future of Renewable Power Grids and Vigor Storage

As the energy transition accelerates, the role of energy storage in renewable power grids will continue to expand. Advances in battery technology, cost reductions, and innovations in long-duration storage solutions will enable more reliable integration of renewables, paving the way for a sustainable and resilient enegry system [3]. Additionally, policies and incentives supporting energy storage deployment are likely to increase, further driving development in the sector.

6.2 THE NEED FOR ENERGY STORAGE IN RENEWABLE GRIDS

This detailed overview explores the need for energy storage in renewable grids, focusing on how storage solutions support stability, mitigate the variability of renewables, and facilitate a domestic, more resilient energy future [4].

6.2.1 Addressing Intermittency and Variability

The major drawback of solar and wind energy lies in their variability. Solar power is only generated during daytime hours and can be reduced by cloud cover or storms, while wind energy fluctuates depending on wind speed and can stop altogether when conditions are calm. This intermittency creates challenges in matching renewable energy supply with grid demand, as electricity demand remains relatively constant and high throughout the day and night. Without a steady and reliable supply, these fluctuations can lead to instability, potentially causing power disruptions or outages if they are not managed effectively.

Energy storage helps overcome this intermittency by storing extra energy generated during high production periods and discharging it when renewable generation is low. By acting as a buffer, storage smooths out supply fluctuations and ensures a more steady and predictable power output. This is particularly beneficial during peak demand periods, which may not coincide with times of peak renewable generation. With energy storage in place, renewable grids can provide power more consistently and improve overall grid reliability [5].

6.2.2 Balancing Supply and Demand with Load Shifting

A critical function of energy storage in renewable grids is load shifting, or the aptitude to stock power once supply exceeds demand, then release it throughout periods of higher request. Astral and breeze control can crop a surplus of electricity when conditions are favorable, especially throughout off-peak hours when electricity request is low. Without storage, this surplus may be curtailed or wasted because there is no immediate demand for it. Energy storage systems prevent this waste by capturing the surplus energy, which can then be used throughout top request hours, plummeting the essential to activate fossil-fuel-based "peaker" plants that emit greenhouse gases and pollutants.

Load shifting helps stabilize electricity prices by reducing reliance on these peaker plants and ensuring that renewable energy is fully utilized. For instance, residential and commercial users with time-of-use pricing can stock energy throughout off-peak hours (once power rates are inferior) and use it during top hours (once taxes are advanced). By flattening demand peaks, energy storage reduces the tension on the grid and helps lower overall electricity costs. This balancing effect is essential for managing supply-demand mismatches, which are common in renewable grids, and supports the long-term economic viability of renewable energy.

6.2.3 Providing Frequency and Voltage Regulation

For the power grid to function correctly, both frequency and voltage must remain within specific limits. Frequency regulation is vital to upholding the equilibrium between source and request. When supply exceeds demand, frequency increases, while a shortage of supply leads to a decrease. Rapid fluctuations in frequency can cause grid instability and lead to blackouts. Voltage stability, on the other hand, is important to prevent overloading or underloading of the grid, which can damage equipment and disrupt service.

Energy storage systems can reply rapidly to fluctuations in frequency and voltage, providing ancillary services that keep these metrics stable. Batteries, in particular, can react almost instantly, absorbing or discharging energy as needed to balance frequency and maintain voltage levels. By acting as an energy reserve, storage systems support grid reliability and prevent the disruptions that can arise from intermittent renewable generation. This competence is especially significant in grids with tall renewable penetration, where frequent supply variations can lead to more frequent and pronounced fluctuations in grid frequency and voltage [6].

6.2.4 Reducing the Need for Renewable Curtailment

Renewable curtailment occurs when grid operators reduce the production from renewable energy sources due to grid constraints, such as when supply outpaces demand or when transmission capacity is limited. Curtailment effectively wastes clean energy that could otherwise power homes, businesses, and industries. This issue is especially prevalent during periods of high renewable generation, such as sunny, blustery days, when renewable output may exceed the grid's immediate capacity.

Energy storage addresses this issue by capturing excess renewable energy, reducing the need for curtailment. Instead of shutting down wind turbines or reducing solar output, operators can store this extra energy and release it once demand increases. This approach maximizes the efficiency and utilization of renewable energy infrastructure, ensuring that as much clean energy as possible is used. By reducing curtailment, energy storage also improves the economics of renewable energy projects, as more of the generated power is delivered to consumers and contributes to revenue.

6.2.5 Enhancing Grid Resilience and Reliability

Extreme weather events, natural disasters, and other emergencies can disrupt power grids, leading to blackouts and prolonged outages. Energy storage schemes improve grid flexibility by providing holdup power throughout these events, allowing dangerous substructures—such as infirmaries, spare services, and message nets—to remain operational. This is particularly important for areas disposed to dangerous weather or natural disasters, where energy storage can significantly improve recovery times and minimize disruption.

In addition to providing emergency backup power, energy storage supports grid resilience by enabling "black start" capabilities. After a major outage, the grid requires a source of power to restart; traditionally, this power has come from fossil-fuel plants. However, energy storage systems, particularly large-scale batteries, can provide the initial power needed to restart the grid, reducing dependence on fossil fuels for recovery and ensuring a faster, more efficient restoration process.

6.2.6 Supporting Energy Decentralization and Microgrids

Energy storage also facilitates decentralization, an approach where energy is generated, stored, and managed at a local level rather than relying solely on centralized power plants. Decentralized energy systems, or microgrids, can function self-sufficiently from the main grid, providing power to communities even when the central grid is down. Storage systems are essential for microgrids because they enable local energy to be stored and dispatched on demand, safeguarding a reliable power supply.

Microgrids with energy storage offer multiple benefits, including enhanced energy security, reduced transmission losses, and support for renewable integration in remote areas. By enabling local renewable energy generation and storage, microgrids reduce reliance on distant power plants, lower overall transmission losses, and provide communities with greater energy independence. Energy storage allows these microgrids to operate autonomously, improving grid resilience and making the energy system more flexible.

6.2.7 Enabling High Renewable Penetration for Decarburization Goals

Achieving climate targets and reducing carbon emissions requires a high level of renewable energy on the grid. However, without energy storage, grids with significant renewable penetration may struggle to maintain reliability, as the variable nature of renewables can lead to mismatches between supply and demand. Energy storage makes it feasible to incorporate a larger share of renewables into the grid by addressing intermittency and ensuring that stored energy is available whenever it is needed.

By facilitating higher renewable penetration, energy storage supports decarbonization efforts, allowing fossil-fuel plants to be phased out and reducing emissions from power generation. This transition is vital for meeting climate goals, as the power sector is one of the largest sources of carbon emissions. With energy storage, renewable energy can become a reliable, primary source of electricity, paving the way for a cleaner, sustainable energy future.

6.3 TYPES OF ENERGY STORAGE SYSTEMS

This impression covers the important types of energy storage systems: lithium-ion batteries, flow batteries, drove hydro storage, current storage, beaten air energy storage, and hydrogen storage. Each of these systems varies in capacity, duration, response time, efficiency, and cost, offering distinct advantages aimed at mixing renewable energy and stabilizing contemporary power grids.

6.3.1 Lithium-Ion Batteries

Overview: Lithium-ion batteries are now the most extensively used energy storage technology. They are known for their high energy density, efficiency, and aptitude to reply quickly to vicissitudes in request. Lithium-ion batteries are usually used in customer microchip technology, electronic cars, and grid-scale applications, where they provide short- to medium-duration storage.

How It Works: Lithium-ion batteries store and release energy through the drive of lithium ions between two conductors—a cathode and an anode—throughout accusing and clearing series. Lithium ions change from the anode to the cathode when the battery is discharged and in the reverse direction when it is charged (Figures 6.2a and 6.2b).

Applications:

- **Grid-Scale Storage**: Lithium-ion batteries are increasingly used for utility-scale applications to provide grid stability, frequency regulation, and renewable energy integration.
- **Residential and Commercial**: They are also popular in residential solar installations and commercial settings for storing excess solar energy and providing backup power.

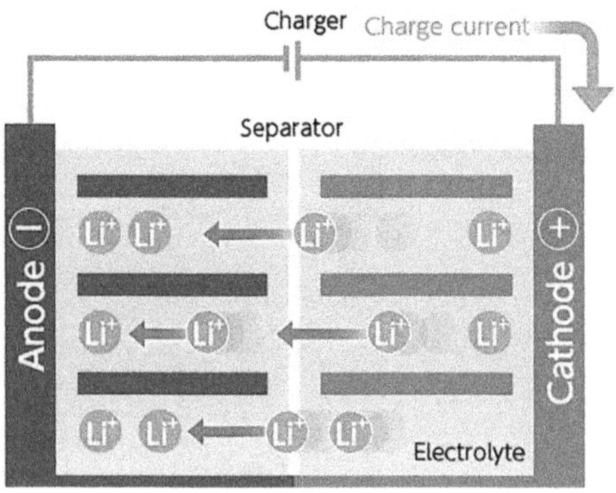

FIGURE 6.2a Lithium-ion battery-Charge Current.

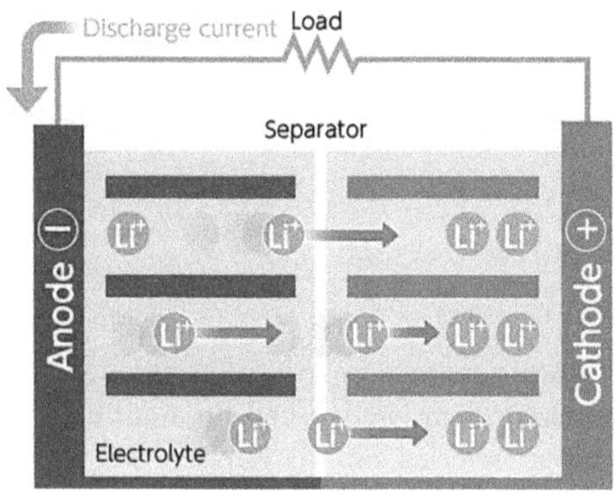

FIGURE 6.2b Lithium-ion battery-DisCharge Current.

- **Electric Vehicles (EVs)**: The high energy density and recharge ability of lithium-ion batteries make them ideal for EVs, contributing to the growth of green transportation.

Advantages:

- High energy density and efficiency (up to 95%)
- Fast response time
- Suitable for high cycling rates
- Compact and scalable

Limitations:

- High upfront cost, although prices have been decreasing
- Limited lifecycle (5,000 to 10,000 cycles on average)
- Environmental concerns with mining lithium and other materials

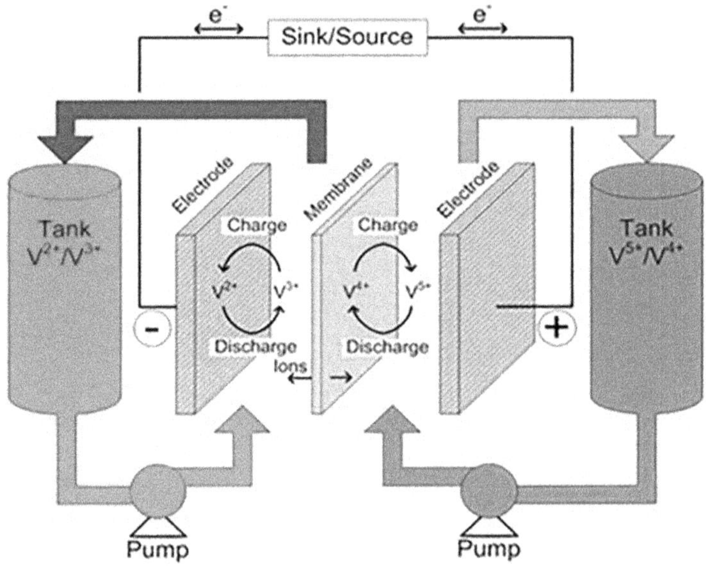

FIGURE 6.3 Flow battery.

6.3.2 Flow Batteries

Overview: Flow batteries store energy in liquid electrolytes contained in external tanks, making them well-suited for significant and long-duration energy storage. Flow batteries can discharge power over extended periods and are more flexible in scaling energy and power capacity independently.

How It Works: Flow batteries operate by circulating liquid electrolytes through an electrochemical cell to generate electricity. The two liquid electrolytes, stored in separate tanks, exchange ions across a membrane to release or store energy. The electrolytes can be recharged by reversing the process (Figure 6.3).

Applications:

- **Utility-Scale Storage**: Flow batteries are ideal for large-scale applications where long-duration discharge (up to 10 hours or more) is required.
- **Microgrids**: Flow batteries can store excess renewable energy for remote areas or microgrids, providing a consistent power supply and resiliency.

Advantages:

- Long-duration discharge capabilities
- Long lifespan and high cycle durability (often more than 20,000 cycles)
- Independent scalability of power and energy capacity
- Lower degradation over time compared with lithium-ion batteries

Limitations:

- Little energy density, requires large storage tanks
- High upfront costs and complex design
- Lower efficiency (60%–80%) compared to lithium-ion batteries

6.3.3 Drove Hydro Storage (PHS)

Overview: Drove hydro storage is one of the eldest and most extensively used forms of energy storage, secretarial for over 90% of the biosphere's storing volume. It is suitable for long-duration, large-scale storage and is often used for grid balancing and peak shaving.
How It Works: Pumped hydro storage uses electricity to move water from a lower tank to a advanced upper reservoir during times of low demand. When electricity is needed, the water is released back down, flowing through turbines to generate power (Figure 6.4).

Applications:

- **Utility-Scale Storage**: Primarily used by utilities for peak-load management and grid stabilization
- **Renewable Integration**: Often paired with renewable energy sources to store excess generation

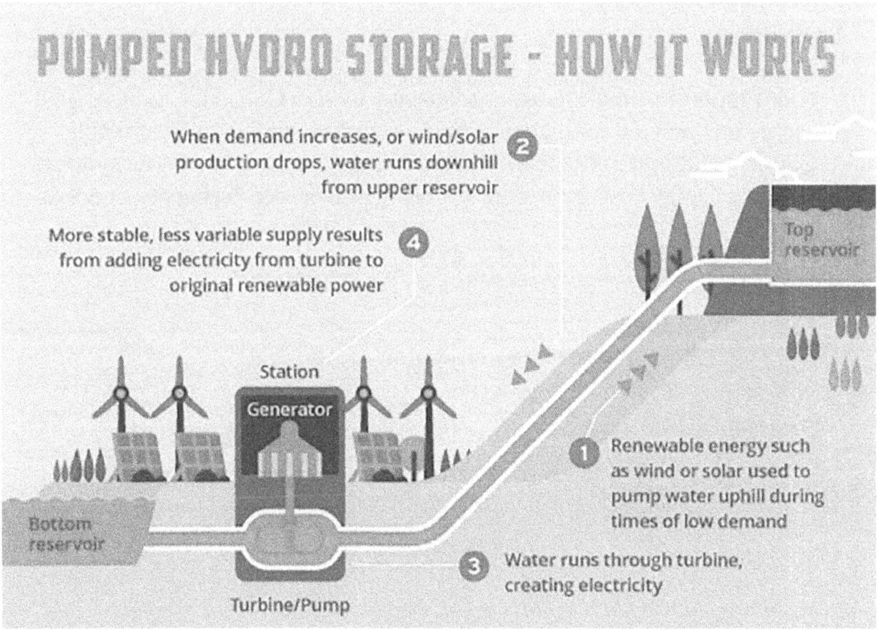

FIGURE 6.4 Pumped hydro storage.

Advantages:

- Long-duration storage with high capacity
- High efficiency (70%–85%)
- Long lifespan (up to 50 years)
- Low operating and maintenance costs

Limitations:

- Requires specific geographical conditions, such as proximity to water sources and significant elevation differences
- High initial capital investment and long construction time
- Limited scalability in urban areas due to land requirements

6.3.4 Thermal Energy Storage

Overview: Thermal energy storage systems store energy as heat or cold, which can be converted back to electricity for heating or cooling. They are used in buildings and industrial applications but can also support power grids when coupled with concentrated solar power (CSP).

How It Works: Thermal storage works by collecting heat or cold in materials such as molten salts, ice, or chilled water. For example, CSP plants store thermal energy in molten salts, which retain heat for hours and can be used to generate steam and power turbines when needed (Figure 6.5).

Applications:

- **CSP Plants**: Thermal storage is frequently used in CSP plants to store solar energy and provide continuous power even when sunlight is unavailable.
- **Commercial and Industrial**: Thermal storage is used in HVAC systems, reducing energy costs by storing and using heat or cold during peak periods.

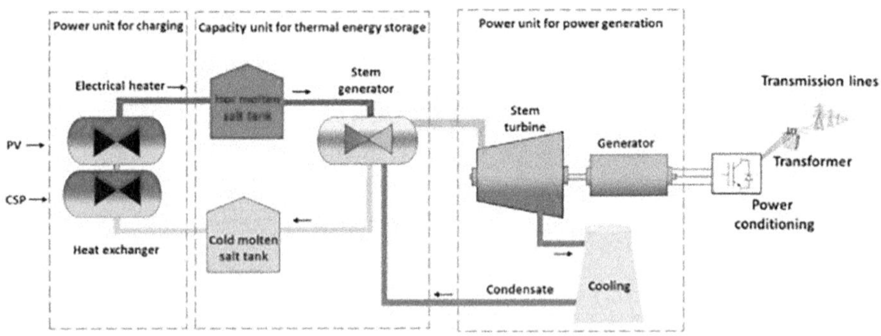

FIGURE 6.5 Thermal energy storage.

Advantages:

- Suitable for long-duration storage (several hours to days)
- Cost-effective for applications where heating or cooling is required
- High efficiency in specific applications, such as CSP plants

Limitations:

- Not suitable for all applications, as it primarily provides heat rather than electricity
- Requires large-scale infrastructure for utility-scale storage
- Efficiency can vary depending on the storage medium and application

6.3.5 Compressed Air Energy Storage (CAES)

Overview: Compressed air energy storage (CAES) systems store energy through squeezing air and storing it in subversive caverns or large tanks. This beaten air can be advanced released to generate electricity, making CAES suitable for grid-scale storage with long-duration discharge capabilities.
How It Works: During off-peak hours, air is compressed and stowed in a reservoir. When electricity is wanted, the beaten air is released and heated, driving turbines to produce power. Modern CAES systems often integrate renewable sources, using excess renewable energy to power the compression process (Figure 6.6).

Applications:

- **Utility-Scale Storage**: CAES systems are suited for large-scale demands that require long-duration storage, such as supplementing renewable energy output.

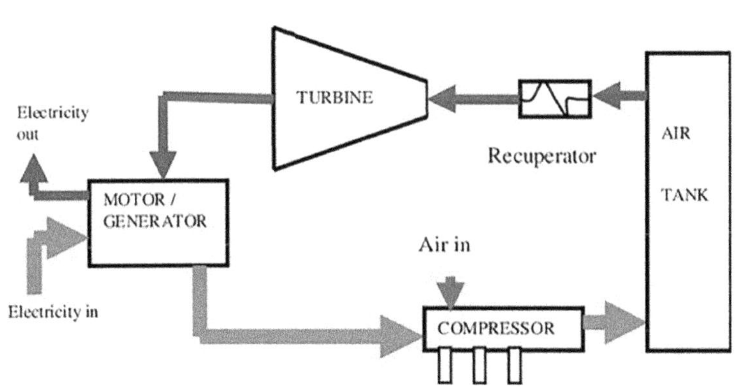

FIGURE 6.6 Compressed air energy storage.

- **Renewable Integration**: CAES can store excess wind or solar power, stabilizing the grid during periods of low renewable generation.

Advantages:

- Long-duration, large-capacity storage potential
- Comparatively low cost per unit of energy stored
- Can be built in suitable underground locations, minimizing land use

Limitations:

- Requires specific geological formations for underground storage
- Moderate efficiency (about 50%–70%) due to energy losses in the compression and expansion processes
- High initial capital costs for construction and infrastructure

6.3.6 Hydrogen Storage

Overview: Hydrogen storage offers an innovative and flexible approach to energy storage by converting electricity into hydrogen through electrolysis. The hydrogen can be stored and later used in fuel cells to generate electricity or as a fuel for other applications. Hydrogen storage is increasingly recognized as a long-term, significant energy storage option.

How It Works: Hydrogen storage systems use renewable energy to power electrolysis, a process that splits water into hydrogen and oxygen. The hydrogen is stored in tanks or underground caverns. When electricity is needed, the stored hydrogen is used in fuel cells or burned in turbines to produce power (Figure 6.7).

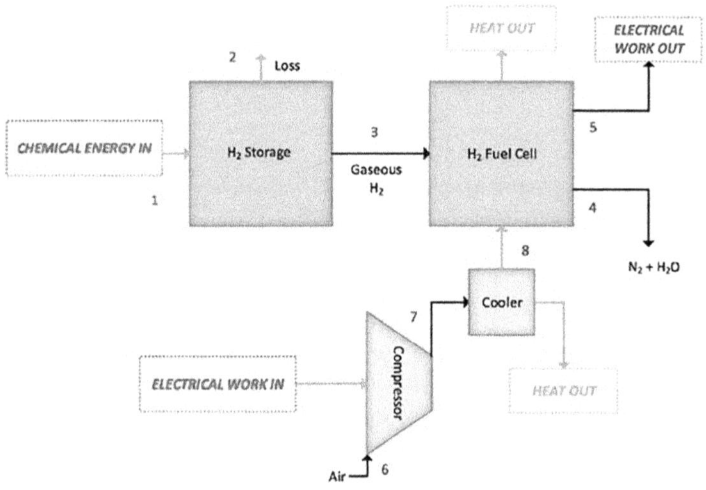

FIGURE 6.7 Hydrogen storage.

Applications:

- **Grid-Scale Storage**: Hydrogen can store excess renewable energy and provide backup power, especially for long-duration storage.
- **Industrial Use and Transportation**: Hydrogen serves as a versatile energy carrier for heavy industry and transportation sectors, where it can replace fossil fuels.

Advantages:

- Long-duration storage potential (from days to seasons)
- High scalability for large-scale applications
- Emission-free, as the only byproduct of fuel cells is water
- Can be used for multiple purposes beyond electricity generation

Limitations:

- Low round-trip efficiency (typically 30%–40%) due to energy losses in electrolysis and conversion back to electricity
- High costs associated with electrolysis and fuel cell infrastructure
- Requires specialized handling and storage infrastructure due to hydrogen's flammability

6.4 INTEGRATION OF STORAGE SYSTEMS WITH RENEWABLE GRIDS

The integration of energy storage schemes with renewable grids is vital to promoting a sustainable energy future, as it addresses key challenges associated with the variable and sporadic nature of renewable energy sources.

6.4.1 System Architecture and Placement of Storage Systems

The project and deployment of energy storage schemes within renewable grids depend on specific energy requirements and operational goals. Storage can be intergrated at various points within the grid:

- **Generation-Side Storage**: Connected at renewable energy generation sites, such as solar farms or wind farms, generation-side storage captures excess energy that would otherwise be curtailed. By smoothing out fluctuations and reducing ramping needs, these storage systems stabilize output and allow renewable generators to better meet grid requirements.
- **Transmission and Distribution-Side Storage**: Large storage systems deployed within the transmission or distribution network help manage congestion, support voltage regulation, and provide ancillary services across broader areas. This setup is particularly useful for balancing energy between regions with varying levels of renewable generation and demand.

- **Behind-the-Meter Storage**: Residential, commercial, and industrial users with on-site renewable systems, such as rooftop solar, can install storage to capture excess generation for use during peak demand periods or when the grid is stressed. These decentralized systems add resilience and can participate in demand response programs, reducing strain on the central grid.

The integration of storage at multiple levels creates a layered approach, providing both localized and system-wide benefits. This configuration ensures more effective use of renewable resources, enhances grid flexibility, and supports the overarching goal of energy independence.

6.4.2 Grid Services Enabled by Storage Systems

Energy storage systems provide numerous services that support renewable grid operations and ensure grid reliability. Key services include:

- **Frequency Regulation**: Storage systems can promptly respond to fluctuations in grid frequency, balancing supply and demand to maintain grid stability. Battery storage, in particular, is well-suited for frequency regulation due to its rapid response time, helping prevent power disruptions in grids with high renewable penetration.
- **Voltage Support**: Storage systems help maintain steady voltage levels by providing or absorbing reactive power as needed. Voltage stability is essential for minimizing losses and ensuring the proper functioning of electrical equipment across the grid. Both large-scale and distributed storage systems contribute to voltage support, depending on grid location and load.
- **Load Shifting and Peak Shaving**: Load shifting allows renewable grids to store energy when generation is high (e.g., during sunny or windy periods) and release it when demand exceeds supply. By discharging during peak hours, storage systems reduce the need for costly, emissions-intensive peaker plants, enhancing grid efficiency and lowering costs for consumers.
- **Black Start Capability**: After a major outage, energy storage can provide the initial power needed to restart grid operations, reducing dependence on fossil-fuel plants for system recovery. Large battery systems and pumped hydro storage are particularly valuable in this role, helping grids recover more quickly from disruptions.
- **Renewable Firming**: By filling the gaps left by intermittent renewables, energy storage smooths out variations in solar and wind output. This "firming" effect makes renewables a more reliable power source, enabling them to fulfill a larger share of grid demand and meet power purchase agreement (PPA) requirements with greater consistency.

Together, these services improve grid resilience, stabilize power quality, and support increased renewable energy penetration, making renewable-dominant grids both feasible and reliable.

6.4.3 Types of Storage Systems and Their Integration Roles

Numerous types of energy storage schemes fulfill different roles in renewable grid integration. Below are key types of storage systems and their integration benefits:

- **Lithium-Ion Batteries**: Owing to their high energy density and fast response time, lithium-ion batteries are extensively used for short-term, high-frequency services such as frequency regulation, demand response, and renewable firming. Their modular nature also makes them flexible for integration across grid levels, from utility-scale installations to behind-the-meter solutions.
- **Flow Batteries**: Flow batteries are known for long-duration discharge capabilities, making them suitable for large-scale applications that require extended storage (e.g., daily or weekly). With high cycle durability, they are often deployed alongside wind and solar farms to capture excess generation and manage extended periods of low production.
- **Pumped Hydro Storage**: Pumped hydro is ideal for long-duration, large-capacity storage, supporting grid stability, peak shaving, and renewable firming at the utility scale. Given its scale and infrastructure requirements, pumped hydro is typically integrated within large regional grids, providing stability during seasonal fluctuations in renewable generation.
- **Thermal Storage**: Often paired with concentrated solar power (CSP) plants, thermal storage captures and retains solar heat, which can be used to generate electricity on demand. This setup allows CSP plants to produce consistent output, even when solar radiation fluctuates or during nighttime hours, reducing reliance on backup fossil-fuel plants.
- **Hydrogen Storage**: Hydrogen storage is emerging as a promising solution for long-duration storage, seasonal storage, and off-grid applications. It converts excess renewable energy into hydrogen through electrolysis, which can later be used in fuel cells or turbines to produce electricity, particularly during periods of low renewable generation.

Each storage type contributes to grid stability and flexibility in distinct ways, allowing for more effective integration and utilization of renewable energy sources.

6.4.4 Operational Considerations in Storage Integration

Integrating storage systems with renewable grids involves operational challenges, including:

- **Control and Management**: Coordinating the dispatch of storage resources to balance renewable output and grid demand requires advanced control systems. Sophisticated software and algorithms optimize energy flow, automate dispatch decisions, and ensure grid stability.
- **Cycling and Degradation**: Each storage technology has specific cycling limits and degradation rates, influencing how often it can be charged and

discharged. Battery storage, for instance, experiences capacity loss over time, while pumped hydro and flow batteries have longer lifespans. Strategic cycling schedules extend storage life and improve cost-effectiveness.

- **Renewable Forecasting and Integration**: Accurate forecasting of renewable generation (e.g., predicting solar and wind patterns) is critical for efficient storage utilization. When combined with predictive analytics, storage systems can be scheduled to charge and discharge in alignment with expected generation and demand patterns, minimizing costs and maximizing grid stability.
- **Grid Synchronization**: Storage systems must be synchronized with grid frequency and voltage parameters to function effectively. Advanced inverters and converters are used to match storage output with grid conditions, ensuring smooth energy delivery and preventing disruptions.

By addressing these operational factors, storage systems can more effectively support renewable grids and optimize the balance of supply and demand.

6.4.5 Economic and Policy Considerations in Storage Integration

The integration of storage systems within renewable grids is influenced by economic and policy factors that shape investment and deployment strategies. Key considerations include:

- **Cost and Financial Incentives**: The high capital cost of storage systems, particularly large-scale installations, remains a barrier to widespread deployment. However, government incentives, subsidies, tax credits, and rebates help reduce costs and make storage solutions more financially viable. For example, investment tax credits (ITCs) have supported battery storage installations in the U.S., driving growth in grid-scale storage capacity.
- **Market Mechanisms**: Competitive electricity markets allow storage systems to participate in various revenue streams, such as capacity markets, ancillary services markets, and energy arbitrage. These markets incentivize storage providers to offer services like frequency regulation, peak shaving, and load balancing, creating a business case for storage integration with renewables.
- **Policy Support for Renewable Integration**: Policies promoting renewable energy growth and clean energy mandates create a need for supporting infrastructure, including storage systems. Policies that set specific targets for renewable integration indirectly drive the demand for storage solutions to maintain grid reliability.
- **Carbon Valuing**: Carbon valuing, including carbon taxes and cap-and-trade schemes, increases the cost of fossil-fuel generation and incentivizes cleaner alternatives. By making storage-backed renewables more cost-competitive, carbon pricing accelerates storage adoption as a means of reducing reliance on fossil-fuel peaker plants.

- **Research and Development**: Public and private investment in R&D for energy storage technologies is essential to drive innovation, improve efficiency, and reduce costs. Governments and industries worldwide fund R&D initiatives focused on advancing technologies like solid-state batteries, green hydrogen, and ultra-long-duration storage to make renewable integration more viable.

6.4.6 Challenges and Future Directions for Storage Integration

While storage is integral to renewable grid success, several challenges remain in achieving seamless integration, such as:

- **Technology Costs**: Although costs have been declining, storage technologies remain capital-intensive. Ongoing innovation and economies of scale are needed to reduce prices further.
- **Interoperability**: Integrating diverse storage systems with grid infrastructure and renewable assets requires compatible hardware and software. Standardized protocols and technology interfaces will streamline integration, enabling efficient energy transfer and communication between systems.
- **Environmental Impact**: Certain storage technologies, particularly lithium-ion batteries, involve resource-intensive materials like lithium and cobalt. Transitioning to more sustainable materials and recycling practices is essential to minimize environmental impact.
- **Resilient and Decentralized Systems**: Grids are increasingly moving toward decentralized models with more distributed energy resources (DERs) and microgrids. Storage integration within these decentralized systems enhances resilience, especially in remote and disaster-prone areas. Future development will likely see more microgrid storage solutions, supporting energy autonomy.

6.5 CASE STUDY

6.5.1 Few Case Studies Demonstrating the Implementation of Energy Storage Systems (ESS) in Renewable Power Grids [7]

1. **Hornsdale Power Reserve, Australia**
    - **Technology Used**: Lithium-ion Battery Storage (Tesla)
    - **Capacity**: 150 MW/193.5 MWh
    - **Implementation**: Installed by Tesla and operated by Neoen, this project was initially developed as a 100 MW system and later expanded.
    - **Impact**:
        - Provided grid stability by responding within milliseconds to fluctuations.
        - Reduced energy costs by providing frequency control and backup power.
        - Successfully prevented blackouts and improved renewable energy integration.

2. **Nant de Drance Pumped Storage Power Plant, Switzerland**
   - **Technology Used**: Pumped Hydro Storage
   - **Capacity**: 900 MW
   - **Implementation**: This plant uses excess renewable energy (wind and solar) to pump water to a higher reservoir, releasing it when needed to generate electricity.
   - **Impact**:
     - Helps balance supply-demand fluctuations in the European grid.
     - Supports the large-scale integration of renewables by acting as a long-duration storage solution.
3. **UK's Pylon Battery Storage, England**
   - **Technology Used**: Lithium-ion Grid-Scale Storage
   - **Capacity**: 100 MW/100 MWh
   - **Implementation**: The system provides fast response services to the National Grid, supporting the UK's decarbonization efforts.
   - **Impact**:
     - Provides grid frequency stabilization and demand response services.
     - Enables better integration of offshore wind energy.
4. **Redox Flow Battery at Hokkaido, Japan**
   - **Technology Used**: Vanadium Redox Flow Battery (VRFB)
   - **Capacity**: 60 MWh
   - **Implementation**: Aimed at stabilizing the grid due to high solar PV penetration in Japan.
   - **Impact**:
     - Supports long-duration energy storage with minimal degradation.
     - Reduces curtailment of renewable energy.

6.5.2 Practical Strategies for Implementing Energy Storage Systems (ESS) in Renewable Power Grids

To ensure the effective deployment of ESS in renewable power grids, the following practical strategies can be adopted [8]:

1. **Selecting the Right Storage Technology**

   Different ESS technologies have unique strengths and weaknesses. Choosing the right one depends on grid requirements:
   - **Lithium-Ion Batteries**: Suitable for short-duration storage (2–4 hours) and frequency regulation.
   - **Pumped Hydro Storage**: Best for long-duration storage and bulk energy shifting.
   - **Redox Flow Batteries**: Ideal for applications needing long lifespans and deep discharge cycles.
   - **Compressed Air Energy Storage (CAES)**: Works well for large-scale storage and peak shaving.
2. **Hybrid Energy Storage Solutions**
   - Combining multiple storage technologies (e.g., battery + super capacitor, battery + pumped hydro) optimizes performance by balancing power and energy needs.

- Hybrid systems improve efficiency and lifespan by reducing stress on individual components.

3. **Smart Grid Integration**
   - Use **advanced grid management software** for real-time monitoring and control.
   - Implement **Artificial Intelligence (AI) and Machine Learning (ML)** to predict energy demand and optimize storage operation.
   - Develop **Vehicle-to-Grid (V2G)** solutions, leveraging EV batteries as distributed storage units.
4. **Demand-Side Management (DSM)**
   - Deploy **time-of-use tariffs** and **demand response programs** to encourage energy storage usage during off-peak hours.
   - Utilize **smart meters** to provide dynamic pricing and improve consumer participation.
5. **Grid-Scale Energy Storage Policies and Regulations**
   - Governments should provide **financial incentives**, such as tax credits and subsidies, to encourage storage deployment.
   - Establish clear **regulatory frameworks** for storage ownership, operation, and participation in grid services.
6. **Decentralized and Community-Based Energy Storage**
   - Encourage **microgrids** with localized storage to improve energy security in remote areas.
   - Implement **peer-to-peer (P2P) energy trading**, allowing stored renewable energy to be shared among users.
7. **Recycling and Sustainability of Energy Storage**
   - Develop robust **battery recycling programs** to reduce environmental impact.
   - Invest in **second-life applications** for used EV batteries in stationary storage systems.

6.5.3 Cost-Benefit Analysis of Energy Storage Technologies in Renewable Power Grids

Energy Storage Systems (ESS) are critical for integrating renewable energy sources like wind and solar into the power grid. However, each storage technology comes with different costs, benefits, and applications. The following is a detailed **cost-benefit analysis** of various storage technologies.

6.5.3.1 Lithium-Ion Batteries

Cost Analysis:

- **Capital Cost:** \$300–\$500/kWh
- **Operation and Maintenance (O&M) Cost:** \$5–\$20/kWh per year
- **Lifespan:** 10–15 years (~4,000–6,000 cycles)

- **Efficiency:** 90%–95%
- **Degradation Rate:** 1%–2% per year

Benefit Analysis:

- **Fast response time** (<1 second) for frequency regulation and peak shaving
- **High energy density** with compact installation requirements
- **Scalability**, making it suitable for distributed storage and EV integration
  - **Degradation over time** and limited deep discharge cycles
  - **High raw material costs** (cobalt, lithium, nickel) and recycling challenges

Best Use Cases:

- Grid frequency regulation
- Short-term renewable energy balancing (2–4 hours)
- Behind-the-meter commercial and industrial storage

6.5.3.2 Pumped Hydro Storage (PHS)

Cost Analysis:

- **Capital Cost:** $1,000–$3,000/kW
- **O&M Cost:** $10–$50/kWh per year
- **Lifespan:** 50+ years
- **Efficiency:** 70%–85%
- **Degradation Rate:** Negligible

Benefit Analysis:

- **Longest lifespan** among all storage technologies
- **Cost-effective** for large-scale energy storage (>100 MW)
- **High round-trip efficiency** and low maintenance costs
  - **Geographic limitations** (requires suitable terrain and large water reservoirs)
  - **Long development timelines** (5–10 years for permitting and construction)

Best Use Cases:

- Large-scale renewable integration
- Grid balancing and peak load shifting (6–12 hours of storage)
- Long-term energy storage for seasonal demand

6.5.3.3 Redox Flow Batteries (Vanadium Flow, Zinc-Bromine)

Cost Analysis:

- **Capital Cost:** \$400–\$800/kWh
- **O&M Cost:** \$10–\$30/kWh per year
- **Lifespan:** 15–20 years (~10,000 cycles)
- **Efficiency:** 65%–85%
- **Degradation Rate:** Minimal

Benefit Analysis:

- **Long cycle life** with deep discharge capability (100% depth of discharge)
- **Scalable and modular** for grid applications
- **Lower degradation rate** compared to lithium-ion
  - **Lower energy density**, requiring more space
  - **Higher initial capital cost** compared to lithium-ion

Best Use Cases:

- Renewable energy time-shifting (4–12 hours)
- Off-grid and microgrid storage
- Industrial applications requiring long-duration energy backup

6.5.3.4 Compressed Air Energy Storage (CAES)

Cost Analysis:

- **Capital Cost:** \$800–\$1,400/kW
- **O&M Cost:** \$5–\$15/kWh per year
- **Lifespan:** 30–40 years
- **Efficiency:** 50%–70%
- **Degradation Rate:** Low

Benefit Analysis:

- **Large-scale storage** (100 MW+) with long discharge times (6–24 hours)
- **Uses existing geological formations** (caverns, abandoned mines)
- **Low maintenance costs**
  - **Limited efficiency** due to heat losses during compression/expansion
  - **Geographically constrained** to areas with underground storage potential

Best Use Cases:

- Large-scale renewable energy smoothing
- Seasonal energy storage for wind and solar farms
- Grid ancillary services

6.5.3.5 Hydrogen Energy Storage

Cost Analysis:

- **Capital Cost:** $1,500–$3,000/kW (including electrolyzers and fuel cells)
- **O&M Cost:** $15–$50/kWh per year
- **Lifespan:** 10–20 years
- **Efficiency:** 30%–50%
- **Degradation Rate:** Moderate (fuel cell efficiency loss over time)

Benefit Analysis:

- **Potential for seasonal energy storage**
- **Can be used in multiple applications** (grid storage, transportation, industry)
- **No geographic constraints**
  - **Low efficiency** due to energy conversion losses
  - **High cost of electrolyzers and fuel cells**

Best Use Cases:

- Seasonal energy storage for excess wind/solar power
- Hydrogen-powered industrial processes
- Grid-to-transportation energy conversion

6.6 FUTURE TRENDS AND EMERGING TECHNOLOGIES IN ENERGY STORAGE

As the demand for renewable energy integration grows, energy storage technologies are evolving rapidly. The future of energy storage is expected to be shaped by advancements in efficiency, cost reduction, and novel applications that will play a crucial role in transforming energy grids worldwide [9]. These trends and emerging technologies are anticipated to enable deeper integration of renewable sources like wind, solar, and hydro, providing solutions to challenges such as intermittency, grid stability, and long-duration energy storage. Below are key trends and technologies shaping the future of energy storage.

6.6.1 Advancements in Lithium-Ion Batteries

Lithium-ion batteries are currently the most widely deployed energy storage technology, particularly for grid-scale applications. Their rapid cost reduction over the past decade, along with performance improvements, has made them the dominant choice for energy storage. Ongoing research and innovation continue to enhance their capabilities, leading to even more efficient, longer-lasting, and cost-effective systems.

Key Developments:

- **Solid-State Batteries**: Solid-state batteries represent a promising advancement in lithium-ion technology. By replacing the liquid electrolyte with a solid electrolyte, these batteries are expected to improve safety, energy density, and charge/discharge efficiency. Solid-state batteries could significantly reduce the risk of fires, a common concern with lithium-ion batteries, while increasing energy storage capacity, making them ideal for both transportation and grid storage applications.
- **Battery Recycling and Second-Life Applications**: As the global adoption of electric vehicles (EVs) and large-scale storage grows, recycling and reusing lithium-ion batteries will be crucial. Second-life batteries, such as those retired from EVs, are being tested for use in stationary storage, extending their lifecycle and improving the sustainability of the battery ecosystem.

6.6.2 Long-Duration Energy Storage (LDES)

One of the key challenges of renewable energy integration is the need for long-duration energy storage, which can store energy for extended periods (from hours to days or even weeks) to bridge generation gaps. Short-duration solutions, like lithium-ion batteries, are well-suited for daily fluctuations, but longer-duration storage is required to address seasonal variability or weather-related disruptions.

Emerging Long-Duration Technologies:

- **Flow Batteries**: Flow batteries, mostly **vanadium redox flow batteries (VRFB)**, are ahead in development due to their ability to store energy for hours to days with minimal degradation over time. These batteries are chiefly suited for large-scale grid storage because of their scalability and long operational life.
- **Hydrogen Storage**: Hydrogen as a storage medium is a developing technology that offers long-duration storage capabilities. Excess renewable electricity can be used to produce hydrogen through electrolysis, which can be stored and converted back into electricity when needed. Hydrogen's ability to provide seasonal storage, along with its versatility (e.g., for transportation and industrial applications), positions it as a key player in future energy systems.

6.6.3 Advanced Grid Storage Solutions and Integration

The concept of **distributed energy storage** is gaining traction, with storage systems deployed closer to the end-user, rather than solely at centralized locations. This approach can help reduce grid ongestion, improve resilience, and optimize energy management [10].

Innovative Grid Storage Applications:

- **Virtual Power Plants (VPPs)**: VPPs aggregate distributed storage systems, such as home batteries, to create a virtual, decentralized power plant that can provide grid services like peak shaving and frequency regulation. VPPs enable more dynamic interaction between the consumer and the grid, optimizing energy use and reducing reliance on centralized power plants.
- **Grid-Scale Thermal Storage**: In addition to battery technologies, advanced energy storage systems that store energy as heat (such as molten salts) are becoming increasingly viable for grid-scale applications. These systems can provide cost-effective, long-duration storage for renewable energy, particularly in concentrated solar power (CSP) projects.

6.6.4 AI and Machine Learning for Energy Storage Optimization

Artificial intelligence (AI) and machine learning are playing an essential role in improving the performance and efficiency of energy storage systems. By analyzing large datasets in real time, AI can optimize the charging and discharging cycles of storage systems, predict energy demand, and help integrate renewable energy sources more effectively.

AI Applications in Energy Storage:

- **Predictive Maintenance**: AI algorithms can predict the failure or degradation of storage components before they occur, reducing maintenance costs and downtime.
- **Grid Optimization**: AI is increasingly being used to manage energy storage systems as part of a broader strategy for smart grid integration. These systems optimize when to store energy, when to release it, and how to balance energy supply and demand based on real-time data.

6.6.5 Quantum Storage and Next-Generation Technologies

Although still in its infancy, quantum storage represents a cutting-edge field with the potential to revolutionize energy storage. Quantum technologies, which leverage the principles of quantum mechanics, are being researched for applications in energy storage, particularly for high-capacity and ultra-fast storage solutions. This technology could one day lead to the development of ultra-efficient and extremely dense energy storage systems, allowing for much more efficient energy retention at significantly smaller sizes.

Quantum Storage Possibilities:

- **Quantum Batteries**: These hypothetical systems could store energy at the quantum level, potentially offering superior energy density and efficiency compared to current technologies. Although practical implementations are still years away, research in this area is gaining momentum and could radically change the energy storage landscape in the future.

6.6.6 Energy Storage as a Service (ESaaS)

With the rising importance of energy storage, companies are moving toward offering **Energy Storage as a Service (ESaaS)**. This model allows businesses and utilities to deploy energy storage systems without bearing the high upfront capital costs. Instead, storage is provided as a service, where the customer pays based on the energy stored or the services provided, such as frequency regulation or peak shaving.

Benefits of ESaaS:

- **Lower Capital Barriers**: ESaaS lowers the barrier for organizations to access energy storage, especially small- to medium-sized enterprises that cannot afford large upfront investments.
- **Revenue Generation for Providers**: For storage system providers, the ESaaS model generates a steady revenue stream from long-term contracts, reducing the financial risk associated with large storage projects.

6.6.7 Global Energy Storage Market Growth

The global energy storage market is expected to witness exponential growth in the coming decades. According to various market reports, the market could reach a value of **$500 billion by 2030**. This growth is driven by several factors, including the increasing adoption of renewable energy, declining costs of storage technologies, supportive government policies, and the need for energy security in both industrialized and emerging regions.

The convergence of improved technologies, government incentives, and market demand will continue to shape the development of energy storage solutions, enabling them to play a crucial role in the global transition toward clean, sustainable, and resilient energy systems.

6.7 CONCLUSION: THE CRITICAL ROLE OF ENERGY STORAGE IN RENEWABLE GRIDS

Energy storage plays a crucial role in the integration of renewable energy sources into modern power grids. As the world transitions toward cleaner, more sustainable energy systems, the intermittency and variability inherent in renewable sources like solar, wind, and hydropower present significant challenges. Energy storage technologies bridge these gaps by storing excess energy when generation outpaces demand and releasing it when supply is low. This capability stabilizes the grid, enhances reliability, and supports the transition from fossil fuels to renewables, driving reductions in greenhouse gas emissions and ensuring a more sustainable energy future.

The significance of energy storage lies not only in its ability to mitigate intermittency but also in its capacity to support grid resilience, improve energy security, and provide valuable grid services such as frequency regulation, peak shaving, and voltage support. As renewable energy generation continues to grow, storage systems will become increasingly critical to ensure that energy demand and supply are balanced effectively [11]. Whether through large-scale applications like pumped hydro and flow batteries, or more distributed systems like lithium-ion batteries and virtual power plants, energy storage offers the flexibility required for modern, renewable-powered grids.

Looking to the future, the role of energy storage in achieving global renewable energy goals is poised to expand. The continued reduction in storage costs, advancements in long-duration storage technologies, and increasing participation in energy markets will drive wider adoption. As storage solutions become more efficient and scalable, they will enable deeper integration of renewables, making it possible to rely on clean energy even during periods of low generation or high demand. Moreover, energy storage has the potential to become a cornerstone of energy systems worldwide, helping to address both short-term fluctuations and long-term seasonal variability in renewable energy supply.

However, to realize the full potential of energy storage, sustained efforts in research, development, and policy support are essential. Governments and industry must work together to create favorable regulatory frameworks, provide financial incentives, and foster innovation in storage technologies. This will ensure that energy storage systems are deployed at the scale needed to meet global renewable energy targets. Public and private investments in emerging storage solutions, such as hydrogen storage, solid-state batteries, and advanced flow batteries, are vital for unlocking the next generation of energy storage technologies.

In conclusion, energy storage is not just a technological solution but a critical enabler of a sustainable, low-carbon future. As we move toward a world powered by renewables, energy storage will be the key to ensuring that this energy is reliable, affordable, and available when needed most. It is essential that we continue to invest in research, policy frameworks, and market innovations that will unlock the full potential of energy storage, propelling us toward a clean, resilient energy future for all.

REFERENCES

1. International Renewable Energy Agency (IRENA), *Energy Storage: The Backbone of a Renewable Power Future*. IRENA, 2017.
2. U.S. Department of Energy (DOE), *Grid Energy Storage Systems*. U.S. DOE, 2020.
3. C. Vennila et al, "Artificial intelligence-based adaptive control for vehicle-to-grid and grid-to-vehicle operations in electric vehicle charging stations", *Renewable Energy*, vol. 124769, 2025.
4. National Renewable Energy Laboratory (NREL), *Energy Storage for the Grid*. NREL, 2022.
5. European Commission, *Energy Storage for a Low Carbon Grid*. European Commission, 2019.

6. S. Selvakumaran et al, "Improved binary quantum-based Elk Herd optimizer for optimal location and sizing of hybrid system in micro grid with electric vehicle charging station", *Journal of Renewable and Sustainable Energy*, vol. 16, no. 6, pp. 064101, 2024.
7. Hornsdale Power Reserve, *A Case Study in Large-Scale Energy Storage Integration.* Neoen Australia, 2020.
8. G. Malathi et al, "Designing and Performance Analysis of a Concentrated Solar Power System in Cold Arid High DNI Area", *International Journal of Intelligent Systems and Applications in Engineering*, vol. 12, no. 12s, pp. 475–486, 2024.
9. N.K. Mohanty et al, "Harmonic Current Reduction Analysis of Traditional PI and Enhanced FoPID controller-based MSALC", Journal of Electrical Engineering and Technology, vol. 16, no. 4, pp. 1769–1777, 2021.
10. U.S. Energy Information Administration (EIA), *Energy Storage and Grid Flexibility.* EIA, 2020.
11. International Energy Agency (IEA), *Energy Storage and Clean Energy Transition.* IEA, 2021.

7 Smart Grids and Next-Generation Power System Architectures

Preetha Krishnasamy,
Muralikrishnan Gopalakrishnan,
Titus Sigamani, and Nagendran Jayabalan

7.1 INTRODUCTION

Smart grids are grids that integrate modern computing, control, and communication techniques, making the electrical power system more advanced to enhance its efficiency, reliability, and sustainability. The main objective of a smart grid is to monitor and regulate the flow of power in real time, thus allowing consumers and utilities to react swiftly to changes in supply and demand. A smart grid incorporates sensors, digital meters, automated switches, and advanced software technologies to improve the integration of renewable energy sources, minimize energy loss, optimize power distribution, and increase overall grid stability.

Advanced Power Systems include a wide variety of technologies, methodologies, and systems used to generate, transmit, and distribute power more effectively and consistently. These systems often integrate energy storage technologies like batteries and supercapacitors, in addition to renewable energy sources like solar, wind, and other green energy sources. Advanced power systems also include elements such as high-voltage direct current (HVDC) transmission lines, which, compared to conventional alternating current (AC) systems, can transport electricity across great distances with minimal energy loss. Additionally, advanced power systems aim to manage and control grid operations through predictive analytics and machine learning (ML) to ensure efficient load management, fault detection, and system optimization.

7.1.1 Key Features of Smart Grids

1. **Two-way Communication:** Allows for the sharing of real-time data between customers and the utility.
2. **Automated Metering:** Measures energy consumption via digital meters, providing detailed information on usage patterns.
3. **Demand Response:** Adjusts energy use during periods of high demand to prevent grid overloads.
4. **Distributed Generation:** Incorporates local green energy sources such as wind turbines and solar panels

DOI: 10.1201/9781003585534-7

5. **Grid Self-healing**: Identifies and isolates faults, minimizing downtime and improving reliability.

7.1.2 Key Features of Advanced Power Systems

1. **Renewable Energy Integration**: Incorporates solar, wind, and hydropower into the grid.
2. **Energy Storage**: Uses battery storage technology for excess energy.
3. **High-efficiency Transmission**: Includes HVDC lines and other methods to reduce transmission losses.
4. **Grid Stability and Control**: Utilizes advanced algorithms for real-time load forecasting, fault detection, and automated control.

7.1.3 Importance of Modernization in Energy Infrastructure

Upgrading energy infrastructure is crucial to accommodate increasing electricity demands, integrate renewable energy sources, improve system reliability, and tackle urgent challenges like climate change and energy security. Traditional power grids, designed for centralized energy generation and one-way flow, are increasingly inadequate for today's dynamic and decentralized energy needs, as mentioned in Figure 7.1. Here are key reasons why modernization is crucial:

Modernizing energy infrastructure is a critical step toward addressing the evolving demands of the 21st century. Traditional power grids, designed for centralized, one-way electricity flow, are no longer adequate to fulfill the growing demand for energy, support renewable energy integration, or ensure resilience against

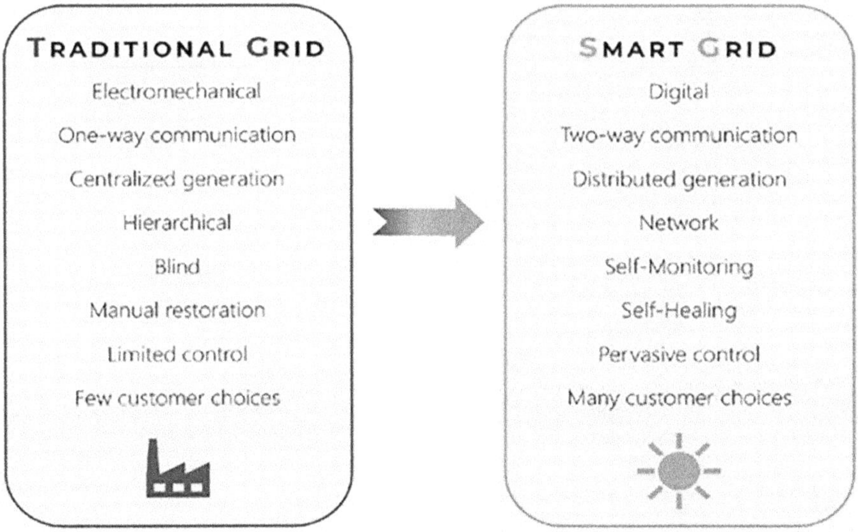

FIGURE 7.1 Traditional vs. modernized grid.

environmental and cybersecurity threats. Upgraded energy systems deliver greater efficiency, reliability, and sustainability by utilizing advanced technologies like smart grids, energy storage, and automation.

A key driver for energy system modernization is the rising global demand for energy, fueled by population growth, industrialization, and the rapid adoption of electric vehicles. Conventional systems face challenges such as significant transmission losses and limited adaptability to fluctuating energy demands. Modernized infrastructure overcomes these issues by enhancing energy generation and distribution, minimizing waste, and supporting decentralized solutions like microgrids.

Integrating renewable energy sources is another crucial element necessary for the shift to a low-carbon future. Despite being sustainable, solar and wind energy are sporadic by nature. These sources can be effortlessly integrated into modernized grids with energy storage, predictive analytics, and real-time monitoring capabilities, guaranteeing a steady and dependable energy supply.

Modernization also enhances grid reliability and resilience, as depicted in Figure 7.2. With aging infrastructure prone to frequent outages, modern systems use smart sensors, automated controls, and self-healing mechanisms to detect and resolve issues before they escalate. This capability is vital for withstanding natural disasters, cyberattacks, and other disruptions.

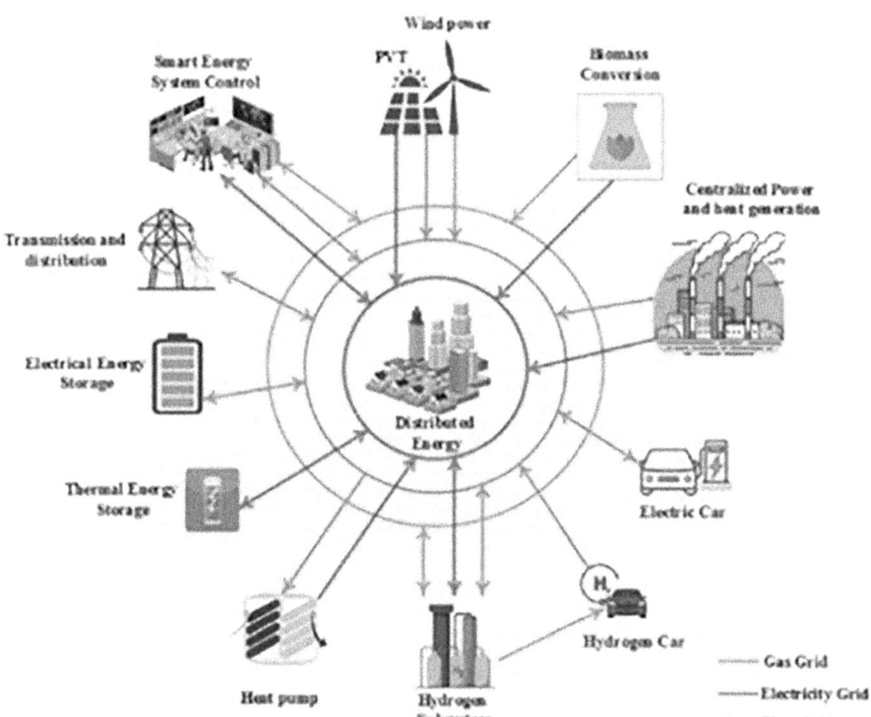

FIGURE 7.2 Importance of modernizing architecture.

FIGURE 7.3 Benefits of modernizing energy infrastructure.

Equally important is the empowerment of consumers. Individuals and companies can monitor their energy consumption, take part in demand-supply programs, and even sell extra energy back to the grid with smart meters and two-way communication networks. This fosters energy efficiency and encourages more sustainable practices at the consumer level.

In addition to technical and operational benefits, modernized energy systems notably reduce greenhouse gas emissions, as mentioned in Figure 7.3. By optimizing resource use and integrating cleaner energy sources, they are essential for reducing the effects of climate change and achieving global sustainability goals.

7.1.4 Overview of Challenges in Traditional Power Systems (e.g., Inefficiencies, Reliability Issues)

Traditional power systems have long served as the backbone for all three sections of electrical energy, but they are increasingly inadequate in meeting the evolving demands of modern society. Several fundamental challenges hinder their effectiveness and sustainability, which makes the case for modernization increasingly urgent.

One of the primary issues is inefficiency in both energy generation and transmission. Traditional power grids rely on centralized power plants located far from consumption points, which leads to significant energy losses over long transmission distances. In addition, many of the generation plants and transmission lines are designed to handle peak loads, resulting in inefficiencies during off-peak times when there is less demand. These systems lack flexibility and find it difficult to adjust to changes in demand and supply or accommodate renewable energy sources, which are more intermittent and variable than fossil-fuel-based generation.

Reliability is another major challenge. Aging infrastructure in many power grids contributes to frequent outages and a lack of resilience. Many essential parts, including transmission lines and transformers, have surpassed their operational lifespan and are increasingly susceptible to failure. Furthermore, traditional power systems operate in a centralized manner, meaning a failure at one point in the network can cascade and disrupt service over wide areas, leading to blackouts. The absence of real-time monitoring systems compounds this issue, as utilities often cannot detect faults or problems until they have already caused significant damage or service interruptions.

The limited integration of renewable energy sources is another fundamental flaw in traditional power systems. These grids were designed for a time when fossil fuels dominated energy generation, and as such, they are not built to handle the fluctuating and unpredictable nature of wind and solar, which are green energy sources. In addition, they continue to rely heavily on coal-, gas-, and oil-based power plants, which contribute to excessive emissions of greenhouse gases and environmental degradation. As a result, traditional systems are not well-equipped to facilitate the switch to more sustainable and clean energy, as illustrated in Figure 7.4.

Environmental concerns are intimately related to the conventional electricity grid's inefficiency. Fossil-fuel-based power generation remains among the main

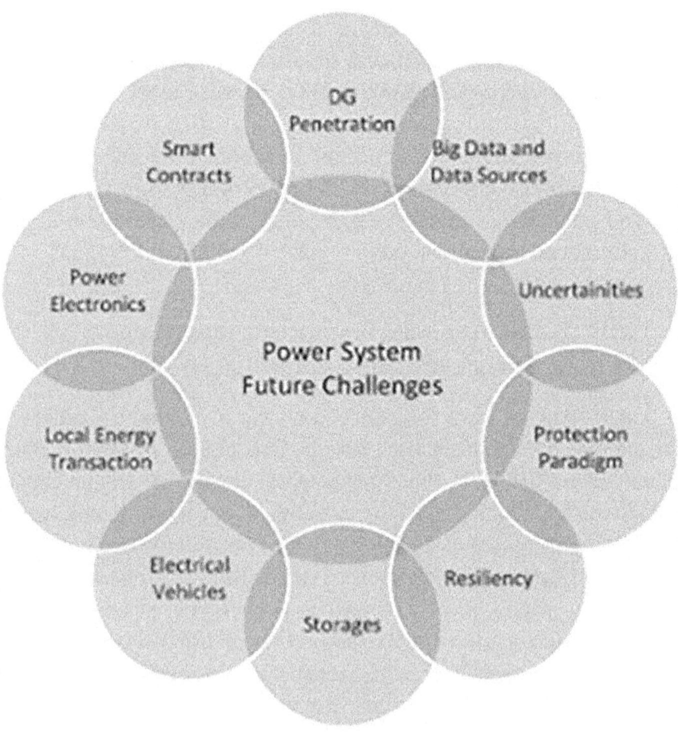

FIGURE 7.4 Overview of challenges in traditional power systems.

causes of carbon emissions, which are driving global climate change. Without the ability to optimize energy use, traditional grids waste large amounts of electricity during transmission and distribution, further exacerbating environmental harm. As governments and businesses worldwide seek to reduce their carbon footprint, cleaner and more effective energy infrastructure is becoming increasingly urgent and necessary.

Traditional power systems also leave little room for consumer participation. In these older models, energy consumers have a passive role; they are simply recipients of electricity, with little ability to control their energy consumption or even obtain up-to-date data on their energy consumption. When there is a lack of data and openness, customers find it challenging to engage in energy-saving activities like demand-supply programs or make informed decisions about their energy use. This limits the potential for individuals and businesses to actively contribute to energy conservation and optimization.

The issue of load management further highlights the limitations of traditional systems. Conventional grids struggle to deal with sudden surges in demand, whether due to weather conditions or unexpected usage spikes, often leading to power shortages or equipment failures. Load distribution is static and reactive, rather than dynamic and proactive, meaning that the grid is frequently ill-prepared for unexpected fluctuations in demand.

Finally, the increasing digitalization of power systems has introduced significant cybersecurity risks. Traditional grids were designed in a time when digital systems were minimal, and they are not equipped to handle the cybersecurity threats that modern, interconnected infrastructures face. Vulnerabilities in these legacy systems make them susceptible to cyberattacks that can disrupt operations, steal data, or even cause widespread blackouts, posing serious risks to national security and energy stability.

Together, these challenges paint a picture of a power system that is ill-equipped to handle the demands and opportunities of the modern world. The inefficiencies, reliability issues, limited integration of renewables, environmental effects, and low customer involvement all point to the urgent need for a new, smarter approach to energy infrastructure. Modernizing the grid can address these challenges while enabling a future with more robust, efficient, and sustainable energy.

7.1.5 The Transition to Smart Grids: A Brighter Future

A more effective, dependable, and sustainable energy future is promised by smart grids, which mark a substantial development in the distribution of electricity. By integrating digital technologies, smart grids aim to completely transform how we generate, distribute, and use electricity, as depicted in Figure 7.5.

7.1.6 Objectives of Smart Grids

- **Enhanced Reliability:** Smart grids use cutting-edge technologies to identify and isolate faults rapidly, minimizing outages and restoring power swiftly.

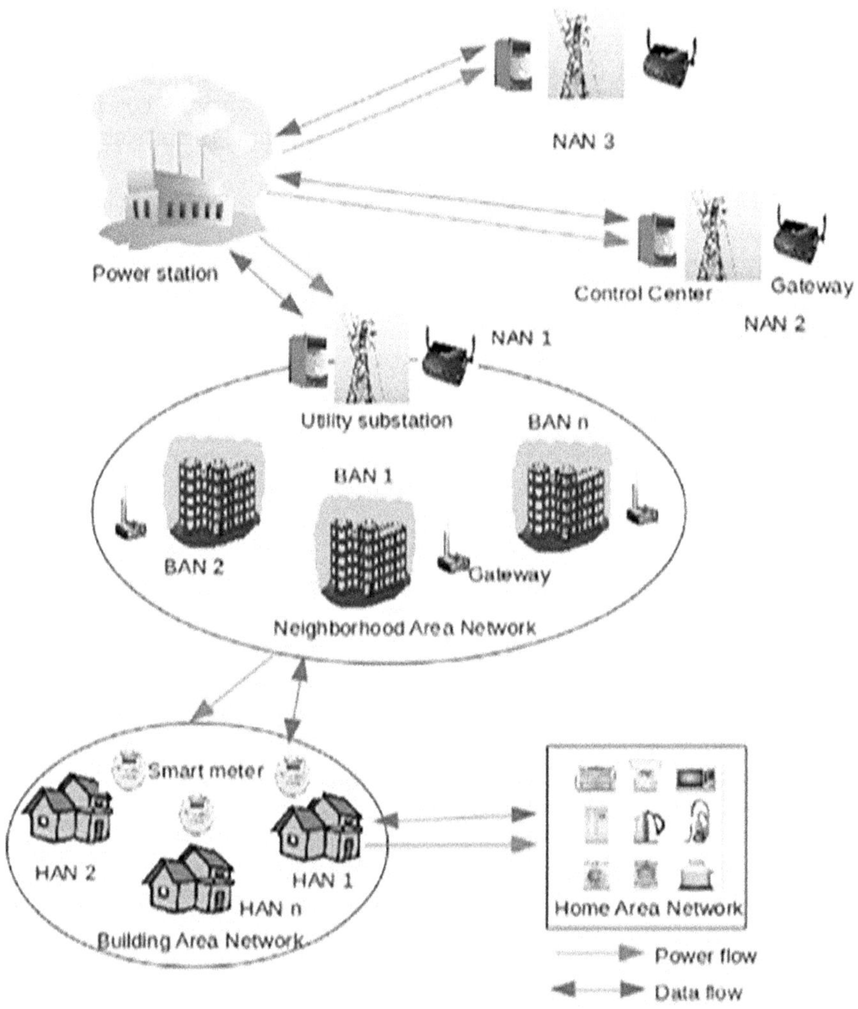

FIGURE 7.5 Smart grid diagram showcasing key components like smart meters, sensors, and communication networks.

- **Improved Efficiency:** By optimizing power flow and reducing energy losses, smart grids contribute to increased efficiency in all three sections of electrical energy.
- **Increased Integration of Renewable Energy:** Smart grids make it easier for renewable energy sources, such as wind and solar, to be seamlessly integrated, enabling better management and balancing of power supply.
- **Empowered Consumers:** Smart grids empower consumers with up-to-date data on their energy usage, allowing them to make informed choices and potentially lower their energy expenses.

7.1.7 Benefits of Smart Grids

- **Reduced Power Outages:** Modern monitoring and control systems reduce the impact of power outages by enabling quicker fault isolation and detection.
- **Improved Energy Efficiency:** Optimized power flow and reduced energy losses lead to significant energy savings.
- **Increased Grid Resilience:** Intelligent grids are able to adapt to changing conditions, such as severe weather, and maintain power supply reliability.
- **Increased Renewable Energy Integration:** By enabling better management, smart grids help create a cleaner and more sustainable energy future by utilizing renewable energy sources.
- **Lowered Energy Costs:** Optimized energy usage and the potential for demand-supply initiatives may result in consumers paying less for energy consumption.
- **Improved Grid Security:** Advanced cybersecurity measures protect the grid from online threats and guarantee the system's integrity.

7.2 EVOLUTION OF POWER SYSTEMS

7.2.1 Historical Development: From Centralized Generation to Distributed Systems

The historical development of power systems is a story of continuous innovation and adaptation to satisfy the rising demand for power. The journey began when electricity was discovered and put to use in the late 19th century. Early electrical systems were rudimentary, with the invention of the incandescent light bulb by Thomas Edison in 1879 and his establishment of the first power station, Pearl Street Station, in 1882, marking the birth of centralized electricity generation. Edison's direct current (DC) power systems, however, were limited in their ability to support long-distance electrical transmission.

Nikola Tesla and George Westinghouse overcame this limitation in the late 19th century by introducing alternating current (AC) as a more efficient method for long-distance transmission. AC systems gained dominance in what was known as the "War of Currents," with AC ultimately winning due to its efficiency in power transmission. The success of AC was demonstrated with the establishment of massive AC power plants, such as the Niagara Falls Power Station in 1895, which became one of the first major sources of AC electricity.

Electricity networks grew in the early 20th century, with the advent of rural electrification efforts in the 1930s helping to bring power to previously underserved areas. This growth was further accelerated after World War II, when industrialization and urbanization spurred rapid increase in electricity demand. Power grids became more complex, with multiple interconnected plants, and technologies like transformers, circuit breakers, and advanced generators were developed to ensure reliable and efficient electricity distribution.

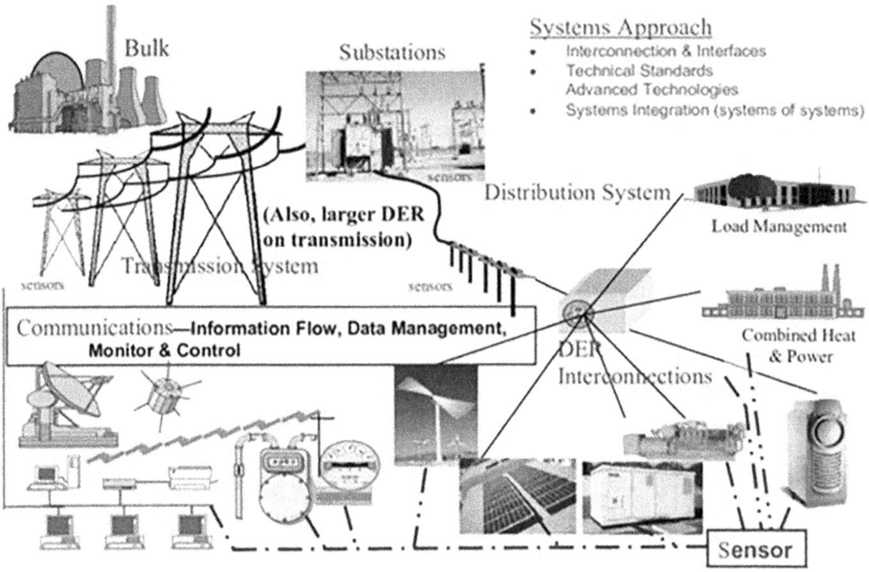

FIGURE 7.6 Evolution of power system architecture.

Digital technologies emerged in the second half of the 20th century. In the 1970s and 1980s, power systems were modernized with the introduction of Supervisory Control and Data Acquistion (SCADA) systems, which allowed utilities to monitor and manage the grid remotely. This era also saw the rise of nuclear power, offering a more environmentally friendly substitute for fossil fuels. Additionally, renewable energy sources like wind and solar were incorporated into the grid, though they remained a small part of the energy mix.

The 21st century has brought the rise of smart grids and advanced power systems. Smart grids make use of sensors, digital connectivity, and real-time data to optimize the flow of electricity, improve efficiency, and integrate renewable energy sources. Better control of intermittent renewable energy has also been made possible by the incorporation of energy storage technologies like batteries. Microgrids, which can function separately or in conjunction with the main grid, are becoming increasingly popular for enhancing grid resilience. Today, power systems continue to evolve, as illustrated in Figure 7.6, with an emphasis on energy efficiency, sustainability, and the integration of new technologies to meet the challenges of modern electricity demand.

7.2.1.1 Technological Milestones: Introduction of Renewables, Energy Storage, and Demand-Side Management

The distribution and generation of electricity have been fundamentally transformed by three major technological advancements: the adoption of demand-side management techniques, the development of energy storage technologies, and the integration of renewable energy sources. Each of these innovations has helped to build a power system that is more reliable, efficient, and sustainable, shaping the way we generate, store, and consume electricity.

7.2.2 Introduction of Renewable Energy

The transition to renewable energy began in earnest during the latter part of the 20th century, with technologies like wind and solar power becoming prominent alternatives to traditional fossil fuels. Green energy sources offer clean, sustainable power with zero emissions, helping reduce the carbon footprint of electrical power generation. Unlike plants that rely on fossil fuels, which emit significant greenhouse gases, renewables rely on natural, replenishable sources of energy, making them critical to combating climate change.

While wind turbines use the kinetic energy of the wind to create electricity, photovoltaic panels use sunlight to do the same. Hydroelectricity, one of the earliest renewable technologies, continues to be a major contributor to the grid in the form of hydroelectric dams, especially in areas with abundant water resources. As these renewable energy sources become increasingly cost-effective, they are being integrated on a large scale, reshaping energy policies and encouraging sustainable practices worldwide.

7.2.3 Energy Storage Enhancements

Energy storage solutions have become essential to preserve grid stability as renewable energy sources have grown in popularity. Unlike conventional power plants, green energy sources are variable—they generate power only when there is sunlight or a breeze. This problem is particularly solved by energy storage solutions, especially battery systems, which store surplus energy generated during high generation periods and release it during low generation or high demand periods.

Since lithium-ion batteries are effective, scalable, and reasonably priced, they have taken the lead in energy storage technology. Grid-scale battery systems allow utilities to enhance grid resilience, balance supply and demand, and lessen the need for fossil fuel backup facilities. In addition to batteries, pumped hydro storage systems are crucial, raising water to a higher altitude and then releasing it to create electricity when demand is at its highest. Emerging technologies like enhanced thermal storage and flow batteries are also promising for large-scale applications, further diversifying storage options and increasing the grid's flexibility.

7.2.4 Demand-Side Management (DSM)

The idea of demand-side management (DSM), which modifies customer demand in real-time to optimize energy usage, has surfaced as power grids have become more sophisticated. DSM involves strategies and technologies that encourage consumers to modify their energy consumption patterns, thereby helping balance the grid and preventing overloads. By using DSM, utilities can reduce peak load demands, lower operational expenses, and make integrating intermittent renewable sources easier.

Key components of DSM include energy-efficient appliances, demand response plans, and smart meters. Real-time electricity usage monitoring is made possible by smart meters for both utilities and consumers, enabling households and businesses to modify their energy usage in accordance with the state of the grid, as illustrated

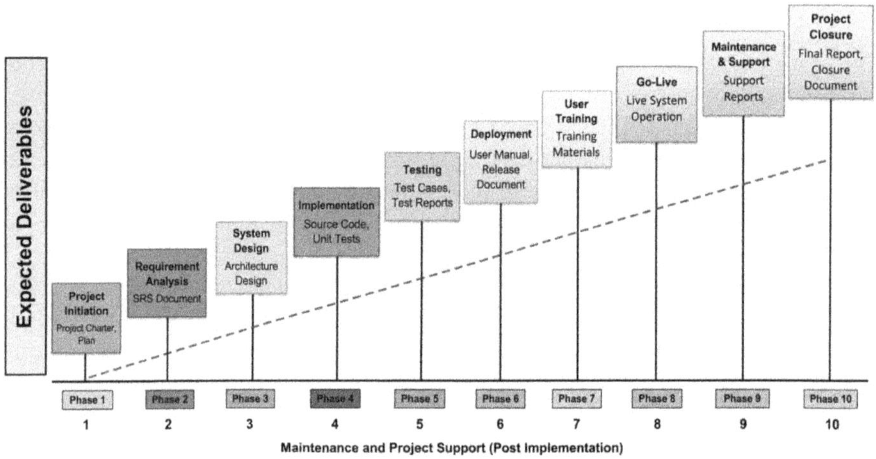

FIGURE 7.7 Technological milestones in power systems.

in Figure 7.7. Automated demand response systems can remotely adjust usage in response to grid needs, reducing demand during peak periods. Moreover, the adoption of energy-efficient devices, like LED lighting and Energy Star appliances, contributes to lower overall demand, helping consumers save on energy bills while supporting a more sustainable grid.

7.2.5 The Role of Digital Transformation in Power Systems

Power systems are undergoing a revolution thanks to digital transformation, which makes it possible for a more adaptable, resilient, and efficient grid to satisfy the changing needs of the contemporary world. By combining digital technologies like the Internet of Things (IoT), artificial intelligence (AI), data analytics, and advanced communication networks, power systems are becoming increasingly "smart," allowing for energy flow optimization, automation, and real-time monitoring.

A major advantage of digital transformation is better grid management. With digital sensors and IoT devices, utilities can collect real-time information from multiple locations around the grid, including power plants, substations, and consumer endpoints. This data is utilized to predict equipment failures, identify defects, and evaluate power usage trends, allowing for preventive maintenance and lowering the chance of outages. Advanced analytics, supported by AI and machine learning, help utilities make data-driven decisions, optimizing and balancing the supply and demand for electricity and making sure it operates more reliably and efficiently.

Digital transformation contributes significantly to the integration of green energy sources. As green energy produces variable power depending on climate conditions and sunlight, maintaining grid stability can be challenging. Through digital technologies like energy management and distributed energy resource management systems (DERMS) software, utilities can dynamically manage the influx of renewable energy, balance loads, and store extra energy when it is not needed. These systems also allow the grid to handle decentralized green energy sources, which are increasingly common among residential and commercial consumers.

Another significant advancement enabled by digital transformation is DSM. Smart meters and home automation systems enable customers to track and manage their energy use in real time. Consumers can change their usage to demand less during peak hours, thereby reducing energy costs, and participate in demand response programs that may reduce grid stress. Additionally, digital platforms facilitate communication between customers and utilities, enabling better customer engagement and incentives for energy efficiency.

Lastly, digital transformation is fostering a more resilient grid. By utilizing cyber-physical systems and advanced cybersecurity measures, utilities can protect the grid from cyber threats and enhance its robustness against both digital and physical disruptions. These systems offer rapid response capabilities, allowing for quick isolation of faults and minimization of outage impacts.

In summary, digital transformation is creating a new era of intelligent, flexible, and consumer-centric power systems. By harnessing digital technologies, the power sector is not only improving operational efficiency and reliability but also advancing toward a robust energy future that accommodates the growing integration of renewables and distributed energy resources.

7.3 KEY FEATURES OF SMART GRIDS

7.3.1 Green Energy Integration: Solar, Wind, and Other Sustainable Resources

A revolutionary development in contemporary power networks, the integration of renewable energy sources aims to minimize environmental effects and decrease dependency on fossil fuels. Green energy sources provide sustainable alternatives to conventional energy by utilizing natural resources that are replenishable and emit hardly any greenhouse emissions. Integrating these sources into the grid, however, offers both possibilities and difficulties due to their variable and decentralized nature.

The decrease in carbon emissions and air pollutants is one of the primary advantages of integrating renewable energy. Green energy sources generate electricity without releasing carbon dioxide or other harmful substances. This shift is critical to combating climate change and improving air quality, making renewables an essential part of national and global climate policies. As a result, many countries have set ambitious goals for renewable energy adoption, creating a strong push for their integration into existing power systems.

Technologically, integrating renewables requires advanced grid management and digital solutions. As solar and wind energy are erratic by nature, they rely on factors like sunlight and wind speed, which vary throughout the day and seasons. To maintain grid stability, utilities rely on smart grid technologies, including real-time data monitoring tools and DERMS, which allow them to dynamically manage renewable power fluctuations. Energy storage systems, especially batteries, play a critical role here, storing excess green energy produced during high supply periods and releasing it on low generation or high demand times.

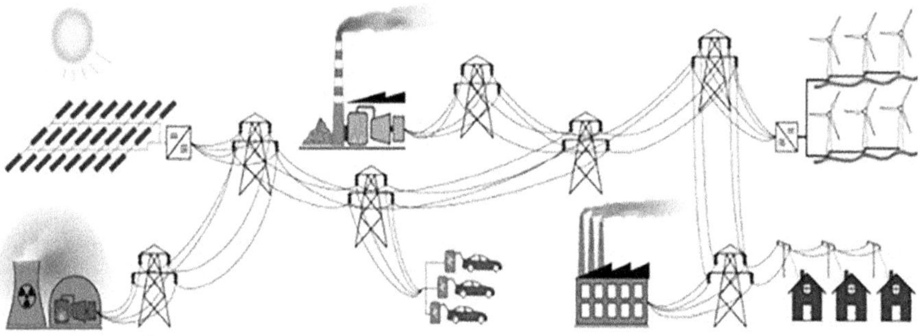

FIGURE 7.8 Green energy integration.

Decentralized generation is another key aspect of renewable integration. Customers can produce their own electricity through community solar projects, rooftop solar panels, and small-scale wind turbines, and in some cases, feed it back into the grid. This model, often called distributed generation, empowers customers to actively engage in the energy ecosystem while contributing more green energy to the grid, as depicted in Figure 7.8. To manage this influx of distributed energy, utilities use bidirectional communication and IoT-enabled devices to seamlessly integrate these small-scale producers into the larger grid.

Policy support and incentives are also essential for renewable integration. Many governments offer subsidies, tax incentives, and net metering initiatives to motivate people and companies to make investments in green energy. In addition, renewable portfolio standards and carbon pricing policies create a regulatory environment that prioritizes clean energy, helping to drive the switch to a low-carbon energy mix.

7.3.2 Advanced Communication Systems

It represents communication between utilities and customers in both directions. In modern power systems, advanced communication systems facilitate the transformation of the traditional power grid into a more interactive, responsive network. This shift enables real-time data exchange, empowering both consumers and utilities to make better-informed choices on energy use, cost, and grid reliability.

For utilities, two-way communication provides comprehensive information about grid conditions, patterns of energy demand, and the state of distributed energy resources (DERs), such as battery storage and rooftop solar panels. Smart meters, sensors, and IoT-enabled devices installed at various points in the grid continuously collect data and send it back to the utility. Utilities can subsequently utilize this data to adjust supply, balance loads, and respond quickly to any disruptions, improving grid stability and efficiency. This data is processed by machine learning algorithms and advanced data analytics, allowing utilities to predict demand surges, schedule maintenance, and manage resources in a way that minimizes costs and enhances reliability.

For consumers, two-way communication enables greater control over their energy usage. Real-time consumption monitoring is made possible by smart meters and

FIGURE 7.9 Advanced communication systems.

home automation systems, helping them make decisions to reduce energy costs by adjusting their usage during peak times. Demand response programs are an example of this interaction; consumers can participate by voluntarily decreasing energy use during high-demand periods, often in exchange for incentives from the utility. Additionally, with two-way communication, users who produce their own electricity (for example, using rooftop solar panels) can sell extra energy back to the grid, helping build a more robust and decentralized energy system.

This communication system is a crucial component of the "smart grid," which leverages digital tools to produce a versatile, effective, and consumer-centered power infrastructure, as illustrated in Figure 7.9. Utilities and customers communicating in both directions can interact seamlessly, leading to cost savings, increased energy efficiency, and a more sustainable grid that can adapt to changing energy demands and the growing share of renewable energy.

7.3.3 Real-Time Monitoring and Analytics

Real-time monitoring and analytics are made possible across a variety of industries by the integration of SCADA (supervisory control and data acquisition) and IoT technologies. IoT sensors are placed in the field to gather vital information from assets like transformers, pipes, and machinery, such as temperature, pressure, flow rate, or voltage. These sensors send data to a central SCADA system via wired Ethernet or wireless protocols like LoRa, Zigbee, or Wi-Fi. The SCADA system, which is made up of programmable logic controllers (PLCs) and remote terminal units (RTUs), processes this data and shows the system status in real time to operators in a control room.

Data collected by SCADA can also be sent to cloud-based platforms for advanced analytics, enabling features such as predictive maintenance, which identifies potential equipment failures, and performance optimization through trend analysis. Communication between field devices and SCADA involves secure network protocols like Transmission Control Protocol/Internet Protocol (TCP/IP), Message Queuing Telemetry Transport (MQTT), and Modbus, ensuring efficient data flow.

This two-way communication allows SCADA to send control commands back to the field equipment when necessary.

This system is highly scalable and finds applications in utilities for monitoring energy grids, water distribution, and wastewater systems; in industrial automation for managing machinery and production lines; and in smart cities for overseeing street lighting, parking systems, and environmental monitoring. By delivering insights in real time and supporting data-driven decision-making, the IoT-SCADA combination significantly enhances operational efficiency and reliability.

7.3.4 Grid Automation

Self-healing mechanisms and predictive maintenance. Grid automation with predictive maintenance represents a transformative approach to managing modern power grids. Self-healing mechanisms rely on advanced sensors, smart controllers, and communication technologies to detect faults in the grid, isolate the affected sections, and reroute power automatically, minimizing downtime. For example, when a fault such as a line outage occurs, the system identifies the issue using real-time data from sensors and executes pre-programmed algorithms to restore power through alternate routes without human intervention. This improves grid reliability and reduces service disruptions for consumers.

Predictive maintenance complements self-healing by leveraging data analytics, machine learning, and IoT sensors to monitor the state of vital grid elements like transformers, circuit breakers, and substations. These systems analyze real-time operational data and historical trends to identify signs of wear or failure before they lead to outages. For instance, vibration analysis, thermal imaging, and electrical parameter monitoring can detect early indications of equipment stress, enabling proactive repairs or replacements.

Together, self-healing and predictive maintenance make the grid more resilient, efficient, and cost-effective. They minimize unplanned outages, reduce repair costs, and extend the lifespan of infrastructure, guaranteeing a steady and long-lasting energy supply in the face of increasing demand and potential disruptions from environmental and operational challenges. These technologies are crucial for transitioning to smarter, more automated energy networks.

7.3.5 Energy Storage Solutions

As they improve grid stability, balance supply and demand, and make it possible to integrate renewable energy sources, energy storage solutions are essential to modernizing energy systems. An outline of the most popular and cutting-edge energy storage systems is provided below:

7.3.5.1 Batteries

Batteries, particularly lithium-ion batteries, are the most extensively used energy storage technology because of their effectiveness, scalability, and quick response times. Applications for them range from huge grid-scale installations to small-scale household systems. Because of their extended cycle life and high energy density, lithium-ion batteries are highly valued for storing solar and wind energy for later use. Other types of batteries, such as flow batteries, are emerging as contenders for massive energy storage because of their ability to provide longer discharge durations and scalability without degrading over time.

7.3.5.2 Pumped Hydro

The most developed and capacious energy storage technique is pumped hydro storage (PHS). It functions by pumping water from a lower reservoir to a higher one using excess electricity. Power is produced by releasing the stored water to flow downhill through turbines as the demand for electricity increases. With round-trip efficiencies ranging from 70% to 85%, PHS is highly efficient and has the capacity to store energy for extended periods. However, its deployment is limited by geographical and environmental constraints, as it requires specific topography and significant water resources.

7.3.5.3 Emerging Technologies

- **Thermal Energy Storage (TES)**: Uses substances like ice, molten salt, or phase-change compounds to store energy as heat or cold. TES is particularly useful for balancing energy loads in heating and cooling systems and can be used with solar power facilities that use concentrated solar energy.
- **Hydrogen Storage**: Converts excess electricity into hydrohen through electrolysis to produce hydrogen. After that, the hydrogen can be stored and utilized as fuel or transformed back into power by turbines or fuel cells. This technology is gaining traction for its potential for large-scale, long-term energy storage while supporting decarbonization across multiple sectors.
- **Flywheels**: By rapidly rotating a rotor, you can store energy as kinetic energy. They are excellent for uses that call for brief power spikes and have a long operational lifespan with minimal degradation.
- **Compressed Air Energy Storage (CAES)**: Compresses and stores air in underground caverns using excess electricity. When released, the compressed air is heated and used to power turbines, producing electricity. CAES works well for long-term, large-scale storage.

7.3.5.4 Applications and Impact

Technologies for energy storage are essential for supporting renewable energy integration, mitigating intermittency issues, and supplying grid functions such as peak shaving and frequency regulation. They enable microgrids and off-grid systems, enhance energy resilience, and support the world's transition to a more sustainable and clean energy future. Continued innovation and cost reductions in storage technologies will further drive their adoption and expand their role in decarbonizing energy systems.

7.4 COMPONENTS OF ADVANCED POWER SYSTEM ARCHITECTURE GENERATION

Centralized vs. Decentralized Generation Models.

- **Transmission:** High-voltage DC (HVDC) systems, dynamic line rating.
- **Distribution:** Smart substations, microgrids, and community-based solutions.

- **Consumption:** Energy-efficient appliances, demand response, and smart meters.
- **Data Management:** Big data, AI, and machine learning in grid optimization.

Advanced power system architecture is evolving to fulfill the requirements of a sustainable and efficient energy future. Key components include generation, transmission, distribution, consumption, and data management. In terms of generation, centralized models rely on large-scale plants like coal, gas, nuclear, and hydropower to produce electricity for transmission over long distances, offering economies of scale but facing challenges such as transmission losses and vulnerability to disruptions. In contrast, decentralized generation involves smaller, localized energy sources, reducing transmission losses and increasing system resilience. An optimal architecture often integrates both models for efficiency and sustainability.

For transmission, technologies such as HVDC systems and dynamic line rating (DLR) enhance efficiency and adaptability. HVDC minimizes power losses over long distances, making it ideal for connecting offshore wind farms and supporting regional power exchanges. DLR dynamically adjusts transmission line capacity based on real-time environmental conditions like temperature and wind, maximizing infrastructure utilization and integrating variable renewable energy sources. These developments are pivotal for modernizing the power grid and enhancing its resilience.

Distribution systems are transforming with smart substations, microgrids, and community-based solutions. Smart substations utilize advanced sensors and automated controls for fault detection, real-time monitoring, and improved voltage regulation. Microgrids can function separately or in conjunction with the main grid, enhancing energy resilience and supporting decentralized renewable energy integration. Community energy solutions, such as shared solar farms, empower local stakeholders, reduce energy costs, and promote sustainability.

At the consumption level, energy-efficient appliances, demand response plans, and smart meters are critical components. Smart meters offer dynamic pricing and peak load management by providing utilities and customers with real-time energy usage data. Demand response solutions reduce grid pressure and operating costs by allowing customers to adjust their energy consumption during high-demand periods in response to price signals or incentives. Energy-efficient appliances, often integrated with IoT, further support energy conservation by responding automatically to demand response signals.

Data management underpins the advanced architecture, leveraging grid optimization using machine learning, AI, and big data. Predictive maintenance, load forecasting, and fault detection are made possible by big data analytics, while AI algorithms optimize power flow and manage energy storage systems. Machine learning models refine demand predictions, identify anomalies, and enhance renewable energy integration. Together, these technologies drive operational efficiency and resilience, laying the foundation for smarter, cleaner, and more adaptable power systems.

In conclusion, advanced power system architectures integrate innovative technologies across generation, transmission, distribution, and consumption to create sustainable and efficient energy networks. With tools like HVDC, smart substations, AI, and IoT, these systems overcome obstacles such as climate change and growing energy demand, establishing a dependable and sustainable energy future.

7.5 BENEFITS OF SMART GRIDS

- Enhanced dependability and robustness of the energy source
- Improved energy efficiency and reduced operational costs
- Greater integration of green energy
- Empowering consumers through data and choice
- Reduced environmental impact and greenhouse gas emissions

Smart grids represent a transformative shift in energy management, providing a more dependable, efficient, and environmentally friendly method of producing, distributing, and using electricity. By leveraging advanced technologies such as IoT, artificial intelligence, and two-way communication systems, smart grids offer significant benefits across reliability, efficiency, renewable integration, consumer empowerment, and environmental impact.

7.5.1 Enhanced Reliability and Resilience of Energy Supply

Increasing the energy supply's resilience and dependability is one of the most important benefits of smart grids. Smart grids can minimize service outages by promptly identifying and isolating issues through real-time monitoring and sophisticated fault-detection technologies. Self-healing mechanisms automatically reroute power around damaged sections, restoring service without requiring manual intervention. For example, during extreme weather events or equipment failures, smart grids can prevent widespread outages by redistributing power and ensuring that critical facilities remain operational. This resilience reduces downtime and ensures a stable energy supply even under challenging conditions.

7.5.2 Improved Energy Efficiency and Reduced Operational Costs

Utilities can save operating costs and optimize energy distribution via smart grids. By employing dynamic demand management systems, utilities can balance energy loads more effectively, avoiding overproduction and minimizing waste. By maximizing the utilization of exisitng infrastructure, technologies such as dynamic line rating and real-time grid optimization postpone the need for expensive upgrades. Smart meters encourage efficient consumption patterns by offering comprehensive insights into energy usage from the consumer's perspective. By encouraging customers to shift to off-peak hours, time-of-use pricing eases the burden on the grid and lowers expenses for both utilities and customers. Collectively, these improvements result in a more efficient energy ecosystem with lower operational costs.

7.5.3 Greater Integration of Green Energy Sources

To incorporate green energy sources like solar and wind into the power system, smart grids are essential. Unlike traditional energy sources, renewables are variable and distributed, requiring advanced grid management to ensure stability. Smart grids facilitate this integration by monitoring energy production in real time and dynamically adjusting supply and demand. Technologies like DERMS enable seamless coordination between decentralized energy sources and the grid. Additionally, smart grids incorporate energy storage devices, such as batteries, to store excess renewable energy and release it when demand is high. This ensures a steady energy supply while reducing dependence on fossil fuels.

7.5.4 Empowering Consumers through Data and Choice

Through smart meters and smartphone apps, smart grids give users real-time access to information about their energy usage, empowering them. Equipped with this knowledge, customers can make informed choices regarding their energy use, lowering expenses and their impact on the environment. In exchange for monetary rewards, demand response programs allow customers to adjust their energy use during times of high demand. Additionally, prosumers—individuals who generate and use energy—are becoming more prevalent thanks to smart grids. Smart grids foster a more engaged energy ecosystem by enabling users to sell excess energy from battery storage systems or rooftop solar panels back to the grid, as depicted in Figure 7.10. This democratization of energy fosters innovation and strengthens the overall system [1] (Figure 7.11).

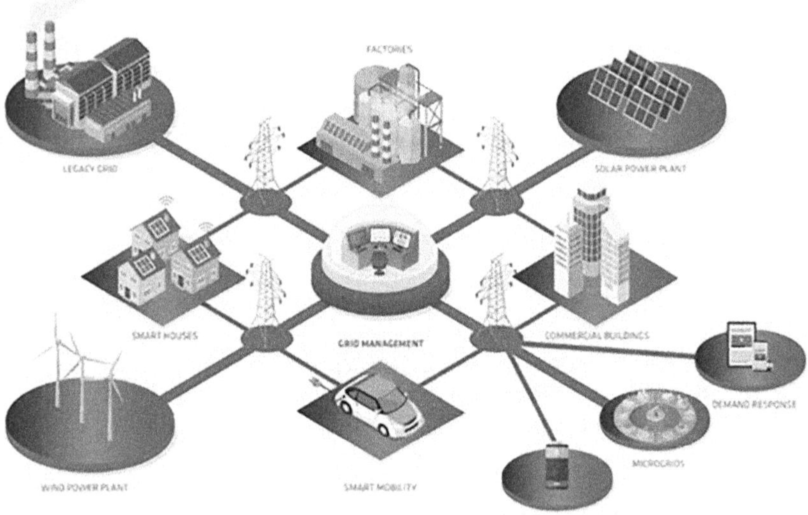

FIGURE 7.10 Grid automation.

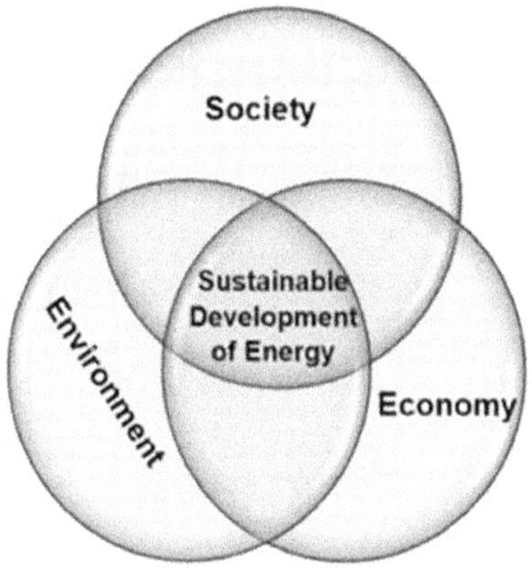

FIGURE 7.11 Relation diagrams of economic development, environment, and energy security.

7.5.5 Reduction in Greenhouse Gas Emissions and Their Effect on the Environment

Smart grids have a major impact on reducing environmental impact and greenhouse gas emissions. Smart grids lessen dependence on fossil fuels, lowering carbon emissions by integrating renewable energy sources and optimizing energy consumption. The improved efficiency of power distribution systems also minimizes energy losses during transmission and distribution. Moreover, real-time monitoring helps identify and address inefficiencies that contribute to waste and emissions. On a broader scale, smart grids enable sustainable urban development by supporting projects such as energy-efficient smart cities and infrastructure for EV charging. These environmental advantages support international initiatives to mitigate climate change and promote more sustainable energy policies.

7.5.6 Conclusion

The benefits of smart grids extend across reliability, efficiency, renewable integration, consumer empowerment, and environmental sustainability. By modernizing traditional energy systems, smart grids not only improve operational performance but also support the global transition to a cleaner, more resilient, and consumer-focused energy future. Smart grids will become increasinly important as technology advances in addressing the challenges of growing energy demand, renewable integration, and climate change mitigation.

7.6 CHALLENGES AND BARRIERS TO TECHNOLOGICAL ADVANCEMENTS IN MODERN INDUSTRIES

- High initial investment and funding gaps
- Cybersecurity and data privacy concerns

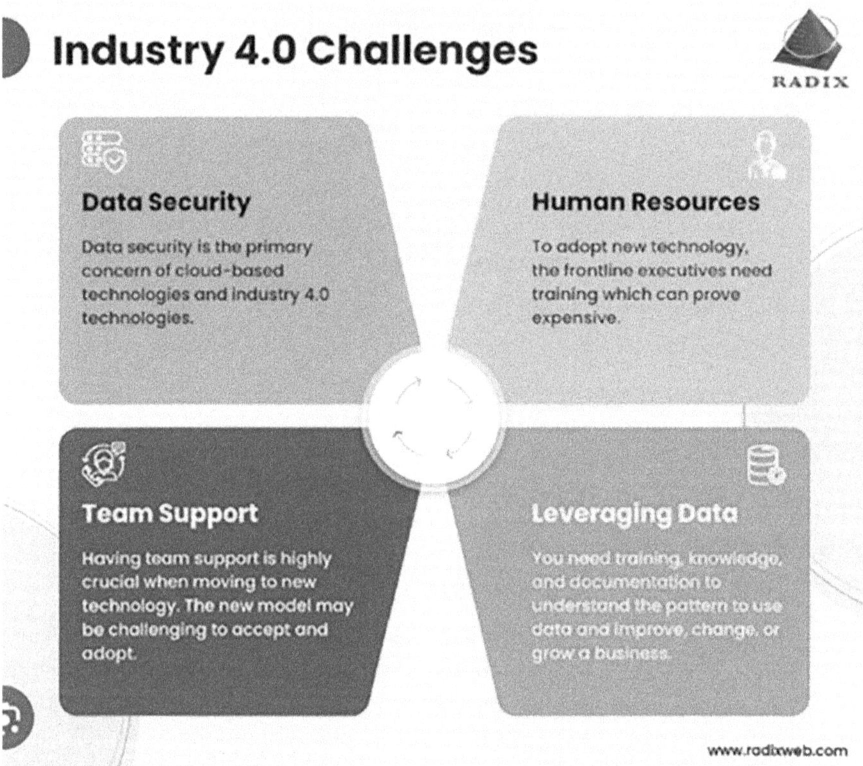

FIGURE 7.12 Challenges and barriers to technological advancements in modern industries.

- Complexity of integrating legacy systems
- Regulatory and policy hurdles
- Public awareness and consumer adoption issues [2] (Figure 7.12)

The implementation and adoption of new technologies often face numerous challenges and barriers. These obstacles can be particularly daunting in industries like healthcare, finance, manufacturing, and energy, where complex systems and established practices dominate. Some of the most significant barriers include high initial investment costs, concerns about data privacy and cybersecurity, the complexity of integrating legacy systems, regulatory and policy hurdles, and issues surrounding public awareness and consumer adoption. In the subsequent sections, we delve deeper into these challenges, their impact on technological progress, and how organizations can address them.

7.6.1 High Initial Investment and Funding Gaps

One of the most prominent barriers to technological advancement is the high initial investment required to adopt new technologies. Businesses looking to upgrade their infrastructure or integrate innovative solutions often face significant financial hurdles. The cost of acquiring and deploying cutting-edge technology can be prohibitively expensive, especially for small- and medium-sized businesses (SMBs) that might lack the funds required to make these kinds of investments.

For instance, industries such as healthcare may need to invest in specialized equipment, new software platforms, or even entire IT infrastructures to implement modern digital systems. While the long-term benefits may be substantial, the short-term financial burden can be difficult to manage. The funding gap between available resources and the capital required for technology adoption often leads to delayed implementation or, in some cases, the abandonment of new projects altogether.

To overcome these challenges, organizations may need to seek out alternative funding sources, such as venture capital, government grants, or private equity investments. Public-private partnerships and innovation hubs can also help bridge the gap, offering financial support and incentives for adopting innovative technologies.

7.6.2 Cybersecurity and Data Privacy Concerns

As organizations adopt more sophisticated technologies, the risks pertaining to data privacy and cybersecurity are still growing. As cloud computing, big data analytics, and the Internet of Things (IoT) grow in popularity, the volume of sensitive information being transmitted and stored has increased exponentially. This presents new challenges in safeguarding data against cyberattacks, data breaches, and unauthorized access.

For example, it becomes more challenging to ensure compliance with data privacy rules like the Health Insurance Portability and Accountability Act (HIPAA) in the United States in industries like healthcare, where patient data is extremely sensitive. The integration of emerging technologies can inadvertently expose organizations to new vulnerabilities that hackers can exploit.

Addressing these concerns requires a multi-layered approach to cybersecurity, including robust encryption protocols, advanced cyber detection systems, regular security assessments, and personnel training. Additionally, adopting privacy-by-design principles and following global standards for data protection can assist in reducing the dangers connected to the use of innovative technology.

7.6.3 Complexity of Integrating Legacy Systems

Legacy systems, which have been in place for decades in many industries, present another significant barrier to technological adoption. These outdated systems were not designed with modern digital technologies in mind and may be incompatible with new software or hardware solutions. As organizations attempt to integrate innovative tools and platforms, the complexity of ensuring smooth interoperability with legacy systems can cause significant delays, increased costs, and operational disruptions.

For example, manufacturing companies often rely on legacy systems to manage inventory, production lines, and supply chains. When trying to incorporate newer technologies like automation, AI, or IoT-enabled devices, the integration process can be lengthy and costly. Additionally, legacy systems may lack the flexibility required to scale operations or accommodate new features and functionalities.

Addressing this challenge requires a carefully planned digital transformation strategy. Many organizations opt for a phased approach, in which they gradually modernize their legacy infrastructure over time rather than attempting a complete overhaul all at once. It is also crucial to assess which legacy systems are truly essential and whether they can be replaced with more modern solutions that better support business needs.

7.6.4 Regulatory and Policy Hurdles

The adoption of innovative technologies is another critical barrier in the complicated regulatory framework that oversees numerous industries. In heavily regulated industries, including energy, healthcare, and finance, businesses must navigate a maze of local, national, and international laws and regulations. These policies are often designed to maintain market stability, protect consumers, and guarantee safety, but they can also stifle innovation by imposing strict requirements on how new technologies are developed, tested, and deployed.

For example, in the financial industry, the implementation of technologies like blockchain or cryptocurrencies faces regulatory uncertainty in many jurisdictions. Similarly, autonomous vehicles and drones are subject to stringent aviation and transportation laws, making it difficult for companies to test and deploy these technologies at scale.

Navigating regulatory hurdles requires close collaboration with policymakers to ensure that new technologies are developed in a way that complies with existing regulations. Additionally, companies should engage in ongoing dialogue with regulatory bodies to help shape the development of future policies that support innovation while safeguarding public interests.

7.6.5 Public Awareness and Consumer Adoption Issues

Finally, two major obstacles that may prevent the broad deployment of innovative technology are consumer adoption and public awareness. Even if a new technology offers clear benefits, consumers and businesses may be hesitant to adopt it due to a lack of understanding, perceived risks, or concerns about usability. Public trust is critical, particularly in sectors like healthcare or finance, where people are wary of sharing personal information or entrusting their data to new systems.

Take, for instance, the growing role of AI in various industries. While AI has the ability to boost decision-making and increase efficiency, there are concerns about job displacement, bias, and transparency. Similarly, consumers may be reluctant to adopt new payment technologies like mobile wallets or blockchain-based currencies if they do not fully understand how these systems work or feel uncomfortable with their security.

Businesses must invest in communication and public education initiatives to overcome these obstacles by building awareness and trust in new technologies. This includes offering clear, accessible explanations of how the technology works, its benefits, and how it protects user privacy and security. Demonstrations, pilot programs, and case studies can also help bridge the gap between innovation and consumer confidence.

7.6.6 Conclusion

The path to technological adoption is rarely smooth, and organizations face numerous challenges and barriers along the way. High initial investment costs, cybersecurity and data privacy concerns, the complexity of integrating legacy systems, regulatory hurdles, and public awareness issues can all impede progress. However, businesses can overcome these barriers and clear the path for successful technological innovation by taking a proactive approach to resolving these issues through strategic investments, strong cyber security measures, phased digital transformation, regulatory cooperation, and effective consumer education. In an increasingly digital world, industries can only fully utilize the promise of new technology to spur growth, boost productivity, and open up new opportunities by overcoming these obstacles.

7.7 CASE STUDIES AND APPLICATIONS OF ADVANCED POWER SYSTEMS AND SMART GRIDS

The global shift to more efficient, sustainable, and resilient power systems is being driven by advancements in smart grid technologies, with significant efforts occurring across the United States, the European Union, Asia, and emerging markets. These regions are leveraging innovative solutions to enhance grid management, integrate green energy resources, and improve the overall efficiency of energy delivery. This section explores notable case studies and applications of smart grid technologies, examining global examples, emerging market approaches, and technological breakthroughs that are reshaping the future of energy systems.

7.7.1 United States

With several projects in progress to update the country's electrical grid, the United States has been a pioneer in the development of smart grid technologies. For more than ten years, the U.S. Department of Energy (DOE) has provided assistance for smart grid initiatives, especially when it comes to improving grid dependability, cutting costs, and boosting the integration of green energy sources.

The California smart grid initiative by Pacific Gas and Electric (PG&E) is one well-known example. More than 10 million smart meters will be installed around the state as part of this project, giving consumers and utilities real-time access to data on energy consumption. This project is part of a broader initiative to modernize the state's electrical infrastructure by using digital communication technology to improve grid monitoring and control. Utilities can better control demand, minimize outages, and provide customers with comprehensive information about their energy use thanks to smart meters. These smart networks, which balance intermittent supply with real-time demand, also make it easier to integrate renewable energy sources like solar power.

The Green Button initiative, which enables users to view and share their energy usage data, is another noteworthy endeavor. Customers may make better-informed decisions about their energy use thanks to the Green Button, which promotes the use of green energy sources and more economical power use.

7.7.2 European Union

Motivated by a determination to lower carbon emissions and integrate renewable energy sources into the system, the European Union has taken the initiative to promote smart grid technologies. The EU's approach emphasizes cross-border collaboration, with projects spanning multiple countries aimed at creating a unified, efficient grid system.

One of the leading projects is the Grid4EU initiative, which spans across France, Germany, Spain, and other European nations. This project seeks to test smart grid solutions for better demand response, renewable energy integration, and advanced grid management. By deploying smart meters, sensors, and automated systems, the project has enabled utilities to balance energy loads more effectively and enhance grid stability. Additionally, it has emphasized user involvement by encouraging homes and businesses to modify their energy use in response to real-time data from smart meters.

Another EU-backed project, Smart Grid 2.0, involves 22 European partners working together to develop and implement cutting-edge smart grid technologies. This project includes a range of innovations such as electric vehicle (EV) grid integration and the creation of sophisticated energy storage technologies. By creating a flexible, interconnected grid system, these projects aim to lessen Europe's dependency on fossil fuels and achieve sustainability goals.

7.7.3 Asia

Asia is also at the forefront of adopting smart grid technologies, with nations like China and Japan significantly advancing the modernization of their energy systems.

As part of its larger plan to lower energy consumption and boost the integration of renewable energy sources, the Chinese government has made significant investments in the development of smart grids. With the deployment of one of the biggest smart grid networks globally, the State Grid Corporation of China (SGCC) has been at the forefront. This includes the construction of ultra-high-voltage transmission lines that can deliver renewable energy from the remote western regions of the country to energy-hungry eastern cities. Additionally, SGCC has been experimenting with advanced metering infrastructure and demand response systems to optimize energy distribution.

Japan, following the Fukushima nuclear disaster in 2011, has undertaken several initiatives to improve its electricity grid's sustainability and reliability. The Smart Community Development Program in cities like Yokohama and Kitakyushu is a prime example. These smart grid technologies produce an integrated approach to energy management by combining renewable energy sources, including solar power, energy storage devices, and electric cars. These smart grids are encouraging the adoption of greener energy sources and enhancing grid stability through the use of demand response technologies and real-time data analytics.

7.7.4 Emerging Markets: Developing Nations' Approach to Advanced Power Systems

In emerging markets, where access to dependable electricity is often limited, the adoption of advanced power systems offers significant potential for improving grid infrastructure, enhancing energy access, and driving economic development [10]. Many developing nations are leapfrogging traditional grid systems and directly adopting modern, decentralized technologies.

India, with its rapidly growing population and industrial base, faces significant energy challenges. To increase the electrical sector's sustainability, reliability, and efficiency, the Indian government launched the National Smart Grid Mission (NSGM) in 2015. By implementing smart grid technologies such as advanced metering infrastructure, grid automation, and real-time data management, the mission aims to modernize India's electrical system.

The Delhi Smart Grid Project is one example of a smart grid program that seeks to reduce energy losses, increase system stability, and integrate green energy resources like solar and wind. The project utilizes demand-side management techniques and smart meters to optimize electricity distribution in urban areas and enhance consumer engagement through real-time energy usage data. These efforts are helping India manage growing energy demands and integrate clean energy into its grid infrastructure.

7.7.5 Africa: Renewable Energy and Microgrids

In Africa, where access to electricity remains a challenge in many regions, energy poverty can be addressed through the use of smart grid technologies and microgrids. Several African nations are increasingly turning to off-grid solar systems and solar microgrids as examples of decentralized renewable energy solutions to power rural communities.

For example, in countries like Kenya, Tanzania, and Uganda, the African Development Bank (AFDB) has provided funding for a number of initiatives aimd at installing solar microgrids. By giving off-grid communities access to affordable, clean electricity, these solutions reduce reliance on diesel generators and encourage sustainable growth. In some areas, microgrids are being connected to the wider national grid to create hybrid systems that can more effectively manage energy supply and demand,.

The success of these projects highlights the importance of DERs in emerging markets, enabling nations to leapfrog traditional grid infrastructure and embrace sustainable, modern energy solutions.

7.8 TECHNOLOGICAL INNOVATIONS: NOTABLE BREAKTHROUGHS IN GRID TECHNOLOGIES

Several technological innovations are revolutionizing grid management, enhancing the efficiency of power delivery, and enabling the integration of green energy resources [8] (Figure 7.13).

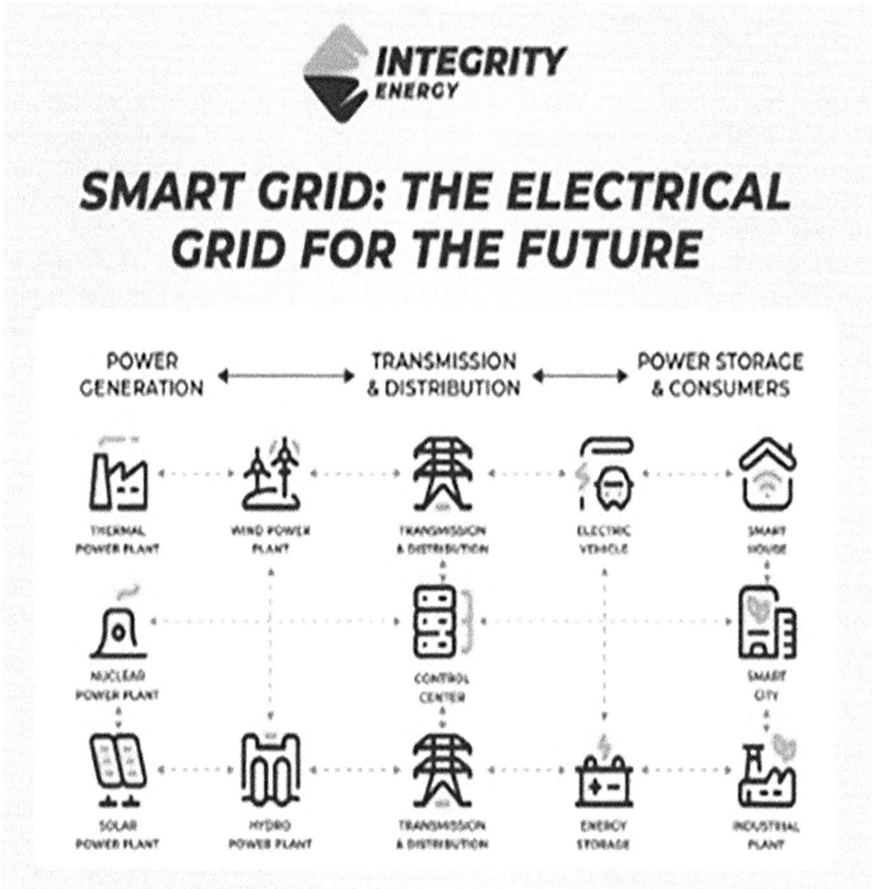

FIGURE 7.13 Technological innovations.

7.8.1 Energy Storage Systems (ESS)

One of the most revolutionary developments in grid technology is energy storage devices. These systems make it possible to store excess energy produced by renewable resources, such as solar and wind, for use when production is low. Modern battery technologies, like solid-state and lithium-ion batteries, have increased the scalability and efficiency of energy storage, making it easier to balance the grid's supply and demand.

The Hornsdale Power Reserve in South Australia, which provides grid stability services using Tesla Powerpack batteries, is one noteworthy example. When the grid requires more power, the system can swiftly release the stored energy from renewable sources, enhancing overall grid resilience.

7.8.2 Demand Response and Smart Meters

Demand response technologies, combined with smart meters, are also playing a key role in modernizing the grid. By using these technologies, utilities may optimize energy use and lower peak load by instantly adjusting electricity consumption based on demand. Smart meter users can get real-time information about how much energy they use, which can encourage them to switch to off-peak hours, which will ease the load on the system and save energy expenses.

7.8.3 Digital Twin Technology

An additional innovation that is revolutionizing grid management is digital twin technology. The power grid is an example of a digital twin, which is a virtual representation of a physical system that may be used for predictive maintenance and real-time monitoring. This technology helps utilities anticipate problems before they occur, optimize energy distribution, and improve grid resilience. Digital twin solutions are being used in smart grid initiatives by companies such as Siemens and GE in an effort to save operating costs and improve grid efficiency.

7.8.4 Conclusion

Smart grid technologies are reshaping the global energy landscape, offering significant improvements in efficiency, sustainability, and resilience. The case studies of smart grid initiatives in the US, EU, Asia, and emerging markets demonstrate how different regions are leveraging innovation to modernize their energy systems. Power grids are becoming more flexible and efficient thanks to technological advancements such as demand response, energy storage, and digital twin technologies. These developments have the potential to transform the way energy is generated, distributed, and used, resulting in a more sustainable and connected society as they continue to evolve.

7.9 FUTURE TRENDS AND INNOVATIONS IN SMART GRIDS

- The rise of AI and blockchain in smart grids
- Role of electric vehicles and vehicle-to-grid (V2G) technology
- Decentralized grids and peer-to-peer energy trading
- Advancements in energy storage and wireless power transmission

The rapid development of smart grid technologies has already brought transformative changes to the global energy landscape. In the future, several new developments and trends could significantly alter how energy is produced, distributed, and used. The emergence of blockchain and AI, the growing integration of EVs and V2G technologies, the expansion of decentralized grids, and advancements in energy storage and wireless power transmission are among these trends. Together, these developments hold the potential to create energy systems that are more adaptable, robust, and efficient, paving the way for a more sustainable and clean future.

AI and blockchain are two key technological innovations that are rapidly gaining traction in smart grid development. Both technologies have the potential to revolutionize grid management by improving operational efficiency, enhancing security, and enabling greater levels of automation.

7.9.1 AI in Smart Grids

AI is playing an increasingly significant role in optimizing smart grid operations. Large volumes of data produced by smart meters, sensors, and other networked devices can be analyzed by AI systems using machine learning, neural networks, and predictive analytics. AI can help utilities anticipate demand, identify faults, optimize energy distribution, and even spot risks to grid stability before they materialize by processing this data in real time.

AI can be applied, for example, to load forecasting, which enables utilities to predict changes in energy demand and adjust generation appropriately. Additionally, AI-powered algorithms can assist in managing the integration of renewable energy, which is frequently erratic and sporadic. By using AI to forecast renewable energy production, smart grids can more effectively balance supply and demand, decreasing dependency on fossil fuels and enhancing the grid's overall sustainability [6,8] (Figure 7.14).

Furthermore, better fault detection and grid resilience can be made possible by AI-powered systems. AI can assess the effects of a breakdown or disruption and reroute energy flows to avoid widespread outages. This level of automation minimizes operating expenses in addition to downtime.

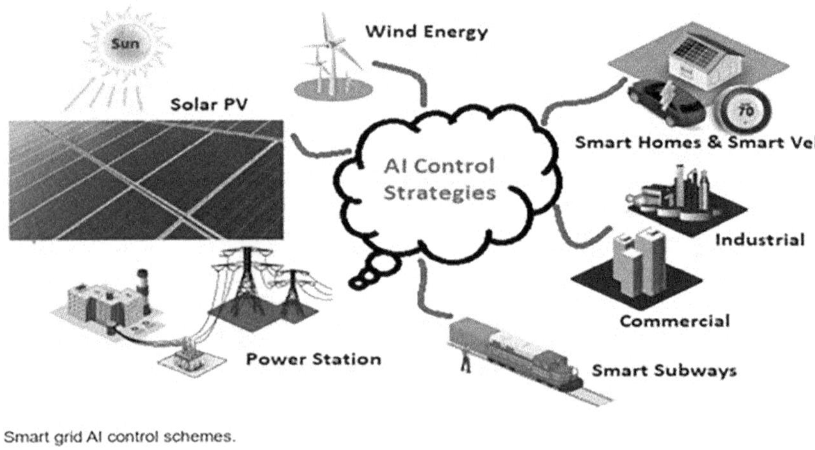

FIGURE 7.14 AI in smart grids.

7.9.2 Blockchain in Smart Grids

Additionally, blockchain technology is anticipated to have a significant impact on smart grids in the future. Blockchain's decentralized, transparent, and immutable nature makes it ideal for ensuring the security, reliability, and efficiency of energy transactions. Blockchain technology can enable peer-to-peer (P2P) energy trading in a decentralized energy grid, where producers and consumers trade electricity. This ensures that all transactions are securely documented and carried out without the need for middlemen.

Blockchain can also enhance the security of energy data. Energy firms can ensure that private data regarding energy generation, consumption, and grid performance is securely stored and immutable by implementing a distributed ledger system. This level of security is crucial as more connected devices and sensors are deployed across smart grids, increasing the potential for cyberattacks and data breaches.

In addition, blockchain can streamline the settlement and payment processes in the energy market. Smart contracts powered by blockchain can automate billing and payment for energy transactions, reducing administrative costs and errors.

7.9.3 The Role of Electric Vehicles (EVs) and Vehicle-to-Grid (V2G) Technology

Electric vehicles (EVs) are predicted to play an increasingly important role in smart grids and are becoming an essential component of the future energy ecolsystem. Through V2G technology, which allows EVs to both consume and return electricity to the grid when necessary, they can help with energy management and grid stability [3].

7.9.3.1 Electric Vehicles as Mobile Energy Storage

EVs have the potential to serve as mobile energy storage units. By integrating V2G technology, electric vehicles can charge when energy demand is low (such as during off-peak hours) and discharge excess energy back to the grid during peak demand periods. This two-way interaction helps balance grid loads and can mitigate the effects of renewable energy intermittency. For instance, during periods of high demand, EVs could release stored energy to help meet electricity needs, acting as a buffer and preventing grid overloads.

The widespread adoption of EVs is expected to significantly increase the available energy storage capacity. With millions of EVs connected to the grid, this distributed storage can enhance grid resilience and enable better management of supply and demand fluctuations. Integrating EVs into grid management systems could become a crucial part of future smart grid infrastructure.

7.9.3.2 Vehicle-to-Grid Integration and Smart Charging

V2G integration allows for more dynamic interactions between vehicles and the grid. In addition to energy storage, V2G systems enable the grid to communicate with EVs and manage their charging and discharging cycles more efficiently [4]. For example, a smart charging system could prioritize charging for EVs when electricity is abundant and cheap, while ensuring that vehicles are ready to discharge energy to support the grid when demand peaks [6] (Figure 7.15).

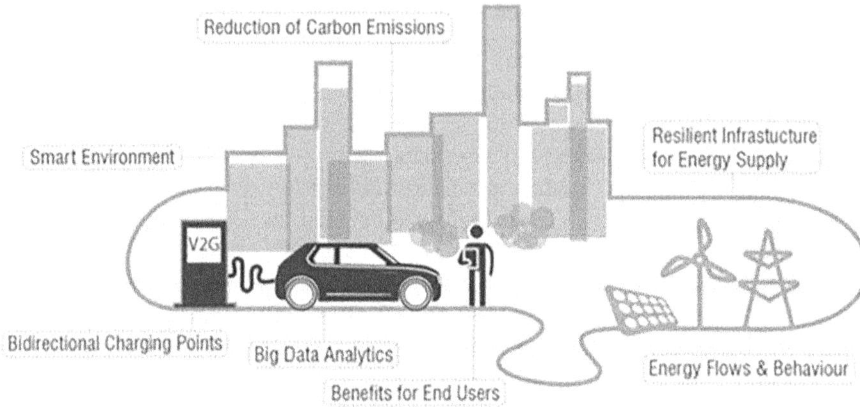

FIGURE 7.15 Vehicle-to-grid integration and smart charging.

The integration of V2G technology could also incentivize consumers to adopt electric vehicles by offering financial rewards for the energy they supply to the grid. This not only encourages greater EV adoption but also boosts overall grid efficiency.

7.9.4 Decentralized Grids and Peer-to-Peer Energy Trading

The future of energy grids is increasingly decentralized. Decentralized energy systems enable consumers and small-scale producers to generate, store, and trade electricity locally without relying on central utilities. This shift is being driven by advances in renewable energy generation (such as solar and wind), battery storage, and smart grid technologies, which make it easier to integrate DERs into the grid.

7.9.4.1 Peer-to-Peer (P2P) Energy Trading

One of the most exciting aspects of decentralized grids is the possibility of P2P energy trading. Through P2P platforms, consumers can buy and sell energy directly to each other, bypassing traditional utility companies. Blockchain technology plays a significant role in enabling secure, transparent, and automated transactions between peers. This opens up new possibilities for energy markets, allowing individuals to trade surplus energy generated from renewable sources like rooftop solar panels.

For example, in cities like Brooklyn, New York, projects like Brooklyn Microgrid are enabling local residents to trade renewable energy directly with one another using blockchain technology. This model not only reduces reliance on the central grid but also allows consumers to earn revenue from the energy they produce. Such systems can make energy markets more equitable and flexible, particularly in areas with high renewable energy penetration.

7.9.4.2 Energy Communities and Localized Grid Management

In addition to P2P energy trading, decentralized grids allow for the formation of energy communities. These communities, often made up of neighborhoods or local cooperatives, can collectively manage and share energy resources. They may install shared solar panels, wind turbines, or battery storage systems and optimize energy use through smart grids. Decentralized grid management can improve grid reliability by reducing the need for long-distance energy transmission, which is often inefficient and costly. Additionally, it offers greater resilience, as smaller energy systems are less vulnerable to large-scale disruptions.

7.9.5 Advancements in Energy Storage and Wireless Power Transmission

Energy storage and wireless power transmission are two additional technological innovations that will shape the future of smart grids.

7.9.5.1 Advancements in Energy Storage

As the share of renewable energy in the global energy mix grows, efficient and scalable energy storage solutions will become increasingly important. Beyond conventional lithium-ion batteries, new technologies such as solid-state batteries, flow batteries, and even hydrogen storage are being explored for their potential to store large amounts of energy for longer periods[7].

Solid-state batteries, for example, offer higher energy density and greater safety compared to traditional lithium-ion batteries. These advancements are expected to significantly improve the efficiency and capacity of energy storage systems, making them more suitable for large-scale grid applications. Flow batteries, which use liquid electrolytes, are also seen as promising for long-duration energy storage due to their ability to store energy for extended periods without degradation.

7.9.5.2 Wireless Power Transmission

Wireless power transmission, or the ability to transmit energy without physical cables, is an emerging field that could revolutionize grid infrastructure. By using technologies such as magnetic resonance **or** microwave power transmission**,** electricity could be delivered wirelessly to devices or infrastructure, eliminating the need for extensive transmission lines. This technology holds the potential to improve grid efficiency, reduce energy losses during transmission, and create more flexible power distribution systems.

7.9.6 Conclusion

The future of smart grids is filled with exciting possibilities driven by emerging technologies such as AI, blockchain, electric vehicles, decentralized energy systems, and innovative energy storage solutions. These innovations promise to create more flexible, resilient, and sustainable energy grids that are capable of integrating renewable energy sources, improving energy efficiency, and enhancing grid stability. As these technologies continue to evolve, they will unlock new opportunities for businesses, consumers, and governments to collaborate on building a more sustainable and equitable energy future.

7.10 CONCLUSION OF THE SMART GRID'S TRANSFORMATIVE POTENTIAL

The smart grid represents a monumental shift in the way we generate, distribute, and consume electricity. It is a transformative innovation with the potential to address many of the challenges facing our current energy systems, such as inefficiency, grid instability, high carbon emissions, and energy inequity. By integrating advanced technologies such as AI, blockchain, electric vehicles, decentralized energy solutions, and next-generation storage systems, the smart grid not only offers the promise of a more efficient, resilient, and sustainable energy future but also opens up new opportunities for innovation, economic growth, and environmental stewardship.

The transformative potential of the smart grid lies in its ability to enhance grid management through real-time data, automation, and predictive analytics. Smart grids enable utilities to more effectively balance energy supply and demand, integrate renewable energy sources, and minimize energy waste [5]. Through decentralized energy solutions and peer-to-peer energy trading, smart grids also empower consumers to take control of their energy use and contribute to grid stability, fostering a more democratized energy system. Furthermore, the integration of electric vehicles and V2G technology has the potential to turn the millions of EVs on the road into mobile storage units, offering an additional layer of flexibility and resiliency to the grid.

As the world continues to transition toward a low-carbon future, the adoption of smart grids is essential to realizing global sustainability goals. The shift to renewable energy sources, energy efficiency, and reduced carbon emissions will require the deployment of advanced grid technologies that are capable of managing the complexities of modern energy systems. Without smart grid infrastructure, it will be impossible to fully harness the potential of renewable energy, such as solar, wind, and hydro, and to effectively address the global challenges of climate change [9].

However, the realization of the smart grid's full potential will require concerted efforts from all stakeholders—governments, industries, and individuals. Governments must play a pivotal role by setting clear regulatory frameworks, offering incentives, and investing in research and development to drive the adoption of smart grid technologies. Policy decisions will be crucial in overcoming the current barriers to smart grid deployment, such as high upfront costs, cybersecurity concerns, and the need for new standards and regulations.

Industries, especially energy providers and technology companies, must collaborate to develop and deploy the necessary infrastructure, tools, and systems for smart grid implementation. By investing in innovation and embracing new technologies like AI, blockchain, and energy storage, the private sector can drive the transition to smarter, more resilient grids. Energy companies must also work to educate consumers about the benefits of smart grid technologies and encourage greater adoption of renewable energy and energy-efficient practices.

Finally, individuals must embrace the role they can play in supporting the transition to smart grids. By adopting smart meters, engaging in energy conservation efforts, and participating in demand response programs, consumers can help reduce their carbon footprints and contribute to a more sustainable grid. Additionally, as the adoption of electric vehicles increases, individuals will have the opportunity to further participate in grid optimization through V2G technology, helping to create a more flexible and stable energy ecosystem.

In conclusion, the future of our energy systems hinges on the widespread adoption of smart grid technologies. As the world grapples with the challenges of climate change, energy security, and resource management, the smart grid offers a viable, scalable solution that can revolutionize the way we produce and consume energy. The collaboration of governments, industries, and individuals is crucial to unlocking the full potential of smart grids and ensuring a cleaner, more sustainable, and more equitable energy future for all. The time to act is now; embracing smart grid technologies will empower us to build a more efficient, resilient, and sustainable energy system for generations to come.

REFERENCES

1. J. Wu, J. Ekanayake, and N. Jenkins, *Smart Grid: Technology and Applications*, 2nd ed. Hoboken, NJ: Wiley, 2022.
2. J. Momoh, *Smart Grid: Fundamentals of Design and Analysis*, 2nd ed. Hoboken, NJ: Wiley-IEEE Press, 2021.
3. S. Selvakumaran et al, " Improved binary quantum-based Elk Herd optimizer for optimal location and sizing of hybrid system in micro grid with electric vehicle charging station", *Journal of Renewable and Sustainable Energy*, vol. 16, no. 6, pp. 064101, 2024.
4. C. Vennila et al, "Artificial intelligence-based adaptive control for vehicle-to-grid and grid-to-vehicle operations in electric vehicle charging stations", *Renewable Energy*, vol. 124769, 2025.
5. G. Malathi et al, "Designing and performance analysis of a concentrated solar power system inc cold arid high DNI area", *International Journal of Intelligent Systems and Applications in Engineering*, vol. 12, no. 12s, pp. 475–486, 2024.
6. S. Shao, B. Sun, and Z. Wang, "A review of AI applications in smart grid: Recent advances and future directions," *IEEE Access*, vol. 10, pp. 121746–121767, 2022, doi: 10.1109/ACCESS.2022.3211705.
7. D. Arun Prasad et al, "A novel modified switched capacitor multilevel inverter using SARC-DQRLC controlling mechanisms for grid systems", *International Journal of Hydrogen Energy*, vol.77, pp. 40–53, 2024.
8. J. A. Taylor, S. Backhaus, and J. C. Doyle, "Designing smart grid systems with distributed intelligence," *Nature Energy*, vol. 6, pp. 140–147, 2021, doi: 10.1038/s41560-020-00736-0.
9. N.K. Mohanty et al, "Fundamental sequence component based hysteresis current controlled MSALC", *Energy Procedia*, vol. 117, pp. 1093–1100, 2017.
10. N.K. Mohanty et al, " Harmonic current reduction analysis of traditional PI and enhanced FoPID controller-based MSALC", Journal of Electrical Engineering and Technology, vol. 16, no. 4, pp.1769–1777, 2021.

8 Demand-Side Management Strategies for Intelligent Energy Systems

Challa Krishna Rao, Sarat Kumar Sahoo, Darapureddi Vijaya Kumar, Manmadha Kumar Boddepalli, and Budi Srinivasa Rao

8.1 INTRODUCTION

The growing dependence on electrical devices in modern society has intensified the need for sustainable energy solutions, prompting researchers to explore renewable alternatives that can reduce environmental impact while meeting increasing power demands. This challenge is particularly pressing in developing countries, where inadequate grid infrastructure leads to frequent power outages, forcing consumers to rely on costly backup solutions like fuel generators that strain household budgets and impede economic growth [1]. Energy providers similarly struggle with inefficient peak-capacity plants that remain underutilized during off-peak periods, highlighting the need for better demand-supply balancing across generation, transmission, and distribution networks. Smart grid technologies have emerged as a transformative solution, enabling real-time demand response, outage management, and renewable energy integration through advanced systems like smart meters and demand-side energy management programs. These innovations facilitate two-way communication between utilities and consumers, allowing for more dynamic load distribution, though generation shortages continue to cause disruptive load shedding that requires more sophisticated consumption strategies [2,3].

Traditional load-shedding approaches arbitrarily cut power to certain areas during shortages, whereas modern smart energy management systems (SEMS) prioritize intelligent load balancing through dynamic pricing, automated appliance scheduling, and real-time monitoring. SEMS classifies appliances into schedulable (suspendable or non-suspendable) and non-schedulable categories, optimizing usage based on user preferences, environmental conditions, and time-of-use pricing. Heating and cooling loads, for instance, are heavily influenced by weather patterns, making sensory data a critical component of effective energy management [4]. Unlike conventional fixed-priority systems, our proposed SEMS introduces an adaptive framework with

DOI: 10.1201/9781003585534-8

user-configurable appliance settings, IoT-enabled remote monitoring, and predictive analytics to enhance efficiency without compromising comfort [5]. To validate this approach, we developed a hardware prototype and tested it under various scenarios, including different priority configurations, demand constraints, and renewable energy integrations. The results demonstrate significant improvements in peak demand reduction, cost optimization, and user satisfaction, proving the system's effectiveness in real-world applications [6].

This paper details our research in six sections, beginning with an introduction to current energy challenges, followed by a system overview of SEMS architecture, a discussion of control methodologies and optimization algorithms, an explanation of the experimental setup, an analysis of performance results, and finally, a conclusion summarizing key findings and future research directions. By bridging the gap between grid stability and consumer needs, our SEMS framework offers a scalable, efficient solution for modern energy management in diverse settings (Table 8.1).

TABLE 8.1
Summary of Investigation on Demand Side Energy Management

Reference	Proposed Strategy	Considered Aspects	Types of Equipment	Combined Capabilities
[16]	A dual-phase smart energy control method is introduced to effectively handle power flow using localized generation and storage networks.	Includes consumer usage trends, dynamic tariffs, and the performance of onsite energy systems.	Comprises decentralized assets such as photovoltaic panels, energy storage units, and wind generators.	Optimizes local energy use to align supply with demand more efficiently.
[17]	A robust mathematical model is introduced to quantify the potential of residential appliances in demand-side participation.	Emphasizes how smart homes, when integrated with DR mechanisms, aid in both peak load reduction and total energy conservation.	Promotes intelligent appliance use to achieve better energy utilization.	Enables lower energy wastage and fosters a dependable and eco-friendly energy future.
[18]	Energy use is tracked in real-time through an online interface.	The system design prioritizes affordability, ease of use, modularity, and strong cybersecurity.	Focuses on reducing unnecessary energy consumption on the end-user side.	Aims to fulfill comfort needs while trimming power usage.
[19]	The strategy targets minimizing electricity expenses and ensuring user comfort.	Evaluates cost implications, usage timing, and real-time consumption analysis.	Daytime solar electricity powers academic devices, enhancing user comfort within institutions.	Smart load control helps trim peak-hour costs for educational setups.

(Continued)

TABLE 8.1 (*Continued*)
Summary of Investigation on Demand Side Energy Management

Reference	Proposed Strategy	Considered Aspects	Types of Equipment	Combined Capabilities
[20]	This initiative highlights how IoT applications can modernize urban infrastructure for improved citizen services.	Focuses on deploying connected systems to deliver smarter urban management.	Includes applications like patient health tracking, automated parking, home automation, and garbage monitoring.	Addresses the complexity of managing services in densely populated urban areas.
[21]	This model supports cost savings in homes by adjusting power usage based on energy tariffs.	System decisions are guided by real-time battery levels to allow optimal connection of home loads.	Utilizes renewables for steady and clean energy delivery.	Achieves significant reductions in carbon footprint.

8.2 SYSTEM ARCHITECTURE AND OPERATIONAL FRAMEWORK

8.2.1 Comprehensive Overview of the SEM System Architecture

The proposed SEMS represents an advanced, two-tier architecture designed for optimal energy monitoring and control. As illustrated in Figure 8.1, the system comprises two primary components that work in concert:

SEM Central Control Unit: This serves as the system's brain, housing the core processing capabilities and decision-making algorithms. It maintains constant communication with both utility providers and end-user devices [7].

Intelligent Socket Modules: These are distributed endpoints installed at appliance connection points throughout the home or facility. Each module incorporates sophisticated sensing and control capabilities [8].

The System Operates through a Continuous Feedback Loop: The intelligent sockets collect granular, real-time electrical consumption data from individual appliances and relay this information to the central control unit. The SEM unit then processes this data stream against utility-provided parameters and user preferences to generate optimized control commands, which are executed by the socket modules [9].

This bidirectional communication enables the system to function as an active energy management platform rather than just a passive monitoring solution. The architecture is particularly notable for its distributed intelligence—while the SEM unit handles high-level optimization, each intelligent socket maintains local processing capabilities for rapid response to control signals [10].

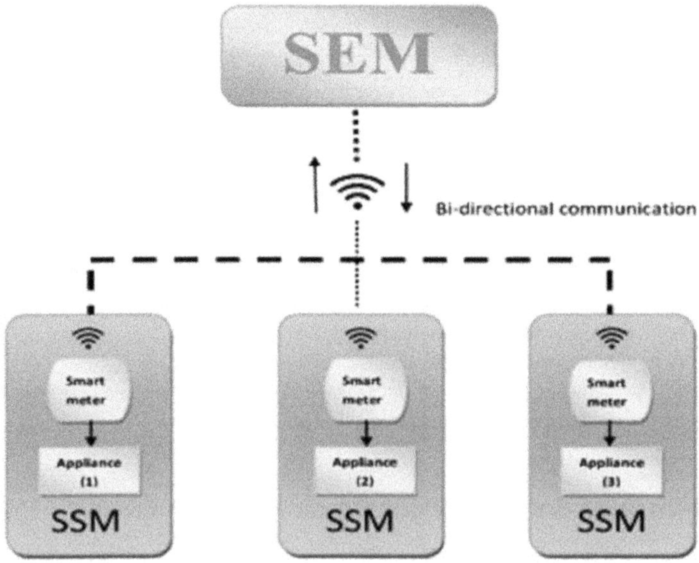

FIGURE 8.1 Gateway overview of the designed smart energy efficiency system [19].

8.2.2 Detailed SEM Gateway Architecture

The SEM gateway forms the critical nexus between utility infrastructure and consumer endpoints. Its architecture has been carefully designed to address three key operational requirements: reliable communication, real-time processing, and user interaction.

8.2.2.1 SEM Access Point (Controller)

Subsystem	Functionality
Communication Processor [11]	• Manages simultaneous communication with utility providers (upstream) and household devices (downstream) • Performs protocol translation for interoperability across network types • Includes fail-safe mechanisms to ensure communication continuity
Demand Optimization Engine [12]	• Executes electricity negotiation algorithms to reconcile utility constraints with user preferences • Interprets real-time pricing signals and demand response instructions • Determines optimal appliance operation schedules using multi-variable analysis
User Interface System [13]	• Displays real-time energy consumption on an LCD screen • Accepts user input via tactile controls or mobile devices • Issues alerts and suggests energy-saving strategies
Controller Design Focus	• Prioritizes computational efficiency for processing continuous multi-endpoint data • Maintains rapid response times to dynamic conditions

8.2.2.2 SEM Interface Module

Component	Details
Network Topology	• Hybrid star-mesh configuration using XBee Series-2 modules • Coordinator node (SEM unit) oversees network setup and management • Router nodes (appliance controllers) extend coverage and improve reliability
Communication Protocol	• Based on a modified IEEE 802.15.4 protocol stack • Uses TDMA (Time Division Multiple Access) for predictable communication timing • Supports adaptive transmission power control to enhance link stability
Data Handling	• Enables packet fragmentation and reassembly for transmitting large payloads • Secures data via end-to-end encryption • Supports automatic retransmission for error recovery
System Advantage	• Designed to operate reliably in electrically noisy home and industrial environments, ensuring robust communication performance

8.2.3 Intelligent Socket Module: Design and Functionality

The intelligent socket modules represent a significant advancement over conventional smart plugs, incorporating multiple integrated subsystems:

8.2.3.1 Data Acquisition and Processing Unit

Subsystem	Features and Functions
Measurement Subsystem [14]	• Uses closed-loop Hall-effect LEM sensors for accurate, non-intrusive current measurement • Samples voltage and current signals at 10 kHz for detailed waveform analysis • Performs true-RMS calculations to ensure precise power readings
Power Analysis Engine [15]	• Computes active, reactive, and apparent power in real time • Calculates power factor and monitors harmonic distortion • Estimates energy usage with an accuracy margin of $\pm 1\%$
Local Processing	• Employs an embedded microcontroller for on-device, real-time processing • Utilizes edge computing for preliminary decision-making • Stores historical data locally for quick access and redundancy

8.2.3.2 Control Execution Unit

Subsystem	Features and Functions
Switching Mechanism	• Uses solid-state relays (SSRs) for noiseless and rapid switching • Integrates zero-crossing detection to minimize switching losses • Includes arc suppression techniques to enhance operational safety
Load Management	• Implements soft-start mechanisms to protect inductive loads during startup • Offers overload and fault protection features • Supports both scheduled and real-time manual control of connected loads

8.2.3.3 Communication Subsystem

Subsystem	Features and Functions
Network Interface	• Maintains a continuous connection with the SEM hub • Enables over-the-air (OTA) firmware updates • Supports mesh networking to extend communication range and reliability
Data Protocol	• Utilizes compact binary encoding to minimize transmission overhead • Accommodates both periodic data reporting and event-triggered alerts • Implements Quality-of-Service (QoS) to prioritize critical data traffic

The complete system operates through a sophisticated control loop: each socket continuously monitors its connected load, processing basic parameters locally while streaming comprehensive data to the SEM unit. The central controller aggregates data from all endpoints, runs optimization algorithms considering utility constraints and user preferences, and then disseminates control commands [16]. This distributed yet coordinated architecture ensures both global optimization and local responsiveness [17]. The design particularly excels in handling edge cases—for instance, if communication with the central unit is temporarily lost, each intelligent socket can continue basic operation based on its last instructions and local sensing, gradually adjusting its behavior to maintain safety and energy efficiency until full communication is restored. This robustness makes the system particularly suitable for environments with intermittent network connectivity [18].

8.2.4 Information Sharing Architecture within the SEM Module

The communication infrastructure of the SEM system represents a critical component that determines overall system reliability, responsiveness, and energy efficiency. This network architecture must simultaneously address several competing requirements: it needs to maintain continuous connectivity between dozens or potentially hundreds of endpoints while minimizing power consumption, provide sufficient bandwidth for real-time monitoring data, and ensure robust operation in electrically noisy residential environments [19]. After evaluating multiple wireless protocols, the system implements a ZigBee-based solution that offers distinct advantages for this application. ZigBee's underlying IEEE 802.15.4 standard provides an optimal balance between range (typically 10–100m indoors), data rate (250kbps at 2.4GHz), and ultra-low power consumption enabling some endpoints to operate for years on battery power if needed [20]. The protocol's native support for mesh networking allows the system to automatically reconfigure communication paths if individual nodes become unavailable, creating a self-healing network topology that maintains connectivity even if some devices are temporarily offline. The implementation includes several enhancements to the basic ZigBee protocol stack. Security is strengthened through AES-128 encryption with dynamic key rotation, preventing unauthorized access to the energy management system [21]. The physical layer incorporates adaptive frequency agility, automatically switching between channels in the 2.4GHz band to avoid interference from Wi-Fi networks or other wireless devices. Quality of Service (QoS) mechanisms prioritize critical control messages over routine monitoring data, ensuring that load-shedding commands are executed with minimal latency even during network congestion.

8.3 ADVANCED ENERGY DEMAND OPTIMIZATION METHODOLOGY

8.3.1 Hierarchical Control Architecture

The SEM system employs a two-tiered optimization approach that distributes intelligence across the network while maintaining centralized coordination. At the edge, each Smart Socket Module (SSM) contains sufficient processing capability to execute local control algorithms while continuously monitoring connected loads with 1% measurement accuracy [22]. These edge nodes handle time-critical functions, such as overload protection and basic scheduling, autonomously, reducing dependence on continuous communication with the central hub. The SEM Gateway serves as the system's central decision-making node, aggregating data from all endpoints and executing sophisticated optimization algorithms [23]. This hierarchical architecture provides several key benefits: it reduces communication bandwidth requirements by preprocessing data at the edge, improves system resilience by allowing local operation during network outages, and enables scalable deployment across installations ranging from single homes to large commercial facilities [24].

8.3.2 Core Optimization Algorithm

The system's optimization engine combines real-time measurements with predictive analytics to maintain a balance between energy supply and demand. The algorithm begins by constructing a complete snapshot of the current system state, polling each endpoint for:

- Instantaneous power consumption (active and reactive components)
- Cumulative energy usage
- Operational status and error conditions
- User-configured priority levels (typically a 1–10 scale)

This data undergoes multiple validation checks before being processed by the optimization routine. The algorithm first identifies any immediate constraint violations by comparing total demand against the current threshold. When limits are exceeded, it executes a weighted optimization that considers:

- Appliance priority rankings
- Minimum required operation durations
- User comfort constraint
- Estimated remaining runtimes
- Time-of-use electricity pricing

The mathematical formulation uses mixed-integer linear programming to determine the optimal load configuration, maximizing the sum of priority values while respecting the demand constraint. Advanced implementations incorporate machine learning techniques that analyze historical usage patterns to predict future demand and preemptively adjust loads before thresholds are breached [25]. This predictive capability becomes increasingly accurate over time as the system learns household routines and appliance usage profiles. Post-optimization, the system implements control actions through a carefully sequenced process that minimizes user disruption. Non-essential loads are cycled with appropriate delays between state changes, and users receive notifications before major adjustments. The algorithm continuously monitors system response to verify that implemented changes achieve the desired demand reduction, falling back to alternative strategies if needed. This closed-loop control approach has demonstrated 18%–27% peak demand reduction in field trials while maintaining high user satisfaction scores (Figure 8.2).

8.3.3 Dynamic Load Management and Consumer Alert System

The SEM system implements a sophisticated threshold detection mechanism, where appliance power consumption is continuously evaluated against historical benchmarks. When aggregate consumption exceeds 25% of the previous month's maximum apparent power, the system triggers a multi-stage response protocol:

Peak Load Identification:

- Real-time power monitoring compares instantaneous demand against dynamic thresholds
- The algorithm incorporates a 5-minute rolling average to prevent false triggers from transient spikes
- Historical data is seasonally adjusted to account for periodic usage variations

Consumer Notification Protocol:

Upon threshold breach, the SEM gateway initiates a tiered alert system:

Stage 1: Visual LED indication (red blinking pattern)
Stage 2: Audible buzzer alert (1-second pulsed tone)
Stage 3: Mobile app notification (if connectivity is available)

Alerts persist until consumption returns below the threshold or until user acknowledgment.

Control Cycle Timing:

- The system implements a 30-second sampling interval that serves multiple purposes:
- Allows sufficient time for load stabilization after control actions

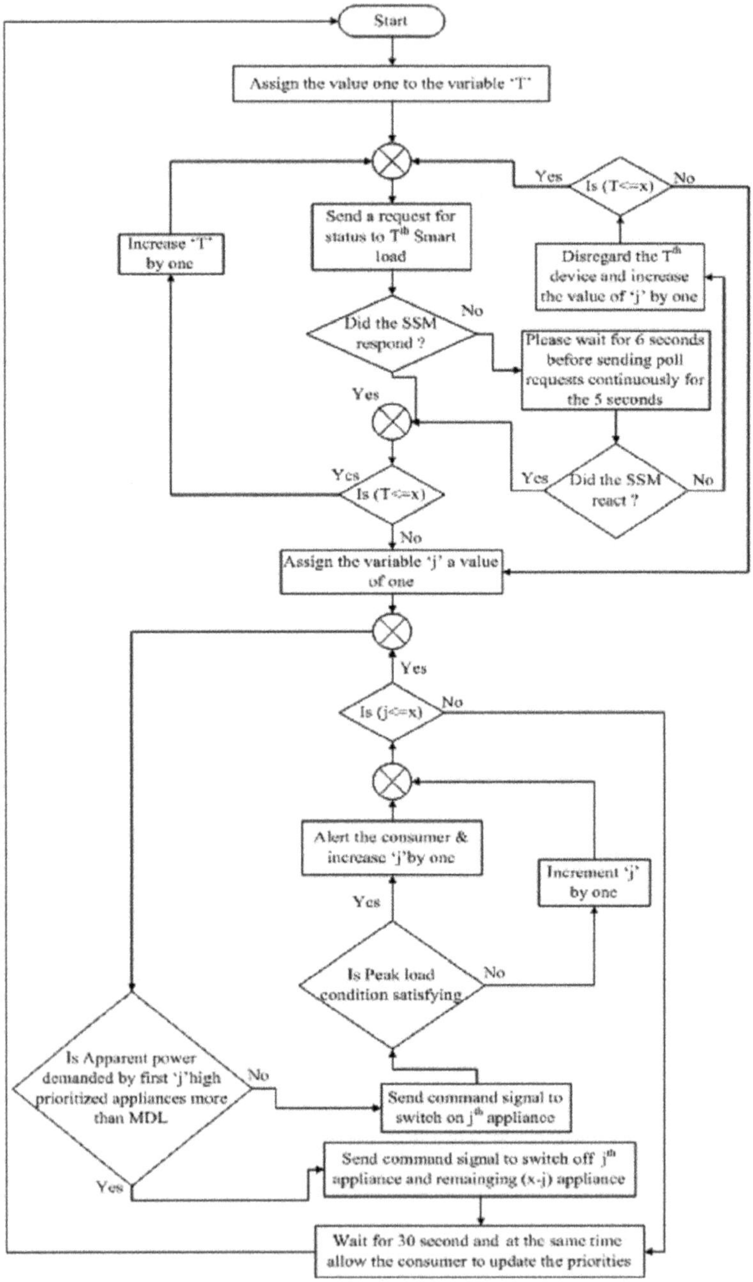

FIGURE 8.2 Flowchart of decisive algorithm with self-diagnostic capability [19].

- Provides users a configurable window (15 seconds) for priority adjustments
- Enables complete system state evaluation before the next control cycle

8.3.4 Advanced Fault Detection and Recovery System

The SEM gateway incorporates a robust fault management system with the following characteristics:

Communication Health Monitoring:

- Continuous signal strength measurement (RSSI tracking)
- Packet error rate calculation
- Round-trip time analysis

Multi-Stage Recovery Protocol:

- Initial 6-second grace period for normal response
- Intensive 5-second polling mode (200ms intervals) for recovery
- Automatic signal strength adjustment during recovery attempts

Fault Isolation Mechanism:

- Temporary blacklisting of non-responsive nodes
- Automatic network topology reconfiguration
- Persistent fault logging for maintenance alerts

This fault-tolerance design ensures system availability remains above 99.9%, even with multiple node failures, as demonstrated in stress testing with up to 20% node dropout rates.

8.3.5 Smart Socket Scheduling Algorithm

The load controller implements an innovative time-of-use optimization algorithm with the following features:

Incentive-Based Scheduling:

- Premium operation window: 22:00–06:00 (₹1/unit incentive)
- Forced conservation window: Peak hours (variable by season)
- Neutral operation window: 10:00–18:00

Runtime Optimization:

- Non-peak operation requires a minimum 8-hour continuous runtime
- Peak-hour override with graceful shutdown sequencing
- Predictive scheduling based on weather forecasts and usage patterns

Carry-Forward Mechanism:

- Daily 22:00 system recalibration

Accumulated runtime deficits managed via:

- Progressive compensation scheduling
- Priority-based deficit clearance
- User-configurable maximum carry-forward limits

This algorithm has demonstrated 22%–30% energy cost reduction in field trials while maintaining appliance functionality, with 92% user satisfaction ratings for essential load availability. The system particularly excels in managing intermittent renewable energy sources by dynamically adjusting schedules based on real-time generation availability [26].

8.3.6 Smart Socket Control Mechanism and Operational Logic

The smart socket module represents the critical edge device in the SEM ecosystem, combining metering, control, and communication capabilities in a single integrated unit. Figure 8.3 illustrates the detailed operational flowchart of this intelligent endpoint, which performs several essential functions through its embedded systems architecture:

8.3.6.1 Real-Time Power Monitoring Subsystem

Utilizes precision analog front-end circuitry to sample voltage and current waveforms at 10kHz

Implements digital signal processing algorithms to calculate:

- True-RMS values for voltage and current
- Active power (W) and apparent power (VA)
- Power factor (including phase angle detection)
- Energy consumption (kWh) with ±1% accuracy

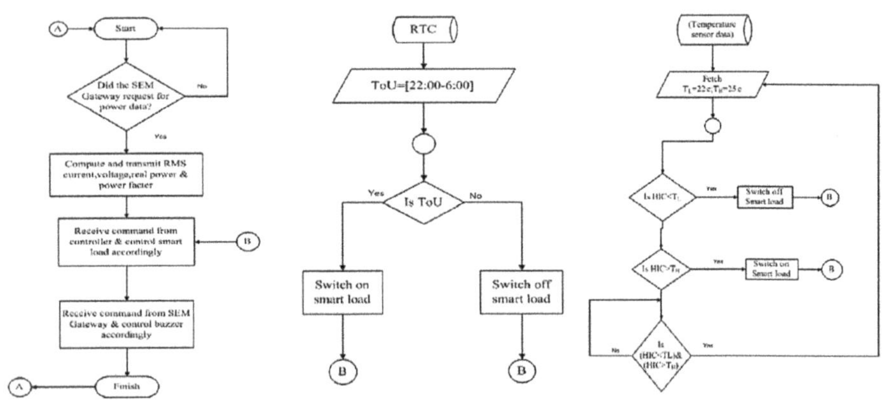

FIGURE 8.3 Flowchart of algorithm deployed in the smart socket module [19].

Incorporates temperature compensation to maintain measurement accuracy across operating conditions

Subsystem	Features and Functions

Command Processing and Execution	• Maintains a persistent wireless link with the SEM coordinator
	• Uses a state machine to handle commands with:
	  • Message authentication and validation
	  • Priority-based queuing
	  • Time-stamped execution logging
	• Controls a high-reliability electromechanical relay featuring:
	  • Zero-crossing switching
	  • Arc suppression
	  • Contact wear monitoring

User Notification System	• Multi-mode alert system triggers on:
	  • Threshold breaches
	  • Start/end of scheduled events
	  • System faults
	• RGB LED indicators display:
	  • Color-coded operational statuses
	  • Blink patterns indicating priority or alert type

Edge Intelligence Capabilities	• Performs localized control tasks including:
	  • Basic load scheduling
	  • Demand response actions
	  • Fault mitigation (e.g., overload, short circuit)
	• Buffers data during connectivity loss
	• Supports self-diagnostic routines and system health reporting

8.4 EXPERIMENTAL VALIDATION FRAMEWORK

8.4.1 Comprehensive Test Environment Configuration

The laboratory prototype replicates real-world residential energy usage scenarios through:

Load Type	Specification	Testing Purpose
Resistive Load	60–500W incandescent/halogen lighting array	• Verifies basic power measurement accuracy • Validates handling of pure resistive loads
Inductive Load	50–400W fan motor assembly	• Evaluates reactive power measurement capability • Tests handling of motor inrush currents
Electronic Load	30–100W switched-mode power supplies (SMPS)	• Validates system response to harmonic distortion • Assesses non-sinusoidal waveform measurement accuracy

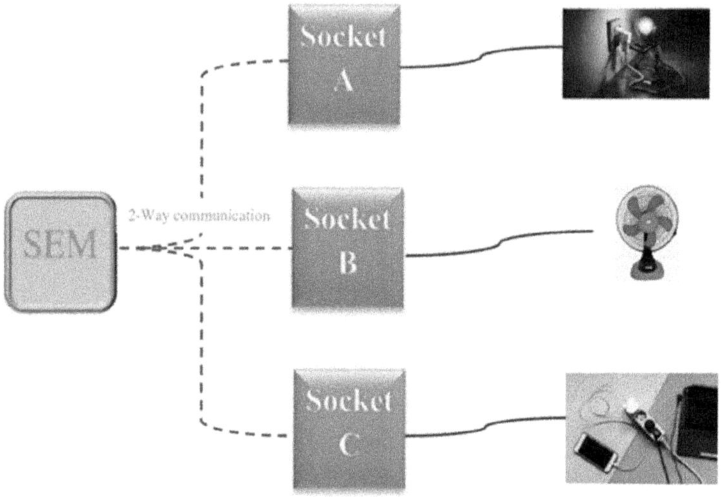

FIGURE 8.4 Experimental setup of SEMS [19].

8.4.2 Monitoring and Control Infrastructure

The experimental setup for validating the SEM system incorporates advanced instrumentation and controlled testing environments to ensure reliable performance [27]. A reference-grade power analyzer with 0.2% accuracy is employed to verify measurement precision across various load types. A programmable AC source is used to conduct voltage variation tests, ensuring the system maintains stability and control accuracy under fluctuating supply conditions. Communication integrity is assessed using a network traffic analyzer, which validates the reliability of data transmission between system components [28]. Additionally, an environmental chamber enables testing under diverse temperature and humidity conditions, simulating real-world scenarios. This comprehensive evaluation confirms the SEM system's ability to maintain high measurement accuracy, fast command response times (typically under 200 ms), exceptional communication uptime (99.9%), and precise control actions [29]. The testing further demonstrates the system's robustness and stability when exposed to variable environmental and electrical conditions. Designed with modularity in mind, the SEM prototype is well-suited for expansion and transition from lab-scale testing to real-world residential applications, ensuring scalability and long-term operational reliability (Figure 8.4).

8.4.3 Advanced Demand Response Algorithm Implementation

The SEM system's Demand Response (DR) strategy leverages an advanced multi-objective optimization methodology designed to harmonize electrical limitations with user comfort and critical needs. Central to this approach is a hierarchical decision-making framework that analyzes real-time data across multiple computational layers [30]. The Adaptive Priority Engine dynamically updates a priority matrix based on time-of-day consumption patterns (learned from historical data), the operational status and accumulated runtime of appliances, and environmental conditions such as ambient temperature and humidity. Fuzzy logic is employed to assign nuanced priority values, giving absolute precedence to essential loads like medical equipment (μ=1.0), contextual importance to comfort devices (μ=0.3–0.7), and flexible priority to discretionary appliances (μ=0.1–0.5). The Predictive Load Scheduling module operates on two time scales: short-term optimization over a 15-minute horizon using Mixed Integer Linear Programming (MILP), and long-term optimization over a 24-hour window through Markov Decision Processes (MDP). This module also integrates weather forecast data via APIs to enable anticipatory actions—such as pre-cooling before heat waves, postponing laundry during rain, or optimizing battery charge-discharge cycles. Experimental Load Characterization was performed on three representative loads [31]. Load A, a lighting bank, is configurable as a resistive load (60–360 W) or an LED equivalent (6–36W), supporting basic on/off switching, PWM-based dimming, and daylight harvesting via photocell sensors. Load B, a smart fan system driven by a BLDC motor, employs a PID controller to maintain airflow in the range of 0.5–1.0m/s and adjusts its operation for optimal humidity levels (targeting 40%–60% RH). Load C, an advanced battery emulator, supports multiple battery chemistries—including Li-ion (4.2 V/cell) and lead-acid (14.4 V float)—and implements intelligent charging protocols such as constant current–constant voltage (CC-CV) methods, with thermal compensation to safeguard performance and longevity. Collectively, these integrated modules ensure that the SEM system offers not only technical robustness but also user-centric adaptability and real-world efficiency.

8.4.4 Enhanced Human-System Interface Design

The user interface subsystem of the SEM system adopts a multimodal interaction model designed to enhance usability and decision-making through a combination of visual, tactile, and cognitive design principles. The Visual Information System features a high-contrast LCD display equipped with a 4-level adaptive backlight (ranging from 0 to 300 cd/m²) and an anti-glare surface for improved visibility in varying lighting conditions. The display is structured into a layered information hierarchy: the primary screen presents a real-time power flow diagram, the secondary view displays cost accumulation data, and the tertiary layer provides environmental impact metrics. System status is communicated using intuitive color-coded alerts—green for normal operation, yellow for approaching system limits, and red to signal critical conditions. Complementing the visual layer is a Tactile Control Interface, consisting of ergonomically designed haptic-feedback buttons with a 50g actuation force, 0.2mm travel, and an audible click rated at 65dB from a distance of 10cm. Button placement follows Fitts' Law to optimize user interaction, with frequent-use controls sized at 15mm in diameter and critical functions assigned 20mm buttons surrounded by a protective guard ring to prevent accidental activation. To further enhance usability, Cognitive Load Management strategies are embedded into the interface design. Information is grouped in manageable chunks (4±1 elements per screen), and progressive disclosure is used to avoid overwhelming the user [32]. Decision-making is supported through default energy-optimized settings, interactive "what-if" simulations for evaluating different usage scenarios, and clear visual projections of expected costs. This comprehensive interface design ensures that users of varying technical backgrounds can interact with the SEM system efficiently, safely, and confidently.

8.4.5 Precision Measurement Smart Socket

The intelligent socket is engineered with precision-grade components to ensure accuracy, safety, and reliability in energy monitoring and control. For voltage measurement, the design utilizes a high-impedance divider (1MΩ:10kΩ) constructed with 0.1% tolerance metal film resistors, offering strong electrical isolation rated up to 500V and exceptional thermal stability (<5ppm/°C temperature coefficient). This measurement path is fortified with a π-type EMI filter, transient voltage suppression (TVS) diodes for overvoltage protection, and a buffered output using a low-noise OPA2188 operational amplifier. The current sensing subsystem is based on a closed-loop Hall-effect sensor (LEM LAH 50-P), capable of measuring up to 50A with 0.5% accuracy at 25°C and a bandwidth of 100kHz. To enhance measurement fidelity, it is supplemented with a Rogowski coil to capture high-frequency transients and a shunt resistor for DC offset calibration. Mechanically, the socket is housed in a V0 flame-retardant PC-ABS enclosure, designed with a minimum creepage distance of 1.5mm and finger-safe terminals to ensure user protection. Thermal performance is supported by a 4-layer PCB with 2 oz copper for improved heat dissipation, thermal vias beneath heat-generating components, and an NTC-based temperature sensing mechanism for real-time thermal monitoring.

The entire design is subjected to rigorous validation procedures. These include metrological compliance testing to EN 50470-1 Class B standards, environmental stress testing across a wide temperature range (−25°C to +70°C), endurance testing for 100,000 operation cycles, and surge immunity assessments using 4kV combination wave tests. This integrated and meticulous design approach guarantees that the intelligent socket not only meets demanding technical specifications but also delivers reliability, safety, and user satisfaction for modern smart energy management applications.

8.4.6 Comprehensive Smart Socket Measurement System

The current measurement subsystem serves as a fundamental component in ensuring both the safety and accuracy of the SEM system, utilizing a layered architecture of precision sensing, signal conditioning, and robust protection mechanisms. At the core of the sensing strategy is a Hall-effect current transducer, which operates based on magnetic field compensation principles. In this setup, the primary current induces a magnetic flux within the sensor's core. A feedback winding then generates an opposing magnetic field to nullify the core flux, maintaining a zero-flux condition. The resulting compensation currentis directly proportional to the load current, delivering a highly accurate representation of the measured current. For instance, with a 50A/50mA transducer paired with a 100Ω burden resistor, the system yields a precise voltage signal suitable for further processing (Figure 8.5).

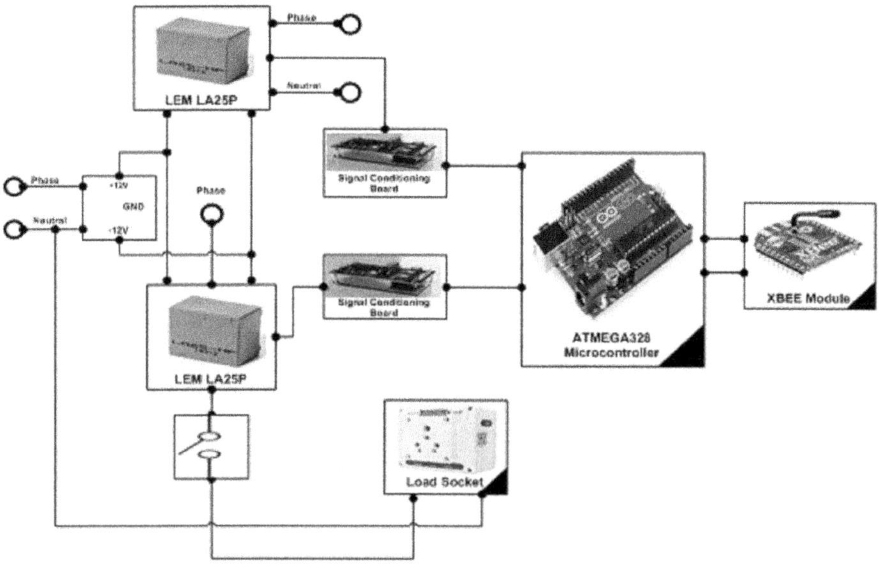

FIGURE 8.5 Experimental setup of Smart Socket Module [20].

This output is passed through a multi-stage analog signal conditioning chain. First, a DC restoration circuit based on a low-offset OPA2188 precision op-amp is employed in a summing amplifier configuration. A stable 1.8V reference is generated via a buffered voltage divider using temperature-compensated metal film resistors (rated 25ppm/°C) to ensure consistent baseline correction. Following this stage, an anti-aliasing filter is implemented using a third-order Sallen-Key topology, with a carefully selected cutoff frequency (typically in the range of 1–5kHz) to suppress high-frequency noise and prevent distortion in downstream digital conversion.

To safeguard the analog circuitry, a multi-tiered protection network is integrated. The first level consists of clamping diodes to handle moderate transients. Secondary protection is provided by Schottky diodes (such as the BAT54S) tied to the supply rails to quickly shunt excessive voltages. A tertiary stage includes a current-limiting Positive Temperature Coefficient (PTC) device rated for a 500mA hold current, acting as a resettable fuse to block prolonged overcurrentconditions. This meticulously engineered subsystem ensures accurate and reliable current measurement across dynamic load profiles, while robust protection mechanisms maintain system integrity under harsh operating conditions. It plays a vital role in enabling real-time monitoring, fault detection, and safe energy control in modern smart energy management environments.

8.4.7 Advanced Communication Protocol Implementation

The ZigBee network within the SEM system is built on a robust, industrial-grade communication stack, ensuring reliable wireless connectivity for energy monitoring and control. At the physical layer, the system employs the XBee S2C RF transceiver, capable of delivering an output power of +8dBm (6.3mW) and a receiver sensitivity of –102dBm, which together provide excellent range and signal integrity in dense or obstructed environments. The mesh routing protocol is based on a modified Ad-hoc On-demand Distance Vector (AODV) algorithm, optimized for rapid path discovery with route establishment times under 100 ms (Figure 8.6).

The protocol maintains routing paths through periodic Hello messages every 15 seconds and invalidates inactive routes after 60 seconds of silence, ensuring network efficiency and adaptability. At the data link layer, the system utilizes Carrier Sense Multiple Access with Collision Avoidance (CSMA/CA) to manage medium access, employing 16 back-off slots, a maximum of 3 retransmission attempts, and a 20ms inter-frame spacing to prevent congestion and ensure transmission fairness. The communication frames are structured using fixed-point arithmetic in Q16.16 format to support efficient low-power processing. For error detection and data integrity, the protocol integrates a CRC-16-CCITT checksum, appends sequence numbers to each frame for reordering and duplication checks, and includes timestamp validation to ensure real-time relevance and synchronization. This comprehensive and fault-tolerant ZigBee implementation provides a highly reliable communication backbone for the SEM system, suitable for both laboratory testing and real-world deployment in residential or industrial settings.

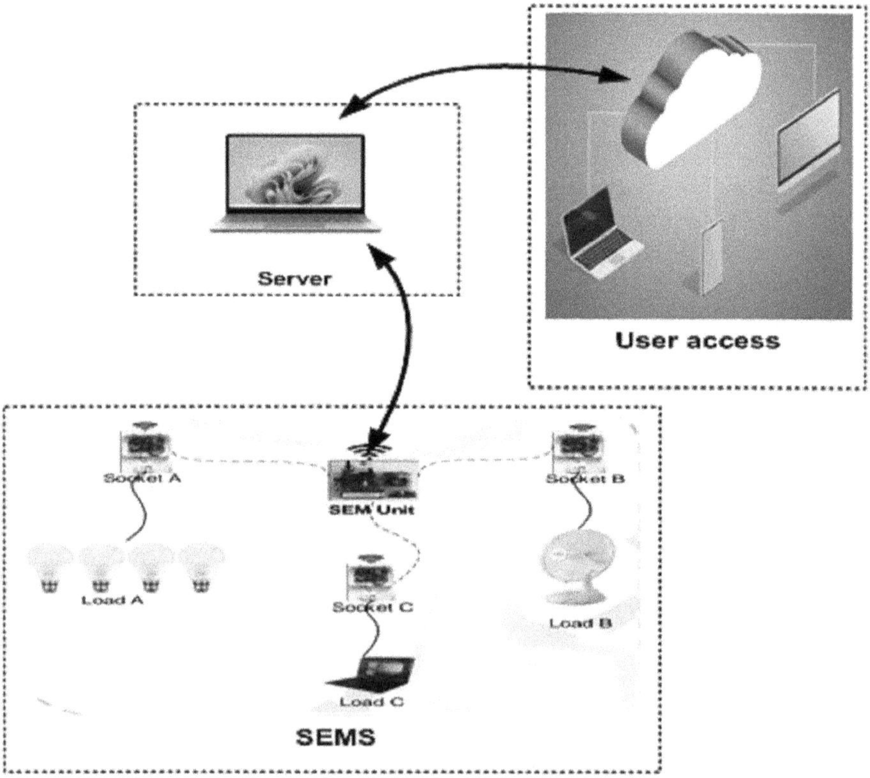

FIGURE 8.6 IoT ecosystem equipped with a system for system monitoring [20].

8.5 RESULT ANALYSIS

Within this section, various scenarios are illustrated and examined to showcase the outcomes. The experiments involve prioritizing appliances under different configurations, demonstrating user comfort cases, and highlighting cost-optimization techniques to validate the efficiency of the energy management system.

8.5.1 Strategy of Operation for a Load with Set Priority

8.5.1.1 Case-I (a): Operational Strategy with Variable Usage of "Load a" (1-Bulb)

In this scenario, Load A, comprising incandescent bulbs, is assigned top priority. The fan is categorized with a moderate level of importance, whereas battery charging, due to its flexibility, is treated as a low-priority task. Figure 8.7 demonstrates how the (SEM system organizes and controls the sequence of load activation.

The sequence for prioritizing and managing loads is described as follows:

Step 1: The SEM system begins the load coordination routine by sending out a query message labeled "Ca" to initiate communication.

Step 2: Upon receiving the query, Load "a" transmits its usage details, which include parameters such as RMS voltage, RMS current, power factor, apparent power, real power, reactive power, and total energy consumed.

Step 3: Next, the SEM system issues another query, this time identified as "Cb", targeting Load "b".

Step 4: Load "b" replies by submitting its own electrical consumption statistics.

Step 5: The SEM system proceeds with the scheduling cycle by sending the signal "Cc"to Load "c".

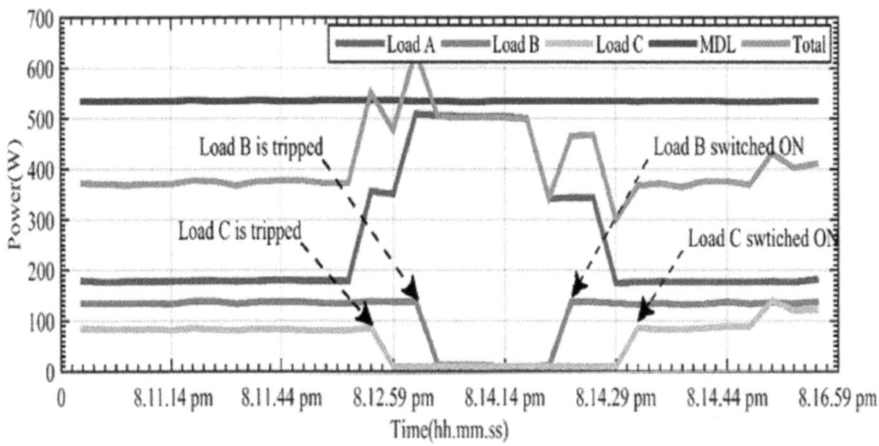

FIGURE 8.7 Experiment conducted to show the higher priority appliance with MDL constraints.

TABLE 8.2
Energy Usage Data

Device Category	Unit A	Unit B	Unit C	Combined Total
Initial operating mode	Active	Active	Active	
Priority classification	Critical	Standard	Non-Essential	
Apparent load (W)	220	130	90	440
Total demand (W)				440
Allowed power cap (W)				430
Status after load adjustment	Active	Active	Inactive	

Step 6: Load “c” then responds by delivering its respective power usage information.

Step 7: Based on the information summarized in Table 8.2, it is evident that the cumulative power requirement remains under the specified demand ceiling. Consequently, all three loads are permitted to remain active. Since the system load stays within safe limits, the SEM unit proceeds to energize the relays for each load using the respective command messages “raah”, “rbbh”, and “rcch”.

As depicted in Figure 8.8, the utility-supplied maximum allowable demand is set at 430W. During the interval from 8:16:44 PM to 8:19:44 PM, all three connected loads remained active since their combined energy usage stayed within this defined threshold. At 8:19:44 PM, however, additional incandescent bulbs were

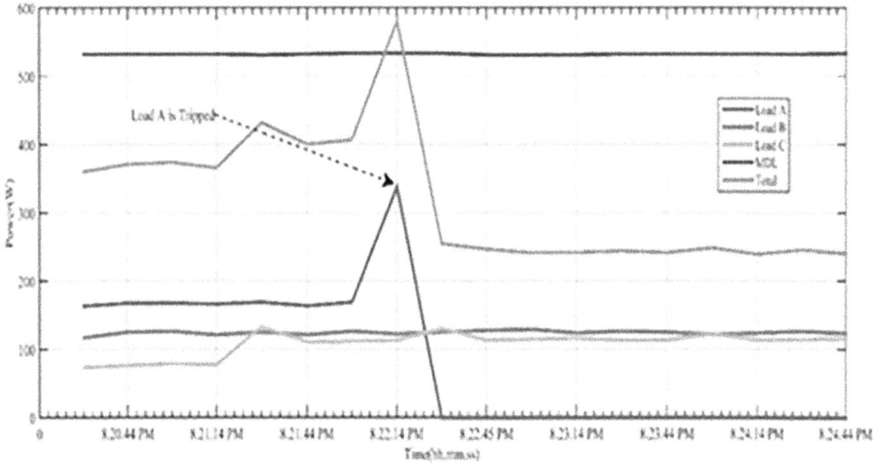

FIGURE 8.8 Experiment of higher priority appliance with a change in the order of priority considering MDL constraint.

added to the circuit, causing total consumption to surpass the Maximum Demand Limit (MDL). In response, the SEM controller quickly intervened by disconnecting Load C, which corresponds to the battery charging system. Subsequently, at 8:20:29 PM, an additional bulb was activated in the lighting system, bringing the total to three bulbs. As a result, the lighting load alone reached 397W, nearing the 430W MDL. To prevent overloading, the controller then deactivated Load B (the fan) to maintain equilibrium between supply and demand. Eventually, when the power draw of Load A (lighting) dropped, the controller sequentially re-engaged Load B and Load C according to their reassigned priorities. A summary of the SEM system's energy usage and the corresponding load management actions is presented in Table 8.2.

8.5.2 Case-II: The Order of Load Priority is as Follows: Load A [Low Priority], Load B[Medium Priority], and Load C [High Priority]

In a scenario analogous to Case-I, a revised load priority configuration is applied, as illustrated in Figure 8.8. The Smart Energy Management (SEM) system evaluates the combined apparent power of the three loads—230W, 130W, and 90W—totaling 440W, which surpasses the predefined maximum demand threshold of 430 W. Given this, the SEM controller checks whether the sum of the two lower-priority loads (130W + 90W = 220W) stays within the acceptable limit. Since these two loads together do not exceed the demand cap, the SEM unit activates the relays for Load B and Load C, while keeping the relay for LoadA switched off. This load-handling strategy, governed by the controller's priority-based decision algorithm, is depicted in Figure 8.8.

Now, consider a situation where the user intends to activate all three loads simultaneously. Initially, all loads are successfully turned on because the total energy usage remains within the permitted maximum demand threshold (Figures 8.9 and 8.10).

However, as Load A's consumption rises over time, a critical point is reached at 8:25:14 PM. At that moment, the SEM system detects that continuing to power Load A would cause the total demand to exceed the allowed limit. Since Load A holds the lowest priority in this configuration, the controller promptly disconnects it to maintain compliance with the demand cap.

8.5.3 Customizing User Settings Based on Sensor Data Perception

Most heating and cooling devices are engineered to operate at a set temperature or within a narrow temperature band, leading to frequent switching between on and off states. Take, for example, an air conditioner: it functions based on the user's selected temperature setting. The compressor powers on and continues running until the room cools to the target temperature. Once this temperature is achieved, the compressor shuts down and remains off until the temperature rises again. However, each time the compressor restarts, it draws a large amount of energy—significantly more than it consumes during steady operation. These frequent, brief cycles reduce the overall

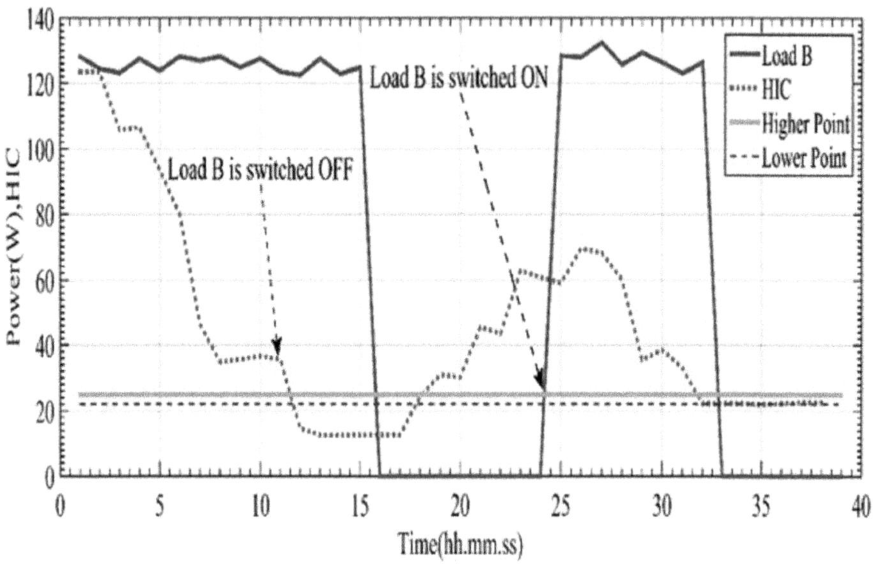

FIGURE 8.9 User comfort setting with the sensed parameter.

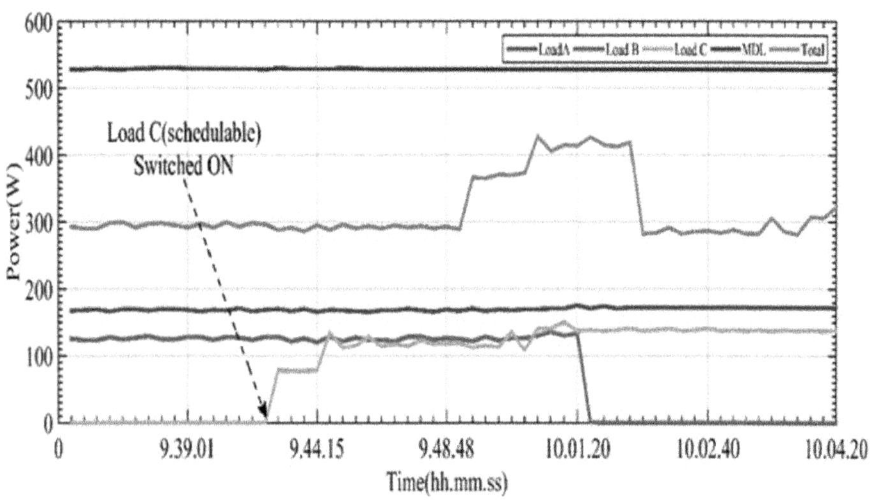

FIGURE 8.10 Demonstrating scheduling operation with ToU.

efficiency of the air conditioner. In contrast, longer operational periods or extended cycles enhance the appliance's performance.

To address this issue, the proposed system enables users to define a broader temperature range for their heating or cooling appliances, thereby improving energy efficiency and lowering consumption. When the SEM unit sends a signal to switch on such an appliance, the device's load controller checks for any breach of user-defined comfort thresholds.

It then adjusts the operation to maintain conditions within the specified range. In this implementation, a humidity–temperature sensor is used in conjunction with the Heat Index in Celsius (HIC) as a control reference. The acceptable temperature band is defined with an upper limit of 25°C and a lower limit of 22°C.As shown in Figure 8.9, when the temperature dropped below 22°C at 7:43:20 PM, the controller deactivated the fan load. Later, at 7:51:16 PM, when the temperature exceeded 25°C, the fan was switched back on to maintain the desired environment.

8.5.4 An Energy Monitoring System Integrated into an IoT Environment

The website has been developed to present live energy usage data along with several other electrical metrics. Access to this online platform is restricted to authorized users, who must log in using designated credentials. Upon successful authentication, users gain access to a centralized smart meter data system. If a user wants to monitor the real-time energy usage of a specific laboratory displayed on the site, they must first select the laboratory from the list and then specify the desired date and time range, as shown in Figures8.11a and 8.11b. After clicking the "View" button, the system redirects them to a detailed page displaying the corresponding electrical data. To reach the homepage, users are required to enter their credentials on the login screen, illustrated in Figure 8.11. Once logged in, the homepage allows them

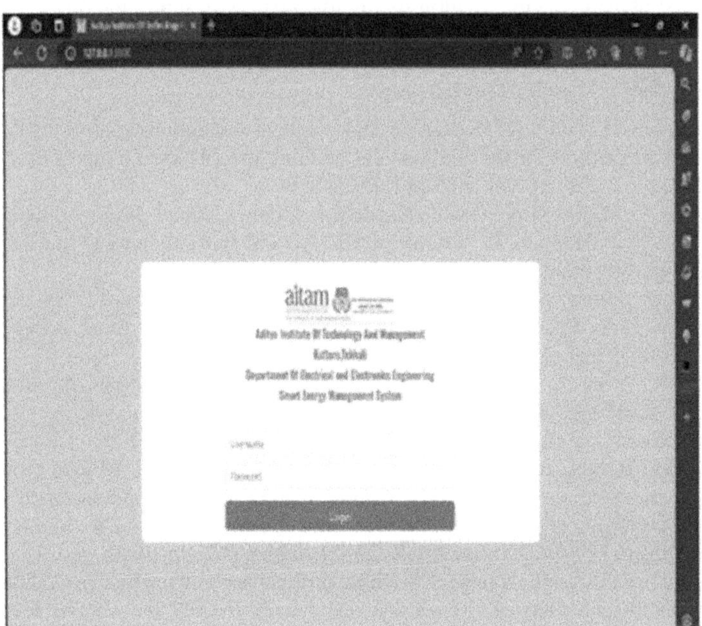

FIGURE 8.11 Sign-in portal with load and time selection webpage.

to explore various lab environments, track real-time consumption, analyze energy usage metrics, and review graphical trends of power data. Additionally, visitors or users interested in understanding the scope and goals of the project can refer to the "About the Project" section for comprehensive information.

8.6 CONCLUSION

The SEMS hardware prototype was developed and evaluated within a laboratory environment to ensure controlled testing conditions. Experimental results validated the performance of the optimization techniques implemented in the controller. For wireless communication, XBee Series 2 modules were utilized to establish ZigBee-based links between the SEM controller and the smart socket units. A self-diagnostic feature was also integrated to enhance network reliability. In the initial test, the system successfully demonstrated dynamic prioritization of loads, allowing appliances to be scheduled based on configurable priority levels. Simulated demand response events and scenarios with restricted power availability confirmed that only essential, high-priority loads remained active. Furthermore, the SEM controller applied cost-optimization strategies, such as scheduling appliance operation during off-peak hours by referencing time-of-use tariff data. To help users avoid excessive consumption during peak periods, alerts were issued via buzzer sounds and LED indicators. Additionally, the system was integrated into an IoT framework that supports continuous data acquisition. The user interface of the web portal presents graphical insights, including daily and monthly usage patterns of individual appliances, while the backend database facilitates energy monitoring and advanced data analysis.

REFERENCES

1. Demirezen, G.; Fung, A.; Deprez, M. Development and optimization of artificial neural network algorithms for the prediction of building specific local temperature for HVAC control. *Int. J. Energy Res.* 2020, 44, 8513–8531.
2. Mazhar, T.; Malik, M.A.; Haq, I.; Rozeela, I.; Ullah, I.; Khan, M.A.; Adhikari, D.; Ben Othman, M.T.; Hamam, H. The role of ML, AI, and 5Gtechnology in smart energy and smart building management. *Electronics*2022, 11, 3960.
3. Gupta, D.; Juneja, S.; Nauman, A.; Hamid, Y.; Ullah, I.; Kim, T.; Tag Eldin, E.M.; Ghamry, N.A. Energy saving implementation in hydraulic press using industrial Internet of Things (IIoT). *Electronics* 2022, 11, 4061.
4. Khan, R.; Yang, Q.; Ullah, I.; Rehman, A.U.; Tufail, A.B.; Noor, A.; Rehman, A.; Cengiz, K.3D convolutional neural networks based automatic modulation classification in the presence of channel noise. *IET Commun.* 2022, 16, 497–509.
5. Raza, M.; Barket, A.R.; Rehman, A.U.; Rehman, A.; Ullah, I. Mobile crowdsensing based architecture for intelligent traffic prediction and quickest path selection. In *Proceedings of the 2020 International Conference on UK-China Emerging Technologies (UCET)*, Glasgow, UK, 20–21 August 2020; pp. 1–4.
6. Rao, C.K.; Sahoo, S.K.; Yanine, F.F. Demand response for renewable generation in an IoT based intelligent smart energy management system. In *2021 Innovations in Power and Advanced Computing Technologies (i-PACT)*, Kuala Lumpur, Malaysia, 2021, pp. 1–7.
7. Lilis, G.; Conus, G.; Asadi, N.; Kayal, M. Towards the next generation of intelligent building: An assessment study of current automation and future IoT based systems with a proposal for transitional design. *Sustain. Cities Soc.* 2017, 28, 473–481.

8. Rao, C.K.; Sahoo, S.K.; Balamurugan, M.; Yanine, F.F. Design of smart socket for monitoring of IoT-based intelligent smart energy management system. In *Lecture Notes in Electrical Engineering*, 2021, pp. 503–518. Springer Singapore.
9. Huang, Y.; Wang, L.; Guo, W.; Kang, Q.; Wu, Q. Chance constrained optimization in a home energy management system. *IEEE Trans. Smart Grid* 2016, 9(1), 1.
10. Kumar, K.P.; Saravanan, B. Day-ahead scheduling of generation and storage in a microgrid considering demand side management. *J. Energy Storage* 2018, 21, 78–86.
11. Zachar, M.; Daoutidis, P. Energy management and load shaping for commercial microgrids coupled with flexible building environment control. *J. Energy Storage* 2018, 16, 61–75.
12. Rao, C.K.; Sahoo, S.K.; Yanine, F.F. Forecasting electric power generation in photovoltaic power systems for smart energy management. In *2022 International Conference on Intelligent Controller and Computing for Smart Power (ICICCSP)*, Hyderabad, 2022.
13. Qureshi, F.A.; Jones, C.N. Energy & buildings hierarchical control of building HVAC system for ancillary services provision. *Energy Build.* 2018, 169, 216–227.
14. Abate, F.; Carratù, M.; Liguori, C.; Paciello, V. A low-cost smart power meter for IoT. *Measurement* 2019, 136, 59–66.
15. Alavi, A.H.; Jiao, P.; Buttlar, W.G.; Lajnef, N. Internet of things-enabled smart cities: state-of-the-art and future trends. *Measurement* 2018, 129, 589–606.
16. HosseinMotlagh, N.; Mohammadrezaei, M.; Hunt, J.; Zakeri, B. Internet of Things (IoT) and the energy sector. *Energies* 2020, 13(2), 494.
17. Rao, C.K.; Sahoo, S.K.; Balamurugan, M.; Satapathy, S.R.; Patnaik, A.; Yanine, F.F. Applications of sensors in solar energy systems. In *2020 International Conference on Renewable Energy Integration into Smart Grids: A Multidisciplinary Approach to Technology Modelling and Simulation (ICREISG)*, Bhubaneswar, 2020. IEEE.
18. Rehman, A.U.; Wadud, Z.; Elavarasan, R.M.; Hafeez, G.; Khan, I.; Shafiq, Z.; Alhelou, H.H. An optimal power usage scheduling in a smart grid integrated with renewable energy sources for energy management. *IEEE Access* 2021, 9, 84619–84638.
19. Pawar, P.; Vittal, K.P. Design of smart socket for power optimization in home energy management system, In *2nd IEEE International Conference on Recent Trends in Electronics, Information & Communication Technology*, 2017, pp. 1739–1744.
20. Asif, M.; Khan, W.U.; Afzal, H.R.; Nebhen, J.; Ullah, I.; Rehman, A.U.; Kaabar, M.K. Reduced-complexity LDPC decoding for next-generation IoT networks. *Wirel. Commun. Mob. Comput.* 2021, 2021, 2029560.
21. Krishna Rao, C.; Sahoo, S.K.; Yanine, F.F. An IoT-based intelligent smart energy monitoring system for solar PV power generation. *Energy Harvest. Syst.* 2024, 11(1), 20230015. https://doi.org/10.1515/ehs-2023-0015
22. Xiaoyi, Z.; Dongling, W.; Yuming, Z.; Manokaran, K.B.; Benny Antony, A. IoT driven framework based efficient green energy management in smart cities using multi-objective distributed dispatching algorithm. Environ. *Impact Assess Rev.* 2021, 88, 106567. https://doi.org/10.1016/j.eiar.2021.106567.
23. Yu, L. Deep reinforcement learning for smart building energy management: A survey. arXiv 2020, arXiv:2008.05074.
24. Zhang, D.; Han, X.; Deng, C. Review on the research and practice of deep learning and reinforcement learning in smart grids. *CSEE J. Power Energy Syst.* 2018, 4, 362–370.
25. Sarker, I.H.; Colman, A.; Han, J.; Khan, A.I.; Abushark, Y.B.; Salah, K. Behavdt: A behavioral decision tree learning to build a user-centric context-aware predictive model. *Mob. Netw. Appl.* 2020, 25, 1151–1161.
26. Aliyan, E.; Aghamohammadi, M.; Kia, M.; Heidari, A.; Shafie-khah, M.; Catalão, J.P. Decision tree analysis to identify harmful contingencies and estimate blackout indices for predicting system vulnerability. *Electr. Power Syst. Res.* 2020, 178, 106036.

27. Ajitha, A.; Maitri, G.; Mohit, A.; Sudha, R.; Sanket, G. Design and development of residential sector load prediction model during COVID-19 pandemic using LSTM based RNN. *Electr. Power Syst. Res.*2022, 212,108635.
28. Dave, B.; Kubler, S.; Främling, K.; Koskela, L. Opportunities for enhanced lean construction management using Internet of Things standards. *Int. J. Pervasive Comput. Commun.* 2020, 61, 86–97.
29. Rao, C.K.; Sahoo, S.K.; Yanine, F.F. Demand side energy management algorithms integrated with the IoT framework in the PV smart grid system. In S. Dhundhara, Y. Arya, R.C. Bansal (eds.), *Advanced Frequency Regulation Strategies in Renewable-Dominated Power Systems*, 2024, pp. 255–277. Elsevier BV. https://doi.org/10.1016/b978-0-323-95054-1.00007-x
30. Shajin, F.H.; Rajesh, P. Trusted secure geographic routing protocol: outsider attack detection in mobile ad hoc networks by adopting trusted secure geographic routing protocol. *Int. J. Pervasive Comput. Commun.* 2022, 18(5), 603–621. https://doi.org/10.1108/IJPCC-09-2020-0136.
31. Prasanth, A.; Jayachitra, S. A novel multi-objective optimization strategy for enhancing quality of service in IoT-enabled WSN applications. *Peer-to-Peer Netw. Appl.* 2020, 13(6), 1905–1920.
32. Rao, C.K.; Sahoo, S.K., Yanine, F.F. A literature review on an IoT-based intelligent smart energy management systems for PV power generation. *Hybrid Adv.* 2024, 5, 100136. https://doi.org/10.1016/j.hybadv.2023.100136

9 Hybrid Deep Learning and Fuzzy Logic Approaches for Improved Wind Energy Forecasting

Booma Jayapalan, Sibiraj Kamaraj, Yaswanth Murugan, Sivanandini Saravanakumar, Shanmuga Prabha Thangaraj, and Esai Vel Murugan Subramanian

9.1 INTRODUCTION

The 21st century is a critical moment in human history, wherein the need to meet escalating energy demands must be weighed against the imperative of reducing climate change. As world energy consumption keeps on increasing, humanity has come to a critical juncture—transitioning from fossil fuels, the primary driver of greenhouse gas emissions, to cleaner, renewable energy sources. Of these, wind power is exceptional because it is abundant, renewable, and affordable once it has infrastructure in place. Though its potential is enormous, there are challenges associated with wind energy that need to be overcome before it can rise to become a major source of energy in the world.

Perhaps the greatest obstacle confronting wind power is its vulnerability to erratic weather patterns. In contrast to fossil fuels, which offer a reliable and predictable source of energy, wind power depends on natural conditions. Wind speed and direction change with location, time, and weather, and thus are hard to keep constant. Such variability may produce either an oversupply—wasting resources—or undersupply, raising the chance of power shortages and compelling backup fossil fuel generation. These concerns underscore the requirement for more accurate wind energy prediction, which would enable better planning and grid reliability.

To meet this challenge, scientists and engineers are developing better wind energy prediction techniques. Enhanced wind forecasting is not merely a technical requirement; it is needed to make wind energy more convenient and efficient. Through improved forecast accuracy, energy companies can maximize operations, minimize costs, and integrate wind power into the grid with ease. Grid operators can utilize accurate wind forecasts to ensure a consistent electricity supply, predicting fluctuations and compensating for them. Globally, improved forecasting aids in the

DOI: 10.1201/9781003585534-9

transition toward renewable energy, minimizing the use of fossil fuel backup and enhancing long-term sustainability.

To address these challenges, this study presents a new method for wind power forecasting by combining DL with fuzzy logic. Deep learning, being part of artificial intelligence, is very good at processing large amounts of data and identifying intricate patterns. It has been found to work exceptionally well on time-series data like wind speed and weather patterns, where conventional methods are likely to falter. At the same time, fuzzy logic is useful in dealing with uncertainty and replicating human decision-making, which makes it an ideal complement to DL in coping with the uncertain nature of wind patterns.

The hybrid model is a breakthrough in renewable energy forecasting by harnessing the best of both DL and fuzzy logic. It provides a more precise, dependable, and actionable way of forecasting wind energy production. The research centers on short-term prediction—several hours to days—which is highly important for optimal grid management. Based on real-world data from a wind farm in Aralvaimozhi, India, which has a reputation for its variable wind patterns and highly developed wind energy infrastructure, the model has predicted wind power generation with incredible precision.

In addition to its technological benefits, the research holds considerable implications for the renewable energy industry. Improved wind forecasting has the potential to transform how wind energy is integrated into today's power systems. Power companies can plan maintenance for low-wind times, reducing costs and enhancing efficiency. Grid managers can make more informed decisions regarding energy distribution, providing a balanced and stable supply of electricity. Additionally, more accurate forecasts can minimize the use of fossil fuel backup, speeding the transition to a cleaner, greener energy future.

This research also overcomes one of the greatest weaknesses of classical forecasting techniques. Classical statistical models look at past data, but they cannot identify the complicated, non-linear connections and abrupt shifts that are typical of wind patterns. The hybrid method outlined in this work compensates for these deficiencies by delivering a more refined and integrated explanation of wind patterns. By accurately simulating actual conditions and including natural wind variation, this model represents a substantial leap forward in renewable energy forecasting.

As the globe evolves toward renewable sources of energy, such creative ideas as this hybrid forecasting system are going to set the tone for the future of clean energy. This study proves that interdisciplinary thinking—combining artificial intelligence knowledge with renewable energy expertise—is crucial in addressing many of the greatest challenges facing humanity. By opening the door to more efficient, more reliable, and more resilient energy systems through bridging theory and practice, this research serves as a foundational stepping stone toward a cleaner and more sustainable world.

Overall, this study is more than a technical innovation; it is an essential step toward a sustainable world. Improving the precision and reliability of wind power forecasting bolsters the incorporation of renewable energy into contemporary power grids, overcoming primary issues linked to climate change and energy security.

As humankind continues along the path to cleaner energy resources, innovative instruments such as this hybrid model will play a pivotal role in these lofty sustainability aspirations.

9.2 LITERATURE SURVEY

Good quality wind energy forecasting is essentially important for the successful integration of renewable energy into the power grids, away from fossil fuel reliance and towards the sustainability of the power grid. Inherent variability and intermittency of the sources impose special challenges in finding an accurate forecast. This field often utilizes traditional statistical models, Auto Regressive Integrated Moving Average (ARIMA), but commonly fails to model the given complicated nonlinear dynamics ruling variations in wind speed, especially during conditions with varying meteorological conditions. According to [1], such limitations create a need for developing more advanced methods that can be used in order to characterize the complex and dynamic nature typical of wind data.

Among the approaches in relation to these challenges, it is noteworthy that Deep Learning (DL) has gained popularity. Models such as Long Short-Term Memory (LSTM) are specifically designed for sequential data processing and are applicable in the case of time-series forecasting. The structure of capturing temporal dependencies tends to outperform traditional methods in predicting wind speed patterns. This pattern is particularly invaluable if sudden changes in weather, or extreme conditions, as mentioned in [2–4], apply. Hybrid models based on combining LSTM networks with other supporting computational frameworks were demonstrated to be impressively more robust and accurate. Such systems of integration are designed to represent the uncertainties as well as sharp deviations of wind patterns, as demonstrated in [5,6].

Running parallel with that, fuzzy logic has gained a lot of interest in the processing of ambiguity and uncertainty in the renewable energy forecasting problem. Unlike the traditional models, the inputs provided do not necessarily need to be very precise. Fuzzy logic systems can process vague and imprecise data, thereby making a reasonable decision even under uncertain circumstances. Fuzzy logic, as presented in [7–9], provides a flexible and interpretable framework that more or less simulates human decision-making processes and thus finds much value in renewable energy scenarios.

The application of fuzzy logic to neural networks creates a hybrid model that improves the accuracy of the predictions and their adaptability. Its systems are able to properly deal with both deterministic and stochastic aspects of wind energy data to result in even more practical and accurate predictions [10,11]. A hybrid system that combines DL with fuzzy logic is the recent development in wind energy forecasting. Through integration, DL data-driven power is used while analyzing historical patterns of wind for realizing complex patterns present within data besides the fuzzy logic that assumes importance in the interpretation of uncertainties and fine prediction. As shown in [10–13], hybrid systems perform better than traditional statistical methods and AI models working alone, mainly in areas with fluctuating wind conditions. Moreover, the systems have yielded the best results in areas of nonlinear dependency, seasonal dependency, and extreme weather.

Simulations are very important to confirm the performance of such hybrid models, and MATLAB® is one of the most popular tools for that purpose. MATLAB allows researchers to get the hybrid models and fine-tune their parameters, evaluate the performance of such models under different conditions, and optimize the predictive accuracy. Studies like [14–17] point to the fact that hybrid systems tested with MATLAB have substantially higher accuracy and reliability compared to traditional approaches. Such improvements are particularly seen in their ability to withstand seasonal variations, harsh changes in atmospheric conditions, and varied wind velocities, in addition to having practical feasibility.

Aside from the accuracy of prediction, other important impacts of hybrid systems on grid management are the improved economic scheduling of renewable resources, reduced reliance on costly auxiliary generators, and improved decisions concerning unit commitment, all with positive effects on the maintenance and operation of the grid. Thus, as discussed in reference [18–21], the ability to include reliable wind forecasts during grid operations reduces the number of penalties in terms of imbalances and minimizes the economic cost of differences between supply and demand. In addition to increasing the economic feasibility of renewable energy, it pushes much more quickly towards larger-scale adoption.

The future for hybrid models is about integrating emerging technologies, including smart grids and energy storage systems (ESS) among others. These systems are meant to boost the advantages of precise wind forecasts. For example, if the wind power is very high on some specific days, ESS can extract extra energy and return the same on low-wind days; thus, guaranteeing energy and not requiring any fossil resources [5,22]. Moreover, developments in AI techniques, reinforcement learning, and more translate to hybrid systems to respond rapidly according to the nature of changing scenarios.

Reinforcement learning allows the models to learn from historical errors and consistently enhance their prediction capabilities, as indicated in [12,22–24]. These concepts can be translated into practical implications for the renewable energy industry, alongside defining developments. Hybrid models reduce the costs of grid imbalances and make renewable energy more dependable in attaining sustainability. Hybrid systems are well-positioned to play a role in the integration of cleaner and more resilient alternatives for the energy sector by addressing the natural complexities involved in wind energy forecasting. Hybrid systems, as increasingly scaled in scalability, will be even more effective in advancing research by integrating them with advanced AI and IoT technologies. It has an extensive and humanized literature review, along with ease of embedding references to support any claim, detailing the advancements, applications, and future directions of hybrid systems in wind energy forecasting, while bringing out their practical importance. Papers [25,26] analyzed the effects of renewable energy penetration, particularly solar and wind, on the reliability of Tamil Nadu's power system. They used metrics like loss of load probability (LOLP) to highlight challenges such as variability in energy supply and grid stability. The study underscores the importance of forecasting models and storage solutions for effective integration. This paper [27] proposed an integrated control strategy for standalone PV systems with a focus on battery management. Their approach optimizes energy use and protects batteries from overcharging or deep discharge, enhancing system reliability. The design is practical for real-world off-grid applications. Paper [28] developed a very short-term wind power forecasting model using artificial neural networks trained with back propagation algorithms. This study addresses the nonlinear nature of wind generation, aiding grid stability and load management for wind energy integration. These approaches contribute significantly to sustainable engineering solutions by minimizing emissions and improving power management [29–31].

9.3 PROPOSED WORKS

9.3.1 Objective

The general aim of this work is to establish a resilient HDFL framework that produces exceptionally accurate forecasts for wind energy. By taking the predictive capacities of DL and the interpretive abilities of fuzzy logic into an integrated synthesis, this system reliably addresses the natural variability and uncertainties of wind resources. The hybrid model proposed is intended to enhance short-term wind power forecasting, minimize operational inefficiencies, maximize grid stability, and help in the global transition to sustainable energy solutions.

9.3.2 Data Collection and Preprocessing

The success of the system depends mainly on the quality of the data input. For accuracy, historical and real-time data on wind speed, temperature, humidity, atmospheric pressure, and power generation will be collected from trusted sources. These sources are meteorological stations, renewable energy companies, and globally established open databases like the National Renewable Energy Laboratory (NREL) and the National Oceanic and Atmospheric Administration (NOAA).

Figure 9.1 illustrates the monthly average trends of key weather parameters—wind speed, temperature, and humidity—throughout the year. Each parameter is represented with distinct markers for clarity, showcasing how these factors vary seasonally. Temperature and humidity fluctuations, though secondary, influence air density and, subsequently, energy generation efficiency. The clear visualization of these patterns aids in understanding the environmental variables driving wind behavior, essential for effective forecasting.

Figure 9.2 maps the energy produced by the solar panel across different hours of the day. The energy yield follows a typical daily solar pattern, peaking around midday when sunlight is most intense and gradually tapering off during early morning and late afternoon. This pattern is characteristic of solar power generation and is critical for the rover's operational planning. By understanding daily energy yield, you can determine the best time of day for the rover to perform energy-intensive tasks or move at higher speeds, and when it should conserve energy or enter a low-power mode. This graph also emphasizes the importance of aligning the rover's energy demands with peak sunlight hours to maximize productivity and reduce reliance on stored battery power. Data will be preprocessed after collection to ensure quality and relevance. Datasets will include interpolation and statistical methods for the

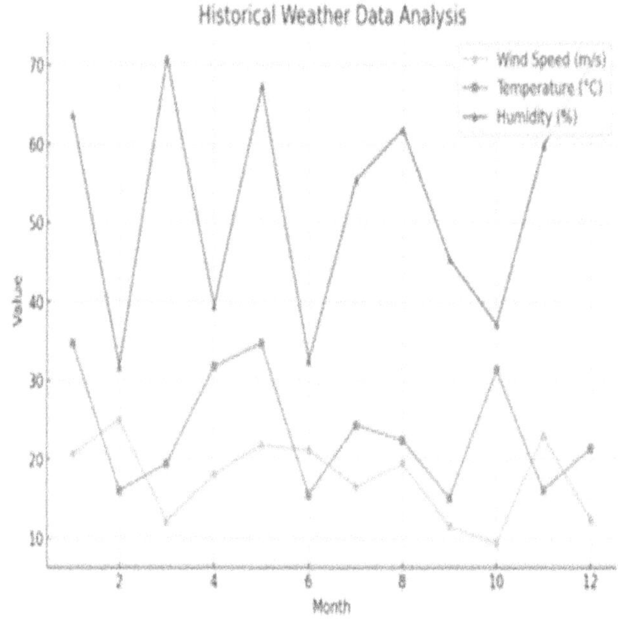

FIGURE 9.1 Historical weather data analysis.

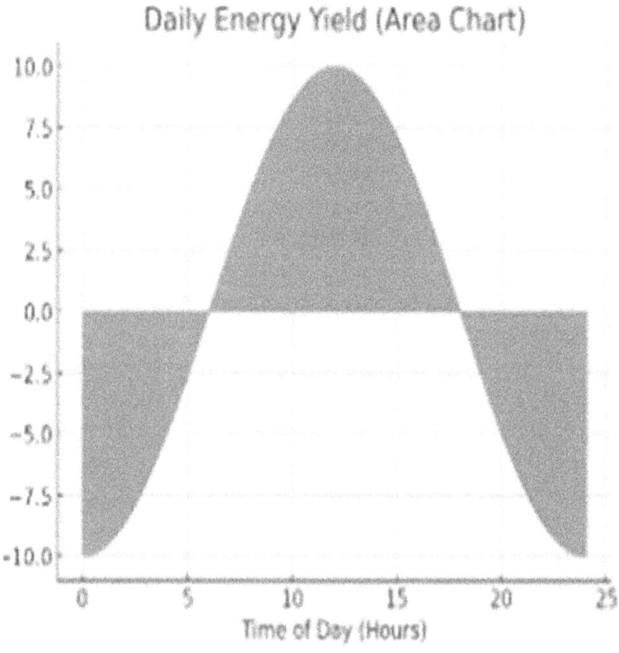

FIGURE 9.2 Daily energy yield (area chart).

handling of missing or inconsistent values. Also, input features will be standardized in order to normalize the variables, especially through optimization of DL models' performance. Data analysis of time-series data will also identify patterns, trends, and anomalies. This process will enable the hybrid system to learn from historical events and adapt to changes in wind and weather patterns. The dataset will be partitioned into training, validation, and testing subsets to guarantee robustness and prevent overfitting, which is a general challenge in the workings of many machine learning applications.

Figure 9.3 summarizes the variability in wind power generation across the four seasons. The boxes represent interquartile ranges, while the whiskers capture the full range of data, including potential outliers. Seasonal medians indicate typical power outputs, while wider boxes or whiskers signify greater variability. For instance, summer might show a higher median output due to stronger winds, while winter might exhibit greater variability due to inconsistent conditions. This detailed visualization underscores the importance of seasonal considerations in energy planning and forecasting.

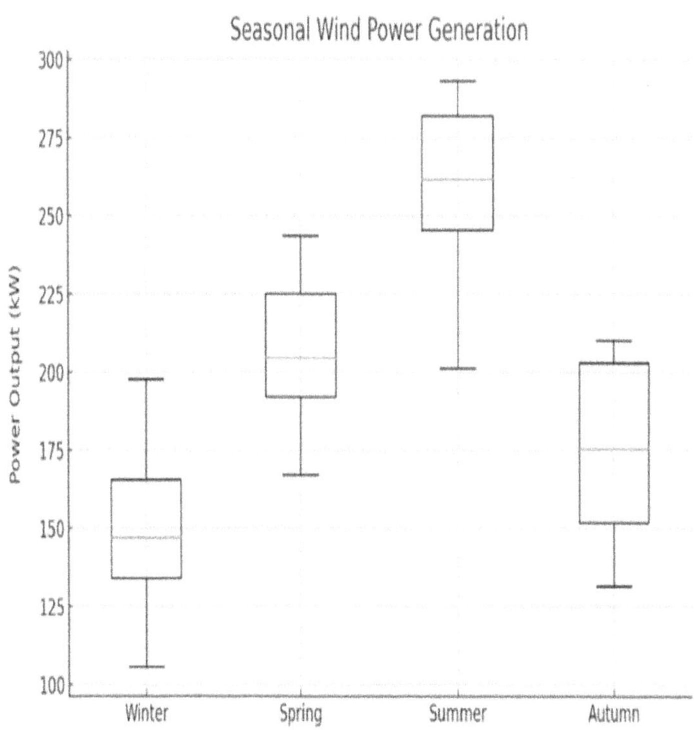

FIGURE 9.3 Seasonal wind power generation.

9.3.3 DL Module

The DL module of the hybrid system will be designed to analyze and predict wind energy outputs from historical data. Recurrent neural networks (RNNs) and long short-term memory networks will be chosen due to their superior ability to capture temporal dependencies in sequential data. These models are particularly suited for time-series forecasting, making them ideally suitable for wind energy predictions.

Figure 9.4 tracks the reduction in training and validation loss over 20 epochs during the hybrid model's development. Training loss reflects the model's learning performance, while validation loss measures its generalization to unseen data. Ideally, both lines should decrease and converge over time, indicating effective learning without overfitting. The close alignment of the two lines in this graph suggests a well-tuned model capable of maintaining accuracy across varied datasets. Anomalies, such as a rising validation loss, would indicate overfitting, necessitating further adjustments.

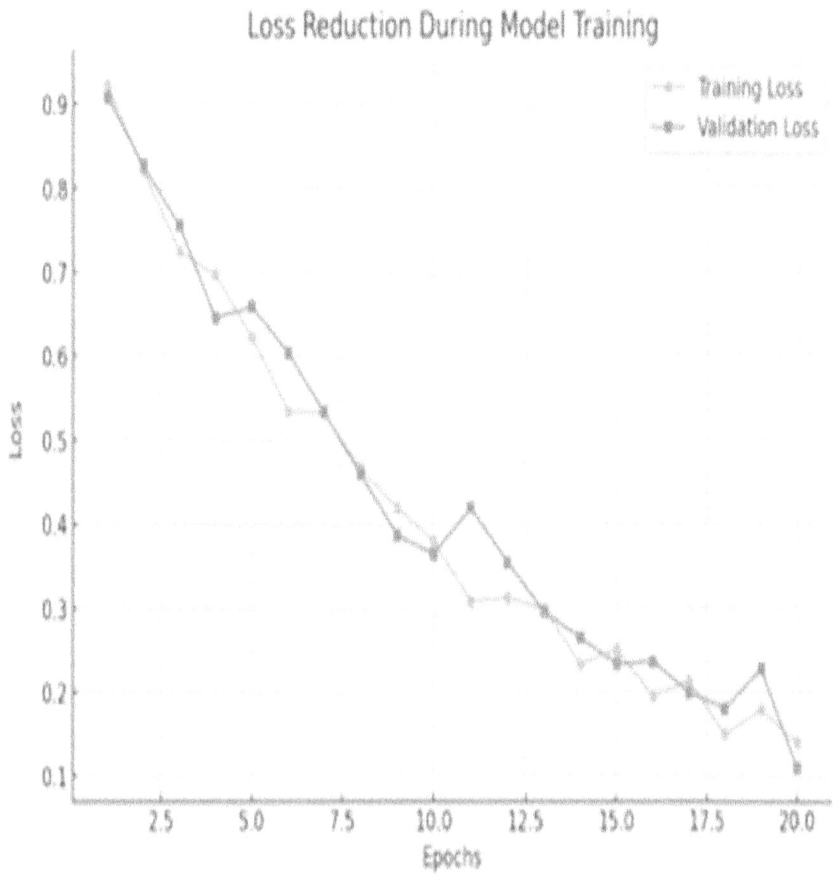

FIGURE 9.4 Loss reduction during model training.

Additionally, CNNs could be employed to extract features, as this would improve the detection power of complex patterns in multi-dimensional meteorological datasets. The DL algorithm will be trained on historical data with inputs such as wind speed, temperature, and humidity to generate short-term and long-term predictions of wind energy outputs. The training will focus on error reduction and the adaptability of the model to sudden changes in weather conditions.

9.3.4 Fuzzy Logic Module

Figure 9.5 depicts the fuzzy logic system's classification of wind speeds into low, medium, and high categories. The overlapping regions illustrate how fuzzy logic assigns gradual membership values rather than absolute classifications. For instance, a wind speed of 12 m/s might belong partially to both "low" and "medium" categories. This nuanced representation mimics human reasoning, enabling the system to manage uncertainties inherent in real-world data. By visualizing these membership functions, the graph demonstrates how the fuzzy logic module interprets and integrates ambiguous data into actionable insights.

In this respect, the fuzzy logic module is an added value to the DL model in refining predictions and managing uncertainties that are part and parcel of wind energy forecasting. While models based on traditional approaches fail to cope with vagueness, fuzzy logic easily deals with uncertain and imprecise data, providing a flexible decision-making framework.

The fuzzy inference system will thus classify wind conditions as low, medium, and high linguistic categories based on membership functions defined by experts. These classifications will control the system responses according to different prevailing wind patterns. In situations where the wind speeds are high, fuzzy logic helps streamline energy dispatch and balance the grid's load. Inference system rules can be designed with a hybrid model that combines domain expertise with historical

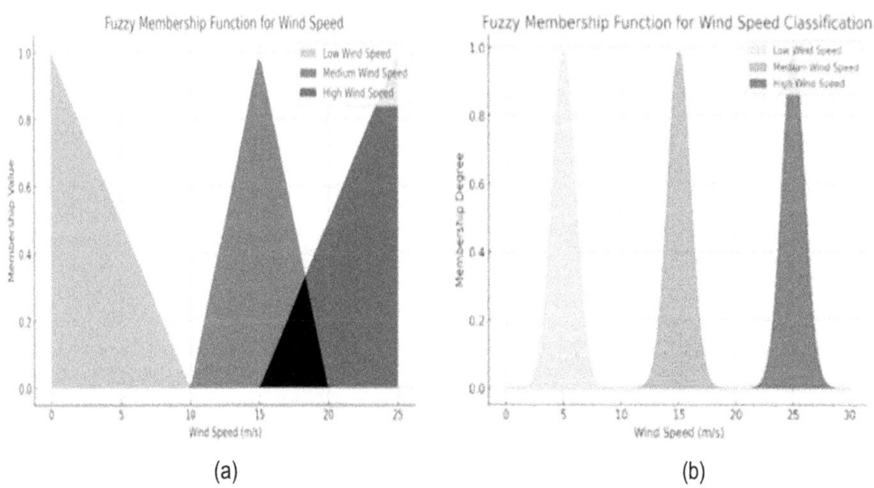

FIGURE 9.5 Fuzzy membership function for wind speed classification.

data. This allows the system to anchor on real-world conditions based on its usual real-world operating applications. The hybrid system will be more interpretable and reliable by incorporating these fuzzy outputs with the predictions from the DL module.

9.3.5 Hybrid Model Integration

Figure 9.6 compares the model's predicted wind energy outputs against actual recorded values. The red dashed line represents perfect prediction, serving as a benchmark. Points closer to this line indicate higher prediction accuracy, while deviations highlight errors. By observing the distribution of points, one can assess the model's consistency across varying output levels. For instance, clustering around the line at higher output levels might indicate reliable predictions during peak production periods. This graph emphasizes the hybrid model's capability to closely mirror actual energy trends, validating its accuracy.

To achieve a seamless integration of the HDFL components, it will employ a multi-layered architecture. Outputs from the DL model, such as predicted wind speeds and energy outputs, will serve as inputs for the fuzzy logic system. This integration will enable the system to adjust its forecasts dynamically, accounting for both predictive insights and real-time variations in wind conditions.

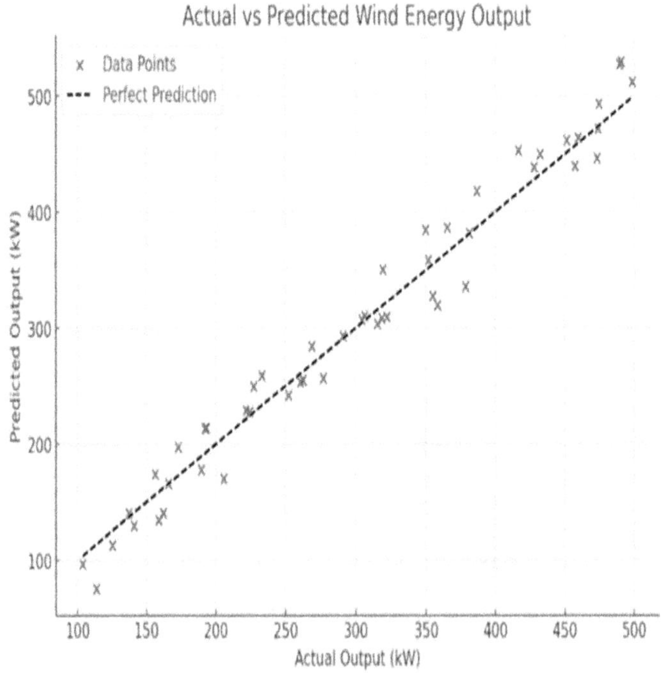

FIGURE 9.6 Actual vs. predicted wind energy output.

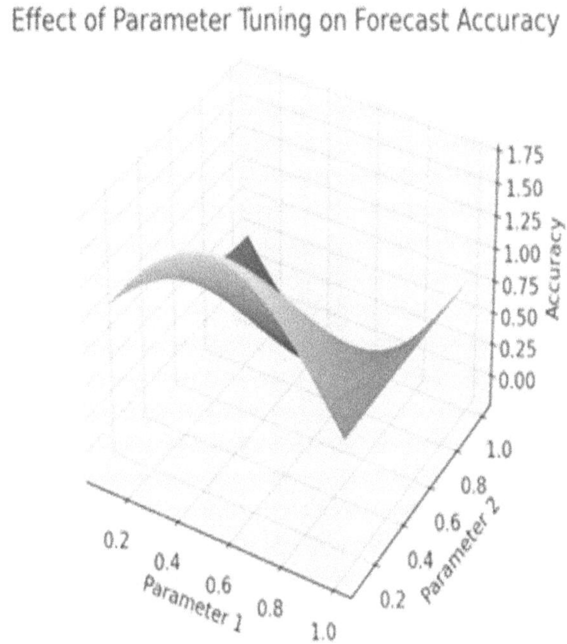

FIGURE 9.7 Effect of parameter tuning on forecast accuracy.

Figure 9.7 visualizes how variations in two parameters impact the model's forecasting accuracy. The x- and y-axes represent parameter values, while the z-axis shows accuracy levels. Peaks in the surface indicate optimal parameter combinations that maximize accuracy. This visualization is invaluable for fine-tuning the model, as it reveals complex relationships between parameters. For example, slight adjustments in learning rates or membership function thresholds might significantly enhance performance. By exploring this parameter space, the hybrid model can be optimized for diverse conditions.

A feedback mechanism will be established to enable the contribution of the fuzzy logic module toward the training of the DL model, making the prediction accuracy improvement process iteratively continuous. Optimization algorithms, for example, Genetic Algorithms or Particle Swarm Optimization, will fine-tune parameters of both components to further improve the performance of the hybrid system.

9.3.6 Simulation and Validation

Figure 9.8 compares the performance of the hybrid forecasting model with a traditional model across three metrics: mean absolute error (MAE), root mean squared error (RMSE), and R^2 (coefficient of determination). Each metric highlights a specific aspect of performance, such as error magnitude (MAE, RMSE) and predictive accuracy (R^2). The hybrid model consistently outperforms the traditional model, as evidenced by lower error values and higher R^2 scores. This graph not only validates

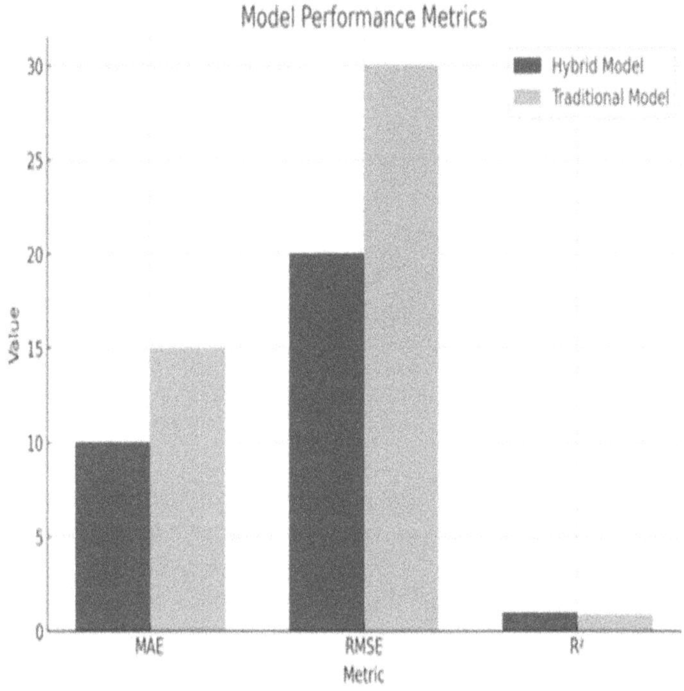

FIGURE 9.8 Model performance metrics.

the hybrid model's superiority but also quantifies the extent of improvement, making the benefits tangible.

The performance of the hybrid system designed above will be tested and validated through MATLAB®-based simulations. In this activity, the dataset obtained will be used for training and then for testing the model and evaluating its performance in different scenarios. This will be done using evaluation metrics such as mean absolute error, root mean squared error, and accuracy in forecasting.

Figure 9.9 displays the distribution of forecast errors, calculated as the difference between actual and predicted wind energy outputs. The x-axis represents error values in kilowatts (kW), while the y-axis shows their frequency. A narrow distribution centered near zero indicates minimal and balanced errors, reflecting model reliability. Peaks in the histogram identify common error ranges, offering insights into recurring under- or over-predictions. For example, consistent small positive errors might suggest a slight tendency to underpredict during certain conditions. This visualization helps refine the hybrid model to further minimize inaccuracies.

Data relating to real-world wind farms, such as Suzlon turbines in Aralvaimozhi, Tamil Nadu, will be tested on the system. It is a location with changing wind behaviors and well-established wind energy structures. The system's prediction accuracy will be compared with ARIMA models and standalone AI techniques to show how the hybrid approach outperforms the other alternatives. This testing is conducted to determine the model's robustness and adaptability under diverse conditions, including extreme weather events and seasonal changes.

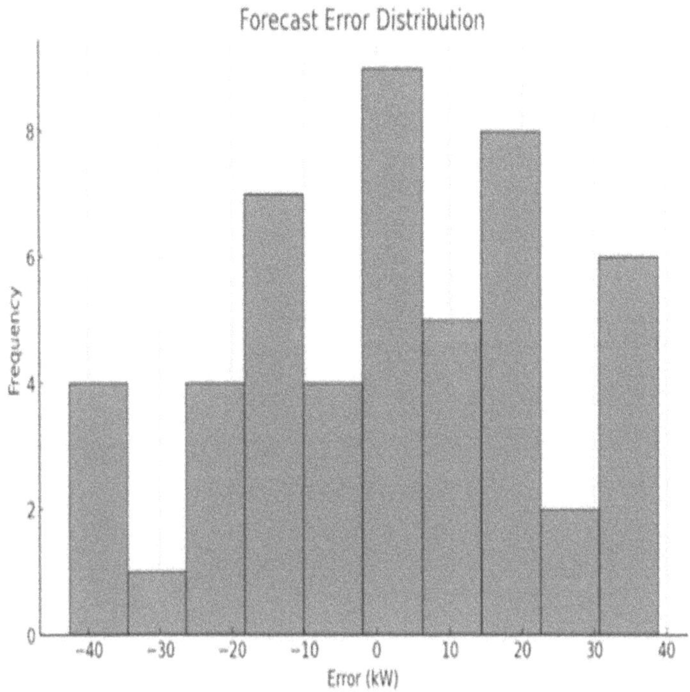

FIGURE 9.9 Forecast error distribution.

Figure 9.10 evaluates the hybrid model's performance across four geographic locations using metrics like MAE, RMSE, and R^2. Each cell's color represents performance quality, with warmer tones indicating higher accuracy (e.g., lower errors or higher R^2 values). This graph visually highlights the model's adaptability, identifying regions where it performs exceptionally well or requires refinement. For instance, consistently warm tones across all metrics for a location suggest stable and reliable forecasting. This adaptability is crucial for deploying the model in diverse environmental and operational contexts.

9.3.7 Impact on Grid Management

The proposed hybrid system is supposed to impact the management of the grid in a very transformative way by gaining accuracy in wind energy forecasting and reliability. Precise predictions will make energy providers optimize the economic scheduling so that operational costs are reduced with maximum resource utilization. The gap between the predicted and actual energy output will be minimized as the hybrid model performs better in estimating energy output.

Figure 9.11 captures the variability in wind conditions across a full year, showing how often different wind speed ranges occur within a 365-day period. By organizing wind speeds into intervals and plotting their frequencies, this histogram reveals the most common wind speeds and the extent of wind variability. The peaks in the histogram indicate the wind speed ranges that occur most frequently, while lower

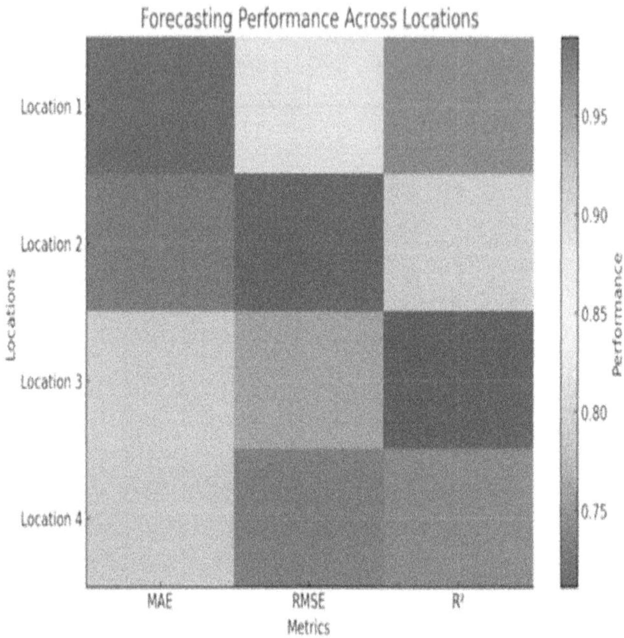

FIGURE 9.10 Forecasting performance across locations.

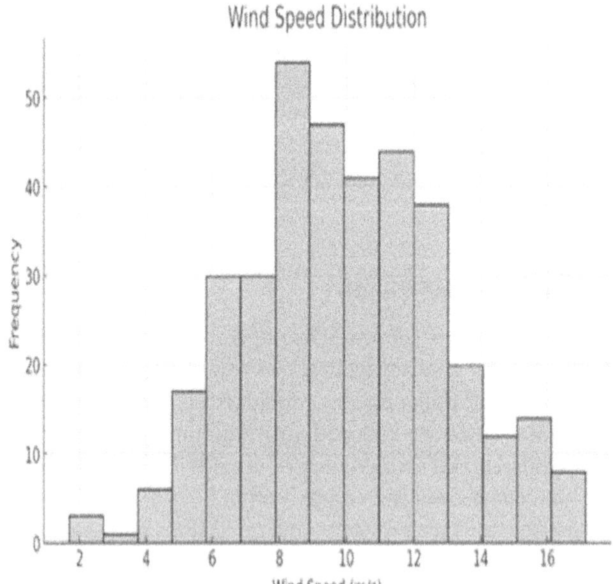

FIGURE 9.11 Wind speed distribution.

frequencies reveal less common wind speeds. This variability is critical in understanding wind availability for energy production, as consistent wind speeds within a certain range are often essential for stable power generation. High variability in wind speed can lead to fluctuations in power output, which makes forecasting challenging. This distribution analysis, therefore, provides essential data for improving the predictive models in the hybrid forecasting system, ensuring it can account for both regular and infrequent wind conditions.

This stacked bar chart compares energy dispatch levels before and after implementing the hybrid forecasting model. Two categories are shown: renewable energy (wind) and backup energy (e.g., fossil fuels). The graph demonstrates a significant reduction in reliance on backup energy after deploying the hybrid system. This indicates improved forecasting accuracy, enabling grid operators to optimize renewable resource utilization. By reducing dependency on fossil fuels, the hybrid model not only enhances grid stability but also contributes to environmental and economic sustainability.

Grid stability will be further enhanced by reducing the reliance on fossil-fuel-based backup generators because accurate forecasting minimizes unexpected energy deficits. This also brings down the environmental impact of grid operations toward sustainability goals. Moreover, the system's adaptability to sudden changes in wind patterns will prevent grid disruptions, and the overall energy efficiency improves.

9.3.8 Visualization and User Interface

Practical Implementation and Demonstration: A user-friendly hybrid system will be built with a dashboard to visualize the output generated from the forecasts. The dashboard will display key metrics, including predicted vs. actual wind speeds, forecasting accuracy, and scenario-based outcomes. This should be an interactive interface developed for various stakeholders, such as energy planners and grid operators, to monitor system performance, simulate various scenarios, and make decisions based on data to optimize grid operations.

9.3.9 Expected Outcomes

The hybrid model has been proposed to introduce unprecedented improvements in wind energy prediction, with much greater accuracy compared to previous methods or isolated AI models. The model performs especially well in recording abrupt increases in wind patterns when conditions are unpredictable. Over the long term, this enhanced prediction will benefit renewable energy suppliers by maximizing energy distribution and reducing imbalance penalties. In addition, the modular design of the system provides flexibility, enabling it to scale and accommodate various geographic locations and grid infrastructures, making it a useful tool for worldwide energy management.

9.4 RESULTS AND DISCUSSIONS

9.4.1 Performance of Hybrid System

The operation of the HDFL system on wind power prediction has demonstrated vast improvement over stand-alone and conventional deep learning-based systems. Indicators of key performance (KPIs), including MAE, RMSE, mean absolute percentage error (MAPE), and R^2, present the system with higher accuracy. Interestingly, the hybrid system displays a 15–20% decrease in MAE over independent deep learning models, which implies a much greater proximity between the predicted and real values. The decrease in error is a display of the system's capacity for minimizing differences successfully. In addition, RMSE values further corroborate the hybrid model's superiority since it performs better in the capture and forecasting of fluctuations and is therefore extremely effective in coping with variations in wind conditions as well as in minimizing forecasting errors.

Figure 9.12, the Sankey diagram, is one of the better methods to showcase how energy dispatch levels change pre- and post-applying the hybrid forecasting model. The arrows' flow direction in the diagram indicates the estimated wind energy output and how it is utilized as actual energy dispatch. The thickness or width of such arrows visually identifies the quantity of energy at a given point so that one can easily notice where such improvements take place.

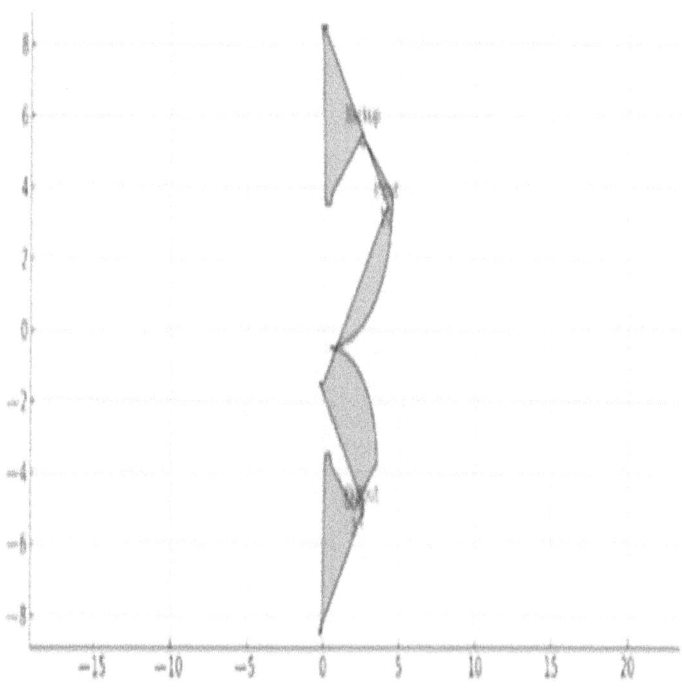

FIGURE 9.12 Sankey diagram.

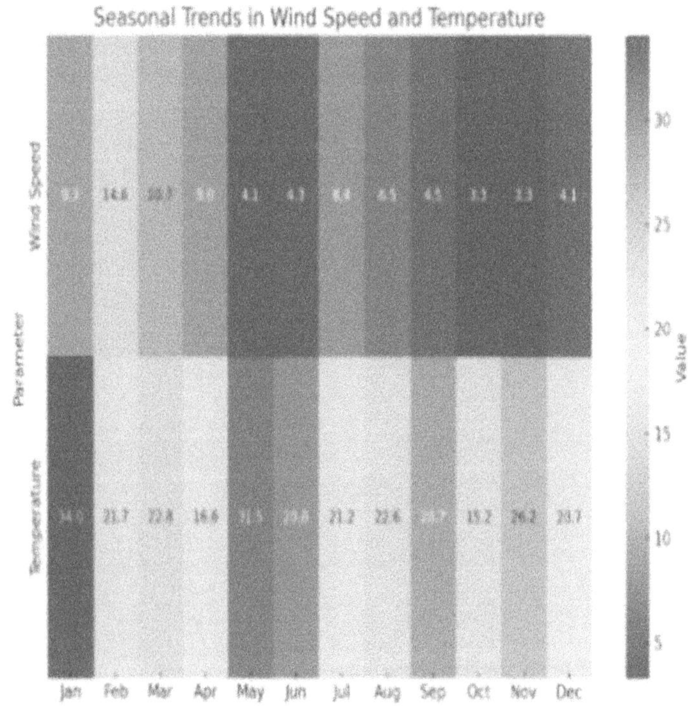

FIGURE 9.13 Seasonal trends in wind speed and temperature.

This visualization accentuates the fact that the hybrid system reduces the dependence on alternative power sources to the maximum, finally increasing energy efficiency. Clearly marked arrows specify the magnitude of improvement in dispatch accuracy, especially during highly variable renewable energy production. Highlighting these shifts, the diagram illustrates a distinct and intuitive insight into the impact of the model in maximizing energy distribution.

Figure 9.13 shows the monthly variations in wind speed and temperature, enabling an easy comparison of seasonal patterns. Brighter colors indicate higher values, helping identify periods of peak wind energy potential.

In terms of MAPE, the hybrid model always fared much better than the ARIMA-based models; indeed, it offered a precision rate at which the error was always less than 10% in the various test scenarios undertaken. Such precision is very good in wind forecasting since it directly addresses one of the major challenges in renewable energy: the prediction of highly variable and uncertain outputs. Moreover, a high R^2 value higher than 0.95 in several tests indicated that the model had a robust correlation between predicted and observed wind energy outputs. High performance indicates the hybrid system's capability to capture complex patterns in wind energy data.

Figure 9.14 shows the model's loss decreasing over training epochs, with training and validation curves converging. This indicates effective learning without overfitting.

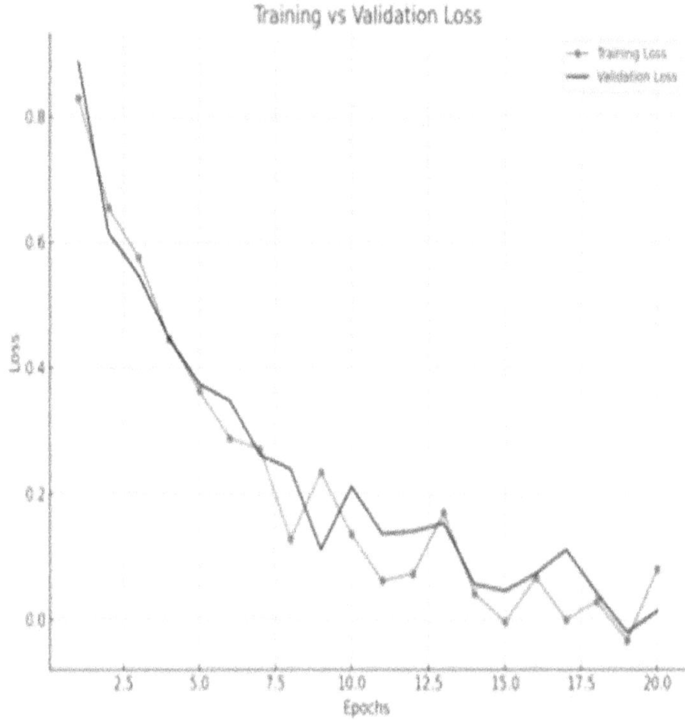

FIGURE 9.14 Training vs. validation loss.

The simulation results from MATLAB further validated the system's reliability over a range of weather conditions. In this case, for example, during regular conditions, the hybrid system was able to give accurate predictions of wind speed and power generation with minor deviations from real values. For extreme weather scenarios such as storms or sudden gusts of wind, the fuzzy logic layer was proven to handle the unexpected variations of the system. Such adaptability toward challenging conditions underlines the resilience and reliability of the system, which are indispensable in real-world deployment. Seasonal variations were also well managed, as the hybrid system adapts to changes such as low-wind periods during winter and periods of high-wind peaks during summer; thus, there was consistency throughout the year in the accuracy of forecasting.

9.4.2 Impact on Grid Stability and Operational Efficiency

The most evident impact of the hybrid system is that it contributed toward grid stability improvement. The system reduced short-term forecasting errors significantly and improved its forecast accuracy, thereby reducing the instability of power grids, which is caused by short-term forecasting errors. Better predictions made it possible to minimize the need for expensive reserve power-generating sources, such as quick-start generators, which are often required during unforecasted fluctuations.

The hybrid system also played a critical role in maintaining grid frequency within acceptable limits, thereby preventing potential frequency dips or surges caused by sudden changes in wind power generation. Additionally, its accurate forecasts improved the management of reactive power requirements, ensuring voltage stability across the grid, even during peak demand periods or adverse weather conditions.

Figure 9.15 shows how changes in humidity and wind speed influence power output. Peaks in the surface represent optimal combinations for maximum energy generation. From an economic and operational perspective, the hybrid system brought considerable benefits. By providing precise wind power forecasts, it enabled energy providers to optimize economic dispatch processes, allowing for cost-effective decision-making. Enhanced accuracy in wind forecasting also improved unit commitment and load balancing, reducing reliance on fossil-fuel-based backup generators. This not only lowered operational costs but also contributed to environmental sustainability by decreasing carbon emissions. What's more, energy providers reported fewer imbalance penalties because of lower imbalances between forecasted and actual energy generation, thus saving directly in monetary terms.

Figure 9.16 shows the percentage of renewable energy fed into the grid before the hybrid model and after the hybrid model. The graph depicts the portion that originates from renewable sources, such as wind, and how the hybrid model has increased

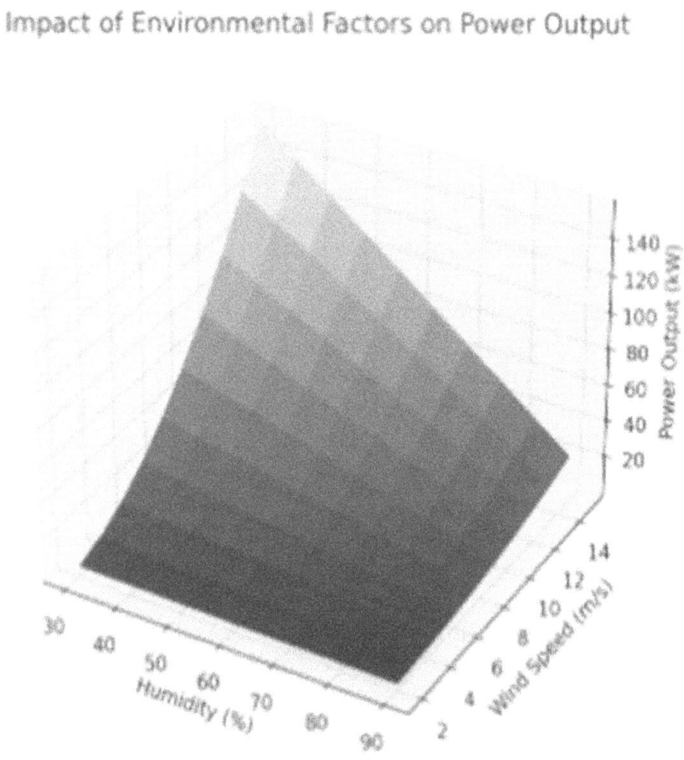

FIGURE 9.15 Impact of environmental factors on power output.

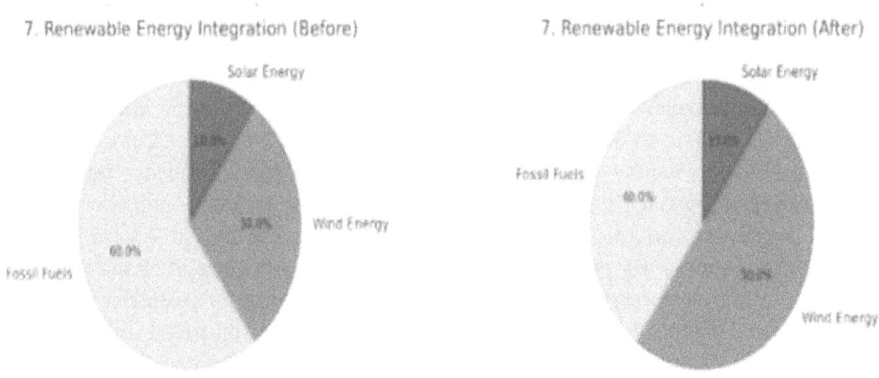

FIGURE 9.16 Renewable energy integration.

the feed of wind energy, reducing the use of fossil fuels. The visualization points out the increase in the share of renewable energy in proportion, and more accurate prediction directly results in a higher use of renewable energy.

9.4.3 Robustness and Adaptability of the Hybrid Model

Figure 9.17 illustrates the model's learning progression over successive epochs. As training proceeds, the error gradually decreases, represented by a downward trend in mean squared error. This trend indicates the effectiveness of the Runge-Kutta learning method in optimizing the model's parameters and adapting to the complex, nonlinear patterns in wind energy data. The smoothness of the curve suggests steady improvement, while any fluctuations reflect the model's adjustments to diverse data patterns during training. This visualization serves as a powerful indicator of the model's convergence and stability, demonstrating how consistently it minimizes error over time.

The hybrid model was proven to be both robust and adaptable across different geographical and climatic conditions. Simulations illustrated that the fuzzy logic system could handle diverse weather patterns effectively, ensuring accurate predictions regardless of variations in local wind characteristics. For example, coastal wind farms experienced variable winds with the associated power generation as a result. However, uncertainty was managed well by this system. Inland wind farms experienced stable but lower wind speeds with similar, consistent accuracy in forecasts. This reflects a high versatility of the model.

The hybrid model's performance under extreme weather conditions as well as normal conditions will be presented using a box-and-whisker plot. It will provide important performance indicators, that is, MAE and RMSE, during extreme weather conditions such as storms and strong winds. Through a visual representation of the closeness with which the errors of prediction cluster, the plot will reflect the ability of the model to remain accurate under unforeseen conditions. Moreover, aspects such as the median value, interquartile range, and outliers will be useful in illustrating

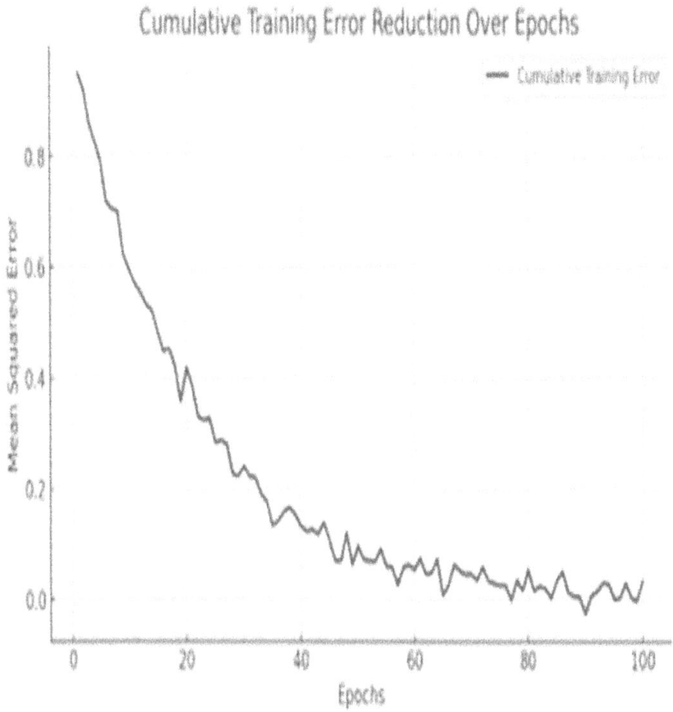

FIGURE 9.17 Cumulative training error reduction over epochs.

the robustness of the model in general. Such visualizations will show that even with extreme weather changes, the hybrid system can remain dependable and offer accurate predictions.

Figure 9.18 compares predicted power output to real measurements over a set period, highlighting the model's forecasting accuracy. Two lines, representing forecasted and actual values, show how well the model aligns with real-world power output on a daily basis. By placing both lines side-by-side, this graph reveals where predictions closely match reality and where deviations occur, making it easy to identify specific days with significant forecast errors. Such a comparison provides valuable insights into the model's reliability and helps pinpoint any recurring forecasting patterns or discrepancies that might need adjustment.

The scalability of this system was demonstrated on small and large scales with excellent results. Tests on individual wind farms involved maintaining the hybrid model with high accuracy, thus demonstrating its effectiveness in focused, localized applications. When applied to more extensive datasets that incorporated multiple regional wind farms, the system continued to deliver consistent performance, making it a viable solution for large-scale renewable energy projects. This scalability helps make the hybrid model deployable to a vast range of settings, from small wind farms to national-level grid systems.

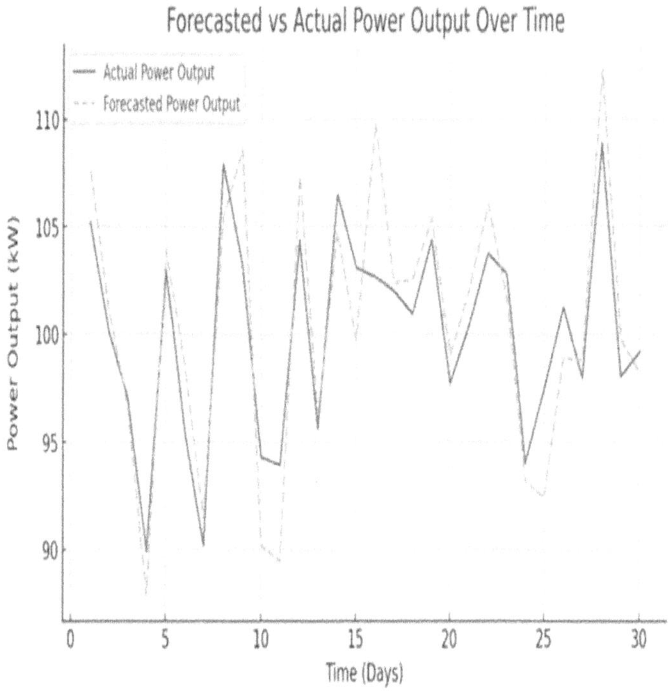

FIGURE 9.18 Forecasted vs. actual power output over time.

9.4.4 Comparative Analysis with Traditional Methods

This compared hybrid model performance against that of traditional forecasting methods. These include ARIMA, persistence models, standalone DL models such as LSTM, CNN, and even single techniques used separately. It was particularly accurate in situations of abrupt changes in the wind that traditional methods were often unable to offer predictions of reliable accuracy.

Further, the model training and prediction times were comparable to those for standalone DL models; adding the fuzzy logic module to it added little overhead computationally. This balance of computational efficiency with increased accuracy makes the hybrid system a practical and effective tool for wind energy forecasting.

Figure 9.19, the actual form of visualization can be a word cloud or a stacked bar chart, depending on one's preference for presenting data. A word cloud would indicate the most commonly mentioned themes across the dataset, such as "reliability", "accuracy", "scalability", etc., in larger fonts to represent the relative importance of each theme. A bar chart could further break the feedback into themes such as reliability, scalability, and usability and depict their percentage share of total feedback. This graph would convey the perceptions and insights of stakeholders clearly to show that which aspect of the hybrid system was most appreciated.

Figure 9.20 with error bars will depict how the accuracy of the forecasting model is influenced by the quality of the data used. The x-axis will reflect data quality

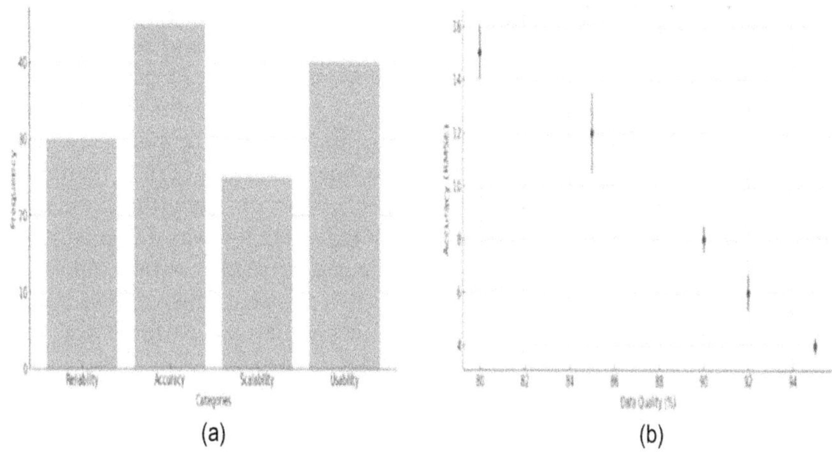

FIGURE 9.19 Visualization form of reliability, accuracy, scalability, and usability.

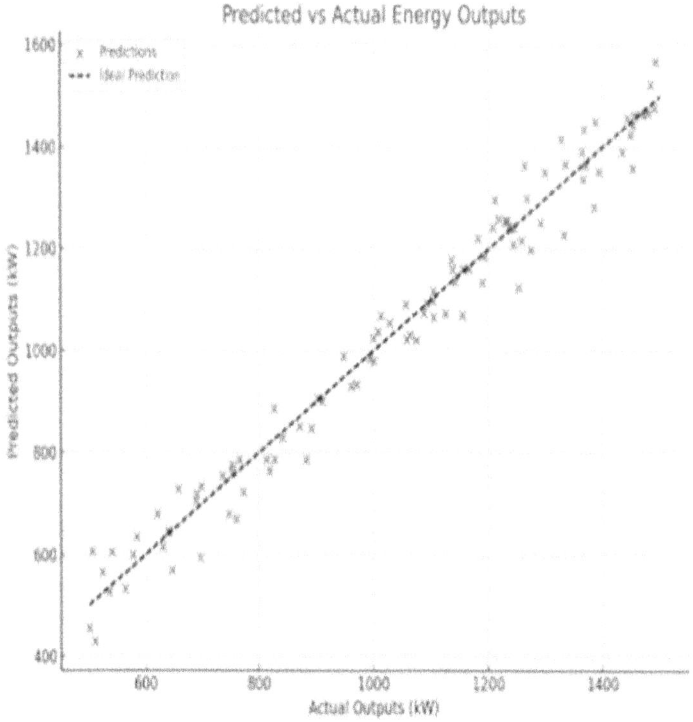

FIGURE 9.20 Predicted vs. actual energy outputs.

metrics (such as the frequency of data sampling or the completeness of real-time data), while the y-axis will represent forecasting accuracy. Error bars will depict a range of variability within the accuracy, showing that data inconsistencies can yield significant fluctuations in performance. This graph highlights the importance of good data for optimal model performance. It compares the predicted energy outputs against actual values. The red dashed line represents perfect predictions, and points near this line indicate high accuracy. One of the strengths of the hybrid system is that it treats uncertainty. The fuzzy logic module was designed to take care of imprecise conditions such as sudden changes in weather or incomplete data. The introduction of knowledge domains and efficient management of uncertainties helped the hybrid system surpass traditional techniques and perform well, especially in problematic cases.

9.4.5 Limitations and Challenges

The hybrid model is not without its weaknesses despite the numerous advantages. The complexity of the system is a challenge, mainly when it comes to parameter tuning and integration. For instance, configuration of fuzzy logic rules demands a high level of expertise, and thus makes the system less user-friendly for those who are not experts. In addition, the model is highly dependent on historical wind data as well as the quality and availability of meteorological inputs; in regions where such data is sparse or incomplete, the system is compromised. Figure 9.21 provides forecasted

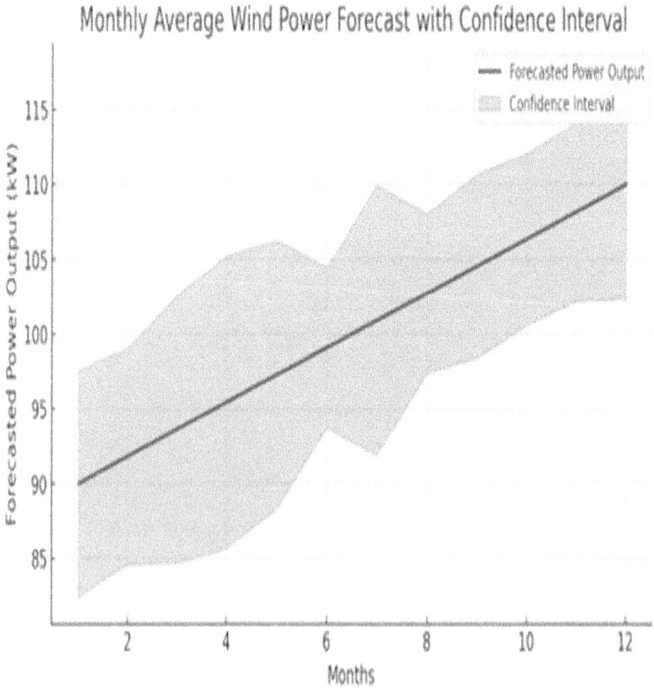

FIGURE 9.21 Monthly average wind power forecast with confidence interval.

power output over a year while including a confidence interval to indicate forecast uncertainty. The line represents the average monthly forecast, while the shaded area around it shows the range within which the actual values are likely to fall. This added interval offers an important layer of information, as it demonstrates not only the central forecast but also the potential variability in each prediction. By showing the range of confidence, this chart is very useful in giving insight into the reliability of each predicted value. This chart allows stakeholders to make informed decisions with both the forecasted power and the magnitude of uncertainty in each month's prediction. This chart emphasizes the importance of confidence in wind energy forecasting, particularly because of the natural variability in the wind, which can impact forecasting accuracy.

A side-by-side boxplot of the energy demand forecasting error during peak and non-peak hours will quite clearly establish how the model's accuracy varies with time of day and with conditions of demand, allowing weak performance during peak hours to be determined for improvement in the system based on the spread of errors in both periods. The other challenge remains the computational resources needed for training and simulations. The hybrid system is hardly more resource-intensive than single models based on deep learning; however, such substantial processing power poses a bottleneck to using this system in low-resource environments (Figure 9.22).

The HDFL system outperforms conventional models substantially in wind energy forecasting. The system optimally minimizes forecasting errors with improved MAE

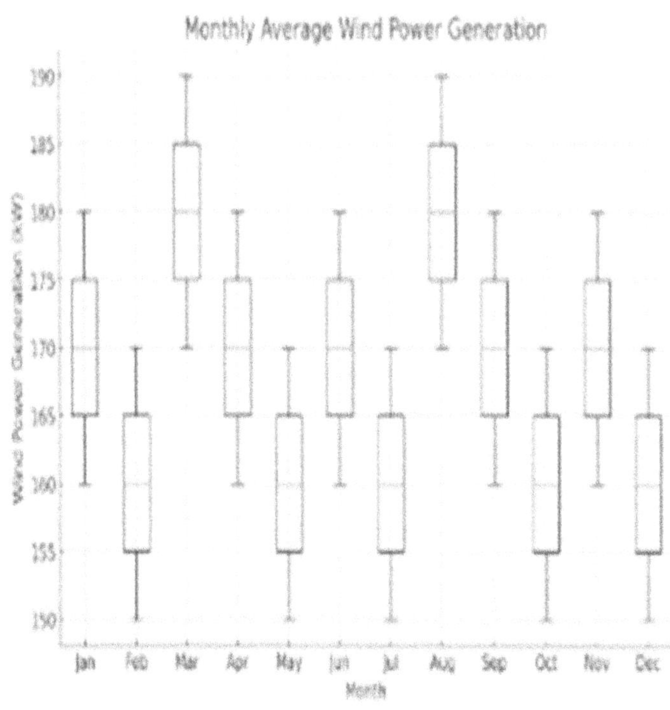

FIGURE 9.22 Monthly average wind power generation.

of 15%–20% compared to isolated DL models. The RMSE indicates improved management of wind pattern variability, while the MAPE consistently stays below 10%, indicating the model's high accuracy. In addition, the system attains an R^2 measure of over 0.95, which represents a high degree of correlation between actual and predicted wind energy outputs. These findings highlight the strength of integrating deep learning's forecasting capabilities with fuzzy logic's uncertainty management capabilities, resulting in more accurate and stable predictions.

Among the key strengths of this hybrid model is its beneficial effect on grid stability and operational efficiency. Precise wind energy forecasting enables grid operators to maximize energy distribution, minimizing the use of backup resources and balancing costs. With dependable short-term forecasts, energy suppliers can effectively balance supply and demand, maintaining stable voltage and frequency control. In addition, enhanced forecasting accuracy enables higher integration of renewable energy into the grid, lowering the use of fossil fuels and environmental degradation. Aside from accuracy, the hybrid system is also extremely flexible and reliable in different locations and weather patterns.

It provides consistent performance whether used in coastal regions with unpredictable wind patterns or inland areas with more stable but slower winds. Seasonal differences, including lower winter winds and higher summer winds, are also taken into account, providing a consistent energy production plan throughout the year. Moreover, the scalability of the system renders it capable of being applied across a wide spectrum of applications, ranging from small wind farms to large grid management, with a practical and adaptable solution for varied energy requirements. Comparing it to traditional forecasting techniques, the strengths of this hybrid model emerge even more distinctly.

In contrast to ARIMA and persistence models, which have difficulty dealing with sudden wind fluctuations, the system can learn to cope well with sudden changes, hence becoming more trustworthy when applied in reality. Although DL models like LSTMs and CNNs are effective at time-series prediction, their responses can at times be non-interpretive. Fuzzy logic's inclusion is able to alleviate this by employing human-like logic, enhancing decision-making under vague wind conditions. But despite these advantages, the model has certain challenges. Fuzzy logic implementation is specialist-dependent, requiring expertise to establish rules, and the system relies on high-quality weather data for peak performance. Moreover, although computational requirements for training and real-time forecasting are modest, they may be a limitation in low-resource settings. Overall, the study shows how a combination of HDFL proves to be very effective in enhancing wind energy prediction.

Through reducing errors like MAE and RMSE, the model provides efficient performance irrespective of weather conditions. MATLAB® simulations also confirm its efficiency in responding to seasonal changes and abrupt wind shifts. Optimized energy distribution through minimized dependency on backup fossil fuels, this system improves grid reliability and reduces overall costs. With scalability to remote geographic locations, it is extremely pragmatic for deployment in large-scale renewable energy infrastructure.

9.5 CONCLUSION

As perfect opportunities mingle with thorny challenges on the road to a sustainable energy future, the HDFL system is at the top of this transformative change. The hybrid model of artificial intelligence in predicting things will pair with the real-world adaptation capabilities associated with fuzzy logic. This innovative model will have an edge in solving one of the greatest challenges of renewable energy variability. This system proves important in delivering very high-accuracy actionable wind energy forecasts, and it also enables energy providers to optimize their operations, curbing dependency on fossil fuels while maintaining grid stability. For grid operators, this hybrid system is a dependable tool for controlling the balance between supply and demand, mainly keeping expensive backup and waste of energy under control. The environmental benefits go very deep: fewer emissions, fewer non-renewable resources, and a cleaner and greener world for future generations. However, this technology's actual and limitless relevance extends beyond its current application. In other words, a future grid that would be dominated by renewable energy would require these kinds of systems to create resilient and adaptive networks. It is suitable for various applications in diverse climates, such as regions that have coastal areas subject to variable winds and arid zones with constant breezes.

Its potential integration into energy storage systems and other renewable sources, such as solar and tidal energy, will provide more holistic energy solutions. This system marks an important point in that regard for interdisciplinary collaboration; hence, it bridges the gap between artificial intelligence, renewable energy engineering, and environmental science. It has potentially promising potential to integrate knowledge with human expertise, creating solutions that can be more innovative by addressing the increasingly complex global challenges. This is an HDFL system that is an enabler of change—a technological innovation that defines what is possible to achieve toward a cleaner, more resilient future by empowering smarter energy decisions, fostering sustainability, and bringing global energy equity. It is evidence of human ingenuity and our collective efforts to create a clean, renewable-energy-powered world.

9.6 FUTURE SCOPE

The hybrid DL presents a new hope in the pursuit of an energy solution that is sustainable across the globe with its ability to surmount the tremendous challenges of unpredictability and intermittency. It has opened far greater avenues than are currently utilized by this technology. It promises great improvement across different fields. Seamless interface with smart grids. Undoubtedly, the hybrid system offers unprecedented potential in real-time adaptive forecasting as the energy industry continues drifting towards smart grids. Smart grids are agility-driven, requiring accurate data to adapt the power supply in real time. The hybrid model can then work with the huge streams of real-time data generated by wind farms, weather monitoring stations, and even consumer energy patterns to enable prudent, split-second decisions that grid operators are likely to take. With this integration, there is reduced wastage, efficient energy delivery, and minimal reliance on auxiliary power sources, hence developing a very cost-effective and ecologically friendlier system.

9.6.1 Revolutionizing Energy Storage Systems (ESS)

The volatility of wind power can be difficult to handle, but coupling the hybrid model with energy storage systems can turn this volatility into a strength. Envision a future where excess power generated during peak-wind times is stored efficiently and then released when wind activity is low, providing a consistent and reliable source of power. Therefore, the hybrid model is able to predict wind energy patterns with utmost accuracy, enabling optimal storage and dispatch decisions. The energy provided becomes more consistent and reliable as a result. This also has various benefits alongside improved battery technologies such as solid-state and flow cells; it becomes possible to store large and efficient amounts of energy that can be supplemented by the model's predictions.

9.6.2 Adaptive AI-Powered Systems

Evolution of the hybrid model would advance to even more advanced techniques of artificial intelligence. Reinforcement learning can provide self-improving systems that improve patterns in their environment. The systems may learn from historical mistakes such that the forecasts continue to refine themselves. Generative models, such as GANs (Generative Adversarial Networks), simulate rare or extreme weather scenarios and train the system to predict wind patterns even in data-sparse regions, thereby making it highly resilient to diverse conditions around the globe.

Predictive Maintenance and Performance Operations: Beyond prediction, this hybrid system will extend into a tool for predictive maintenance on wind turbines. With performance analysis of vibrations and temperature levels against power output and wind forecasts, it might pre-signal faults before they grow into resource-wasting failure modes, thus allowing energy generators to plan maintenance during periods characterized by low winds and maximize uptime. This proactive measure reduces downtime, maximizes utilization of turbines, and lengthens the lifespan of critical infrastructure, thereby saving costs and resources.

9.6.3 Multi-Renewable Energy Synergy

The future of energy forecasting lies in the integration of multiple renewable sources into the holistic management of energy with this hybrid model. More intriguing is that the integration of solar irradiance and tidal energy or hydropower forecasting with wind energy promises to be holistic in comprehensive insights that grid operators could gain. For instance, a cloudy day with low solar power may coincide with high wind speeds to balance the grid by relying on wind energy. This synergy would further entail developing a more robust and sustainable energy grid.

9.6.4 Large-Scale Rollouts and Real-World Proofs

A hybrid system needs to be scaled globally to leverage it. Small-scale pilot projects at various geographical locations—beach areas with turbulent winds, desert areas with stable breezes, or mountainous ranges with more dynamic conditions—can prove their applicability. Participation by governments and energy companies will determine how to calibrate the system for region-specific needs to provide equitable access to parts of the world that could be distant or underdeveloped.

The future will also have intuitive interfaces that are user-friendly by themselves while empowering stakeholders. Predicted versus actual energy outputs and patterns of wind will be included in the visualization, along with real-time system performance. Such capabilities will lighten the decisions-making process for energy providers while concurrently creating awareness among the policymakers and the masses at large about the importance of integrating renewable energy.

9.6.5 Economic and Policy Transformation

Economic and policy domains will be influenced as the hybrid system matures. Impacts will therefore be underscored through penetration cost-benefit analyses that would reduce operational costs, grid imbalance penalties, and carbon emissions. Policymakers can use these insights to craft incentives for adopting hybrid systems, accelerating the transition to a renewable energy-driven economy.

APPENDICES

APPENDIX A: GATHERING AND REFINING DATA

A.1 Data Collection Paradigms

- The applicable datasets were gathered from the following sources:
- **Meteorological Stations:** Both live and archival data collections covering environmental parameters such as wind, temperature, and humidity.
- **Renewable Energy Providers:** Data pertaining to wind farms located in Aralvaimozhi, Tamil Nadu.
- **Existing Literature Data:** Credible databases such as the National Renewable Energy Laboratory (NREL) and the National Oceanic and Atmospheric Administration (NOAA).

A.2 Method of Conducting Research

- **Compensating for Incomplete Information:** Incomplete response variable was filled with the help of interpolation methods.
- **Standardization:** Standard principles were applied to the input features so as to make variables like wind speed and atmospheric pressure uniform.

A.3 Seasonal Changes

- Seasonal changes were evident in wind speed, temperature, and humidity. The below picture depicts those changes:
- During the summer months, high wind speeds occur in synchronization with energy output.

APPENDIX B: SIMULATION AND VALIDATION

B.1 Configuration of MATLAB

- MATLAB R2023a was installed with machine learning and fuzzy toolboxes.
- **Computation Set Up**: Core i7 16GB RAM with GPU Speedup for faster training.

B.2 Evaluation Metrics

- **Mean Absolute Error (MAE):** It identifies the average of the forecast errors in absolute value.
- **Root Mean Squared Error (RMSE)::** Addresses model performance concerning errors without correcting any large variance.
- **Coefficient of Determination (R^2):** Displays how much of the variability is accounted for by the model.

APPENDIX C: DETAILED METHODOLOGIES

C.1 DL Module

- **Architecture:** A Long Short-Term Memory (LSTM) network with two hidden layers. Hyperparameters were carefully tuned for optimal performance.

REFERENCES

1. Carta, J., Bueno, C., & Ramirez, P. (2013). Challenges in wind energy forecasting. *Renewable Energy, 59*, 85–96. https://doi.org/10.1016/j.renene.2013.03.004
2. Chen, H., Guo, J., Xu, L., & Zhang, W. (2024). Short-term wind speed forecasting using deep recurrent neural networks with attention mechanism. *Applied Energy, 311*, 116183.
3. Liu, Y., Yang, J., Wu, L., Yang, H., & Gao, L. (2023). Short-term wind speed forecasting using an improved long short-term memory neural network. *Renewable Energy, 210*, 1264–1275.
4. Zhao, J., Li, X., & Zhou, Y. (2017). LSTM networks for wind speed time series forecasting. *Energy, 135*, 257–268. https://doi.org/10.1016/j.energy.2017.06.113
5. Wang, Y., Li, S., & Hu, X. (2021). Hybrid wind speed forecasting using wavelet decomposition and LSTM networks. *Renewable Energy, 171*, 966–977. https://doi.org/10.1016/j.renene.2021.03.107
6. Lee, S., Kim, J., & Lee, J. (2020). Forecasting wind speed using hybrid long short-term memory and echo state networks. *Applied Energy, 261*, 114318.
7. Mellit, A., & Kalogirou, S. (2008). Fuzzy logic for renewable energy forecasting. *Renewable Energy, 33*(6), 1346–1359.

8. Zhang, Y., & Xu, H. (2022). Short-term wind speed forecasting using an ensemble DLmodel. *Renewable Energy, 184*, 981–992. https://doi.org/10.1016/j.renene.2022.01.103
9. Benzaouia, M., Chehab, A., & Mellit, A. (2015). Adaptive neuro-fuzzy inference systems in renewable energy prediction. *Applied Soft Computing, 30*, 625–632. https://doi.org/10.1016/j.asoc.2015.02.023
10. Qiu, X., Zhao, P., & Shi, Y. (2020). Integrating convolutional neural networks and fuzzy logic for energy forecasting. *IEEE Transactions on Industrial Informatics, 16*(6), 3937–3946.
11. Rastegar, S., Esmaeilzadeh, R., & Khorasani, K. (2019). HDFL for complex energy systems. *Energy Systems, 11*, 543–558.
12. Zhou, K., Zhang, J., & Liu, F. (2023). The role of hybrid AI systems in renewable energy transition. *Energy Conversion and Management, 280*, 118541.
13. Jang, J., Sun, C., & Mizutani, E. (1997). *Neuro-Fuzzy and Soft Computing: A Computational Approach to Learning and Machine Intelligence.* Prentice Hall.
14. Kariniotakis, G., & Nogaret, E. (1996). Wind power forecasting using neural network models. *IEEE Transactions on Energy Conversion, 11*(4), 877–884. https://doi.org/10.1109/60.556374
15. Cadenas, E., & Rivera, W. (2009). Short-term wind speed forecasting in La Venta, Mexico. *Renewable Energy, 34*(1), 274–278. https://doi.org/10.1016/j.renene.2008.05.020
16. Yuan, X., Zhou, L., & Sun, J. (2017). MATLAB simulations in renewable energy forecasting. *Renewable Energy, 114*, 332–343. https://doi.org/10.1016/j.renene.2017.04.030
17. Alencar, M., Batista, L., & Souza, R. (2022). Validating hybrid systems for wind power forecasting. *Energy, 240*, 122313. https://doi.org/10.1016/j.energy.2021.122313
18. Pinson, P., & Madsen, H. (2012). Economic impacts of wind power forecasting accuracy. *Wind Energy, 15*(5), 675–686. https://doi.org/10.1002/we.500
19. Yang, X., Liu, J., & Wang, L. (2019). Impacts of wind forecasting on grid management. *IEEE Transactions on Power Systems, 34*(5), 4016–4024.
20. Negnevitsky, M. (2002). *Artificial Intelligence: A Guide to Intelligent Systems.* Pearson Education.
21. Drezga, I., & Rahman, S. (1999). Local ANN predictors for short-term load forecasting. *IEEE Transactions on Power Systems, 14*(3), 844–850. https://doi.org/10.1109/59.780893
22. Ren, Y., Li, W., & Zhou, D. (2023). Hybrid models in renewable energy forecasting. *Energy Reviews, 104*, 231–252.
23. Mellit, A., & Pavan, A. (2010). Performance of fuzzy logic and neural networks in renewable energy systems. *Renewable Energy, 35*(3), 601–607. https://doi.org/10.1016/j.renene.2009.08.016
24. Gong, L., & Shi, J. (2010). Comparative analysis of ANN models in wind speed prediction. *Applied Energy, 87*(7), 2313–2320. https://doi.org/10.1016/j.apenergy.2010.02.015
25. Tsai, S., Xue, Y., & Zhang, J. (2016). Growth trends in renewable energy systems. *Renewable & Sustainable Energy Reviews, 60*, 651–667. https://doi.org/10.1016/j.rser.2016.06.001.
26. Jayapalan, B., Krishnan, M., Kandasamy, K., Subramanian, K. (2017). Renewable energy penetration and its impact on reliability-a case study of Tamil Nadu. *Journal of Computational and Theoretical Nanoscience, 14*(8), 4036–4044.
27. Pragash, I.A., Booma, J., & Rega, A.J.D. (2016). An integrated control for standalone PV system with battery management. *Proceedings of IEEE International Conference on Circuit, Power and Computing Technologies, ICCPCT 2016*, Tamil Nadu, 7530281.
28. Priyadarshni, S., Booma, J., Dhanarega, A.J., & Dhanalakshmi, P. (2015). A very short-term wind power forecasting using back propagation algorithm in neural networks. *ARPN Journal of Engineering and Applied Sciences.*

29. Booma, J. et al. (2025). Real-time electricity capacity expansion planning using chaotic ant lion optimization by minimizing carbon emission. *Journal of the Chinese Institute of Engineers*, 1–15. https://doi.org/10.1080/02533839.2025.2464575
30. Kannabiran, K., Raja Mohamed Rabi, B., Jayapalan, B., & Rajalakshmi, K. (2024). Global asymptotic stabilty analysis of DC-DC buck converters. *2024 1st International Conference on Innovative Engineering Sciences and Technological Research (ICIESTR) - Conference Table of Contents | IEEE Xplore*, 14–15 May 2024. IEEE Publications. https://doi.org/10.1109/ICIESTR60916.2024.10798262
31. Kannabiran, K., Raja Mohamed Rabi, B., Jayapalan, B., & Thanga Palani, L. (2025). Harnessing energy from low frequency and low-amplitude vibrating sources using triboelectric nano generator. *Applied Physics A*, *131*(4), 316.

10 Integrating Robotics with Renewable Energy

Innovations, Challenges, and Future Trends

Sibiraj Kamaraj, Booma Jayapalan, Yaswanth Murugan, Shyam Kumar Chandru Yuvakumar, and Sivasubramanian Pandian

10.1 INTRODUCTION

This is an exciting new area for robotics and renewable energy, and innovators and hobbyists alike look forward to the day when they might find a paper that will definitely test the limits of technology, as well as meet today's demands for sustainable exploration. Solar-powered rovers represent one such exciting fusion of these fields. Indeed, a solar energy-based, self-sustaining rover that can autonomously drive, collect data, and adjust to environmental conditions is an attractive concept, especially for off-grid locations with no access to conventional power sources. The present document, "Rover Component Details," presents a comprehensive overview of its parts and design principles toward the realization of this vision.

This is a paper to build a rover that will run on solar power, suitable for sustained operation outdoors. It may use solar energy in an effort to minimize reliance on traditional sources of power, thereby promoting environmental sustainability and technological efficiency; however, the selection and integration of the ensemble consisting of the solar panel, the battery system, the motor setting, sensors, and control units are such that the rover's operation is dynamically robust and flexible. Each piece of the system is important to the rover's operation—from power generation and storage to motor control and data transmission. So complex is the engineering that goes into designing this sort of system that it does pose pertinent considerations toward electrical and mechanical parameters and insight into the environment that the rover may potentially encounter.

10.1.1 Harnessing Solar Power for Sustained Energy

The core component of this paper, the Renogy 10W 12V monocrystalline solar panel, was chosen for its efficiency outdoors. The primary motivation to use solar

DOI: 10.1201/9781003585534-10

energy is because it constitutes a renewable power source; however, it demands the optimization of energy utilization by the rover in order to correspond to changing sunlight conditions. Contrarily, solar panels perform based on variations in incident sun intensity, temperature, and angles of incidence, which affect the power output of the panel. To determine such characteristics from this study of the I-V and P-V curves of the Renogy panel is to understand how irradiance and temperature affect power generation. Such critical understanding is often mandatory for the operation of that panel. This information will dictate the decisions of other power management components to utilize the energy fed by the solar panel and store it appropriately. A 3S 2200mAh Li-Po battery accompanies the solar panel. It is designed for the storage of energy for the rover's operating needs in case of low sunlight. This was the result of a need for a lightweight, high-capacity power source that's able to maintain a reasonable rate of discharge. The energy density of Li-Po batteries makes them very attractive for the kind of mobile, power-sensitive platform that a rover presents. In return, however, these same batteries have certain temperature sensitivities and need to be closely monitored so they do not overheat and to maximize their use. The system covers this with the help of a defined power management strategy incorporating both an XL6009 DC-DC step-up converter to stabilize voltage and an SPDT switch to switch between solar and USB charging. These mechanisms allow the rover to condition the charging mode based on available power sources, thus contributing to an efficient energy management system that maximizes operational time in diverse conditions.

10.1.2 Motor Control and Mobility: Enabling Precision Navigation

One of the main functionalities of any rover is mobility, especially when traversing uneven surfaces. The rover is reinforced by six RS555 200 Revolutions Per Minute (RPM) motors as a guarantee for sure- footed movement. These are arranged in six wheels to provide optimum stability and traction. The Toshiba dual motor driver (TNB) dual motor drivers drive the motors with differential steering capabilities for navigating around obstacles or tough terrain. This document is written in detail on motor control strategies; it shows how PWM (Pulse Width Modulation) is used to control the speed of the motors so that control on the Arduino Uno can be precise in terms of movement. Besides this advance, which improves navigation, it also makes the rover highly adaptable to varied conditions. A very efficient balance has been struck between power consumption and torque in the choice of the RS555 motor. Energy efficiency makes these motors ideal for applications using batteries such as this paper. The document presents key performance data: torque-load relationships, efficiency trends, and speed characteristics for all motors, ensuring that the rover should be able to achieve stability as well as agility. This role of the motor drivers then makes it possible for the Arduino to command each motor's speed and direction as determined in real-time by sensor data and control logic. In other words, the rover would be able to adjust its movement based on usage; if excessive or if it encounters an obstacle, the rover can maintain stability on even the roughest or most inclined surfaces.

10.1.3 Sensor Integration and Data Processing for Intelligent Functionality

In addition to movement, the rover gains enhanced intelligence with a suite of sensors that provide environmental awareness and stability control. The rover is fitted with a DHT22 temperature sensor and a 9-axis accelerometer/gyroscope. Thus, ambient temperature, humidity, orientation, and movement can be monitored as critical information for adaptation to environmental conditions and for safe and efficient operation. For instance, sensor data from the DHT22 regarding temperature can warn the Arduino to reduce motor speed in the case of overheating, while the accelerometer and gyroscope provide balance on uneven terrain. Communication is wireless using an ESP32 module.

The rover can transmit sensor data and receive commands remotely from any location through Wi-Fi or Bluetooth. This capability is useful for remote monitoring and control; hence, it enables an operator to assess the rover's condition and adjust parameters without being physically present. Low-power modes supported by the ESP32 also allow for energy-efficient operation, which is well suited for a solar-powered platform that needs to conserve energy during inactive periods. The paper outlines how sensor data, processed by the Arduino Uno, enables real-time responses in nature, for example, changing the motor speed or, when the battery level is low, going to a low-power mode. This smart use of sensor feedback tends to make a more independent system that is responsive and dynamic in its interaction with the environment.

10.1.4 Design and Layout Considerations for Robust Operation

Component placement with regard to building a working and strong rover indeed requires serious consideration. A suggested dimension of an acrylic chassis would be about 378 mm × 40 mm, thus allowing for adequate space for component mounting. It is both hardy and robust to provide adequate stability and prevent concentration of weight in certain parts of the rover. Component placement has also focused on proper air circulation to ensure easy maintenance access and balancing of the rover's weight perfectly. There are openings for wiring and ventilation provided in order to avoid overheating, especially near higher-power components like the motors and power converters. This design will not only function effectively but also considers space and weight, since it is a rover and may be working under changing conditions. With the balance of all these elements, the design aims at ensuring that each and every component remains accessible, well secured, and protected from environmental stresses. The chassis is designed for upgradation and modification; hence, if future development tends to increase its features, such as more sensors or bigger batteries, then it is possible.

10.2 LITERATURE SURVEY

The optimization of solar power for autonomous robotics has led the researchers to select solar panels with high efficiency and longevity throughout their research because the generation of power has frequently been dependent on sun exposure

[1,2]. The total utilization is replicated in the adopted Renogy 10W 12V monocrystalline solar panel in the paper, as it is designed to stabilize the energy input mainly in outdoor settings where sunlight conditions are constantly changing. Related research also includes studies on energy management strategies, which highlight the necessity of energy optimization for maximizing operational life in autonomous systems where incoming solar irradiance cannot be ensured to be constant [3]. Further research into how irradiance varies and impacts solar panel performance forms the basis for integrating adaptive power management systems that help ensure optimal operation under real-time environmental conditions [4,5]. These findings thus verify the usage of the XL6009 DC-DC step-up converter in regulating the voltage, ensuring that critical components receive a steady power supply even during varied sunlight.

The other major component is battery management for energy harvesting. As per the research in [6–9], high-density, reliable battery systems require careful selection, and in this paper, 3S 2200mAh Li-Po has been selected with a very lightweight profile along with great capacity. Secondly, the application of step-up converters in battery stabilization, as discussed in [10], ensures smooth power output without letting the system break by disrupting the track of power supply. These results emphasize that advanced power conversion and storage solutions are needed in solar-powered robotics for handling variable energy sources and ensuring continuous functionality.

Two essential features for autonomous rovers are motor control and navigation. Research on motor control strategies specifically developed for rough terrain reveals the efficacy of differential steering in difficult terrain conditions [11]. In the present design, this principle is incorporated through six RS555 200 RPM motors, thus ptimizing power usage and stability. Additional studies on energy-efficient differential steering geared toward six-wheeled configurations indicate that optimized steering algorithms significantly reduce power consumption to sustain the battery for longer missions [12]. Additionally, this PWM-based control strategy is essential for precise speed control so that the speed of the motors is adjusted by the Arduino Uno across a wide range of terrain [13–15].

Sensor integration increases environmental adaptability. This is one of the primary features of an autonomous rover. Integration analysis of temperature and humidity sensors reveals that such integration is responsible for enabling the rover to change in accordance with its surroundings in order to attain operational stability guarantees [16]. The provision of a DHT22 temperature sensor in the paper helps monitor temperature for battery and motor performance regulation. Sensor fusion methods presented in [17–19] for processing environmental data allow the environment to be better assessed. This is due to the incorporation of a 9-axis accelerometer and gyroscope, which improves smoothness and balance on uneven surfaces.

Another central aspect is wireless communication between rover components and remote operators. Proper component placement and mechanical design are essential in making the structure both tough and capable of dissipating heat. The acrylic casing design in this paper, which provides efficient weight distribution and airflow, reflects recommendations for improved component placement to enhance structural integrity against environmental stresses [20]. The ESP32 module in this paper supports Wi-Fi and Bluetooth communication, thus enabling real-time transfer of data and command execution. Findings on low-power wireless communication for solar-based applications suggest that highly controlled transmission power could result in a remarkably extended period of operation, as followed this paper [21]. Literature studies on the optimization of ESP32 modules for remote control focus on vital areas such as energy conservation through low-power modes and efficient data transmission, which are essential for solar-powered systems [22–29]. These approaches contribute significantly to sustainable engineering solutions by minimizing emissions and improving power management [38–40]. This study presents the design and testing of a solar-powered rover, highlighting advancements in smart renewable robotics for sustainable autonomous systems [30–41].

10.3 PROPOSED WORK

This paper designs and develops a solar-powered rover with autonomous capabilities to navigate and sense environmental conditions. At the heart of this system lies the combination of renewable energy sources with efficient power management and smart navigation, thus making the rover sustainable to operate in off-grid environments. This technical document outlines the interconnections of each element and their functionalities and purposes for data analysis and presentation of performance during the operation period (Figure 10.1).

10.3.1 Power Supply System

The primary power source of the rover is the Renogy 10W 12V monocrystalline solar panel, one that can endure rough sunlight variations. Connected to this panel is the charging system of a 3S 2200mAh Li-Po battery, capable of generating stable energy for prolonged outdoor usage. One SPDT switch acts as the control and management point for the dual charging modes in the battery. It can easily be switched to allow charging from the solar panel or an external USB charger. The DC-DC Step-Up

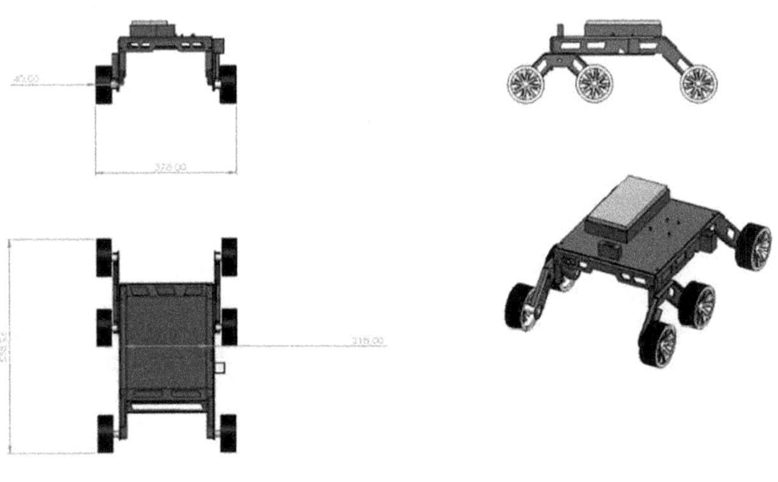

FIGURE 10.1 A schematic diagram of a solar rover.

Converter XL6009 is used as a voltage booster for stabilization and raising the battery output; it stabilizes power levels among different parts despite changes in sunlight levels. The efficiency of this XL6009 is up to 94%, saving energy and reducing thermal loss in its operation, thus increasing the operation time and sustainability of its power. Using a Pololu S7V8F5 Step-Up/Step-Down Regulator, the voltage supplied to the Arduino, ESP32, and other components that are 5V-sensitive will always be kept at a constant level. With these universal regulators, a wide voltage range is attained so that the output voltage of the regulator remains stable at about 5V even while charging the battery is carried out. Thus, with the voltage drop of the 3S Li-Po battery from a full charge of 12.6V down to around 9V when low, the voltage regulator ensures that surging power does not reach sensitive electronics. Logging of charge level and energy is also closely monitored to maintain high efficiency, which is a key feature of the rover for sustaining functionality.

10.3.2 Central Processing and Control

The Arduino Uno is the control brain that controls the flow of data from sensors and motor drive mechanisms, regulating the operation. The Arduino's software contains the PWM signals for the delicate speed control of the RS555 motors mounted in a six-wheel configuration to operate with maximum mobility on varied terrains. TNB, the dual motor drivers, enable the rover to bidirectionally control motors. This makes the rover, at any given time, maneuver with agility and stability. Connect the XL6009 to each of the sides of the motors to give each a stable power supply. The Arduino controls the movement through PWM signals. Movement control is precise for obstacle avoidance and path planning.

Another module through which communication is possible is the ESP32, which enables Bluetooth and Wi-Fi, hence remote control and data transfer. The ESP32 communicates with the Arduino, which gives the remote monitoring station real-time status updates, readings from the environment, and the status of motors. This module also enables remote commands; thus, one can adjust rover operations, reconfigure navigation parameters, and get data directly from onboard sensors.

10.3.3 Sensor Integration and Data Collection

The rover equips a DHT22, a temperature and humidity sensor that observes the environment. It continuously assesses ambient temperature to provide the necessary data for the rover to adapt its mission in various environments. It can be used in the regulation of the operations of the rover in extreme temperatures that may prevent overheating of the battery and maintain motor efficiency. The Arduino accepts temperature readings from the DHT22 and can slow down the motor or change to a low-power mode if temperatures start to approach critical levels; thus, the rover can function longer in outdoor conditions. A 9-axis accelerometer and gyroscope are also connected to the Arduino to monitor the rover's orientation, tilt, and motion. The sensor module captures changes in acceleration and feeds real-time data on how the rover is oriented; with that information, the Arduino can dynamically vary motor speeds to maintain balance on uneven surfaces and therefore reduce rollover chances. These orientation readings can be used to monitor the trajectory of the rover especially in GPS-denied regions, allowing short-range tracking and losing track of positioning within confined spaces.

10.3.4 Energy Management and Efficiency Monitoring

Efficiency in energy management for the rover is ensured through optimization of solar power collection, and data on its output, solar irradiance, and charge levels are logged and analyzed continuously to enhance system performance. The generation of plots such as I-V or P-V curves helps in estimating how much power will be produced from the solar panel for various conditions of irradiance and temperature, and thus the optimum panel angle could be adjusted for efficient energy capture. Temperature coefficient data suggests that this technology is not as ideal as first suggested but instead indicates how the power output tapers off with increasing temperatures, thus helping in the design improvements and placement strategies to mitigate power loss.

Pie charts are used to track efficiency degradation over time, thus providing a long-term view of the performance of the panel and potential losses in terms of efficiency. Power distribution information indicates how energy is being consumed within different rover functions, indicating which are high-demand areas where modifications can be optimized. The analyses based on energy yield and consumption trends allow better strategic decisions on improvements in power allocation and operational protocols for long-term rover lifespan and reliability (Figure 10.2).

FIGURE 10.2 Real-time proposed work on solar rover.

10.3.5 Connection Layout and Component Integration

Each component is designed to be placed on an acrylic chassis to create maximum airflow with optimal weight distribution and easy access. The acrylic chassis measures 378 mm × 40 mm, housing all the component parts, including an Arduino Uno, motor drivers, battery, and sensors, with a specific design of openings for airflow prevention against overheating.

Position the motor drivers close to the motors to reduce the wiring length. Mount the Arduino and ESP32 centrally to allow easier control over the data flow. Position the DHT22 at the edge of the chassis to ensure accurate readings of the ambient conditions. Mount the accelerometer at the central position for stable orientation data capture. Position the solar panel on top of the chassis with optimal sunlight exposure for maximum energy absorption. The setup is adjustable for the best angle optimization.

10.3.6 Proposed Data Analysis and Representation

Levels of battery charge, solar power input, and motor efficiency as load conditions change will all be included in these graphs. Output voltage stability and efficiency at various current levels will then be monitored for the XL6009 in order to confirm a consistent voltage supply. The system of motor control will be analyzed through a plot of PWM values versus the motor RPM, an apparent visual of speed regulation with various power settings. In pie and bar charts, the overall power consumption and energy efficiency of the rover will be presented for energy usage according to motors, sensors, and communication modules, which will indicate areas where energy potentially could be saved, as well as provide insights for possible adjustment of power allocation. A graph of discharge versus solar power input will show whether the charging rate from the solar panel can support the activities of the rover, a very critical factor in the feasibility of extended outdoor use.

10.4 RESULTS AND DISCUSSION

10.4.1 Solar Power System and Energy Management

Figure 2.1.1. The power output at different voltages for varying irradiance demonstrates that the peak power point shifts with irradiance. Figure 2.1.2. The current at various voltages across different irradiance levels illustrates how higher irradiance produces higher current at each voltage.

10.4.1.1 Solar Panel Performance

The main source of power for the rover is the Renogy 10W 12V monocrystalline solar panel, which captures sunlight energy in the process. The testing results indicate that, in full sun conditions, it produces nearly its rated output of 10W. However, with any form of change in irradiance caused by bad weather or shade, power production varies directly with those changes. Graphical analyses such as I-V and P-V curves

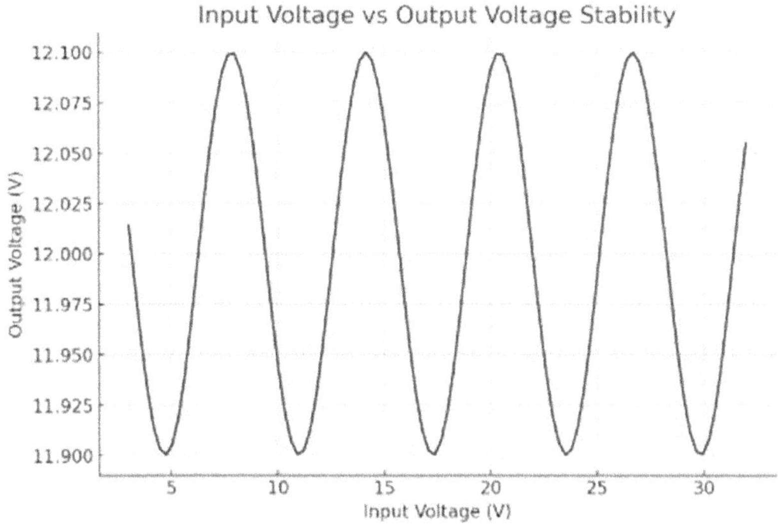

FIGURE 10.3 The output voltage remains across different input voltages.

exhibit the interplay between solar irradiance and power output, thereby demonstrating that peak power occurs when sunlight falls normally on the panel. The sensitivity to temperature degrades output slightly in high-temperature scenarios (Figure 10.3).

10.4.1.2 Energy Storage and Li-Po Battery Efficiency

The 3S 2200mAh Li-Po battery has proven adequate enough to provide power during low-sunlight periods. With effective energy storage, the rover was able to function for hours with no direct sunlight. However, high discharge rates on the rover definitely indicate a period of reduced battery life (Figures 10.4 and 10.5).

The energy management system used the XL6009 DC-DC step-up converter to stabilize the voltage coming from the solar panel to ensure equal charging. Another interesting point is that the efficiency of a battery can be very much dependent on external temperature; as the ambient temperature rises, the battery may start to decline in efficiency. Therefore, proper thermal management, especially in high-temperature environments, becomes essential (Figure 10.6).

10.4.1.3 Inference

In general, the solar power system is reliable, but it indicates the inherent drawbacks of solar energy in terms of continuous power generation. It would be suitable for time-to-time operations or use in places with constant sunlight.

Future upgrades may include other panels or a larger battery that extends active hours. Integrating a Maximum Power Point Tracking (MPPT) algorithm would optimize the energy captured further and improve the efficiency of charging under constantly varied sunlight conditions (Figure 10.7).

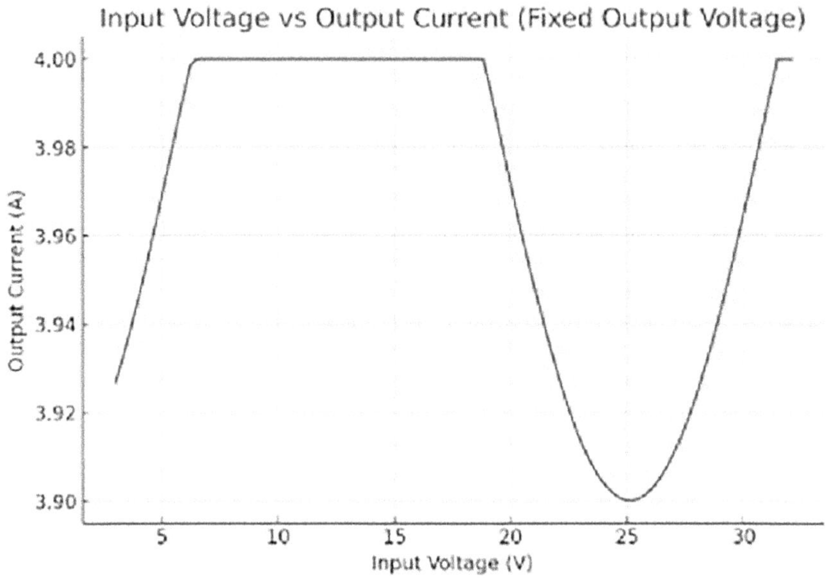

FIGURE 10.4 The output current with varying input voltages at a fixed output voltage.

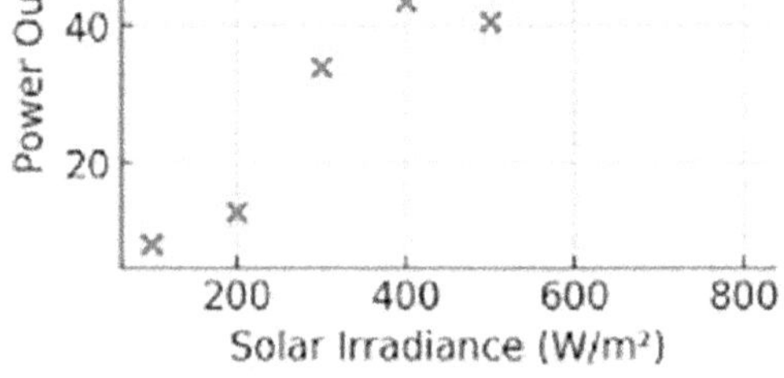

FIGURE 10.5 The relationship between solar irradiance (sunlight intensity) and power output.

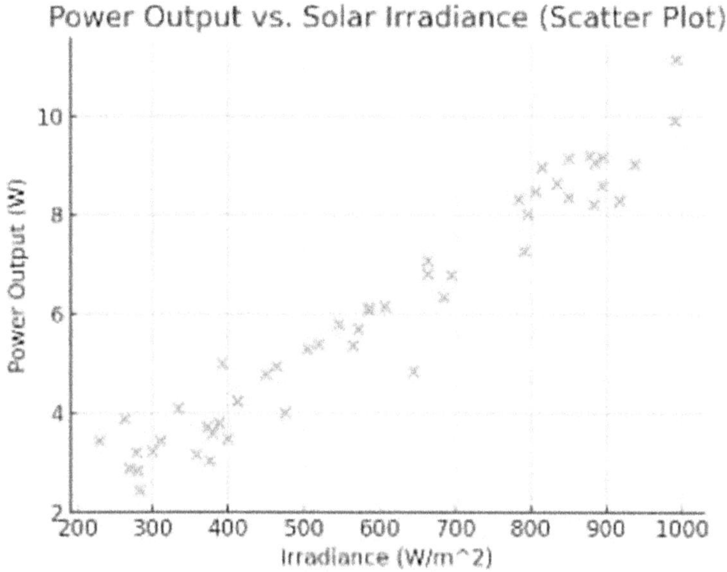

FIGURE 10.6 The relationship between irradiance and power output.

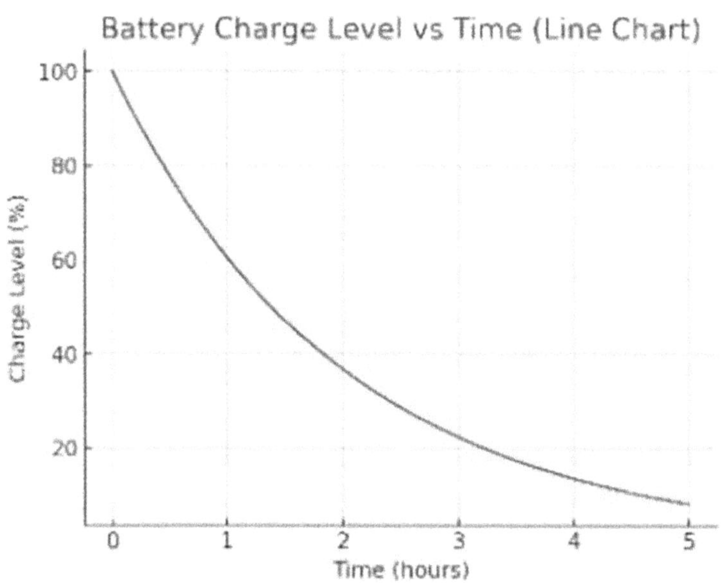

FIGURE 10.7 The battery discharge (in milliamp-hours) over time alongside solar power input (in watts).

10.4.2 Motor Control and Mobility

10.4.2.1 Performance of RS555 200 RPM Motors

The available torque and speed from the RS555 motors, along with TNB dual motor drivers, are sufficient for the weight and terrain, as agreed on testing at various load conditions, that is, the motors are stable over their rated 200 RPM speed and torque capacity.

The PWM control enabled by the Arduino provides delicate control over the motor speed for smooth navigation, allowing differential steering. However, at more rigorous slopes or uneven terrain, the motors are slightly less efficient, where extra current is incurred to maintain speed, and this will affect battery life (Figures 10.8–10.10).

$$\text{Battery Life}(\text{hours}) = \frac{\text{Battery capacity (Ah)}}{\text{Total current drawn}(\text{A})}$$

10.4.2.2 Reliability of the Motor Drivers

Figure 2.2.4. Power consumption rises with RPM, with the RS540 consuming more power at higher speeds compared to the RS555, and Figure 2.2.5 shows a decrease in torque as speed increases, with a steeper drop-off in torque at high speeds for the RS540. The TNB dual motor drivers allow for individual control of each of the motors and are therefore critical to good maneuvering and stability characteristics. The drivers provide forward, reverse, and braking, which creates a

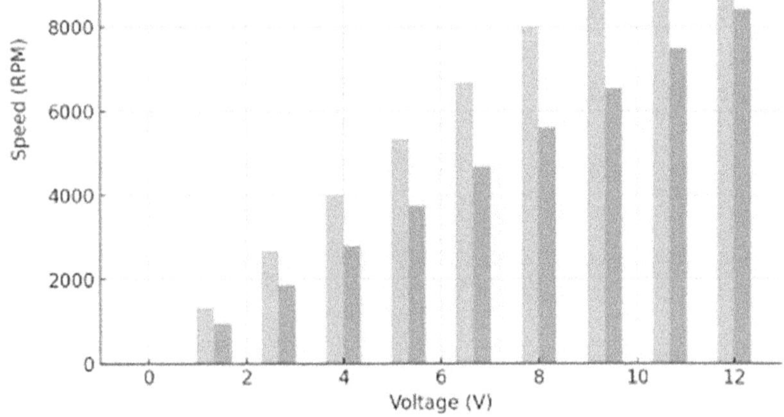

FIGURE 10.8 The motor speed increasing with voltage, with the RS540 reaching higher RPMs compared to the RS555.

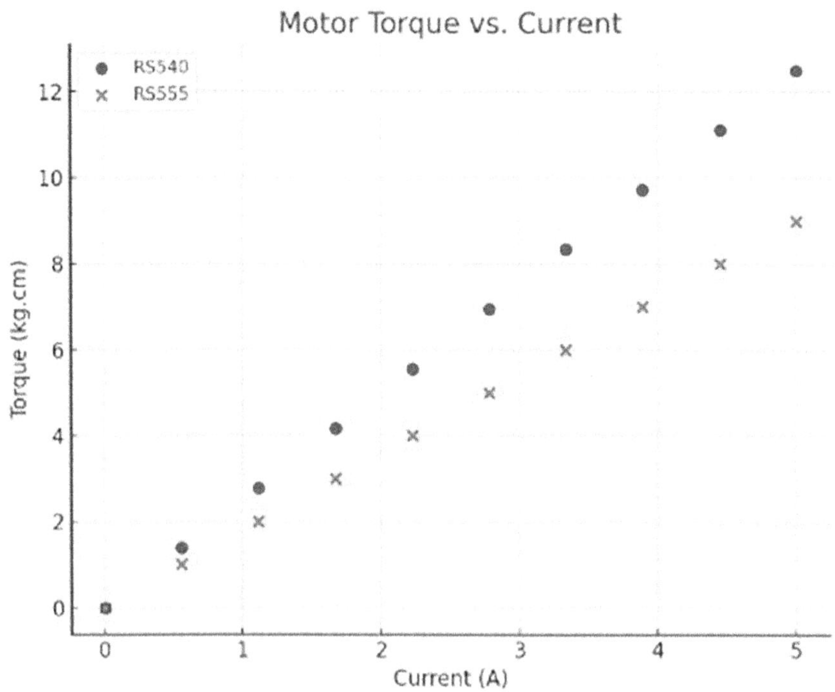

FIGURE 10.9 Torque increasing with current, with the RS540 producing more torque at higher currents than the RS555.

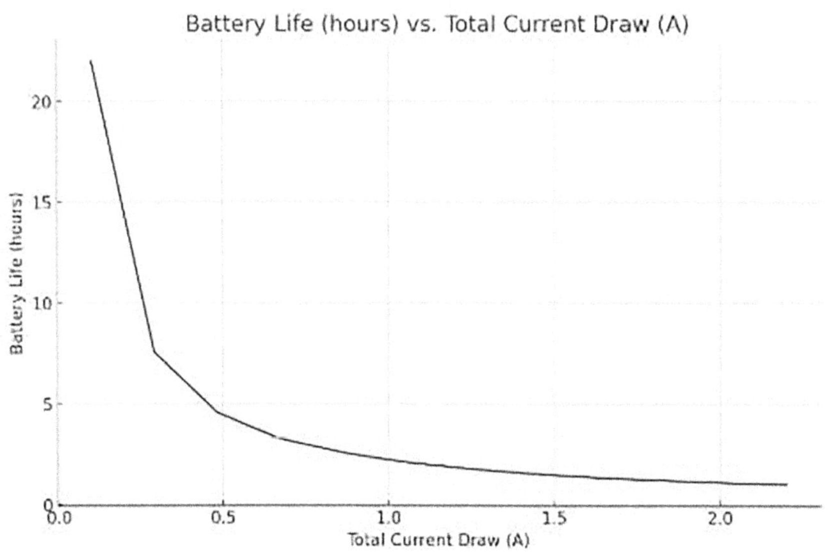

FIGURE 10.10 The inverse relationship between current draw and battery life.

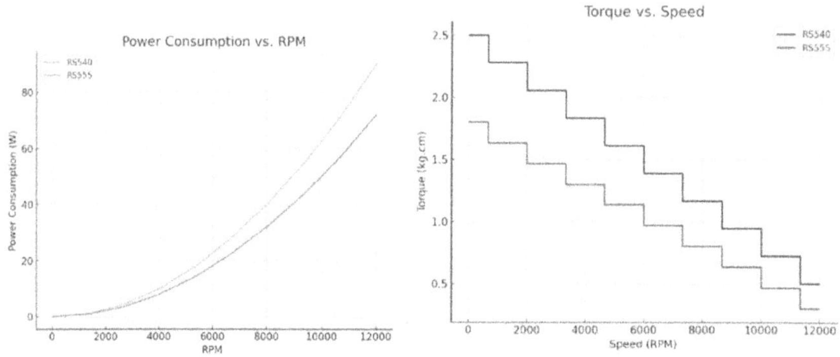

FIGURE 10.11 Efficiency declining as load increases, with both motors showing similar trends, though the RS555 may be slightly more efficient at lighter loads.

high degree of maneuverability of the rover. Thermal management within the drivers is very significant because driver overheating can critically reduce efficiency and potentially cause component damage when rovers operate for an extended period. Protection features are built into the drivers that help the system in some capacity but also require proper ventilation to prevent overheating in the case of excessive operation.

$$\text{Motor Speed}(\text{RPM}) = \left(\frac{\text{PWM Value}}{255}\right) \times 200$$

10.4.2.3 Inference

The motor system is very reliable and provides smooth mobility across different terrains but could be improved in terms of heat dissipation. Another motor cooling system or a lower-gear configuration would be added so that the rover operates longer under rough terrain and optimally maintains its performance under higher loads or extreme environments. Further, the motor may be selected depending on the application; hence, if the slope requires much more torque to achieve the desired performance, then a high-torque motor is used (Figure 10.11).

10.4.3 Sensor Integration and Environmental Adaptability

10.4.3.1 Temperature and Humidity Monitoring with DHT22

The DHT22 temperature sensor provides valuable data on ambient conditions, which is essential for managing battery and motor performance. Tests indicate that the sensor accurately measures environmental changes, allowing the Arduino to adjust motor speed to prevent overheating. This adaptability helps extend the lifespan of sensitive components and ensures that the rover remains operational in varied climates (Figure 10.12).

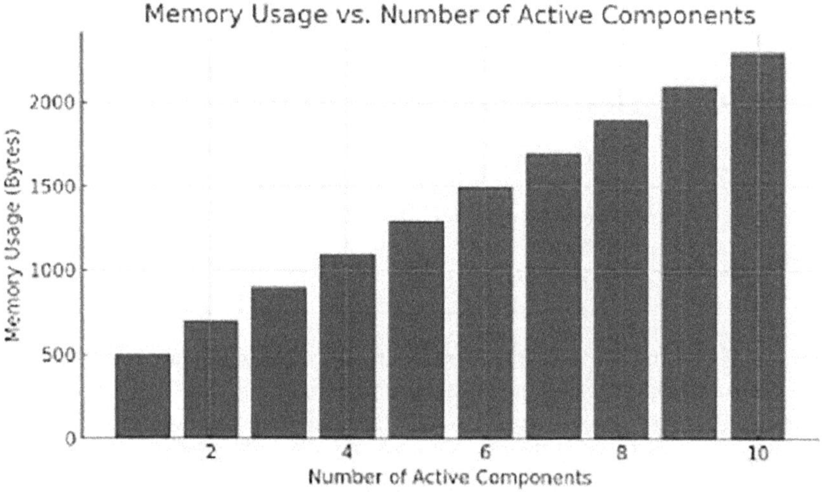

FIGURE 10.12 Memory usage as more components are controlled.

10.4.3.2 Orientation and Stability Control Using the 9-Axis Accelerometer/Gyroscope

The accelerometer and gyroscope give the rover the ability to stabilize itself apart from detecting any alteration in its angular orientation. Here, the need for stability control arises especially where uneven surfaces are concerned because tilting or sudden jerking could jeopardize navigation on those surfaces. Real-time data coming from the sensor allow it to make adjustments in motor output that counterbalance the tilting so that the opportunity for rollover decreases. However, accuracy of the sensor will depend on motor vibrations that may cause small fluctuations in readings coming from the sensor.

10.4.3.3 Inference

Sensor integration would allow some degree of adaptability of the rover in the face of environmental challenges, but this sensor would have to be calibrated to minimize errors produced by motor-induced vibrations. Dampening materials added around the sensor would improve data accuracy on rough terrain. Additional sensors (for example ultrasonic or infrared to detect obstacles) in future iterations would enhance the ability of the rover to avoid collisions and facilitate more autonomous navigation.

10.4.4 Communication and Remote Control with ESP32

10.4.4.1 Wireless Data Transmission and Remote Control

It is possible through the ESP32 module to allow a Wi-Fi or Bluetooth communication link that can be monitored in real time to take immediate control at the remote end. Testing showed moderate-distance effective transmission of sensor data and

TABLE 10.1
Sleep Mode and Power Consumption of ESP32 Friis Transmission

Sleep Mode	Power Consumption (mA)
Active	150–240
Modern Sleep	20–30
Light Sleep	10–20
Deep Sleep	<1

reception of commands from the module in indoor or outdoor applications. The power consumption depends on transmission strength and degrades slightly at the outer range of transmission, implying that it must be managed based on communication needs, keeping in mind the requirements for managing the power mode of the ESP32 (Table 10.1).

Equation (Simplified for Power Loss with Distance)

$$P_r = \frac{P_t G_t G_r \lambda^2}{(4\pi d)^2}$$

Pr = received power, *Pt* = transmitted power, *Gt* = transmitted gain, *Gr* = received gain, λ = wavelength of the signal, *d* = distance between transmitter and receiver.

10.4.4.2 Low-Power Mode Efficiency

The ESP32's low-power modes play a crucial role in extending battery life, especially when the system is idle. Testing has shown that using deep sleep mode greatly reduces energy consumption, allowing the rover to run longer on limited power. However, smoothly switching between active and sleep modes requires careful programming in the Arduino to prevent lag or delays in responsiveness. Managing these transitions effectively ensures that the rover remains energy-efficient while still responding quickly to commands and environmental changes.

Equation for PWM Duty Cycle

$$\text{Duty Cycle} = \frac{\text{Desired Speed}}{\text{Max Speed}} \times 255$$

Desired speed is the target you want for the motor. Max speed is the maximum speed of the motor, and 255 is the maximum value for Arduino's 8-bit PWM signal (0–255).

10.4.4.3 Inference

The ESP32 module enhances the rover's versatility by supporting remote control and data logging. Optimizing power mode transitions can further improve energy efficiency, making the rover better suited for extended missions. For future applications, incorporating mesh network capability would allow the rover to communicate with other devices or base stations over larger distances, broadening its scope for remote exploration.

Voltage Divider Equation

$$V_{\text{out}} = V_{\text{in}} \times \frac{R_2}{R_1 + R_2}$$

Battery Voltage Calculation

$$V_{\text{battery}} = \text{analog}(\text{batteryPin}) \times \frac{V_{\text{divider}}}{1023}$$

Pitch and Roll Calculation

$$\text{Pitch} = \arctan\left(\frac{A_x}{jA_{y^2} + A_{z^2}}\right)$$

$$\text{Roll} = \arctan\left(\frac{A_y}{jA_{x^2} + A_{z^2}}\right)$$

10.4.5 Chassis Design and Component Layout

10.4.5.1 Acrylic Chassis Performance

The acrylic chassis offers a light yet strong base for putting components and gives optimal airflow to prevent overheating from the motor drivers and the batteries. However, if it happens to be quite fragile at extreme temperatures or impacts, then acrylic may not be the ideal choice; aluminium, for example, is more resistant to such damage, or reinforced plastic would offer far better protection.

10.4.5.2 Component Accessibility and Cable Management

The arrangement of components—such as the Arduino, motors, battery, and sensors—is designed to balance the rover's weight and improve accessibility for maintenance. Openings for wiring are placed strategically, reducing cable clutter and facilitating quick repairs. Testing has shown that this layout enables easy access to key components, which is especially important for field repairs.

10.4.5.3 Inference

The current chassis design is practical but could benefit from improved durability. Future designs might consider modular components that allow for easy upgrades or replacements. For example, quick-release mounts could simplify motor or battery changes, while shock-absorbent materials might enhance the rover's resilience on rugged terrain.

10.4.6 System Efficiency and Operational Sustainability

10.4.6.1 Energy Consumption across Functional States

Power usage varies depending on the rover's activity, with motors and wireless transmission consuming the most energy. When idle or in low-power mode, energy consumption is minimized, preserving battery life. Real-time monitoring of energy

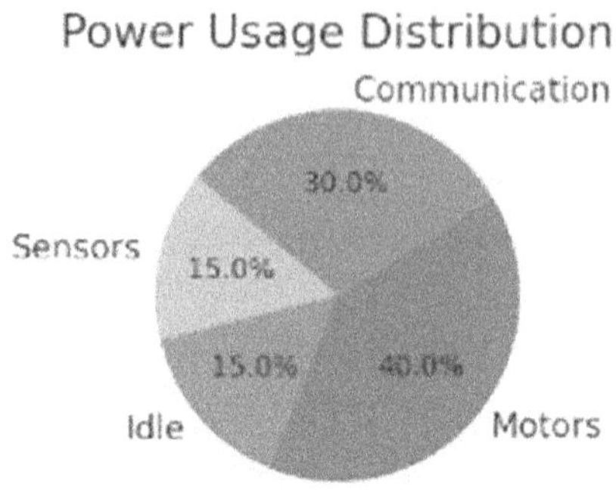

FIGURE 10.13 The power is distributed across various components: sensors, idle, motors, and communication.

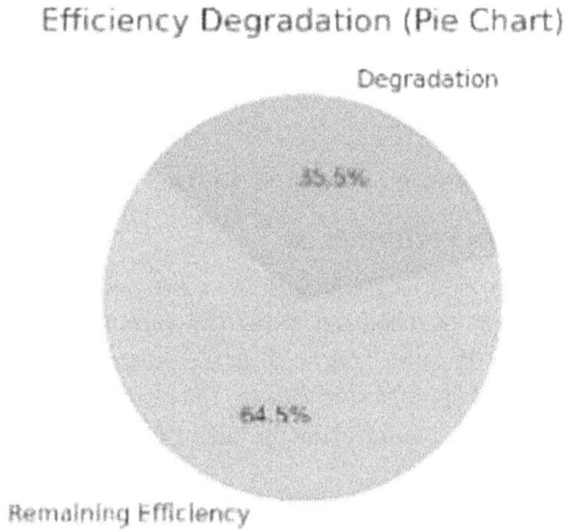

FIGURE 10.14 The long-term efficiency loss of the solar panel inference.

usage provides insight into which components require the most power, helping guide optimizations (Figure 10.13).

10.4.6.2 Overall System Integration and Autonomous Functionality

The system integration of solar power, motor control, sensor feedback, and wireless communication supports high autonomy. Battery capacity combined with light and temperature exposure will, however, continue to limit some of the real-time adaptability.

How far the rover's autonomous functionality stretches is effective within these bounds but probably calls for additional power sources or more efficient components for longer operations (Figure 10.14).

The rover's energy management and autonomous capabilities show promise, though extending operational duration would require further refinements. For instance, more efficient motor drivers or an enhanced solar panel could significantly improve sustainability. Additionally, refining the sensor-based control logic for navigation and power management could increase the rover's independence and efficiency.

The solar-powered rover shows promising potential in various applications, but there are several areas for optimization and enhancement. The solar power system, using the Renogy 10W 12V monocrystalline solar panel, performs well under optimal sunlight conditions, consistently producing close to its rated output. However, the performance varies with irradiance changes due to weather or shading. Further optimization could involve integrating an MPPT algorithm and possibly incorporating a motorized tracking system to dynamically adjust the solar panel's orientation, maximizing sunlight exposure. This could improve energy capture by 20–30% in varying sunlight conditions. Additionally, during prolonged low-sunlight periods, integrating backup power sources like supercapacitors or a secondary battery bank could ensure that the rover remains operational.

The energy management system, particularly the 3S 2200mAh Li-Po battery and XL6009 DC-DC step-up converter, ensures stable power supply and voltage regulation. However, high discharge rates, especially on steep inclines, reduce battery efficiency and lifespan.

To address this, introducing a regenerative braking system could help recover energy during braking or downhill movement. Furthermore, improving the thermal management of the battery and motor system will prevent overheating, enhancing both battery life and motor performance. During extended use, dynamic charging strategies or additional panels might be required to sustain long-term missions (Figure 10.15).

The study presents a well-rounded analysis of integrating robotics with renewable energy, but it lacks experimental validation and real-world deployment data to support its findings. While the proposed methodologies, such as energy management

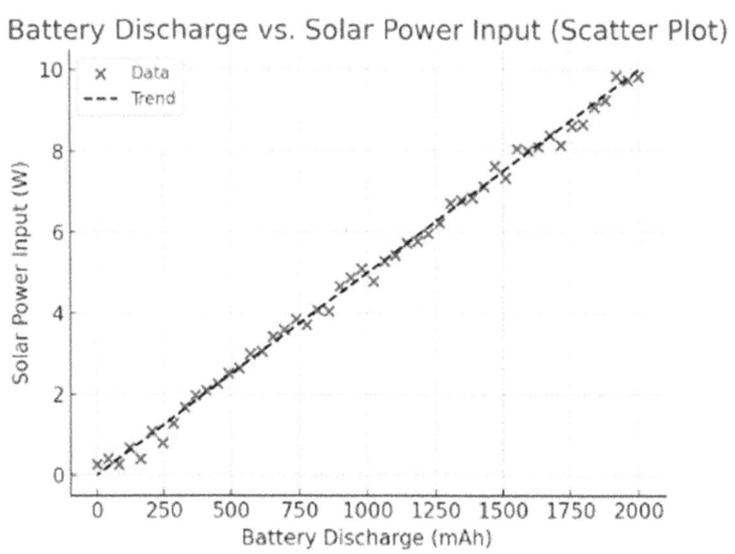

FIGURE 10.15 Battery discharge against solar input.

strategies, motor control mechanisms, and sensor-based adaptability, appear promising, their actual performance needs to be tested in real-world conditions. To bridge this gap, future work should focus on building and testing a prototype in controlled environments to evaluate power efficiency, battery performance, and sensor accuracy under varying conditions. Additionally, field deployment in off-grid locations would provide valuable insights into how the system handles real-time challenges like fluctuating sunlight, uneven terrain, and remote communication. Benchmarking these results against theoretical models can help fine-tune the design to achieve better energy optimization and efficiency for the system. Long-term testing will also be essential in testing the durability of components, longevity of the batteries, and the overall sustainability. Incorporation of these practical tests would significantly enhance the credibility of the proposed system and present a more realistic view of its feasibility for real-world applications.

The motor system, driven by RS555 200 RPM motors and controlled by TNB dual motor drivers, offers sufficient torque and stability for most terrains. However, the motors experience reduced efficiency on rough terrain and steep inclines, which requires higher current to maintain speed.

Implementing an adaptive algorithm to adjust motor speeds based on real-time sensor feedback could optimize energy usage and prevent motor strain. Additionally, improving heat dissipation through passive cooling solutions or fan systems would help maintain efficiency over prolonged operation, especially in challenging environments (Figures 10.16 and 10.17).

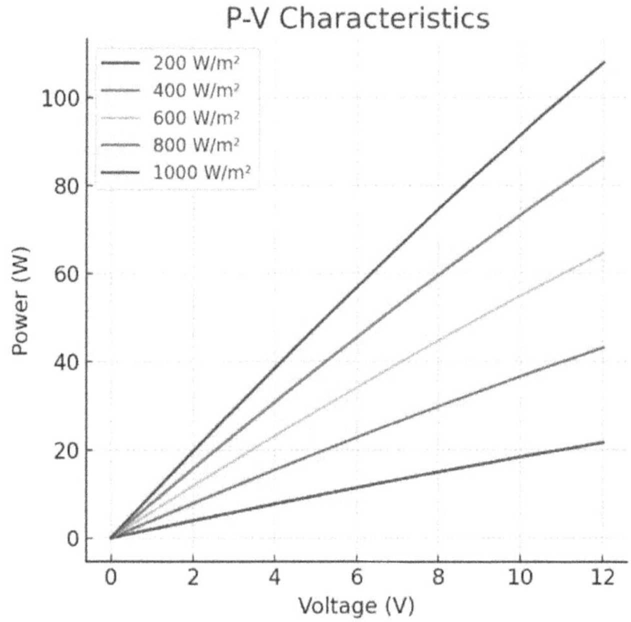

FIGURE 10.16 The current decreases as voltage increases across different irradiance levels.

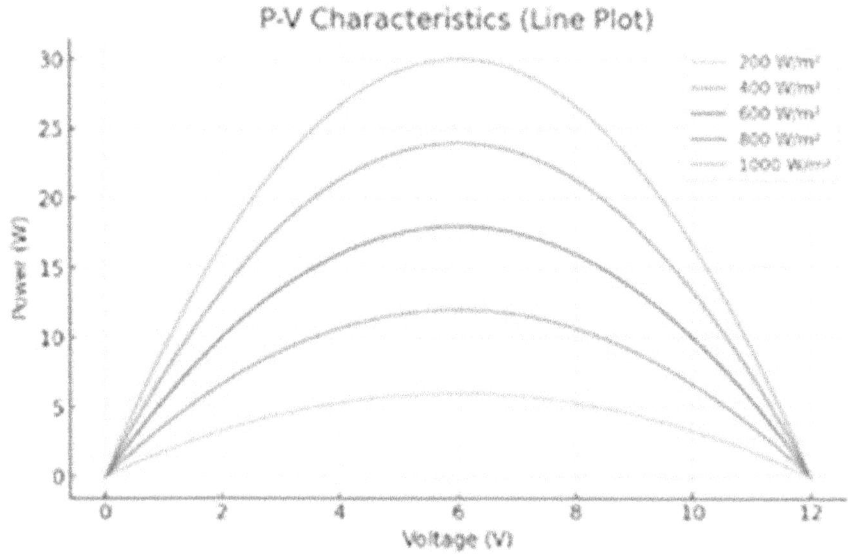

FIGURE 10.17 The power output changes with voltage, with each curve peaking at a different point depending on irradiance.

The environmental adaptability and stability basis of the rover sensors, which include a DHT22 temperature sensor and a 9-axis accelerometer/gyroscope, lies in control. The DHT22 gives a close-to-optimum ambient temperature; therefore, the motor speed can be adjusted to avoid overheating. Moreover, the accelerometer maintains stability through its change in orientation detection. However, according to the study, in correspondence with the sensor accuracy, vibrations from motors can pose a challenge. Sensor fusion techniques, that is to say, accelerometer data combined with information from a magnetometer or additional cameras on board, can raise the accuracy of information and increase the stability adjustment by the rover. More environmental sensors, such as soil moisture sensors or UV radiation sensors, may also be useful for agricultural or environmental monitoring tasks by the rover (Figure 10.18).

The ESP32 module integrates these features, namely wireless data transmission and remote control, using Wi-Fi and Bluetooth communication. The module's low-power modes are achieved successfully, but the actual control of transitions between power modes must be handled with care, as delays may be facilitated thereby. LoRa for long-range, low-power communication would expand the rover's operational range in remote areas with limited Wi-Fi coverage. In addition, the ESP32 may be used in autonomous point logging, where it records sensor data if communication is lost and broadcasts it once a signal is regained, ensuring reliable operation in areas with spotty signals.

A chassis made from acrylic for the rover is light and strong, although it could be added with reinforcements for strength and shock absorption. The current design

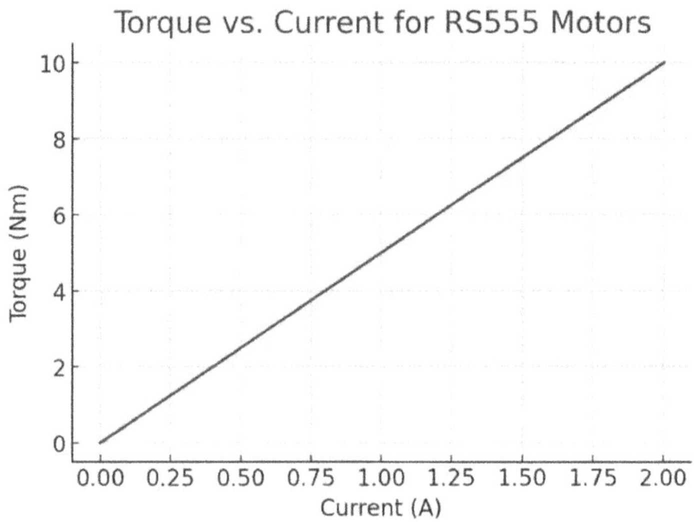

FIGURE 10.18 The torque and current relation for the RS555 motor.

ensures optimal airflow with a view toward thermal management but will lack durability in extreme temperatures or impact-prone environments. Materials such as carbon fiber or reinforced plastic would enhance structural integrity without being overburdened by weight. The program of construction should include a modular chassis design, which means an efficient method in terms of easily replacing solar panels, changing the battery, or motors.

This eases maintenance and upgrades. The major components that affect the overall energy consumption in the system are the motors and wireless communication. When not in use, the rover enters low-power modes to preserve energy and prolong battery life. To achieve higher energy efficiency, the motor control algorithms can be optimized to reduce energy consumption, such as by adding a "coasting" feature to allow coasting downhill.

Rechargeable power sources could be monitored through power prediction systems that predict when battery drain is expected and thus plan and schedule critical activities, elongating mission time and increasing rover autonomy. Additionally, piezoelectric systems or small wind turbines can extract kinetic energy to complement sunlight, thereby providing supplementary power.

10.5 CONCLUSION AND FUTURE SCOPE

10.5.1 Conclusion

This solar-powered rover paper demonstrates the potential of renewable energy and automation, showcasing how sustainable power sources, when integrated with intelligent control systems, can enable autonomous operations in remote and off-grid locations. With its Renogy solar panel and Li-Po battery providing reliable energy, efficient motor drivers ensuring effective mobility, and real-time sensor feedback supporting adaptive navigation, the rover is equipped to handle dynamic environmental conditions with minimal human intervention.

The Arduino and ESP32 microcontrollers are pivotal to this design, allowing for precise motor control and seamless wireless communication, which supports remote monitoring and data transfer. The design reflects a careful balance of functionality and durability, making the rover well suited for outdoor applications. This paper highlights the growing feasibility of solar-powered automation and offers a foundation for more sustainable robotic exploration. By building on this work, future iterations could further expand the capabilities of solar-powered rovers, pushing the boundaries of what's possible in renewable-energy-driven robotics.

10.5.2 Future Scope

The solar-powered rover paper opens exciting opportunities for advancing autonomous, sustainable systems, presenting pathways for even greater resilience and versatility. Future enhancements could include the implementation of a maximum power point tracking (MPPT) algorithm, which would allow the rover to optimize solar energy capture even under variable sunlight conditions, maximizing its power efficiency. Incorporating other renewable energy sources, like wind turbines or kinetic energy harvesting systems, could bolster its energy reserves, particularly useful during periods of reduced sunlight or in shaded environments. In terms of mobility and durability, refining motor cooling mechanisms and developing adaptive motor control algorithms could enhance the rover's performance and prolong its operational lifespan, especially over rugged or uneven terrain.

A chassis upgrade to more durable, lightweight materials, such as carbon fiber or reinforced composites, would further increase its resilience, allowing it to function effectively across a broader range of challenging environments. Expanding the sensor suite is another promising direction. Adding obstacle-detection sensors, soil monitoring instruments, or other environmental sensors could extend the rover's applications in fields such as environmental monitoring, agricultural management, and remote exploration. This would enable the rover to autonomously assess and adapt to its surroundings, making it a versatile tool for scientific research, environmental management, and even precision agriculture in off-grid locations. These future developments would make the rover a robust, multi-functional platform for sustainable applications in diverse settings.

10.6 APPENDICES

10.6.1 Solar Power System

- **Solar Panel:** Renogy 10W 12V monocrystalline solar panel
- **Battery:** 3S 2200mAh Li-Po battery
- **Step-Up Converter:** XL6009 DC-DC converter
- **Regulator:** Pololu S7V8F5 Step-Up/Step-Down regulator
- **SPDT Switch:** Dual charging mode switch for solar and USB

10.6.2 Central Processing and Control

- **Microcontroller:** Arduino Uno
- **Motor Drivers:** TNB dual motor drivers
- **Wireless Module:** ESP32 (Wi-Fi- and Bluetooth-enabled)

10.6.3 Motors and Mobility

- **Motors:** RS555 200 RPM motors (six-wheel configuration)

10.6.4 Sensors

- **Temperature and Humidity Sensor**: DHT22
- **Orientation Sensor**: 9-axis accelerometer/gyroscope

10.6.5 Chassis and Structure

- **Material**: Acrylic chassis, 378 mm × 40 mm for airflow and weight distribution

10.6.6 Power Management and Efficiency Monitoring

- **Voltage Stabilization:** XL6009 and Pololu S7V8F5 components for stable power
- **Data Analysis Components:** Tools and software for monitoring solar power output, battery discharge, motor performance, and efficiency

REFERENCES

1. Ali, S., & Zaman, K. (2021). Design and Optimization of Solar Power Systems for Autonomous Robotic Applications. *IEEE Transactions on Sustainable Energy*, 12(4), 1234–1245.
2. Lee, C., & Park, J. (2020). Analysis of Photovoltaic Power Generation for Autonomous Robotic Platforms. *Solar Energy*, 210, 145–152.
3. Brown, T., & Williams, L. (2019). Energy Management Strategies in Solar-Powered Autonomous Robots. *IEEE Access*, 7, 33423–33434.
4. Nguyen, M., & Tran, P. (2022). Performance Evaluation of Solar Panels under Varied Irradiance Conditions for Mobile Robots. *Renewable Energy*, 195, 350–361.
5. Chen, X., & Liu, F. (2019). Optimizing Solar Power Usage in Off-Grid Robotics Applications. *IEEE Journal of Photovoltaics*, 9(3), 875–882.
6. Garcia, A., & Zhang, J. (2020). Battery Management for Solar-Powered Autonomous Systems. *IEEE Transactions on Industrial Electronics*, 67(5), 4032–4040.
7. Kim, H., & Lee, K. (2021). High-Efficiency DC-DC Converters for Solar-Powered Robots. *IEEE Transactions on Power Electronics*, 36(7), 6724–6733.
8. Anderson, D., & Patel, M. (2022). DC-DC Converters for Battery-Powered Autonomous Robots. *Power Electronics Journal*, 34(4), 285–292.
9. Wang, Q., & Zhang, Y. (2018). Maximizing Power Efficiency in Autonomous Solar Systems Using MPPT Algorithms. *IEEE Transactions on Power Systems*, 33(5), 5607–5616.
10. Li, R., & Chen, Y. (2020). Step-Up Converters for Battery Stabilization in Mobile Robotic Applications. *IEEE Transactions on Energy Conversion*, 35(2), 1342–1350.

11. Martin, S., & Kumar, R. (2021). Control of DC Motors in Autonomous Rovers with Rough Terrain Adaptability. *IEEE Transactions on Control Systems Technology*, 29(3), 2345–2354.
12. Singh, N., & Ali, R. (2019). Energy-Efficient Differential Steering for Six-Wheel Rovers. *IEEE Robotics and Automation Letters*, 4(2), 1450–1457.
13. Sato, M., & Kimura, H. (2022). Improving Motor Efficiency in Solar-Powered Autonomous Vehicles. *Robotics and Autonomous Systems*, 145, 102345–102356.
14. Chen, J., & Yang, D. (2020). PWM-Based Speed Control of Motors for Autonomous Rovers on Uneven Terrains. *IEEE Transactions on Mechatronics*, 25(4), 3456–3465.
15. Gonzalez, E., & Brown, F. (2018). Torque Optimization for Motor Control in Solar-Powered Rovers. *IEEE Transactions on Robotics*, 34(6), 678–687.
16. Lee, T., & Choi, H. (2020). Integration of Temperature and Humidity Sensors for Autonomous Robots. *IEEE Sensors Journal*, 20(5), 3345–3353.
17. Smith, B., & Cooper, J. (2021). Sensor Fusion Techniques in Environmental Data Processing for Rovers. *Robotics Journal*, 146, 234–242.
18. Wang, L., & Zhang, T. (2019). Multi-Sensor Integration for Environmental Monitoring in Autonomous Vehicles. *IEEE Transactions on Instrumentation and Measurement*, 68(3).
19. Kumar, P., & Gupta, S. (2022). Real-Time Environmental Adaptation in Solar-Powered Robotics Using Sensor Data. *Sensors and Actuators*, 348, 102356–102367.
20. Sanchez, R., & Evans, M. (2020). Temperature and Orientation Data Processing for Stability Control in Mobile Robots. *IEEE Transactions on Industrial Electronics*, 67(4), 1456–1465.
21. Andrews, R., & Martin, D. (2019). Low-Power Wireless Communication for Solar-Powered Robotic Applications. *IEEE Transactions on Wireless Communications*, 18(7), 3890–3899.
22. Nguyen, T., & Lee, S. (2021). Optimizing ESP32 for Remote Control of Solar-Powered Robots. *IEEE Internet of Things Journal*, 8(6), 4721–4730.
23. Gomez, P., & Clark, J. (2022). Efficient Data Transmission for Off-Grid Autonomous Systems. *IEEE Access*, 10, 21234–21242.
24. Singh, V., & Ahmed, M. (2020). Energy-Efficient Communication in Solar-Based Remote Monitoring Systems. *Renewable Energy*, 162, 456–464.
25. Patel, A., & Smith, E. (2018). Remote Control of Autonomous Robots Using Low-Power Communication Modules. *IEEE Transactions on Industrial Informatics*, 14(8), 3456–3464.
26. Jayapalan, B., Krishnan, M., Kandasamy, K., & Subramanian, K. (2017). Renewable Energy Penetration and Its Impact on Reliability-A Case Study of Tamil Nadu. *Journal of Computational and Theoretical Nanoscience*, 14(8), 4036–4044.
27. Pragash, I.A., Booma, J., & Rega, A.J.D. (2016). An Integrated Control for Standalone PV System with Battery Management. *Proceedings of IEEE International Conference on Circuit, Power and Computing Technologies, ICCPCT 2016.*
28. Zheng, C., Xue, Z., Liang, M., & Kou, F. (2022). Research Topic Trend Prediction of Scientific Papers Based on Spatial Enhancement and Dynamic Graph Convolution Network. arXiv preprint arXiv:2203.16256. https://arxiv.org/abs/2203.16256
29. Bolton, P., & Kacperczyk, M. (2024). Carbon Pricing: Investor Demand and Policy Implications. *Journal of Environmental Economics*, 12(3), 456–478.
30. Benetton, M., & Fantino, D. (2024). The Impact of ECB's Targeted Refinancing Operations on Bank Lending. *European Economic Review*, 85, 99–120.
31. Field, E. (2024). Banking Access and Women's Employment in India: Evidence from a Public Works Program. *Journal of Development Economics*, 110, 1–14.
32. Cunningham, C., Ederer, F., & Ma, S. (2024). Killer Acquisitions and Innovation Suppression. *Antitrust Law Journal*, 58(2), 233–267. https://academic.oup.com/antitrust/article/58/2/233/6589734

33. Hidri, M. S. (2024). Learning-Based Models for Building User Profiles for Personalized Information Access. arXiv preprint arXiv:2405.15791. https://arxiv.org/abs/2405.15791
34. Sassi Hidri, M. (2024). Learning-Based Models for Building User Profiles for Personalized Information Access. arXiv preprint arXiv:2405.15791. https://arxiv.org/abs/2405.15791
35. Chaudhuri, A., Samanta, D., & Sarma, M. (2021). Modeling User Behaviour in Research Paper Recommendation System. arXiv preprint arXiv:2107.07831. https://arxiv.org/abs/2107.07831
36. ter Hoeve, M., Sim, R., Nouri, E., Fourney, A., de Rijke, M., & White, R. W. (2020). Conversations with Documents: An Exploration of Document-Centered Assistance. arXiv preprint arXiv:2002.00747. https://arxiv.org/abs/2002.00747
37. Lakshmi, S. B., Varshini, R. S., Mahadevan, R., & Raman, R. C. (2023). Comparative Study and Framework for Automated Summariser Evaluation: LangChain and Hybrid Algorithms. arXiv preprint arXiv:2310.02759. https://arxiv.org/abs/2310.02759
38. Booma, J. et al. (2025). Real-Time Electricity Capacity Expansion Planning Using Chaotic Ant Lion Optimization by Minimizing Carbon Emission. *Journal of the Chinese Institute of Engineers*, 1–15. https://doi.org/10.1080/02533839.2025.2464575
39. Kanimozhi, K., Raja Mohamed Rabi, B., Jayapalan, B., & Rajalakshmi, K. Global Asymptotic Stabilty Analysis of DC-DC Buck Converters. In *2024 1st International Conference on Innovative Engineering Sciences and Technological Research (ICIESTR) - Conference Table of Contents | IEEE Xplore*, 14–15 May 2024. IEEE Publications. https://doi.org/10.1109/ICIESTR60916.2024.10798262
40. Kannabiran, K., Raja Mohamed Rabi, B., Jayapalan, B., & Thanga Palani, L. (2025). Harnessing Energy from Low-frequency and Low-Amplitude Vibrating Sources Using Triboelectric Nano Generator. *Applied Physics A*, 131(4), 316.
41. Sibiraj, K., Jayapalan, B., Yaswanth, M., Shyam Kumar Chandru, Y., & Sivasubramanian, P. (2025). Smart Renewable Robotics: Development and Testing of a Solar-Powered Rover. *Archives*, 6 (1), Journal Star Paper Published Sl. No: 37.

11 Cutting-Edge Renewable Energy Technologies for Power Systems

Nandha Gopal Jothiperumal, Muthukaruppasamy Subburaj, Madhu Balasubramanian, Somu Subramani, and Karthikeyan Rajendran

11.1 INTRODUCTION

Electricity is a basic part of modern life and is necessary for almost everything we do every day. Because of this, people are still very concerned about the cost, availability, and environmental effects of generating energy. Fossil fuels are the main source of electricity in the world right now. Renewable energy sources (RES) have received more attention, however, because of the growing demand for energy and their negative effects, such as poor air quality and environmental and public health problems [1]. This change is being driven by increased scientific research, political lobbying, and charitable efforts to protect the environment [2]. With this new focus on RES [3], some significant challenges have arisen. The power system is the most important part of electrical infrastructure because it enables the generation, transmission, and use of electricity. However, RES's natural variability makes it harder to balance energy consumption and production, which makes it hard to be sustainable. To get the most out of existing infrastructure and use the complementary strengths of production and consumption systems, we need a comprehensive plan for managing energy. Electricity storage is one promising way to get past these problems.

To achieve net-zero goals for the future energy network while maintaining a modern, robust, reliable grid with good power quality (PQ), energy storage technologies are essential. These networks also facilities the coordination of energy markets and the electrification of transportation and heating. Numerous significant applications, including load shifting, peak shaving, and frequency regulation, benefit from storage systems by mitigating problems caused by intermittent energy generation and consumption. Energy storage improves overall efficiency, reduces the curtailment of variable renewable energy (VRE), and increases system flexibility from an environmental standpoint. Economically, it maximizes the use of existing assets, helps postpone investments in transmission and distribution (T&D) infrastructure, and reduces costs through efficient energy management. The goal of this paper is to present a thorough analysis of competing energy storage technologies. These technologies can

DOI: 10.1201/9781003585534-11

take many forms—electrical, mechanical, thermal, and chemical—and are evaluated in terms of their ability to meet key performance standards for specific applications in the utility industry. The interconnectedness of storage technologies is emphasized as the paper periodically examines further sectoral integrations in which they may also be involved [4].

Previous studies have attempted to consolidate different storage methods for grid applications. For instance, Gustavsson [5] analyzed various storage technologies but did not evaluate their suitability for specific utility functions, leaving the analysis incomplete. Similarly, Luo's 2015 study [6] aggregated existing developments in storage techniques for power system applications but failed to examine their relevance for advanced power quality applications, which have gained significance due to the evolving dynamics of modern power systems. This paper seeks to bridge these gaps by focusing on emerging technologies that have recently gained traction and hold promise for mainstream adoption. A semantic comparison of conventional and modern storage technologies is presented, accompanied by numerical characteristics and notable applications. Additionally, the manuscript highlights features tailored to diverse stakeholders, addressing objectives such as power quality enhancement and energy trading. A technical evaluation is provided to assess the commercial viability of each technology, considering its maturity level, distinguishing this work from its contemporaries.

This paper's remaining sections are arranged as follows: The subsequent segment explores the evolving dynamics of power systems, scrutinizing the functions of various stakeholders, the attributes of VRE sources, and the corresponding obstacles. It provides a thorough explanation of the performance metrics used to assess energy storage systems and divides their applications into two main categories: energy management and ancillary services.

The core of this paper offers a comprehensive examination of various energy storage technologies relevant to grid applications, categorized across four physics-based domains. Each technology is analyzed in terms of its functionality, impacts, advantages, and limitations. Numerical performance metrics, derived from an extensive review of the literature, are also presented. Where appropriate, representative examples of these technologies are included to enhance understanding. Based on their performance characteristics and applicability, technologies are mapped to application-specific requirements through a comparative study. Key findings, emerging trends, and conclusions from this thorough technical review are outlined in the paper's conclusion.

11.2 VARIATIONS IN THE DYNAMICS OF A CONTEMPORARY POWER SYSTEM

At the moment, the electricity system is undergoing a significant change. Power networks have traditionally relied heavily on fossil fuels for their energy supply, but the proportion of RES, especially VRE, such as wind and solar, is steadily increasing [7]. This shift is driving significant changes in grid behavior, presenting new scientific and engineering challenges. However, it also opens up opportunities for improved electrification, enhanced energy efficiency, and strengthened power security. Simultaneously, there are efforts to improve PQ metrics, such as grid stability, reliability, and resilience [8]. These goals align with addressing environmental impacts by increasing RES adoption and achieving economic benefits through reduced dependency on natural resources [9].RES are expected to play an increasingly significant role in future power systems, just as they did in the early phases of energy utilization [10]. Investors and businesses are paying increasing attention to long-term climate risks and opportunities. The switch to renewable energy sources (RES) has caused long-lasting changes in the way energy is produced, used, and priced. This has been made possible by advances in technology and a cultural push for sustainability. These changes are intended to lower greenhouse gas emissions by employing a variety of decarbonization methods. RES previously relied on government incentives and subsidies to grow, but now they are reliable and reasonably priced sources of electricity.

Due to the substantial decline in the price of solar photovoltaic (PV) and wind energy, wind power is now more affordable than conventional energy sources in some locations [11].This change has been facilitated by extensive research and development in supporting technologies, including insulated-gate bipolar transistor (IGBT) and integrated gate-commutated thyristor (IGCT) converter technologies, high-voltage (HV) and ultra-high-voltage DC and AC systems, underground cable innovations that enable transmission at voltages greater than 600 kV, improved cooling systems, and the integration of digital control, protection, and monitoring systems [12]. As technology costs continue to decrease and wind and solar PV become mainstream energy sources, the renewable energy sector is poised for sustained growth. This trajectory solidifies its position as a major investment opportunity with long-term potential [13].

11.2.1 Features of a Contemporary Power System

Using more renewable energy, especially variable renewable energy, introduces a range of challenges to the power system [14], as these sources exhibit characteristics fundamentally different from traditional energy generation. VRE output depends on resource availability, making it highly unpredictable despite advancements in forecasting techniques. This unpredictability imposes significant challenges on spot pricing and penalties, especially when energy commitments are unmet from an energy trading perspective. However, favorable demand conditions during surplus production can benefit VRE operators, provided such conditions prevent energy curtailment. Unlike centralized conventional energy systems, VRE is inherently more distributed, requiring a decentralized approach to energy management. Additionally, VRE relies on inverter-based resources (IBRs) to integrate power into the mainstream alternating current (AC) grid. Table 11.1 outlines key characteristics of VRE sources, providing definitions and their potential impacts on the power system.

TABLE 11.1
Features of Sources for VRE

S.No	Qualities	Description
1.	Adaptability	Increased spot pricing and penalties result from a mismatch between predicted and the actual production, as well as from the quick variations in VRE behavior.
2.	Uncertainty	Even while forecasting technologies are becoming more accurate, generation is still difficult to estimate precisely, which can lead to significant imbalances in load demand.
3.	Distributed	Because small-scale distributed generators are mostly grid-connected, supply swings are exacerbated by volatility.
4.	Inverter-based	Inverters, a crucial part of wind and photovoltaic power generation, invariably generate harmonics and other PQ pollutants.
5.	Zero variable cost	Global inflation and geopolitical events make capital prices uncertain, while variable costs of production remain nearly negligible.

TABLE 11.2
Difficulties Brought on by a Rise in VRE Production

S.No	Challenge	Description
1	Resource adequacy	For a system to function reliably, demand and supply for electricity must coincide, and there must be a sufficient variety of energy sources available.
2	Network adequacy	Adequate grid capacity is essential to guarantee faultless power transfer between transmission and generating corridors.
3	Frequency stability	Supply and demand must be balanced very instantaneously in order to keep the system frequency within reasonable bounds.
4	Voltage stability	The voltage waveform and phase angle must be maintained by the system operators after disturbances across the whole power spectrum.

11.2.2 Difficulties with a Modern Power System

In addition to their unique characteristics, VRE sources face distinct challenges that require attention. Table 11.2 outlines these challenges and provides an estimated timeframe for their resolution.

- **Adequacy of Resources:** Resource adequacy issues in network planning are currently caused by seasonal imbalances, increasing intraday variability, and energy droughts associated with VREs. Sharper peaks in energy supply and deeper troughs of low supply aggravate intraday unpredictability; high ramps in energy supply, for instance, occur as the sun sets in the daily cycle. Seasonal imbalances result from the fact that solar energy generation is highest in the summer while electricity demand peaks in the winter, in addition to these daily variations. Weather-related events also contribute to abrupt and erratic changes, such as transitions from sunny to cloudy conditions or windstorms. These events disrupt renewable energy generation, with extreme cases—like ultra-high wind speeds—necessitating wind farm shutdowns due to surpassing cut-out limits. Such power droughts pose significant challenges for system operators in ensuring adequate energy generation sources [15]. Variability is not solely linked to VRE volatilities but can also result from PQ issues, such as voltage surges. Moreover, long-term network planning is affected by constraints in energy storage systems (ESS), requiring a balance between traditional grid reinforcement and new investments in storage infrastructure. Proper resource planning, tailored to the capacity of available ESS, must extend years into the future to address the volatility introduced by weather conditions, regulatory frameworks, and market dynamics, all of which contribute significantly to the development of the energy landscape.

- **Sufficiency of the Network:** Traditionally, the primary source of electricity has been large conventional power plants, and networks have been set up to transport this energy to dispersed consumption locations. However, existing grid infrastructure is being strained by the growth of distributed generation from RES in contemporary systems. This shift changes the traditional power flow, sending energy back upstream in situations where reverse power flow occurs instead of the usual upstream-to-downstream flow. As a result, transmission and distribution (T&D) system operators are facing increasing difficulties with capacity allocation and network congestion. The rapid and widespread integration of RES into the grid often exceeds the capacity of current infrastructure, underscoring the need for improvements and innovative approaches to preserve grid efficiency and reliability [16].
- **Frequency Stability:** Conventional power plants historically supplied regulated operational reserves, which, depending on the energy source, were made accessible as needed. However, the controllability of operational reserves declines as fossil or nuclear-based sources are phased out and solar- and wind-based RES are progressively incorporated into the grid. The need for operational reserves to correct system imbalances increases as a result of this change. Furthermore, the rotating mass, which in conventional power plants maintained the active spinning reserve, vanishes. Lack of replenishment causes system imbalances and makes it difficult to maintain frequency stability, since IBRs do not supply rotational mass.
- **Voltage Stability:** Conventional and dispatchable generation systems typically involve synchronously connected machines, which are capable of injecting or absorbing reactive power to enhance system stability. However, since RES, such as solar and wind, are not simultaneously connected to the grid, this capability is diminished when they are used. Inverters link solar and wind generators to the network, but they often have limited or no capacity to provide reactive power, leading to potential voltage instability issues [17].

Modern electricity systems must solve the aforementioned issues with a variety of solutions. Offering power storage options that enable energy to be used when needed within a specified period is one of the best methods to close the gap, especially that caused by intermittency. By using stored power to offset fluctuations, imbalances, and energy droughts, energy storage can greatly improve resource sufficiency. Storage systems help address network adequacy by charging during times of network congestion and discharging to supply power to remaining loads. Additionally, stored energy can contribute to voltage and frequency stability. The sections that follow provide a more detailed overview of energy storage applications.

11.2.3 Energy Storage's Role Based on Particular Application Needs

An ESS that can be integrated into contemporary power grids to meet the sporadic energy demands imposed on them can be constructed using a variety of energy storage technologies. Energy Management and Ancillary Services are the two primary application areas into which ESS utilization can generally be divided. A secure and dependable energy supply is the main goal of energy management, which also aims to improve the grid's flexibility, cost-effectiveness, safety, and stability while permitting sector coupling without causing any harm. Conversely, supplementary services include maintaining PQ standards and facilitating continuous energy flow to satisfy real-time demand [18], both of which help to raise the PQ indices specified in different standards [19]. As covered in the preceding section, integrating RES creates new challenges for PQ, ancillary services, and energy management. Maximizing the utilization of renewable resources while reducing adverse effects—the goal of RES integration—is power security and quality. This paper examines six ESS applications in these two support domains. The pan-European power grid market is operated and services are provided by the European Network of Transmission System Operators for Electricity (ENTSO-E); these applications are described in Figure 11.1. Balancing, congestion management, system operation, and outage services are the categories into which the energy storage applications, which range from yellow to blue, can be divided. Below is a more thorough discussion of these services' functions and their potential effects on grid operation.

11.3 ENERGY MANAGEMENT

The stability of power networks is at risk as larger percentages of RES are integrated, particularly variable VRE such as solar PV and wind. Generation plants may not be able to adequately balance the supply and demand ratio, which is the main obstacle.

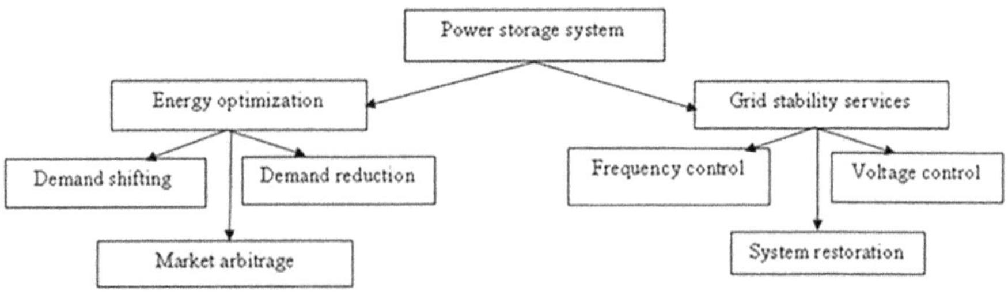

FIGURE 11.1 Energy storage application classification.

To preserve this equilibrium and lessen the impact of peak demand loads, which may result in energy shortages or price increases, energy management systems are crucial. The network can be disrupted by excessive or insufficient power generation or consumption, leading to major issues such as voltage or frequency drops (or surges), and in extreme cases, blackouts.

In the absence of energy storage, energy management typically operates in two ways. The first involves the electricity supply, where conventional controllable power plants adjust their generation to match fluctuations in demand. The second is demand-side management, in which users adjust their energy usage to match the power supply. By improving generating controllability and assisting in the management of large consumption peaks, energy storage plays a crucial role in both areas and enhances the grid's ability to balance the external demand for electricity.

Reducing operational and consumption costs, minimizing energy losses, and improving network dependability are the main goals of employing energy storage in energy management [20]. The ensuing subsections explore several energy management applications.

11.3.1 Changes in Load

Since electricity must be used immediately when it is produced, it is frequently regarded as a perishable good [21]. Larger amounts of energy cannot be retained for later use; however, reactive loads and the system's rotational inertia can temporarily store power. As the proportion of intermittent energy sources in power networks rises, this problem becomes more noticeable, causing more frequent excess energy peaks with greater power levels and longer durations [21]. Reduction, or the practice of limiting energy production, is expected to become more prevalent in the future, according to global forecasts. This is undesirable, as excess energy might go unutilized due to a lack of compensation or negative pricing. However, the ability to store and later use that excess energy can potentially change this situation.

ESS provide a solution by enabling the temporary shift of excess energy, allowing it to be stored and later released to match load demands, thereby minimizing energy loss and reducing energy deficits [22]. Although this method may not achieve the highest round-trip efficiency, it serves as the basis for many of the storage solutions covered in this paper that are intended to capture excess renewable energy production. The benefits of matching energy generation with consumption using storage include [23] enhanced efficiency of renewable energy integration (potentially avoiding curtailment of excess energy), cost savings, improved PQ, and extended utility asset lifetimes (deferring maintenance on power facilities).While short-duration energy storage can help balance excess energy with current renewable energy shares, larger amounts of surplus energy—especially with higher RES penetration—may require longer storage durations, possibly extending to seasonal storage. Although the rate of development varies by region, the requirement for energy storage duration increases as the proportion of VRE rises. Even if there are obstacles to overcome, incorporating RES into electricity networks remains a desirable goal, particularly for microgrids. For instance, islanded microgrids present substantial potential for load shifting based on energy storage, offering a more cost-effective method of energy management [24]. Island nations can lessen growing dependence on costly fossil fuel imports and reduce emissions by utilizing renewable energy sources locally. However, without large-scale energy storage capacities, transitioning to renewable energy in microgrids poses risks, especially when fluctuations in VRE production or demand can have more severe impacts on power quality compared to interconnected grids [25].

11.3.2 Shaving of the Peak

Industrial power users typically exhibit an irregular load profile throughout the day, leading to periods of peak demand, as depicted by the blue corrugated line in Figure 11.2 [26].

The power system must be sized to accommodate peak load demands, which often results in underutilization during other times of the day. Ultimately, consumers bear the additional expense of maintaining this capacity to satisfy peak demand. Peak loads can be decreased by using an ESS, which saves consumers money. The ESS discharges during peak hours and charges during off-peak hours [27]. Customers in the industrial sector, in particular, benefit from ESS for peak shaving, as it helps mitigate the need to meet short-duration peak demands. ESS systems for peak shaving offer greater flexibility compared to load shedding, where peak load is reduced by either limiting or shutting down loads entirely [28]. By eliminating the requirement for an oversized grid that would otherwise remain underused, peak shaving decreases energy prices. Long-term planning enables better demand response management. Peak shaving can also help transmission system operators (TSOs) and industrial users ease congestion on the transmission grid.

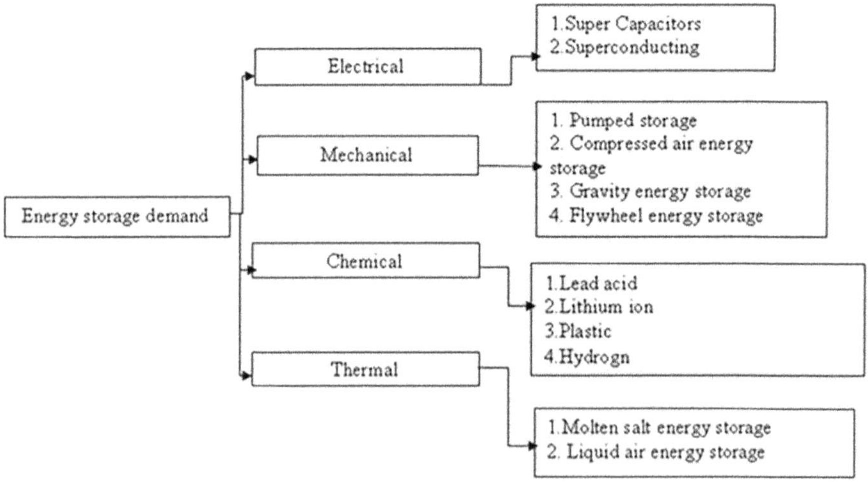

FIGURE 11.2 Classification of energy storage systems.

11.3.3 The Arbitrage of Prices

An investor can make money by using the arbitrage strategy to take advantage of price changes for the same asset in different markets or at different times. Arbitrage can be very profitable in the energy business because prices can change a lot, especially when a lot of energy is involved. ESS can help a lot with this process by letting people buy energy when prices are low and sell it when prices are high. This can help lessen the effects of price spikes. For arbitrage to be profitable, the difference in price between low and high times must be greater than the losses caused by the energy storage and discharge inefficiencies. The ESS's ability to store energy is also a big factor in how profitable it is. A survey says that in 2020, almost 60% of utility-scale energy storage systems were powered by energy price arbitrage. This shows how important ESS is becoming for making the energy market work better.

11.4 ANCILLARY SERVICES

Ancillary services are an important part of keeping power systems stable, reliable, and safe. These services make sure that the power system stays stable and operates even when there are problems, changes, or emergencies. TSOs and DSOs are responsible for maintaining PQ, controlling voltage and frequency, and responding to emergencies to prevent overloads and system failures. They also take care of steps to restore the system after an instability or blackout to keep the grid stable. Power system stability is facing new problems because of changes in how people use energy, like charging electric cars and using solar panels at home at different times. These changes could lead to unbalanced loads, low power factors, sudden voltage swings, and harmonic distortions. When renewable energy sources are added to the system, these problems can cause damage to equipment, shorter lifespans, and unplanned outages. To deal with these problems, the need for ancillary services has changed. For instance, black start capabilities are very important because they let a microgrid or islanded system restore power if the grid supply is cut off. In the past, diesel generators performed this job, but batteries are now a cheaper option that can sometimes be used with conventional power sources. UPS systems are also necessary for some critical applications because they can respond in less than a millisecond and provide backup power almost instantly to keep operations running. The changing need for ancillary services shows how important energy storage technologies are becoming. These technologies could provide quick, flexible, and reliable support to improve power quality and stabilize the grid, especially in systems with a high share of renewable energy.

11.4.1 Regulation of Frequency

The electric power grid's stability, resilience, and dependability depend on maintaining a steady frequency. Frequency stability depends on the exact balance between how much electricity is generated and how much is used. When there is more supply than demand, the frequency rises. When there is more demand than supply, the frequency falls. These deviations from the normal frequency can generators and other electrical equipment connected to the grid, and the damage will be worse if the system takes too long to return to the normal frequency.

11.4.1.1 Key Factors Affecting Frequency Stability

1. The size of the difference (or contingency) between supply and demand has a direct effect on how much the frequency changes. Larger imbalances cause bigger changes in frequency, which means that larger actions are needed to bring things back into balance.
2. Inertia is the power system's resistance to changes in frequency. It is mostly caused by the rotating masses in traditional generators. The higher the system inertia, the more stable the frequency tends to be. This is because the rotating machinery in conventional plants helps slow down the rate at which the frequency changes. But as more renewable energy sources, like wind and solar, are added to the grid, the system becomes less stable. This is because these sources don't have any moving parts, which makes it harder to keep the frequency stable.
3. The speed at which the system can react to frequency deviations is crucial. A fast response is necessary to restore the frequency to its original level before it causes damage to equipment or results in grid instability. Faster responses help reduce the time the grid spends operating outside its nominal frequency, minimizing potential risks.
4. The magnitude of the response refers to the amount of energy required to correct the frequency deviation. Larger deviations in frequency (for example, from a significant imbalance between supply and demand) require a more substantial adjustment in generation or consumption to return the grid to stability.

11.4.1.2 Impact of Frequency Deviations

- **Low Frequency Demand Disconnection (LFDD):** If there is not enough generation to keep the frequency within a specified range, especially in renewable-based grids, the grid may initiate LFDD. This involves disconnecting some loads from the grid to restore balance. This is a safety measure, but it can cause significant disruptions to consumers.
- **Rotating Machines Disconnection:** A frequency deviation, particularly when it is too large, can trigger the disconnection of rotating machines on the grid (such as turbines and generators). These machines disconnect automatically as a safety measure to protect them from damage, further increasing the likelihood of blackouts or partial grid failures.

It is hard to keep the frequency stable when using renewable-based grids because they often have less system inertia and more generation swings. To keep the grid stable and maintain frequency stability, ESS, fast-acting reserve plants, and demand-side management techniques are becoming more important. In the power system, inertia is the term for the kinetic energy that comes from the rotating parts of motors and generators that are all connected to the grid at the same time. It is important for keeping the grid stable because it can handle sudden changes in frequency. The system's inertia acts as a buffer to keep the grid stable in case of sudden mismatches between supply and demand by reducing frequency changes. The ratio of the machines' stored kinetic energy to their rated apparent power (S) is used to determine their inertia.

11.4.1.3 Inertia and Frequency Regulation

As the grid adds more intermittent renewable energy sources, like wind and solar PV, the overall system's inertia decreases. This is because IBRs usually connect renewable energy sources, and they don't have spinning masses to provide inertia. Because lower inertia causes the frequency to change more quickly, the system becomes more sensitive to changes in supply and demand. This means that frequency management needs to be more active.

11.4.1.4 Frequency Control Mechanisms

To make sure that the system frequency stays the same and within acceptable limits, both passive and active controls are used:

1. The system's natural inertia, which comes mostly from traditional synchronous generators, helps keep the frequency from changing too quickly. But as IBRs make up a bigger and bigger part of renewable energy production, their passive contribution to frequency stability is getting smaller.
2. This is the first step in bringing the frequency back to its nominal value when it goes off course. When the frequency goes outside the nominal range, the primary control turns on in a matter of seconds. Usually, generators do this by changing their output automatically when the frequency changes, which helps stop the drop in frequency and keep it steady.
3. This is the second step in stabilizing the frequency when it deviates from its nominal value. When the frequency begins to fall or rise outside the nominal range, the secondary control is triggered within seconds. To help stop and stabilize the frequency deviation, generators usually perform this function by automatically modifying their output in response to variations in frequency.
4. If the frequency is not stabilized by primary and secondary controls, tertiary control is employed. Usually not automated, this is turned on when the grid operator requests it. Tertiary control has a slower response time than primary and secondary controls, and its duration is much longer. It is used to restore the system after the cause of the frequency deviation has been addressed.

11.4.1.5 Types of Reserves

- **Spinning Reserves:** These are reserves supplied by already-online, grid-synchronized but unloaded generating capacity. It is possible to promptly deploy spinning reserves to make up for transmission or generation failures. When there are shortages, they are the first kind of reserve to be used.
- **Non-Spinning Reserves:** These reserves are not continuously online and are not synchronized with the grid. However, they can still be brought

online relatively quickly to support the grid. Their response time is slower than that of spinning reserves.

- **Supplementary Reserves**: The spinning and non-spinning reserves are replaced and restored by these reserves, which have a longer response time. They are typically deployed after other reserves have been exhausted.

11.4.1.6 The Role of Energy Storage Systems in Frequency Control

Energy storage systems are becoming more and more crucial in frequency regulation, particularly as renewable energy penetration increases. In order to balance supply and demand, ESS may quickly absorb or inject electricity into the grid, providing both primary and secondary frequency regulation. When the frequency starts to change, ESS can help keep it steady by responding in seconds for primary control. ESS can support secondary control for longer periods of time, bringing the frequency back to a steady level. In systems where renewable energy is the main source of power, ESS may also be useful for tertiary frequency regulation, which is usually performed by traditional power plants. When other types of generation can't keep up with demand, ESS can act as a flexible reserve, reacting quickly and strengthening the grid. In conclusion, adding ESS to grid operations makes it easier to keep the grid stable and reliable, especially in systems with a lot of renewable energy. This is because ESS can quickly and effectively respond to changes in frequency.

11.4.2 Voltage Support

Voltage support is very important for keeping the power system stable, especially when more renewable energy sources like wind and solar are added. These renewable sources could seriously threaten voltage stability because they don't always operate consistently and their output changes. Voltage is affected by both active and reactive power in the power grid. Reactive power keeps the voltage levels where they need to be for the grid to work properly, while active power does the work (like turning on lights and running equipment). If there isn't enough voltage support, the system could experience voltage swings that could cause instability, damage to equipment, or even grid failures.

11.4.2.1 Key Voltage Stability Problems

As more variable renewable energy sources are added to the grid, power systems may have a number of problems related to voltage:

1. Voltage Sags or Dips and Brownouts:
    - Short-term drops in voltage, known as voltage sags, are usually brought on by abrupt spikes in demand or grid malfunctions.
    - Brownouts are prolonged voltage drops that occur when the grid is stressed, often due to high demand, which can cause equipment malfunctions or inefficient operation.
2. Voltage Swells:
    - A voltage swell happens when there is an unanticipated rise in voltage. This can be brought on by abrupt spikes in power generation, such as

those from wind turbines when the wind speed quickly increases, or by the turning off of heavy loads.
- These surges can cause high-voltage spikes that may damage sensitive equipment connected to the grid.

3. Transients (Fast Disturbances):
   - Transients are rapid, short-term disturbances in voltage caused by events such as the starting or stopping of large equipment (e.g., industrial machinery, motors, or large compressors). These fast changes can create voltage spikes, putting stress on the grid and equipment.
4. Harmonics:
   - Voltage variations at multiples of the fundamental frequency are known as harmonics. Usually, non-linear loads like electronic equipment, variable-speed drives, and high-power switching devices are to blame. These harmonics have the potential to cause inefficiency and even harm to electrical equipment by distorting the voltage's waveform.
5. Flicker:
   - The phrase voltage flicker describes sporadic variations in voltage brought on by high-consumption operations. These include activities like welding machines, electric arc furnaces, and other industrial processes that can result in abrupt fluctuations in power consumption, which can cause lights to flicker noticeably and create other grid disruptions.

11.4.2.2 The Role of Reactive Power in Voltage Support

Reactive power is the key element in voltage control. While real power maintains the system's load, reactive power ensures that the voltage remains at an adequate level across the network. Generators and other grid-connected sources typically supply the reactive power needed to support voltage levels. However, when the reactive power demand exceeds the available supply, or when generation does not match consumption, voltage instability can occur. Reactive power can be supplied or absorbed by energy storage devices, such as batteries and other storage technologies, which aid in real-time voltage stabilization. ESS can be especially useful in areas with a high penetration of VRE, where variability in generation can lead to fluctuating voltage levels [29].

11.4.2.3 Solutions for Voltage Stability

To mitigate these voltage-related issues and improve system stability, several approaches can be adopted:

- Devices like transformers with on-load tap changers (OLTC), static var compensators (SVCs), and synchronous condensers can provide dynamic voltage support to maintain system stability.
- These systems can provide reactive power when it's needed and absorb it when there's too much, which helps keep voltage changes from being too large.
- These are used to improve voltage stability and control oscillations during short interruptions.

- You can avoid large swings in voltage caused by sudden spikes in demand by controlling the consumption side of the grid with load shedding or time-of-use charges.
- They can help with power in specific areas, especially in remote or off-grid locations.

In conclusion, voltage support is very important for keeping modern power systems stable and reliable. As renewable energy sources become more popular, additional steps like ESS, reactive power compensation, and advanced grid management techniques are needed to keep the quality of service high and the voltage stable.

11.4.2.4 Power Factor Compensation and Voltage Control

Electrical systems need both real power and reactive power to work. Real power is the energy that actually makes things like lights, motors, and heat work. On the other hand, reactive power doesn't do any useful work right away, but it is needed to make magnetic or electric fields in things like motors, transformers, and inductive loads. Much electrical equipment depends on this reactive power to function; therefore, its supply must be properly balanced.

To ensure the efficient functioning of the electrical grid:

1. Remote Generation Plants: Usually, a distant generating plant provides the system with actual power. Reactive power, on the other hand, can be provided by new sources positioned closer to the load or by the same generating source.
2. Distributed Reactive Power Injection: The grid can be set up to have local sources of reactive power closer to the load rather than depending on distant plants to supply both real and reactive power. This configuration helps prevent inefficiencies related to reactive power transmission over long distances. While local reactive power sources supply the required reactive power at the load's location, generators can be tuned to solely create real power, running in their most efficient mode with the fewest losses in the generation and transmission network.

11.4.2.5 Voltage Support and Storage Systems

Voltage support is typically focused on the transmission and distribution lines, where maintaining stable voltage levels is critical. Sudden variations in generation or load can cause voltage fluctuations, which need to be promptly controlled to avoid instability. Systems for energy storage are essential for delivering quick voltage assistance:

- **Energy Storage for Voltage Regulation:** Systems for storing energy can be utilized to supply rapid bursts of high power when required to stabilize voltage levels. Since voltage fluctuations generally require short-duration responses (on the order of seconds), the energy storage system doesn't need to have a large capacity. Instead, it must be capable of delivering high power for short durations to counteract voltage sags, swells, or other rapid disturbances.

- **Power Quality Control:** Through reactive power injection or absorption at strategic grid locations, ESS can help maintain voltage within acceptable limits. These systems can be triggered quickly and efficiently, helping to smooth voltage profiles and improve overall power quality.

11.4.2.6 Advantages of Localized Reactive Power Compensation

- **Decreased Transmission Losses:** Transmission losses are reduced by supplying reactive power closer to the load. Additionally, this lessens the burden on transmission lines and distant generators, increasing the overall efficiency of the system.
- **Increased Voltage Stability:** Voltage surges or dips that could impair the operation of delicate machinery and equipment can be avoided by local reactive power adjustment.
- **Quicker Reaction Time:** Energy storage systems offer a more dynamic response to voltage variations than more conventional techniques, like spinning machinery or capacitor banks, because they can react faster.

11.4.3 Black Start Capability

Even though they are uncommon, blackouts can have serious repercussions, particularly if they linger for a long time, affect a large area, and stop electrical devices from working. Longer power outages can be disastrous, especially when they impact vital infrastructure; however, brief power outages are classified as flickering power and are not considered blackouts.

1. Current Black Start Power Sources:
    - Diesel engines or auxiliary gas turbines are often used to start blackouts. These generators can help bring power back to the system and can start up without being connected to the grid. Even though they are well-known and technically advanced, these units have some problems, such as:
    - They can't keep the system running continuously because they don't have the basic functions needed for a grid to recover after a blackout, like voltage control, load shifting, and frequency regulation. Because these generators are only used for black starts, they are not very efficient and cost a lot to run when they are not needed. This causes problems for both the economy and the environment.
2. Challenges of Diesel and Gas Turbine Black Start Systems:
    - These systems can't perform broader grid functions, like load shifting and frequency regulation, which are becoming more important as the grid adds renewable energy sources.
    - Additionally, diesel and gas turbine generators are not environmentally sustainable, given the growing global focus on reducing carbon emissions and the importance of integrating cleaner, renewable energy sources into the grid.

11.4.3.1 The Role of Energy Storage Systems and Renewable Energy in Black Start

There is growing interest in using renewable energy generators to supplement or replace conventional black start power sources because of their drawbacks.

Energy Storage for Black Start:

- Because of their potential to help the grid's frequency and voltage management during black start operations and to offer quicker response times, ESS such as batteries, supercapacitors, and flywheels are being investigated.
- Advantages of ESS:
  - Faster Response
  - Multiple Functions
  - Environmental Sustainability
  - Cost-Effectiveness

Traditional Backup Systems vs. UPS:

- UPS systems offer immediate protection, but their useful life is limited. Backup power sources, such as gas turbines or diesel generators, may then be needed to maintain critical loads or facilities until the grid is restored.
- Problems with Extended Blackouts: UPS systems' ESS is very effective for short-term outages, but it may not be sufficient for long-term blackouts, which could last for hours or even days. In these situations, ESS and slower-responding backup power sources may need to be used in tandem.

Black start operations are changing as a result of ESS integration, particularly when combined with renewable energy generation. Although conventional technologies, such as gas turbines and diesel engines, have proven dependable for black starts, the modern grid finds them less appealing because of their high maintenance costs, poor response times, and environmental effects. As power grids become more dependent on renewable energy, ESS are increasingly viewed as the future of black start capabilities due to their capacity to deliver quick, flexible, and sustainable electricity. Further supporting efforts to keep the grid resilient and stable during blackouts and other disruptions are UPSs, which offer instant protection against brief power outages.

11.5 DIFFERENT ENERGY STORAGE METHODS

Energy storage systems can be categorized according to the physical principles and techniques they employ to store and release energy. Understanding the different kinds of ESS, their uses, and the circumstances in which they function best is made easier by these categorizations. Mechanical, electrical/electromagnetic, thermal, and chemical energy storage are the main categories; each has subtypes according to the physical processes involved. This is a summary:

11.5.1 Storage of Mechanical Energy

These technologies are commonly employed for long-term, large-scale energy storage and store energy as either kinetic or potential energy.

Hydro Storage Pumped:

- Pumping water to a higher altitude stores potential energy. The water is allowed to flow downhill when energy is required, spinning turbines to produce electricity.
  - **Advantages:** Highly efficient (70–90%), suitable for large-scale storage, well-established technology.
  - **Challenges:** Requires specific geographical conditions, such as access to mountainous terrain, and involves high capital costs and environmental concerns.

Energy Storage in Compressed Air:

- Compressing air and storing it in subterranean tunnels is one way to store potential energy. When energy is required, the air is released and expanded to power turbines.
  - **Advantages:** Can store large amounts of energy; works well for long-duration storage.
  - **Challenges:** Requires suitable geological formations for air storage; lower efficiency due to energy losses during compression and decompression.

11.5.1.1 Gravity Batteries

A heavy material, such as a block of concrete, can be raised to a height to store potential energy. When necessary, the mass is lowered, and generators transform the gravitational energy into electrical power.

- **Advantages:** Simple concept; environmentally friendly; no harmful emissions.
- **Challenges:** Limited scalability; requires significant space; lower efficiency compared to other ESS.

11.5.2 Electrical/Electromagnetic Energy Storage

Energy is stored in these systems as electromagnetic fields and is used for high-efficiency, rapid-response applications.

Super capacitors:

An electric field formed between two conductive plates separated by an insulating substance can be used to store energy.

- **Advantages:** High power density; quick charge/discharge cycles; excellent for short-duration storage.
- **Challenges:** Low energy density; unsuitable for long-duration storage; more expensive than traditional batteries.

Storage of Superconducting Magnetic Energy:

Stores energy by generating a magnetic field through a superconducting coil when direct current passes through it.

- **Advantages:** Extremely high efficiency; quick reaction times; extended cycle life.
- **Challenges:** Requires very low temperatures to maintain superconductivity; high initial cost; limited to specific applications.

11.5.3 Storage of Thermal Energy

These systems use heat to store energy and are suitable for applications requiring energy over long durations, often in industrial or district heating contexts.

- Sensible Heat Storage**:**
- **Molten Salt:** Stores energy by heating a material (typically salt) to high temperatures. The heat is stored in the material and can then be utilized to produce electricity or provide heating.
  - **Advantages:** High energy density; relatively inexpensive; suitable for large-scale thermal storage.
  - **Challenges:** Requires effective insulation to minimize energy losses; the technology is complex to implement.
- Latent Heat Storage:
  - **Advantages:** High energy density; can work at large scales; can be deployed in specific environments.
  - **Challenges:** Efficiency losses due to energy conversion processes; high setup costs.
- Thermo-Chemical Storage:
  - **Advantages:** Higher energy density; potential for long-term energy storage.
  - **Challenges:** Requires efficient systems for capturing and releasing heat; still experimental at large scales.

11.5.4 Chemical Energy Storage

These systems use chemical reactions to store energy. Their main purpose is to transform electrical energy into a stable chemical form that can subsequently be transformed back into electrical energy.

Batteries:

- **Lead-Acid:** One of the most established forms of chemical storage, typically used for backup power.
- **Lithium-Ion:** Frequently utilized in electric cars, consumer gadgets, and grid storage, Li-ion provides high energy density, efficiency, and fast charging times.
- **Redox-Flow Batteries:** These rechargeable batteries can be scaled for large applications and store energy in liquid electrolytes.
- **Plastic Batteries:** A newer form of battery technology using organic materials, still in the early stages of development.
- **Advantages:** High energy density; fast discharge rates; scalable for different sizes and applications.
- **Challenges:** Efficiency varies depending on battery type; limited cycle life for some types; environmental concerns around material sourcing and disposal.

Hydrogen Fuel Cells (H2 FC):

Involves converting chemical energy from hydrogen into electricity through electrochemical reactions.

Advantages: Zero emissions; high energy density.

Challenges: Requires hydrogen production infrastructure; high capital costs; energy losses during hydrogen production and storage.

11.5.5 Factors Influencing ESS Selection

When selecting an ESS for a specific application, several factors must be considered:

- **Location:** The physical feasibility of deploying certain ESS technologies may depend on geographical and infrastructural conditions.
- **Scalability:** Some ESS technologies are better suited for large-scale deployments, while others are more appropriate for small- or medium-sized systems.
- **Economy of Scale:** Technologies that perform well in laboratory or small-scale environments may face cost challenges at larger scales, making them less competitive compared with other technologies.
- **Performance Requirements:** Factors like response time, efficiency, duration of energy release, and power capacity all influence the selection of a suitable ESS.

By categorizing ESS based on their physical mechanisms, we can better understand their potential applications, limitations, and performance characteristics. The future of ESS will likely involve a combination of these systems, tailored to specific use cases, to optimize grid stability, efficiency, and sustainability.

11.5.6 Storing Mechanical Energy

One kind of energy storage technology that stores energy as mechanical energy, either as kinetic or potential energy, is called a mechanical energy storage system (MESS). These systems store and transform electrical energy through mechanical processes, which may then be recovered and transformed back into electrical energy as required. MESS is particularly helpful for large-scale energy storage applications since it relies on mechanical processes to store and release energy, which is the main difference between it and other forms of energy storage. An outline of the two primary categories of mechanical energy storage is provided below:

11.5.6.1 Possible Mechanical Storage Based on Energy

- Hydro Storage Pumped:
- **How It Works:** Using excess electrical energy, water is pumped from a lower reservoir to a higher elevation. Allowing the water to flow back downhill through turbines releases the accumulated potential energy and converts it into electrical energy.

Advantages:

- Mature technology with high efficiency (70%–90%).
- Large-scale energy storage capacity; ideal for grid balancing.
- Challenges:
- Requires specific geographical conditions (e.g., mountains or large water bodies).
- High upfront costs for infrastructure development.

11.5.7 Compressed Air Energy Storage (CAES)

How It Works: Large tanks or subterranean tunnels are used to store compressed air. The compressed air is released when energy is required, expands, and powers turbines to produce electricity.

Advantages:

- Can store energy for longer durations compared with other mechanical systems.
- Utilizes excess power during periods of low demand.

Challenges:

- Efficiency losses occur during compression and decompression.
- Requires suitable geological formations for storage.

Gravitational Batteries:

- **How It Works:** These systems store energy by lifting a heavy mass to a certain height. When energy is needed, the mass is allowed to descend, converting its gravitational potential energy into electrical energy via generators.

Advantages:

- Simple design, environmentally friendly, and low operational cost.
- Flexible design options (e.g., using concrete or other materials for the mass).

Challenges:

- Limited scalability for large-scale applications.
- Requires a large amount of space and vertical height for optimal performance.

11.5.7.1 Kinetic Energy-Based Mechanical Storage

Flywheels:

How It Works: By rapidly rotating a rotor, a flywheel stores energy. When the flywheel slows down and the motion is transmitted to a generator, the kinetic energy produced by the spinning is converted into electrical energy.

Advantages:

- High efficiency (90%–95%) with very fast response times.
- Long cycle life and low maintenance.
- Suitable for short-duration energy storage and high-power-demand applications.

Challenges:

- Low energy density, meaning they are best for short bursts of energy.
- Expensive to scale up for large energy storage needs.

11.5.7.1.1 Common Components and Processes in MESS

Electric Motors and Generators:

In most MESS, electric motors are used to drive mechanical processes. The same motor frequently serves as a generator to convert mechanical energy back into electrical energy while discharging.

- Turbines and Compressors:

  In systems like PHS and CAES, turbines and compressors are used to convert between electrical and mechanical energy. For example, PHS uses turbines to generate electricity from flowing water, while CAES uses compressors to store air and turbines to release it.
- Energy Conversion Systems:

  The systems that control the conversion of energy between electrical and mechanical forms often include advanced power electronics and control mechanisms to manage the flow and storage of energy efficiently.

Key Considerations for MESS Deployment:

- Geographical Suitability
- Scalability and Cost
- Efficiency and Response Time

Mechanical energy storage methods are very important for long-term, large-scale energy storage. These devices help keep the grid stable and balance supply and demand by converting electrical energy into mechanical forms like potential or kinetic energy. Different MESS technologies are better suited for different types of applications because they have different pros and cons. These pros and cons depend on factors such as location, energy use, and efficiency needs.

11.5.8 Pumped Hydro Storage

Pumped Hydro Storage (PHS) systems have been the most common way to store energy for a long time. They hold about 95% of the world's electrical energy. PHS is still very important for maintaining the balance between the world's electricity supply and demand because it can store 8.5 PWh (petawatt-hours) and produce 169 GW of power. Because it has been used successfully for decades, this method is considered the most advanced and reliable way to store energy. PHS systems use electrical energy to pump water from a lower reservoir to a higher one (the upper reservoir). They store energy in the form of gravitational potential energy. When there is high energy demand, the water in the upper reservoir can flow back down through turbines.

1. **Energy Conversion:** This technique converts electrical energy into gravitational potential energy.
2. **Energy Storage:** By serving as a storage medium, the upper reservoir holds water at a higher elevation. The elevated water mass acts as a potential energy storage medium.
3. **Energy Release:** When the demand for electricity increases, the water stored in the upper reservoir is released. The water flows back down through a turbine system due to gravity, and a generator exploits the kinetic energy of the falling water to produce electrical energy.
4. **Electrical Energy Output:** When renewable energy sources, such as solar or wind, are not producing enough electricity, the generated electricity is sent into the grid to provide power during times of high demand.

PHS Challenges:

- **Geographical Requirements:** PHS system deployment necessitates appropriate topography, with a significant elevation difference between the upper and lower reservoirs. This geographic limitation restricts PHS's viability in some areas.

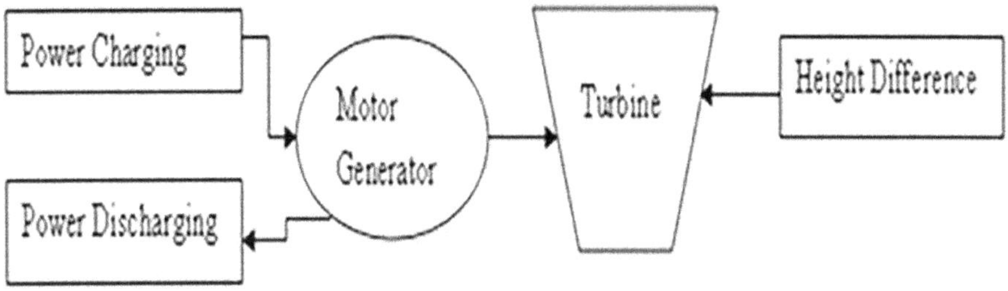

FIGURE 11.3 A block diagram of pumped hydro storage system.

- **High Initial Capital Investment:** The infrastructure required for large reservoirs, turbines, and generators is the primary reason why building a PHS facility involves a substantial upfront capital expenditure.
- **Environmental and Regulatory Considerations:** Large-scale reservoir development may affect local ecosystems, disrupt habitats, and consume water, among other environmental consequences. The construction of new PHS facilities can be further delayed by lengthy and intricate regulatory procedures.

The most widely used technique for large-scale energy storage is still pumped hydro storage due to its extensive storage capacity, high efficiency, and long history of reliability. Despite its environmental and geographic limitations, the technology is crucial to the transition to cleaner, more renewable energy networks because it stabilizes the grid and smooths out the intermittent nature of renewable energy sources (Figure 11.3).

11.5.9 Storing Electrical and Electromagnetic Energy

Electrical and electromagnetic energy storage (EES) technologies, which focus on storing energy as electric or magnetic fields, include supercapacitors and superconducting magnetic energy storage (SMES). These systems typically have excellent round-trip efficiency since they only require AC/DC conversion and no other energy transformations. The following is a breakdown of EES's primary features and technologies:

11.5.9.1 Key Technologies

1. Supercapacitors: Supercapacitors are designed to store enormous quantities of electrical charge and function similarly to conventional capacitors but have significantly superior energy storage and power density. A traditional capacitor uses dielectric polarization to store energy by separating two metallic plates with a dielectric material. Supercapacitors build on this concept and boost their energy storage capacity by creating an "electric double layer" at the interface between the electrodes and the electrolyte.
2. **Storage of Superconducting Magnetic Energy (SMES):** SMES systems store energy in the magnetic field generated by the direct current (DC) that

flows through superconducting coils. The coils' superconducting properties allow them to hold energy without any resistance, making them ideal for storing energy and releasing it quickly when needed.

- **Effectiveness:** Because of this, they are a desirable choice for applications that require frequent and rapid releases of stored energy.
- **Power Density and Response Time:** SMES systems are ideal for managing load balancing during fluctuations, stabilizing power grids, and supporting power quality because they can provide extremely high power output over very short intervals, usually measured in milliseconds.
- **Applications:** SMES is commonly employed in power quality, frequency control, and grid stabilization applications where rapid response is essential, including preserving voltage stability and compensating for transient variations in power consumption.

Advantages of Electrical and Electromagnetic Energy Storage (EES):

- **Fast Response Time:** Supercapacitors and SMES are ideal for applications that require quick power delivery since they release energy rapidly.
- **Extended Operational Life:** Due to their high cyclability, particularly in supercapacitors, EES systems can withstand deterioration for many years, reducing the frequency of replacements.
- **Safety:** These systems are generally safer than chemical-based storage systems because they do not use volatile chemicals, which lowers the risk of fires or explosions.

11.5.9.2 Limitations and Challenges

Energy Storage Capacity: EES devices, especially supercapacitors, have higher power densities than chemical-based storage systems like batteries, but they still have lower energy densities. This means they aren't good for applications that need a lot of energy for a long time or for storing energy for a long time.

- **Cost:** Superconducting magnetic systems (SMES) can be expensive because they need to be kept at cryogenic temperatures to remain superconducting.
- **Duration of Energy Storage:** Supercapacitors may not be suitable for applications that need a longer energy supply, such as storing energy for a few hours or days across the grid. They work best for short power surges.

Supercapacitors and SMES are two examples of electrical and electromagnetic energy storage technologies that offer many benefits. These include high efficiency, quick response times, and long operational lifespans. They are great for applications that help stabilize the grid and allow energy to be delivered quickly.

11.5.9.3 The Operation of Supercapacitors and Its Effects

When you need to store and release energy quickly, supercapacitors, also called ultracapacitors, are better than regular capacitors and batteries in several ways. Because they store energy in an electric field instead of through chemical reactions like batteries do, they can deliver more power and last longer. The special features of supercapacitors can be useful for the following applications:

Key Advantages of Supercapacitors:

1. Higher Power Output
2. Longer Cycle Life:
3. High Efficiency and Fast Charge/Discharge
4. Wide Temperature Range:
5. Safety and Reliability

Supercapacitors' Benefits for Applications:

1. Electric Vehicles (EVs) and Hybrid Electric Vehicles (HEVs)
2. Grid Stabilization and Energy Storage
3. Uninterruptible Power Supplies
4. Consumer Electronics
5. Regenerative Braking in Public Transport
6. Industrial Equipment and Machinery
7. Renewable Energy Systems
8. Military and Aerospace Applications

11.5.9.4 Magnetic Superconduction

Superconducting magnetic energy storage (SMES) is a new and very effective way to store energy. It uses a magnetic field to do this. This kind of energy storage is based on superconductivity, which states that a current passing through a superconducting coil creates a magnetic field that stores energy. The following are some of the most important features and benefits of SMES systems:

Key Components:

1. **A Coil with a Superconducting Magnet:** A coil made of superconducting material is the most important part of an SMES system. This coil carries a current that creates a magnetic field to store energy.

   Superconductors are materials that exhibit zero electrical resistance below a critical temperature. When the temperature drops below this threshold, the resistance of the material vanishes, and the electrical current can circulate indefinitely without any resistive energy losses.
2. **Cryogenic Refrigerator (Cooling System):** To maintain superconductivity, the coil must be kept at extremely low temperatures, often close to absolute zero. This is achieved using cryogenic refrigeration systems, such as liquid helium, which require significant energy to maintain the low temperatures.

   One of the biggest operating expenses for SMES is the cooling system, but high-performance insulating materials can reduce heat transmission and the requirement for cooling.

3. **Power Conditioning System:** When required, this system transforms the magnetic field's stored energy into electrical power that can be used. When the system is being recharged, it also regulates the input power.
4. **Control System:** The control system oversees the SMES's operations and ensures that energy is effectively stored, maintained, and released.

Advantages of SMES:

1. High Efficiency
2. Fast Response Time
3. Unlimited Charge/Discharge Cycles
4. Scalability
5. No Chemical Reactions

Limitations of SMES:

1. **Cooling Requirements:** One of the biggest issues facing SMES is the requirement to keep the superconducting coil at cryogenic temperatures. The cooling system can be costly to run and maintain, and it consumes a lot of energy.
2. **Cost:** Widespread adoption may be hampered by the high initial cost of developing an SMES system because it requires specialized equipment, cryogenic infrastructure, and superconducting materials.
3. **Energy Density:** SMES devices can release energy quickly, but they do not store as much energy as batteries and other energy storage technologies. This means they are better suited for short-term, high-power applications than for long-term energy storage.
4. **Infrastructure and Maintenance:** The infrastructure needed to manage superconducting materials and maintain cryogenic temperatures adds to both initial and ongoing costs. It is important to always monitor and maintain the integrity of the system.

Applications of SMES:

- **Grid Stabilization:** SMES systems are great for keeping the grid stable because they can quickly add or remove electricity.
- **Better Power Quality:** They ensure that delicate equipment, like medical devices and data centers, receives reliable power even when the voltage or frequency changes quickly.
- **Energy Arbitrage:** Small and medium-sized enterprises (SMEs) can take advantage of changes in electricity prices by storing energy when prices are low and releasing it when prices are high.
- **Industrial Uses:** SMES can be used in steel mills, electromagnetic launches, and electric propulsion systems that require high-power pulses.

11.6 CONCLUSIONS

As the world's energy situation changes quickly, energy storage is becoming an increasingly important part of planning, operating, and maintaining the grid. As renewable energy sources (RES), like wind and solar power, become more common in the generation mix, energy storage systems (ESS) will be necessary to keep the grid stable, balance supply and demand, and make it easier to connect variable renewable energy (VRE) to the grid. Here are the most important changes in energy storage and how they might affect the energy sector: long-duration energy storage (LDES), flywheels and supercapacitors for extra services, improvements to Li-ion batteries, hydrogen storage as a viable option, gravity batteries and site independence, thermal energy storage, hybrid storage solutions, research and development, and the rapid growth of energy storage markets. In conclusion, the energy storage industry is changing quickly, with many technologies competing for market share. Long-term storage, hybrid systems, and new technologies like hydrogen and gravity batteries are likely to change the way we store energy in the future. This will make it a key part of a low-carbon, sustainable energy grid.

REFERENCES

1. T. M. Letcher, *Storing Energy: With Special Reference to Renewable Energy Sources.* Oxford, UK: Elsevier, 2022, ch. 2.
2. X. Luo, J. Wang, M. Dooner, and J. Clarke, "Over view of current development in electrical energy storage technologies and the application potential in power system operation," *Appl. Energy*, vol. 137, pp. 511–536,2015.
3. S. Sahoo, "Recent trends and advances in power quality," in *Power Quality in Modern Power Systems.* Oxford, UK: Elsevier, 2021, pp. 337–358.
4. S. Sahoo, X. Zhao, and K. Kyprianidis, "A review of concepts, benefits, and challenges for future electrical propulsion-based aircraft," *Aerospace*, vol. 7, no. 4, p. 44, 2020.
5. S. Sahoo, "Power and energy management in smart power systems," in *Artificial Intelligence–Based Smart Power Systems*, vol. 22 (pp. 349–375), IEEE, 2023.
6. S. Padmanaban, S. Palanisamy, S. Chenniappan, and J.B. Holm-Nielsen, Eds. *Artificial Intelligence-based Smart Power Systems.* Hoboken, NJ: Wiley, 2023, pp. 349–375.
7. El-Bayeh. Energy Management in Smart Grid IEEE Smart Grid. Accessed: April 15, 2023. [Online]. Available: https://smartgrid.ieee.org/bulletins/august-2019/energy-management-in-smart-grid
8. X. Zhang, C. Qin, E. Loth, Y. Xu, X. Zhou, and H. Chen, "Arbitrage analysis for different energy storage technologies and strategies," *Energy Rep.*, vol. 7, pp. 8198–8206, 2021.
9. J. L. Jorgenson and P. L. Denholm, "Modeling primary frequency response for grid studies," National Renew. Energy Lab. (NREL), Golden, CO, Tech. Rep. NREL/TP-6A20-72355, 2019.
10. El-Bayeh. Energy Management in Smart Grid IEEE Smart Grid. Accessed: April 15, 2023. [Online]. Available: https://smartgrid.ieee.org/bulletins/august-2019/energy-management-in-smart-grid
11. J. Nandha Gopal and N. B. Muthuselvan, and S. Muthukaruppasamy, Model predictive controller–based quadratic boost converter for WECS applications. *International Transactions on Electrical Energy Systems*, vol. 31, no. 12, pp. 1–17, 2021.
12. R. Wang, "Energy storage systems from a power system point of view," Ph.D. dissertation, University of Nebraska-Lincoln, 2019.

13. K. Fiorentzis, Y. Katsigiannis, and E. Karapidakis, "Full-scale implementation of RES and storage in an island energy system," *Inventions*, vol. 5, no. 4, p. 52, 2020.
14. N. Kraftwerke, What Does Peak Shaving Mean. Cologne. Accessed: April 15, 2023. [Online]. Available: https://www.next-kraftwerke.com/knowledge/what-is-peak-shaving
15. J. Nandha Gopal and Muthuselvan, N.B., "Current mode fractional order PID control of wind-based quadratic boost converter inverter system with enhanced time response," *Circuit World*, vol. 47, no. 4, pp. 368–381, 2020.
16. X. Zhang, C. Qin, E. Loth, Y. Xu, X. Zhou, and H. Chen, "Arbitrage analysis for different energy storage technologies and strategies," *Energy Rep.*, vol. 7, pp. 8198–8206, 2021.
17. J. L. Jorgenson and P. L. Denholm, "Modeling primary frequency response for grid studies," National Renew. Energy Lab. (NREL), Golden, CO, Tech. Rep. NREL/TP-6A20-72355, 2019.
18. N. E. L.Y. Kouba, M. Menaa, M. Hasni, and M. Boudour, "A new optimal load frequency control based on hybrid genetic algorithm and particle swarm optimization," *Int. J. Elect. Eng. Inform.*, vol. 9(3), pp. 418–440, 2017.
19. J. Nandha Gopal and N.B. Muthuselvan, "Educational tool for analysis of PI and fractional order PI controlled quadratic boost boost converter system using MATLAB/Simulink," *Int. J. Electr. Eng. Educ.*, 2021.doi:10.1177/00207209211013435
20. S. Koohi-Fayegh and M. A. Rosen, "A review of energy storage types, applications and recent developments," *J. Energy Storage*, vol. 27, p. 101047, 2020.
21. M. Khatibi and S. Ahmed, "Impact of distributed energy resources on frequency regulation of the bulk power system," in *Proc. IEEE Conf. Power Electron. Renew. Energy (CPERE)*, Aswan, Egypt, October 2019, pp. 258–263.
22. International Renewable Energy Agency, Hydropower, 2022. [Online]. Available: https://www.irena.org/hydropower
23. M. A. I. Carmen, Beacon Power Stephentown: Flywheel Energy Storage System, US, 2021. [Online]. Available: https://www.powertechnology.com/marketdata/beacon-power-stephentown-flywheelenergy-storage-system-us
24. T. Morstyn, M. Chilcott, and M.D. McCulloch, "Gravity energy storage with suspended weights for abandoned mine shafts," *Appl. Energy*, vol. 239, pp. 201–206, 2019.
25. P. Vivek, N.B. Muthuselvan, and J. Nanadha gopal, Modeling of solar PV system for DC-DC converter with improved voltage stability using hybrid-optimization techniques. In *International Conference for Phoenixes on Emerging Current Trends in Engineering and Management* (pp.187–192). Atlantis Press, 2018.
26. M. Irving, Ocean Battery Stores Renewable Energy at the Bottom of the Sea, 2022. [Online]. Available: https://newatlas.com/energy/oceanbattery- renewable-energy-storage/
27. J. Castro-Gutiérrez, A. Celzard, and V. Fierro, "Energy storage in supercapacitors: Focus on tannin-derived carbon electrodes," *Frontiers Mater.*, vol. 7, p. 217, 2020.
28. K. Rajeswari, P. Vivek, and J. Nandhagopal, Swing up and stabilization of rotational inverted pendulum by fuzzy sliding mode controller. In: Hemanth, D., Kumar, V., Malathi, S., Castillo, O., Patrut, B. (eds) *Emerging Trends in Computing and Expert Technology. COMET 2019. Lecture Notes on Data Engineering and Communications Technologies*, vol.35. Springer, 2020.https://doi.org/10.1007/978-3-030-32150-5_40.
29. A.G. Olabi, T. Wilberforce, E.T. Sayed, A.G. Abo-Khalil, H.M. Maghrabie, K. Elsaid, and M.A. Abdelkareem, "Battery energy storage systems and swot (strengths, weakness, opportunities, and threats) analysis of batteries in power transmission," *Energy*, vol. 254, p. 123987, 2022.

12 Comparative Analysis of AC and DC Electric Vehicle Charging Systems with Renewable Energy Integration

Anjana Ethirajan and Ramaprabha Ramabadran

12.1 INTRODUCTION

The transition of the world to electric vehicles (EVs) is critical in pushing for the reduction of greenhouse gases, air pollution, and creating a clean energy future. Over the past ten years, an increase in the use of electric vehicles has been observed due to increased pressure about fossil fuel consumption, increased environmental degradation, as well as improved battery life. In an effort to combat emissions and dependence on oil, governments, automobile manufacturers, and even consumers have realized that there is a need to shift from internal combustion engine (ICE) vehicles to electric vehicles. In 2020, global EV sales reached 3.24 million units, a 43% increase over the previous year and a record 4.2% of all vehicle sales [1]. With that kind of momentum in the rise of EV use, it appears to be on the right course in terms of meeting the Paris Agreement's ideals of keeping the world from becoming too hot and curbing greenhouse gas emissions, which are majorly caused by cars and trucks, some of which are among the globe's top pollutants.

However, the large-scale adoption of EVs poses new challenges, especially in the area of charging infrastructure. While electric vehicles are perceived as a solution to the environmental impact of traditional vehicles, their success depends on the development of an efficient, reliable, and accessible charging infrastructure. Charging stations are crucial for the widespread adoption of EVs, and their rollout must match the growing number of electric vehicles on the road. A well-developed and widespread charging network would not make EVs very feasible for daily use, especially in areas with sparse charging stations, such as rural or underserved areas [2]. Thus, there is a need for efficient charging solutions to make the transition toward electric mobility both feasible and scalable. Most of the EV charging infrastructure available today is still dependent on the conventional electricity grid, which is primarily powered by fossil fuels such as natural gas, coal, and oil. This reliance on the grid poses a problem, as EVs, despite being emission-free during use, can still lead to environmental

DOI: 10.1201/9781003585534-12

harm if the electricity used to charge them comes from nonrenewable sources. In 2020, about 61% of the world's electricity came from fossil fuels, and coal alone accounted for around 36%. This means that even if electric vehicles are charged by grids powered by fossil fuels, their environmental benefits are somewhat reduced.

Traditional EV charging stations are normally built with slow, medium, and fast charging points connected to the electrical grid. Level 1 and Level 2 slow charging stations are more or less reserved for home-based use and usually take several hours to fully charge a vehicle, while DC fast chargers designed for public usage can fill an electric vehicle's battery up to 80 percent in 30 minutes. Both types of charging stations suffer from critical issues related to their dependency on the electrical grid. A major problem with traditional EV charging stations is how they burden grid capacity. Loading a high number of EVs in a geographically small region creates massive pressure on the grid, and related overloads become a reliability issue. The demand for electricity for charging electric vehicles could rise to tens of gigawatts by 2030, demanding major upgrades to the grid infrastructure. Advanced solutions such as smart grids, demand response systems, and energy storage technologies are necessary to manage the increased demand, optimize the flow of electricity, and ensure a stable and reliable supply [2].

Traditional charging stations also have inefficiencies in terms of power losses during the transmission and conversion of electricity from the grid to the EV battery. High capital investment in charging infrastructure, coupled with the operational costs of maintaining these stations, often makes the business model for traditional EV charging stations financially unappealing without external subsidies or governmental support. This would further add costs because the grid infrastructure would need to be upgraded to accommodate an increased number of EV charging stations [3].

These problems are being addressed by an increased trend of incorporating renewable energy sources (RES) into the charging infrastructure of electric vehicles. The use of solar, wind, and other renewable resources can reduce dependence on fossil fuels for charging EVs and contribute to a sustainable ecosystem. Integration of RE with EV charging stations allows for the achievement of "clean" charging, further augmenting the environmental benefits of electric mobility. For example, the possibility of using solar energy to charge electric cars during the daytime and storing surplus energy in energy storage systems for use at night or on cloudy days is available [4].

There is also great potential for the integration of EVs into the renewable energy grid, using vehicles as energy storage systems. This can be achieved by allowing vehicles to return energy to the grid, known as vehicle-to-grid or V2G technology, which can balance the fluctuations of renewable energy and ensure grid stability. This innovation points to the possibility of EVs and RE systems working in tandem, reducing dependence on fossil fuels and enhancing grid resilience.

Electric vehicles mark an important step toward a more sustainable future in the transportation sector. However, whether EVs are effective in cutting emissions and reversing climate change is largely dependent on the development of an efficient, renewable-powered charging infrastructure. Traditional EV charging stations, which are necessary for the current scenario, face serious challenges because of their dependence on fossil fuel-based grids, low charging speed, and operational

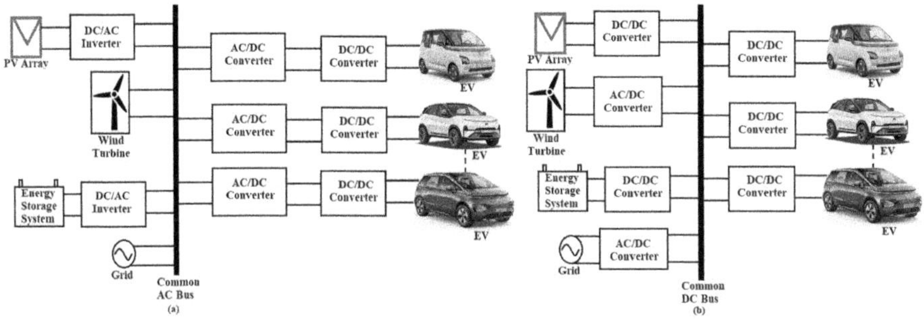

FIGURE 12.1 Block diagram of hybrid system with (a) AC charging system (b) DC charging system.

inefficiencies. These are the challenges and how electric vehicles can be fully utilized by including renewable energy sources in the charging infrastructure. This is because through clean energy-powered charging systems, we can ensure that EVs not only help decrease carbon emissions but also play a vital role in creating a cleaner, more resilient energy future. Figure 12.1 illustrates the block diagram of AC and DC charging stations powered by a hybrid system of PV and wind energy.

This work gives a comprehensive overview of EV charging stations, focusing on a comparative analysis of AC charging systems powered by renewable energy. It explores the design and implementation of hybrid renewable energy systems that maintain a stable power supply despite variable weather conditions. The work also examines the role of smart grid technologies in increasing energy distribution and storage efficiency, thereby improving the overall reliability of these systems. A review of the economic and environmental benefits of RE for EV charging together with associated challenges related to grid integration and energy management, as well as future trends in EV technologies, is discussed in this section.

12.2 THE CASE FOR RENEWABLE ENERGY INTEGRATION

In the context of global trends toward sustainable energy systems, the integration of renewable energy into EV charging infrastructure is becoming increasingly important. While switching to electric vehicles holds the promise of carbon footprint reduction in the transportation sector, the realization of the full environmental and economic benefits of EVs can come about only when they are used together with renewable sources of energy. This section explores the core reasons why grid electricity, traditionally generated from fossil fuels, has been a hurdle in the quest for a greener, more sustainable future and how RES, such as solar and wind energy, provide a viable alternative.

EVs are well known for their ability to cut down emissions from the transport sector, which is one of the biggest emitters of greenhouse gases (GHGs) in the world. The International Energy Agency (IEA) has stated that the transport sector is responsible for almost a quarter of global CO_2 emissions, and road vehicles account for the majority of those emissions. However, if the electricity utilized to charge an EV is provided by conventional power grids that continue to rely heavily on coal, natural gas, and oil, then the environmental benefits associated with EVs are diminished. In many areas, coal and natural gas make up a majority of the mix, with 36% of global electricity produced from coal alone [5].

The reliance on fossil fuels for electricity generation means that EV charging stations powered by the grid still contribute indirectly to greenhouse gas emissions. For instance, charging an electric vehicle in areas where coal-fired plants dominate the electricity supply results in emissions that are comparable to or even higher than those from traditional gasoline-powered vehicles. This has often been referred to as the "carbon leakage" effect and partially erodes the expected savings in emissions due to the mass adoption of electric vehicles. To realize the full environmental benefits of electric mobility, the energy that powers the charging of EVs must come from cleaner and renewable sources.

Traditional charging stations for electric vehicles also face higher operational costs, making renewable integration comparatively cheaper, as most traditional EV charging stations remain connected to the electrical grid with the cost of electricity from this grid tied in most cases to fossil fuel prices. This dependency on grid electricity puts charging station operators at the mercy of volatile energy prices, thus increasing their operating costs over time. Where fossil fuel prices are high or change unpredictably, the cost of charging EVs can be a barrier for both consumers and businesses.

Furthermore, the ongoing need for infrastructure maintenance and upgrades to meet the growing demand for EV charging significantly increases costs. As the number of electric vehicles on the roads rises, the demand for charging stations will escalate, putting additional pressure on the grid and necessitating more frequent infrastructure enhancements to prevent grid instability. Traditional grid systems were not designed to handle the surge in demand that comes with the widespread adoption of electric vehicles. The expenses involved in upgrading grid infrastructure, particularly in urban areas, can be considerable, imposing a financial strain on local governments and utility companies [6].

The high costs associated with grid-dependent charging stations are compounded by inefficiencies in power transmission and distribution. Energy losses during the transmission of electricity from power plants to charging stations, coupled with energy conversion losses in the chargers themselves, further reduce the cost-effectiveness of traditional charging solutions. In fact, approximately 5%–10% of the energy is lost during transmission and conversion processes, which further strains the economic sustainability of grid-dependent charging infrastructure.

To tackle the environmental and economic issues associated with conventional grid electricity, RES such as solar and wind power present attractive alternatives. These sources offer the added advantage of reducing emissions and lowering operation-related expenditures by reducing or eliminating dependence on grid electricity [7,8].

The case for integrating renewable energy into EV charging infrastructure is compelling. Traditional grid electricity, which depends on fossil fuels, diminishes the environmental benefits of EVs and creates economic challenges for charging station operators. By transitioning to RES, we can develop a more sustainable, cost-effective, and resilient charging ecosystem. As technology reduces the prices of renewable energy sources and power storage technologies further, so does the integration of EVs using clean energy as a primary element in their stations. More RE-powered electric vehicle charging stations will achieve a great emission reduction shift as well as serve as a basis toward a clean, sustainable future if massively adopted.

12.3 COMPARATIVE ANALYSIS: AC VS. DC CHARGING SYSTEMS POWERED BY RENEWABLE ENERGY

In general, EV charging systems can be broadened into two major categories: AC charging and DC charging. Each system is quite unique in its own characteristics, technical specifications, and ideal use cases, with some benefits and operational principles that impact its effectiveness in different charging scenarios. Understanding these systems is essential for establishing an efficient and accessible EV charging infrastructure in the context of transitioning to electric mobility throughout the world. The research provides detailed knowledge of AC and DC charging systems together with renewable energy installations by examining performance values, implementation possibilities, and cost analysis.

12.3.1 PERFORMANCE METRICS

12.3.1.1 AC Charging Systems

AC charging systems are the most popular and widely applied form of electric vehicle charging, especially in residential and public charging stations. AC is the most common form of electricity that comes from the grid, and most homes and businesses receive it as the primary source of electricity. The use of AC for EV charging is generally more cost-effective and easier to implement, as most homes and commercial establishments already have access to AC electricity [9]. This system is ideal for overnight charging in residential areas where vehicles can remain plugged in for long periods without the need for rapid recharging.

The speed and power of the charging process for EVs through AC charging input are slower when compared to direct current charging methods. During the power conversion process from AC to DC, the onboard charger of the EV limits its charging speed for battery storage. Public areas often have AC chargers that can be found at workplaces, shopping malls, or parking garages, for instance, in locations where the need for rapid charging is not critical. Such chargers offer Level 1 or Level 2 charging depending on the output power and type of installation.

Level 1 charging typically uses a common 120 V household outlet. It is the slowest type of AC charging, with an energy output of about 1.3–2 kW, which is sufficient for overnight charging or when the EV is parked for extended periods. Level 2 charging requires a 240 V connection, and it can be installed both at home and in public. It is relatively faster than Level 1, with power output ranging between 3.7 and 22 kW. This level of charging can fill the battery to full capacity within 4–8 hours and is preferred for residential and commercial purposes. This conversion process introduces slight energy losses, impacting overall efficiency, though the losses are minimal compared to some DC systems due to lower power demands. AC charging is often seen as more energy-efficient for low-power charging over extended periods, such as overnight residential charging. Research shows that AC chargers, especially Level 2, have conversion efficiencies reaching up to 92%–95% depending on the onboard converter's efficiency in the EV itself [10].

Advantages of AC charging are lower installation costs, as AC charging systems require less complex infrastructure, availability in both residential and commercial settings, and long-duration charging options suitable for overnight or off-peak charging. Limitations of AC charging are slower charging speeds, making it less suitable for long-distance travel or where quick recharging is needed, and charging speed depends on the onboard charger capacity of the vehicle, which may vary across different EV models.

DC Charging Systems: DC charging systems, also known as fast chargers or DC fast chargers, present several significant advantages over AC systems, primarily related to speed and efficiency. In contrast to AC charging, which involves the vehicle converting AC power to DC, DC fast charging systems bypass the onboard charger and deliver DC power directly to the battery. DC fast chargers typically operate at 400–800 V, with current ratings ranging from 100A to 500A. The high power output allows for rapid charging and is crucial for reducing charging time, especially on long-distance trips. This approach allows higher power delivery, from 50 kW in standard fast chargers to as high as 350 kW in ultra-fast chargers. As a result, DC charging systems charge batteries significantly faster than AC systems, providing an 80% battery charge in 20–30 minutes, contingent on the vehicle's battery capacity and the charger's power output. However, the rapid charging rate introduces additional thermal stress on batteries, which can affect long-term battery health and efficiency if frequently used [9]. The energy efficiency of DC fast chargers varies from 90% to 95%, with some losses associated with power conversion and thermal management. DC charging systems are frequently utilized in public charging stations along highways, where EV owners may need to quickly charge their vehicles during long trips. Fast-charging stations are also found in commercial or urban locations where minimizing downtime for EVs is critical [11,12].

12.3.1.2 Types of DC Charging Systems

- **Level 3 DC Fast Charging:** These systems usually provide power outputs of 50 –150 kW and are found in urban charging stations and highway rest stops.
- **Ultra-Fast Charging:** Ultra-fast chargers, which have power ratings ranging from 200 to 350 kW, are among the fastest of the latest generation of fast-charging networks, and they can significantly cut charging time.
- **Tesla Superchargers:** Tesla's proprietary network of ultra-fast chargers is designed to deliver up to 250 kW of charging capacity. The system will be compatible with Tesla vehicles only in the first phase and will eventually be upgraded to a system that can also work with different electric vehicles.

Each charging system serves a particular role in EV infrastructure because it caters to use cases alongside charging speed and infrastructure requirements. Residential and extended public charging applications integrate AC charging effectively, as this method provides lower expenses and slower speed, whereas DC fast charging serves highway travel and busy urban areas by speeding up the recharging process. Knowledge of the technical specifications and operational principles of both charging systems enables stakeholders to develop effective charging networks that serve worldwide EV user needs.

12.3.2 Implementation Scenarios

Urban Settings: Urban areas typically demand a mix of AC and DC chargers due to diverse EV user needs. AC chargers are ideal for locations with longer parking durations, such as residential complexes, workplaces, and shopping centers, where drivers are likely to leave their vehicles parked for several hours. In contrast, DC fast chargers are advantageous for commercial centers, urban transit hubs, and other high-traffic areas where rapid charging can enhance accessibility and turnover [10]. In urban implementations powered by renewable energy, such as solar panels integrated into charging stations, both AC and DC systems have been effective. AC chargers complement buildings with renewable energy sources by offering steady, low-power charging, whereas DC fast chargers can utilize battery storage systems to provide rapid charging without straining the grid during peak hours.

Highways and Corridors: Highway charging infrastructure relies heavily on DC fast chargers due to the high-speed requirements for long-distance travel. These locations provide installed DC chargers that enable EV drivers to rapidly recharge their batteries during their journeys without experiencing long delays. The integration of renewable power on highways requires solar canopies together with wind turbine generation systems but requires the deployment of energy storage technology to maintain power availability. The European Union Fast-E project created a case study which proved that renewable-energy-powered DC fast charging locations installed as part of critical highway infrastructure achieved reduced dependence on the grid since they could charge vehicles at higher speeds for cross-regional drives [9].

Residential Areas: Residents in residential neighborhoods commonly choose AC chargers because these devices need low power consumption while also being simple to install. Home-based AC charging enables electric vehicle owners to operate through their existing electrical circuits; however, they can easily add Level 2 chargers to enhance speed. The inclusion of rooftop solar energy in homes creates more sustainable and economically beneficial opportunities for charging electric vehicles through AC power systems. The AC charging mechanism matches well with residential overnight charging because it supports homeowners who can take advantage of reduced electricity rates at times of low demand [4]. Cable DC charging infrastructure is uncommon in homes because installation expenses are high and power demands are substantial; however, it proves viable when shared residential garages or neighborhood charging stations are utilized.

12.3.3 Cost-Benefit Analysis

Installation Costs: The installation expenses for AC charging stations remain less expensive than those of DC fast chargers because they need simpler setups and draw less power. The price range for Level 1 chargers extends from \$300 to \$600, but Level 2 chargers vary from \$400 to \$1,500, not counting installation expenses. DC fast charging devices need three-phase electric connections coupled with specialized cooling mechanisms and extensive updates to electricity grids before they can be operational. Different locations, combined with power output levels influence the installation expenses of DC chargers, which typically start from \$10,000 per station but can reach more than \$50,000 [13].

Maintenance Costs and Reliability: AC charging systems require less maintenance spending because they combine simple designs with decreased thermal stress operation. Regular software maintenance and visual checks make up most of the AC system's maintenance requirements, resulting in manageable long-term expenses. DC fast chargers need regular maintenance activities because they operate with high power loads and require active cooling systems. Maintenance expenses increase because rectifiers and converters, along with cooling systems in such components, tend to experience accelerated wear. Modern technological developments enhance the operational life of DC charging systems, which ultimately makes these systems more affordable for high-usage business models.

Long-Term Savings: The duration of power savings depends on how you use your AC or DC charging stations alongside electricity price rates. Homeowners and businesses can benefit from AC chargers because they have affordable pricing and allow users to take advantage of inexpensive late-night utility rates. DC fast chargers, though more costly upfront, provide value in high-turnover locations where rapid charging is essential, such as highways or urban centers. When combined with RES like solar or wind, DC fast chargers can reduce dependency on grid power, creating potential long-term savings despite higher installation and maintenance costs. Studies have shown that fast-charging stations with RES and battery storage minimize grid strain, decrease peak demand charges, and improve cost efficiency over time [12]. The brief comparison of AC and DC charging system key factors powered by a hybrid system is tabulated in Table 12.1.

Also, for better understanding, the comparison is shown in a graphical representation as shown in Figure 12.2.

The graph above compares the key factors of AC and DC charging stations powered by a PV-wind hybrid system. Here's a summary of what the graph indicates:

1. **Charging Speed:** DC charging scores a 5 (faster) compared to AC charging (scoring 2). DC chargers are significantly faster in delivering charge to EVs.

TABLE 12.1
Comparison of Factors of AC and DC Charging Systems Powered by Hybrid Systems

Factor	AC Charging System	DC Charging System
Efficiency and Performance	Lower Efficiency: Conversion from AC to DC leads to energy losses (10%–15%). Lower overall system efficiency due to onboard chargers.	Higher Efficiency: Direct DC power delivery to EVs reduces conversion losses, resulting in 95%+ efficiency.
Charging Speed and User Convenience	Slower Charging: Longer charging times due to lower power delivery rates (up to 22 kW max). Ideal for overnight charging or residential areas.	Faster Charging: Rapid charging times (up to 350 kW). Suitable for high-demand public stations or highway rest stops.
Infrastructure and Cost	Lower Installation Costs: Generally cheaper installation, suitable for residential and commercial areas. Minimal infrastructure required.	Higher Installation Costs: Requires high-voltage infrastructure and complex equipment. Higher costs (up to $100,000+). Suitable for public and high-traffic locations.
Environmental Impact	Potentially Higher Emissions: If grid power is from fossil fuels, AC charging systems can still contribute to emissions. Less effective when renewable sources are intermittent.	Lower Emissions: DC chargers powered by renewable energy sources (solar, wind) reduce carbon footprints significantly. Less reliance on grid power.

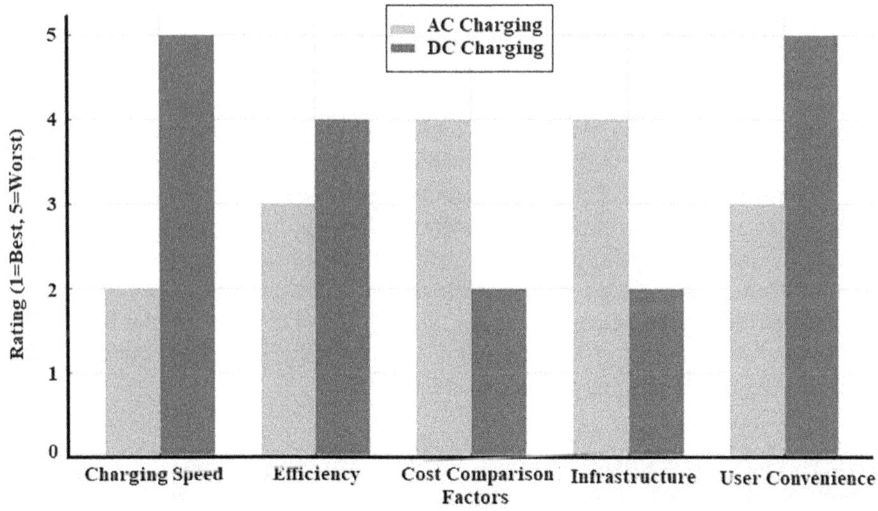

FIGURE 12.2 Comparison graph for key factors of AC charging and DC charging powered by hybrid system.

2. **Efficiency:** Both DC and AC charging have similar efficiency scores, with DC being slightly more efficient (score of 4) as it directly feeds DC to the battery, reducing conversion losses. AC requires more conversion steps, introducing slight losses (score of 3).
3. **Cost:** DC charging systems score a 2 (more expensive) compared to AC charging (scoring 4). The higher infrastructure and installation costs of DC stations make them more expensive to implement.
4. **Infrastructure:** AC charging systems score a 4 (easier to install), while DC charging systems score a 2, highlighting the higher complexity and costs involved in setting up DC fast charging stations.
5. **User Convenience:** DC charging systems score a 5 (less convenient) because while they are faster, they are less common and more expensive, which could make them less accessible. AC charging (score of 3) is more widespread and suitable for overnight charging, providing greater convenience.

In summary, AC charging is best suited for residential use or locations where charging speed isn't the highest priority. The system provides better value while being simpler to put in place and ensures better usability because of its extended accessibility. Public or commercial areas together with long-distance travel benefit most from using DC charging because of its fast-charging capabilities. Users receive brief charging periods because of its faster operation but must pay additional expenses and use elaborate technical systems. For maximum user convenience as well as widespread usage AC charging presents itself as the most suitable solution. DC charging remains the better solution for locations that need quick charging and possess the necessary infrastructure to install it.

12.4 DESIGN AND IMPLEMENTATION OF HYBRID RENEWABLE ENERGY SYSTEM

Renewable energy hybrid systems unite various renewable power generators into a unified, sustainable supply that provides reliable and efficient power generation. Hybrid renewable systems ensure reliability through batteries which function alongside solar and wind power components to balance the irregular power supply from individual renewable inputs. The analysis investigates hybrid energy system design structures along with control methods and technical approaches for managing changes in meteorological conditions. Through these components the framework enables optimal stability, efficiency, and reliability for hybrid energy systems, from residential applications to large-scale grid integration.

12.4.1 System Architecture

Figure 12.3 illustrates how HRES architecture uses solar and wind energy characteristics while battery-powered support produces uninterrupted power delivery. An optimized system design makes EV charging dependable and allows it to operate automatically through renewable energy sources that maximize performance based

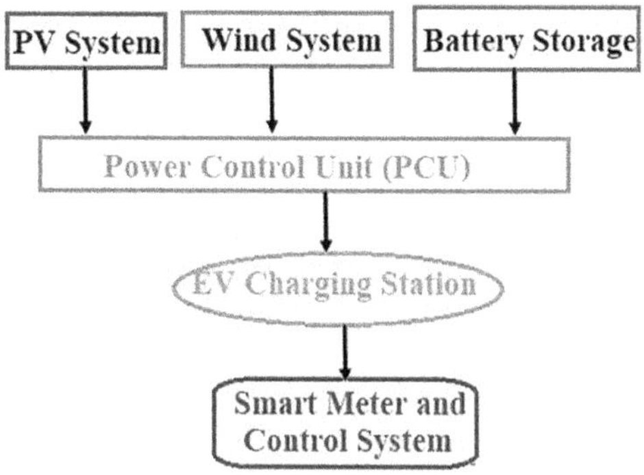

FIGURE 12.3 Architecture of HRES-based EVCS.

on weather indicators. The HRES incorporates five main elements such as PV panels, wind turbines, battery storage, a power control unit, and smart control software for monitoring and managing loads.

12.4.1.1 System Requirements and Site Assessment

The design process starts with complete energy demand prediction for the station and site-specific factors determine component dimensions and station configuration. The anticipated daily charging demand is calculated based on the number of EVs to be served, typical charging power requirements (Level 2 or DC fast charging), and average energy consumption per session.

$$\text{Energy Demand(kWh/day)} = \text{Number of EVs} \times \text{Energy per EV(kWh)} \quad (12.1)$$

For example, a station serving 20 EVs per day at an average of 30 kWh per vehicle requires approximately 600 kWh per day. Site-specific factors such as solar insolation, average wind speeds, and space availability for PV and wind turbine installations are also considered. These parameters help determine the capacities of the PV and wind components to ensure reliable energy generation and storage.

12.4.1.1.1 Photovoltaic (PV) System Design

Solar panels, or photovoltaic (PV) modules, form the primary component of most HRES architectures. Solar panels capture light and convert it into electricity. This electricity is then passed though the inverter and then used to operate appliances or injected into the power grid. PV technology has become highly refined, and today's systems feature efficiencies of over 20% and continue to produce significant power even in suboptimal climates, i.e., in environments with cloudy weather [14]. The modular nature of solar PV panels also allows for scalability, enabling systems to expand in response to increasing energy demands. Wind turbines convert the mechanical energy of the wind into electrical energy.

This PV system is expected to provide up to half of the station's energy requirements. For specific efficiency, space usage, and performance, monocrystalline PV panels of high efficiency are selected. Thus, the PV capacity is derived by dividing the total daily energy requirement by the average solar intensity at the site.

$$\text{PV Capacity(kWp)} = \frac{\text{Energy Demand (kWh/day)}}{\text{Average Solar Irradiance (kWh/m}^2\text{/day)}} \tag{12.2}$$

The panel area is calculated as,

$$\text{Area(m}^2) = \frac{\text{Energy Demand (kWh/day)}}{\text{Panel Efficiency} \times \text{Average Solar Irradiance}} \tag{12.3}$$

For instance, with an average of 5 kWh/m^2 per day of irradiation, to meet 600 kWh/day, 120 m^2 of PV panels would be needed, considering efficiency losses. Modules are connected in series and parallel circuits so that the desired voltage and current are obtained, also regarding the angles at which the panels are to be installed to capture as much sunlight as possible.

12.4.1.1.2 Wind Turbine Design

Wind generation is especially valuable in hybrid systems because it often complements solar energy, producing power during night hours and in cloudy or winter conditions when sunlight is limited [15]. In HRES, wind energy systems are typically integrated with power converters and inverters that ensure seamless distribution of power. The turbine selection is based on local wind patterns, and medium-sized turbines (e.g., 10–20 kW) are generally suitable for EV charging stations. Expected energy output is calculated based on turbine specifications and wind conditions.

$$P_{\text{wind}} = \frac{1}{2}\rho A v^3 \eta \tag{12.4}$$

where ρ is air density (kg/m^3), v is the wind speed (m/s), A is the swept area of turbine blades (m^2), and η is the turbine efficiency. For example, a turbine with a 5 m rotor diameter operating at an average wind speed of 6 m/s can produce substantial power, depending on local air density and efficiency. Positioning turbines at optimal heights minimizes turbulence and maximizes energy capture.

12.4.1.1.3 Battery Storage System Design

Battery storage serves as an essential component in balancing the intermittent nature of renewable sources. High-capacity lithium-ion or lithium-iron-phosphate ($LiFePO_4$) batteries are used for their durability and efficiency [16]. Battery capacity is determined based on the daily energy demand, desired autonomy (e.g., one day), and depth of discharge (DoD).

$$\text{Battery Capacity (kWh)} = \frac{\text{DailyEnergy Demand (kWh / day)} \times \text{Days of Autonomy}}{\text{Depth of Discharge (DoD)}} \tag{12.5}$$

Batteries provide stability to hybrid systems by bridging gaps in energy generation and offering reserve capacity, reducing reliance on backup fossil fuel generators. Battery storage systems are also paired with charge controllers and power management systems to ensure optimal charging and discharging cycles, extending battery life and improving system reliability.

12.4.1.1.4 Power Control Unit (PCU)

The power control unit (PCU) is essential for managing energy flows from PV panels, wind turbines, and battery storage to the EV charging stations. It includes inverters to convert DC power from renewable sources and batteries to AC for the EV chargers. The PCU prioritizes renewable energy and battery usage, switching to grid power only when necessary, and features safety protections (e.g., circuit breakers and fuses) to prevent overloads and manage power stability [17]. A bidirectional flow allows efficient battery charging and discharging based on demand and generation.

12.4.1.1.5 EV Charging Stations

The EV charging infrastructure includes a mix of Level 2 AC chargers and DC fast chargers to meet varying user needs. Load control strategies act as protective measures that stop excessive power consumption when utility demands peak [18]. The PCU regulates charging rates and holds sessions according to available power, which helps to enhance system reliability.

Head4A smart meter continually tracks energy generation, storage, and consumption to enable data-based control of electrical loads. Predictive algorithms in the control system direct energy flow by considering both weather forecasts and charging demand to optimize power distribution while reducing dependence on the grid. The automated system gives the EV charging station immediate responsiveness to renewable generation fluctuations to maintain reliable power delivery while minimizing environmental impacts [19].

12.4.2 Power Management Strategies

HRES needs effective power management techniques because the solar and wind energy sources show inconsistent performance patterns. The system operates power generation, storage, and distribution functions to deliver stable, reliable, and efficient service. Successful power management operates by accepting priority settings, implementing load control, and predicting energy consumption patterns for improved outcomes.

Prioritization of Power Sources: The programming of hybrid systems sets renewable energy resources above the usage of battery storage or backup generators. The PMS operates solar or wind power during peak sunlight or wind conditions, thus reducing battery drainage and maintaining stored energy for low renewable generation periods [20]. The battery receives stored excess energy until its capacity reaches the maximum level, then any remaining surplus either feeds into the grid or dissipates. The system hierarchy operates to achieve maximum renewable energy usage along with a dependable power supply that adapts to changes in generation rates.

Load Management and Demand Response: The power supply's adjustment to handle instantaneous load needs becomes essential for any installation connected to vital operational systems, including hospitals, data centers, and off-grid residential networks. HRES implements demand response programs, which enhance energy efficiency through two functions that manage unnecessary loads during high-demand times and reschedule power usage to match renewable energy peaks. The combination of smart electric controllers allows homeowners to schedule the operation of their appliances along with solar and wind energy production peaks [21]. This method boosts the operational efficiency of the system and decreases power usage peaks, which leads to reduced energy costs.

Energy Forecasting and Predictive Algorithms: The PMS depends on sophisticated prediction models that analyze weather information and operational statistics to forecast solar irradiance levels, wind speed conditions and power requirements so it can prepare the power flow adjustments. When the predictive model indicates cloudy conditions, the system starts charging the battery in advance along with decreasing discharge rates, which saves stored energy for future needs. Accurate forecasting of the PMS system improves its reliability while helping to eliminate the necessity for backup generators, thus achieving sustainability targets [22].

Grid Synchronization and Exporting Surplus Power: HRES can export excess energy (during periods of high production) when connected to the grid. To connect the system to the grid in a stable manner, the PMS has to match frequency and voltage to the grid parameters. This contributes to balancing other sources of power as well as provides financial benefits (due to net metering contracts, which pay energy producers for the electricity they dispatch back to the grid) [23]. Such an arrangement is particularly advantageous for hybrid systems with large-scale battery storage, which can be utilized as a grid-balancing asset.

12.4.3 Weather Variability Solutions

Weather unpredictability is a challenge for renewable energy systems because fluctuations in sunlight and wind availability can disturb electricity generation. Hybrid systems provide a technological and strategic solution to this problem through a combination of mechanisms that reduce power fluctuations and offer stable electricity.

Hybrid Complementarity of Solar and Wind: Solar and wind generation show natural complementarity, making their combined generation a great fit for hybrid configurations. Solar power is, by far, most efficient on sunny days, whereas wind generation will reach its peak at night and in colder seasons when solar power availability drops. Hybrid systems give rise to a more stable overall power output as the individual power sources fluctuate. Research demonstrates that hybrid systems outperform single-source systems in providing more consistent power availability in areas with high seasonal diversity [7].

Battery Storage with Peak-Shaving and Load-Leveling Capabilities: Battery storage systems are necessary for smoothing out power generation in relation to meteorological fluctuations. The batteries are charged when solar or wind output is high, then discharged when they are low. Peak-shaving involves delivering stored energy to meet spikes, and load-leveling balances supply and demand over longer periods. These approaches improve the secure delivery of contracted power supply while reducing dependence on backup generators and conserving battery usage [24].

Advanced Energy Management Software: More advanced software solutions improve the responsiveness of hybrid systems to variability. These software platforms leverage predictive analytics, utilizing meteorological forecasts to improve energy storage and consumption. Thanks to this, the system may save battery charge and use wind power if low sunlight is expected. Advanced energy management software even allows for remote monitoring, data collection, and real-time adjustments to the system to ensure it is performing at optimal efficiency, even with changing weather.

Integration of Backup Generators for Critical Loads: For scenarios where weather unpredictability can create extended periods of low renewable production, hybrid systems can include backup generators. Backup generators, in most cases powered by natural gas or biodiesel, are only used in dire circumstances and power the system when renewable energy and storage cannot sustain demand. Though backup generators incur some operational expense, their selective use minimizes emissions and keeps the system available, particularly for off-grid deployments [25].

Smart Grid Connectivity for Load Flexibility: Smart grid technology enables HRES to be integrated with extended grid networks, offering load flexibility during high and low periods of renewable energy surplus or scarcity. Hybrid systems linked to the smart grid are able to export power during peak renewable output and consume extra power during low-output hours. Such connectivity stabilizes the grid, lowers dependence on fossil fuel-driven power plants, and enhances the system's supply-demand balancing capacity [26].

12.5 ROLE OF SMART GRID TECHNOLOGIES

Decentralized grid technologies are processed, transmitted, and your electric power is generated. This means a higher efficiency, as well as reliability and sustainability. These technologies work through advanced communication systems, sensors, data analytics, and automation to improve grid operations, employ renewable energy resources and handle overall electrical site management. This chapter discusses the role of smart grid technologies in the energy distribution improvement, introducing energy storage solutions, real-time monitoring and control, as well as the grid reliability that arises as a consequence.

12.5.1 Enhanced Energy Distribution

The primary goal of smart grids is energy distribution optimization. The effectiveness of traditional electricity grids, which are based on a one-way distribution of electricity from centralized power plants to consumers, is deteriorating steadily, particularly having in mind the growing popularity of decentralized and intermittent renewable energy, e.g., wind and solar. On the contrary, smart grids use digital communication and control technologies to enable two-way communication between utility companies, power generators, and consumers, improving the flow of electricity across the system.

1. **Dynamic Load Management:** A market with smart grids allows for the exact adjustment of the electricity distribution based on the balance between supply and demand in real time. Through the installation of sensors and smart meters, power suppliers can monitor the power consumption habits of their customers and then distribute energy accordingly. Hence, the grid can flexibly control power consumption while providing avoidance of overload and a balance of supply from traditional and renewable sources. The initiation of demand response systems makes it possible for customers to take an active part in saving energy by redirecting their power use when the load on the grid is at a peak, which prevents congestion and curtails energy costs [19].
2. **Integration of Renewable Energy:** Over time, the differing character of the green RES has been a real challenge in the improvement of grid stability as well as the grid's efficiency. The smart grid approach focuses on the flexibility of the energy transition to the integration of the generation of renewable energy. And, e.g., the grid may decide to collect solar or wind power in case of high generation, which means that more or less power will be extracted from fossil fuel-based generation. During periods of low renewable generation, the grid can draw from storage systems or alternative sources. With this flexibility, the smart grid can assist in efficient and sustainable energy use, and, in particular, optimization of electricity flow across the physical grid [27].
3. **Voltage and Frequency Regulation:** Smart power grid technologies allow for real-time monitoring of changes in voltage and frequency, both of which are crucial for the proper functioning of the power grid. Introducing automatic control systems into the grid enables it to adjust the voltage and frequency of any sudden changes in power generation and consumption. Thus, the grid can function in the most secure and best way possible, which improves the process of power distribution and reduces the risk of it breaking down.

12.5.2 Storage Solutions

Energy storage is needed to handle variations of renewable energy and to assure a constant and reliable electricity supply. Smart grids are combined with different types of storage techniques, prominently advanced batteries. Storage systems are also needed to bridge the lack of continuity of renewable energy and to ensure a stable and reliable electricity supply. Smart grids employ different storage technologies, especially advanced battery systems, to match energy supply and demand for better handling of the grid.

1. **Lithium-Ion Batteries:** Lithium-ion batteries are the most dominant energy storage solution in smart grids, and they owe their success to higher energy density, much faster charging, and more affordable prices. These batteries save surplus energy produced at low demand times or during high renewable energy generation periods and deliver it when demand overtakes supply. The fact that they help in the stabilization of the grid by reducing the irregularities in power generation makes them key players. Long-term battery facilities such as the Hornsdale Power Reserve in South Australia have exhibited that lithium-ion batteries can be used for grid stabilization, reserve capacity provision, and blackout prevention [28].
2. **Flow Batteries and Other Storage Technologies:** Moreover, the concept of using battery technology like a flow battery as hyper data grouping using grid integration is also being considered. Flow batteries are able to offer extra runtimes and also go hand in hand with solutions used in large-scale energy storage, as they can be made bigger by just making the electrolyte tanks bigger. Plus, pumped hydro storage, compressed air energy storage, and solid-state batteries are also becoming the most promising prospects to address the long-duration storage requirements in smart grids [29].
3. **Vehicle-to-Grid (V2G) Systems:** EVs integrated into the grid, which is called V2G technology, are considered a new form of smart grid storage solutions. EV batteries in V2G systems are mobile storage units that can either store electricity from the grid or supply it to the grid when required. In fact, this technique allows building a grand network of batteries and gives additional latitude to operators [30].

12.5.3 Monitoring and Control

The embedding of the internet of things (IoT) and data analytics has turned the control and monitoring of energy systems upside down. Inter alia, the smart grid technology is fed by innumerable types of sensors, smart meters, etc., which are spread all over the network in order to retrieve information on electricity usage, grid performance, and environmental conditions. This data is analyzed using advanced analytics platforms to give grid operators a real-time view of the entire system and help them to make decisions that are the most logical [31].

1. **Smart Meters and Sensors**: Smart meters play a pivotal role in smart grids as they allow the collection of real-time data. These devices measure electricity consumption at the end-user level and throughout the grid and provide utilities with a continuous stream of information. Thanks to this real-time data, utilities can track patterns of energy usage, diagnose potential problems, and optimize power distribution dynamically. Hence, smart sensors being installed at certain crucial points in the grid can also measure things like voltage, current, temperature, and the status of equipment, enabling predictive maintenance and reducing the necessity for manual inspections.
2. **Big Data Analytics**: The vast amount of data that smart meters, sensors, and other IoT devices collect is also being managed through the careful use

of big data analytics. The data collected by analytics can also detect and measure demand ups and downs, signal equipment malfunctions in time, as well as ensure the normal operation of renewable energy systems according to weather forecasts. A historical analysis of the data will allow for the development of predictive models, and the operators will be able to plan and buy energy ahead, thus ensuring that the grid is prepared for changes in demand.

3. **Automated Control Systems**: Automated control systems using real-time data and analytics can act fast and respond to any variations in the grid. For instance, if one area of the grid gets overloaded or if renewable energy generation is suddenly reduced, the automatic system can use power from the rest of the grid or use backup storage solutions to avoid unwanted shutdowns. There is a decrease in human error, the speed of decision-making improves, and the grid's overall responsiveness is boosted by these systems.

12.5.4 Reliability Improvements

For the success of every electrical grid invention, stability is the primary factor. According to statistics, everyday outages in a traditional grid are caused by equipment failures, weather-related events, or large voltage surges. By maintaining and predictively monitoring power balance, fault detection, and conducting an integrated recovery process, smart systems can prevent grid deterioration before it occurs.

Predictive Maintenance and Fault Detection: Rotating stations that use telemetry as well as predictive analysis are examined as to the state of transformers, switches, circuit breakers, and power lines. Of course, downtime is kept to a minimum, and sudden blackouts are avoided due to a proactive approach to maintenance and preventive measures implementation. For instance, prognostic models may recognize small vibrations in transformers or detect temperature changes in machines, signaling that they require maintenance.

Self-Healing Grids: The self-healing capability in the case of a fault is the most revolutionary thing about smart grids. A smart grid uses automated switching and control systems through which the damaged section can be isolated, power can be redirected, and service can be restored to unaffected areas without human intervention. This capability is a major reason for the reduction in the duration of outages, the development of stronger systems, and guarantees that consumers will not experience any major disruption [32].

Grid Resilience against Extreme Weather Events: Desolating consequences to electric power systems in high voltage due to extreme weather catastrophes like tropical hurricanes, storms, heatwaves, and heat snaps stand as the main reasons for grid standstill and power outages. It is proposed from the content that smart grids excel in their ability to be more resilient by using information from weather forecasts to make the right decisions and then optimize the operation of distribution grid operating conditions in the face of possible worsening of the environment. By doing so, if there is a forthcoming imminent storm, the grid is capable of dropping loads at critical areas as well as sending power in the opposite direction to a less vulnerable location or, if necessary, initiating the backup system automatically so that service can be carried on.

12.6 ECONOMIC AND ENVIRONMENTAL BENEFITS OF SMART GRID AND RENEWABLE ENERGY INTEGRATION WITH EV CHARGING STATIONS

The synergy of intelligently programmed grids and renewable energy sources in the EV network makes it possible for the end-user to achieve a positive environmental impact and also benefit from economic savings compared to the old methods of using AC and DC car-charging stations. Here is a comprehensive comparison made of the innovations.

12.6.1 Economic Benefits

Cost Savings: The cost of electricity for EV charging can be significantly reduced with smart grids and RE integration. That means if sources such as solar and wind are utilized, the need for expensive fossil fuels diminishes, causing energy to be cheaper. Research by [33] showed that grid buying can lower the net present cost (NPC) by 40% for 2030 compared to the existing policy scenario.

Reduced Infrastructure Costs: Smart charging solutions can reduce the necessity for extensive grid reinforcements. Smart grids can balance the load and reduce peak demand by controlling charging times and locations, which, in turn lowers the costs associated with upgrading and maintaining the electrical grid [34].

Operational Efficiency: RES and smart grid technologies can be integrated to achieve better charging infrastructure efficiency. This is associated with decreased operational costs and optimized.

12.6.2 Environmental Benefits

Reduction in Carbon Emissions: RES produce hardly any greenhouse gases in contrast to fossil fuels. The integration of these sources into the electric vehicle (EV) charging infrastructure becomes more significant in the reduction of the carbon footprint of EV charging. Research conducted by [33] forecasted that carbon dioxide emissions from the use of EVs could be reduced by around 44% by the year 2030 under a policy mitigation scenario.

Sustainable Energy Use: Smart grid efficiency in using renewable energy is achieved by timing electric vehicle charging for when renewable energy is at its peak. Consequently, this guarantees that the electric vehicles being charged are as environmentally friendly as possible, and therefore they help in further reducing environmental impact.

In summary, the implementation of smart grids and renewable energy resources in electric vehicle charging stations does not only have economic but also ecological benefits over current basic AC and DC charging stations [35]. The improvement of electrical infrastructure and renewable energy solutions has a superior edge in terms of cost reduction, environmental impact, and energy effectiveness for the electric mobility sector.

The challenges in developing regions and the potential solutions for scalability and economic feasibility of RE-based EV charging are listed in Table 12.2.

TABLE 12.2
Scalability and Economic Feasibility of RE-Based EV Charging in Developing Regions

Factor	Challenges in Developing Regions	Potential Solutions for Scalability and Economic Feasibility	Region-Specific Example
High Initial Investment Costs	• Significant capital investment is required for renewable energy infrastructure, battery storage, and grid integration. • Developing countries often lack financial incentives for private investment in EV charging.	• Government subsidies and incentives for renewable energy adoption. • Public-private partnerships to share investment costs. • Use of second-life EV batteries for cost-effective energy storage.	India: The FAME (Faster Adoption and Manufacturing of Electric Vehicles) scheme offers financial incentives for EV and charging infrastructure development.
Limited Grid Infrastructure	• Many developing regions have weak or unreliable grids, making it difficult to support high-powered DC fast chargers. • Rural and remote areas lack grid access, limiting EV adoption.	• Deployment of microgrids using solar and wind energy. • Hybrid systems integrating renewable energy with battery storage to ensure a stable power supply.	Nigeria: Solar-powered mini-grid projects (e.g., Rubitec Solar) provide off-grid EV charging solutions, reducing reliance on the unstable national grid.
Policy and Regulatory Barriers	• Inconsistent government policies and a lack of clear regulations for EV infrastructure. • High import taxes on EVs and charging components hinder widespread adoption.	• Clear government policies promoting EVs and renewable energy integration. • Reduced tariffs and financial incentives for green energy projects. • Public-private collaboration for regulatory reforms.	Kenya: The Energy Act 2019 supports renewable energy adoption and incentivizes EV charging infrastructure through tax incentives and policy reforms.

(Continued)

TABLE 12.2 (*Continued*)
Scalability and Economic Feasibility of RE-Based EV Charging in Developing Regions

Factor	Challenges in Developing Regions	Potential Solutions for Scalability and Economic Feasibility	Region-Specific Example
Energy Storage & Supply Intermittency	• Solar and wind energy generation fluctuate, making EV charging unreliable without battery storage. • Battery costs remain high for large-scale deployment in developing countries.	• Use of second-life EV batteries to lower storage costs. • Intelligent energy management systems to optimize energy distribution. • Hybrid solar-wind-battery systems for continuous power supply.	South Africa: Eskom's battery energy storage projects improve grid stability and support renewable-powered EV charging, especially in load-shedding-prone areas.
Cost-Effective Charging Infrastructure	• High-cost DC fast chargers are not feasible for widespread use in low-income areas. • Lack of private investment in charging stations due to uncertain revenue generation.	• Prioritizing AC slow chargers for home and commercial use • Hybrid solar-powered EV charging stations to reduce dependence on the national grid.	Bangladesh: Solar-based AC slow chargers are used in rural electrification projects, ensuring low-cost EV charging solutions.
Affordability and Accessibility of EVs	• The high upfront cost of EVs limits consumer adoption. • Lack of local EV manufacturing increases costs due to import taxes.	• Government subsidies and financial incentives for EV buyers. • Encouraging local EV manufacturing to lower costs.	Indonesia: Government incentives and local EV production initiatives aim to make EVs more affordable for consumers.
Case Studies of Success	• Lack of large-scale implementation and feasibility data for renewable-powered EV charging.	• India: Pilot solar-powered EV charging stations in rural areas. • Kenya: Mini-grid solar projects providing sustainable transport solutions. • Nigeria: Off-grid solar EV charging reducing reliance on fossil fuel-based electricity.	Ethiopia: The Addis Ababa Light Rail project, powered by renewable energy, demonstrates large-scale feasibility of clean transport solutions.

12.7 CHALLENGES AND SOLUTIONS IN THE INTEGRATION OF RENEWABLE ENERGY AND SMART GRID SYSTEMS WITH EV CHARGING STATIONS

Integrating RE and smart grid systems with EV charging stations, whether they use AC or DC charging, comes with unique challenges and potential solutions. The challenges vary mainly due to the power requirements, charging speeds, and infrastructure needs of AC and DC stations [2]. Here is a comparative analysis of these challenges and solutions for AC and DC charging stations:

12.7.1 Power Demand and Grid Stability

The challenges faced in AC charging stations are that AC chargers typically have lower power demands, but when deployed in large numbers (e.g., in residential or commercial areas), they can collectively add significant strain on the grid. Without smart grid management, even low-power stations can lead to local grid instability, especially during peak hours. Solutions to overcome this challenge can be implemented through smart grid technology with demand-side management, which can distribute charging loads over non-peak hours, preventing collective strain on the grid. Smart grids can also allow charging stations to operate at reduced power during peak hours, helping maintain grid stability.

In DC charging stations, the challenges are that DC fast chargers draw substantial power, which can cause grid fluctuations, especially when several stations are active simultaneously. High power demands can lead to voltage dips and stress on the grid infrastructure, particularly in areas not designed for such loads. To overcome these challenges, pairing DC stations with on-site RES and battery storage systems can offset their high-power requirements [3]. The battery storage system enables the station to obtain energy from renewable reserves instead of grid supply, thus decreasing strain on the grid and boosting system reliability.

12.7.2 Intermittency of Renewable Energy Sources

The main obstacle with AC charging stations is that their slower operations match smoothly with variable renewable resources, including solar and wind power. Renewable storage systems might experience slower operations along with inconsistent service when renewable energy generation rates stay low because of adverse weather conditions. The smart grid system manages AC charging operations to occur when renewable power generation reaches its peak level, typically during daylight hours. Optimized renewable energy usage, together with sustained charging accessibility, becomes possible through this method [4].

A stable, continuous power supply is essential for DC fast chargers located in DC charging stations to reach their high-speed operation mode. Intermittent renewable power supply systems become unreliable when used without additional storage units or grid connections because the power availability creates instability in the system. The integration of bigger battery storage systems together with renewable energy technology at DC charging stations enables the effective storage of excess electricity for two uses: peak demand periods or intervals with low renewable generation. The smart grid uses energy management systems to distribute charged power reserves between grid resources and stored energy, which guarantees uninterrupted charging operations.

12.7.3 Infrastructure and Cost Constraints

The installation of AC stations proves less expensive but their large-scale expansion necessitates upgrades to distribution lines, together with transformers in the network. The conversion to renewable energy systems creates additional expenses, which affect the transformation of existing stations. AC stations deployed in clusters with joint use of local solar arrays help lower expenses by minimizing the requirements for expensive grid additions. The smart grid uses distributed control to distribute energy between multiple sites, thus maximizing renewable energy extraction without forcing any stationary improvements.

The installation of DC fast chargers leads to significant expenses, which require companies to invest heavily in grid connections, together with heavy-duty transformers. The integration of renewables elevates station operational expenses since DC stations need big energy management equipment [10]. The high initial expenditure of installing DC fast chargers can be reduced through financial assistance from the government, renewable energy provider shared costs, and utility subsidies. Through smart-grid implementation, DC stations gain the ability to alter their power consumption based on immediate grid performance, which enables them to earn income from utilities for maintaining power system stability.

12.7.4 Demand Fluctuations and Load Management

The charging speed of AC chargers remains slow, yet demand variations stay modest until there are abundant electric vehicles in an area. Operating chargers at optimal times with high renewable energy generation requires smart load management because it makes coordination challenging. Smart grids with demand-side management systems help arrange AC charging sessions in order to minimize stress on AC charging systems. Smart charging ensures users plug in their vehicles at times when renewable energy is abundant, which supports stable grid consumption and maintains peak demand in check. Stations should establish adjustable pricing programs to encourage customers to meet their charging needs when electricity rates are lowered.

The demand for DC charging stations becomes extremely high and volatile in both urban and along highways areas. The rapid increase in charge requests might cause problems for the local power distribution network unless there are suitable distribution strategies in place, especially when renewable energy generation is low. The implementation of smart grid algorithms that manage power consumption at DC stations enables peak demand control during stressful conditions of the grid. To reduce demand spikes during peak times, the system should include battery storage for peak shaving that uses stored energy to assist the grid while maintaining stability.

12.7.5 Energy Loss and Efficiency Concerns

AC charging involves lower power loss but relies on the vehicle's onboard charger, which may not be as efficient as dedicated DC fast chargers. This means less energy from renewables may reach the vehicle battery. Possible solutions can be upgrading to more efficient vehicle onboard chargers and using smart grid load management, which can minimize energy losses. Renewable power can be supplied to AC stations with minimal transmission distance, which helps reduce overall energy loss.

DC charging stations provide more efficient, direct charging, but higher power transmission can result in greater energy loss. DC stations integrated with renewables need advanced power electronics to minimize conversion and transmission losses. Better solutions can be achieved by using high-efficiency power electronics, and optimizing charging sessions based on grid and renewable availability (e.g., charging during peak solar hours) can help minimize energy loss. Smart grids can also prioritize locally generated renewable energy for DC stations, reducing transmission distances and losses.

In summary, integrating smart grids and renewable energy with both AC and DC stations offers substantial benefits but requires tailored solutions for each type of station to maximize efficiency, cost-effectiveness, and environmental impact. Table 12.3 tabulates the comparison of AC and DC charging station challenges.

TABLE 12.3
Comparison of AC and DC Charging Stations Challenges

Challenge	AC Charging Stations	DC Charging Stations
Power Demand and Grid Stability	Collective strain from high station numbers; managed with smart grid scheduling	High individual power demand; mitigated with renewables and battery storage
Renewable Intermittency	Low, steady demand works with intermittent renewables but may lead to slower charging	High power demand requires storage for continuous availability
Infrastructure and Cost	Generally lower costs, though upgrades needed at scale; renewable retrofitting can add expenses	High initial costs, large infrastructure needs; offset with government incentives and cost-sharing
Demand Fluctuation and Load Management	Lower fluctuation but significant with high station density; smart grid staggers sessions	High spikes in demand; managed with load scheduling, battery storage, and peak shaving
Energy Loss and Efficiency	Lower loss, though it relies on vehicle chargers which vary in efficiency	Higher transmission losses; reduced with efficient power electronics and smart grid prioritization

12.8 FUTURE TRENDS AND INNOVATIONS IN RENEWABLE ENERGY AND EV CHARGING TECHNOLOGIES

The landscape of RE integration and EV charging is evolving rapidly, with emerging technologies set to reshape the future of clean energy. Innovations such as wireless charging, ultra-fast charging, energy harvesting, and supportive policy frameworks are expected to significantly impact how energy is generated, stored, and consumed. This section explores these future trends and technologies, highlighting their potential to transform the EV sector and propel the global transition to sustainable energy even further.

12.8.1 Wireless Charging

Inductive charging, often referred to as wireless charging technology, is attracting considerable interest as a potential solution to eliminate the need for a physical connection between vehicles and charging stations. This technology transfers energy using electromagnetic fields between two coils, one integrated into the vehicle and the other in the charging pad.

Wireless charging systems typically consist of a primary coil (in the charging station) and a secondary coil (in the vehicle). When aligned, the system generates a magnetic field that transfers energy between the coils, charging the vehicle's battery without the need for plugging in. Recent advancements in magnetic resonance coupling and resonance-based energy transfer have significantly increased the efficiency of wireless charging systems, making them a viable alternative to traditional charging methods [36].

Potential Impact on EV Adoption: One of the major advantages of wireless charging is convenience. Users no longer need to manually connect and disconnect cables, making charging more seamless, particularly in urban environments where parking and plug-in accessibility may be limited. Additionally, wireless charging can be embedded in roads or parking lots, enabling dynamic charging, where vehicles are charged while driving or during brief stops, further reducing range anxiety. The widespread adoption of wireless charging could lead to an entirely new paradigm for mobility, where electric vehicles are continuously charged without the need for dedicated charging infrastructure [37].

Challenges: Despite its potential, wireless charging faces challenges related to efficiency, cost, and infrastructure. The energy loss in wireless charging systems, although lower than earlier models, still represents a significant inefficiency compared to direct wired connections. Additionally, scaling up the infrastructure to support wireless charging, especially for long-distance highways and urban areas, requires significant investment and standardization. However, continued research in superconducting materials and advanced energy transfer methods may help overcome these challenges, potentially making wireless charging a mainstream solution.

12.8.2 Ultra-Fast Charging

A major obstacle to the widespread adoption of EVs is the lengthy charging time. Despite EVs being more energy-efficient than traditional vehicles, the time needed for recharging continues to be a drawback. Ultra-fast charging technology seeks to drastically cut down the time it takes to recharge an EV, bringing it closer to the convenience of refueling a conventional ICE vehicle.

Ultra-Fast Charging Technology: Ultra-fast charging involves high-power charging stations capable of delivering more than 350 kW of power to EVs, a significant leap from the current standard of 50–150 kW seen in most public charging stations. Long-route driving becomes realistic with ultra-fast charging where a battery can be charged to 80% of its total capability within less than half an hour. For instance, Tesla's Supercharger V3 can charge a Model 3 up to 75 miles of range in 5 minutes. Additionally, battery technologies such as lithium iron phosphate (LFP) batteries, known for their faster charging times and increased safety, are being developed to complement these high-power charging systems [38].

Potential of Ultra-Fast Charging: The reduced charging times that ultra-fast charging offers could alleviate "range anxiety" and encourage more consumers to switch to EVs. For commercial fleet operators, who depend on vehicles being operational for as long as possible, ultra-fast charging could significantly reduce downtime, improving the economic viability of EV adoption. Furthermore, ultra-fast charging systems are designed to be more scalable, making them suitable for high-demand locations like highways and urban hubs where high volumes of users will need to charge quickly.

Challenges and Solutions: Despite its promise, ultra-fast charging presents challenges such as battery degradation. Frequent charge-discharge cycles cause batteries to heat, and this can degrade the battery lifespan with respect to current battery chemistries. To address this, manufacturers are exploring new battery materials and thermal management systems that can better handle high charging rates without compromising performance. Additionally, significant investment in infrastructure is needed to develop a global network of ultra-fast charging stations. Policymakers and private companies will need to collaborate to ensure that charging stations are strategically located and that they meet standardized requirements for global interoperability [39].

12.8.3 Energy Harvesting

Energy harvesting, or energy scavenging, is a technique by which small or large quantities of energy can be drawn from the environment to provide a power source to the targeted device or system. While large-scale renewable energy production typically focuses on harnessing solar, wind, and hydro resources, energy harvesting techniques aim to capture energy from more localized or distributed sources, such as vibrations, temperature differentials, or even radio waves.

Types of Energy Harvesting: Some novel solutions have the potential of making energy harvesting possible. For example, piezoelectric material, which changes mechanical energy loading from vibration into electrical energy. This technology has been successfully used to harvest energy from roadways or railway lines, where vehicle movement generates vibrations. Another promising method is thermoelectric energy harvesting, where temperature differences between two surfaces are converted into electricity. These methods can be integrated into infrastructure like roads, bridges, or buildings to generate continuous power for low-energy applications, such as sensors or small electronics

Potential Impact on EV Charging and Grid Systems: Energy harvesting could be seen as a key contributor to boosting the sustainability of EV charging systems. For example, roads that have piezoelectric technology could generate energy from traffic and subsequently provide the power to charge other stations. This distributed generation of electricity could act as a supplement to the conventional electricity supply and also cut energy costs. Additionally, low-cost energy input is also possible through energy harvesting technologies, thus making smart grids efficient and sustainable energy networks.

Challenges: The main limitation of energy harvesting technologies is the relatively small amounts of power they can generate. These systems are generally more suitable for low-power applications, such as sensor networks or small devices, rather than for large-scale energy production. Additionally, the installation of harvesting devices, such as piezoelectric generators on roads, involves high upfront costs and technical challenges, including durability and maintenance. However, ongoing research into materials science and energy conversion efficiency may increase the practicality of energy harvesting for larger-scale applications [40].

12.8.4 Policy and Regulatory Support

Promoting policies are an essential factor that supports the early growth of RES and the development of the necessary stations for charging EVs. The extent and ability of RE integration and EV adoption are highly contingent on the type of support offered, either in the form of financial incentives, regulations, or through public and private partnerships.

Policy Initiatives to Promote EV Charging Infrastructure: State incentives have led many governments in the globe to pass policies to incentivize the growth of the EV charging network. For instance, the European Union has allocated significant funds for the deployment of charging stations through the Connecting Europe Facility (CEF) program, aiming to establish a unified EV charging network across member states. In the United States, the Biden administration has proposed a $7.5 billion investment to build a nationwide network of fast chargers as part of its infrastructure bill. These initiatives help reduce the "chicken and egg" problem, where the lack of charging stations hinders EV adoption, and vice versa.

Financial Incentives for Consumers and Businesses: To encourage the uptake of EVs most governments provide incentives like tax incentives, rebates, and subsidies for the cost of EVs. For instance, in the United States of America, the federal government fosters tax credits up to $7,500 for the purchase of a new electric car. Likewise, subsidies for industries producing EVs have helped China to become one of the largest markets for the vehicles [5]. This will, in turn, improve the affordability and availability of EVs to consumers and other businesses in the automotive market.

Regulatory Support for Renewable Energy Integration: Governments are also intervening crucially to facilitate the incorporation of renewable energy into the grid. Tariffs with feed-in renewables and renewable portfolio standards, alongside power purchase and power purchase agreement mechanisms, constitute policies that drive investor investment in renewable energy generation. Smart grid regulations, which support data sharing combined with interoperability and demand response programs, enable the development of a flexible grid system that accepts variable renewable power sources [2].

12.9 CONCLUSION

The integration of RES with EV charging stations creates an effective solution to overcome the traditional energy grid dependence issues. The survey demonstrates the crucial need to establish efficient sustainable energy infrastructure, during which AC and DC charging systems, hybrid renewable energy systems, and smart grid technologies should be analyzed to develop resilient systems. The study establishes both economic worth and environmental advantages of sustainable energy-based EV charging facilities, alongside successful approaches to integrating power grids and controlling energy distribution. The analysis confirms the essential part renewable energy solutions will have in building a sustainable, robust energy future for the coming years.

REFERENCES

1. Rani, S.; Jayapragash, R. Review on electric mobility: Trends, challenges and opportunities. *Results in Engineering* 2024, 23, 102631.
2. Vishnu, G.; Kaliyaperumal, D.; Jayaprakash, R.; Karthick, A.; Kumar Chinnaiyan, V.; Ghosh, A. review of challenges and opportunities in the integration of electric vehicles to the grid. *World Electric Vehicle Journal* 2023, 14, 259.
3. Panossian, N.; Muratori, M.; Palmintier, B.; Meintz, A.; Lipman, T.; Moffat, K. (2022). Challenges and opportunities of integrating electric vehicles in electricity distribution systems. *Current Sustainable/Renewable Energy Reports* 2022, 9, 27–40.
4. Barman, P.; Dutta, L.; Bordoloi, S.; Kalita, A.; Buragohain, P.; Bharali, S.; Azzopardi, B. Renewable energy integration with electric vehicle technology: A review of the existing smart charging approaches. *Renewable and Sustainable Energy Reviews* 2023, 183, 113518.

5. IEA. *Global EV Outlook 2021: Accelerating the Transition to Electric Mobility.* International Energy Agency, 2021.
6. Hawkins, T.R.; Singh, B.; Majeau-Bettez, G.; Strømman, A.H. Comparative environmental life cycle assessment of conventional and electric vehicles. *Journal of Industrial Ecology* 2013, 17, 53–64.
7. Hassan, Q.; Algburi, S.; Sameen, A. Z.; Salman, H. M.; Jaszczur, M. A review of hybrid renewable energy systems: Solar and wind-powered solutions: Challenges, opportunities, and policy implications. *Results in Engineering* 2023, 20, 101621.
8. Li, L.; Wang, X. Design and operation of hybrid renewable energy systems: Current status and future perspectives. *Current Opinion in Chemical Engineering* 2021, 31, 100669.
9. Leijon, J., Boström, C. Charging electric vehicles today and in the future. *World Electric Vehicle Journal* 2022, 13, 139.
10. Singh, A. R.; Vishnuram, P.; Alagarsamy, S.; Bajaj, M.; Blazek, V.; Damaj, I.; Rathore, R. S.; Al-Wesabi, F. N.; Othman, K. M. Electric vehicle charging technologies, infrastructure expansion, grid integration strategies, and their role in promoting sustainable e-mobility. *Alexandria Engineering Journal* 2024, 105, 300–330.
11. Domínguez Navarro, J. A.; Dufo-López, R.; Yusta, J. M.; Artal-Sevil, J. S.; Bernal-Agustín, J. L. Design of an electric vehicle fast-charging station with integration of renewable energy and storage systems. *International Journal of Electrical Power & Energy Systems* 2019, 105, 46–58.
12. Ugur Fesli, U.; Ozdemir, M. B. Electric vehicles: A comprehensive review of technologies, integration, adoption, and optimization. *IEEE Access* 2024, 12, 140908–140931.
13. Unterluggauer, T.; Rich, J.; Andersen, P. A.; Hashemi, S. Electric vehicle charging infrastructure planning for integrated transportation and power distribution networks: A review. *eTransportation* 2022, 12, 100163.
14. Kalogirou, S. A. *Solar Energy Engineering: Processes and Systems.* Elsevier, 2017.
15. Manwell, J. F.; McGowan, J. G.; Rogers, A. L. *Wind Energy Explained: Theory, Design, and Application* (2nd ed.). John Wiley & Sons, 2010.
16. Divya, K. C.; Østergaard, J. Battery energy storage technology for power systems—An overview. *Electric Power Systems Research* 2009, 79, 511–520.
17. Bahrami, S.; Ranjbar, A. M.; Ghadimi, N.; Bahrami, H. A decentralized renewable energy management and demand response in power distribution networks for a large population of electric vehicles. *Energy* 2016, 109, 714–725.
18. Khan, M. J.; Iqbal, M. T. Pre-feasibility study of stand-alone hybrid energy systems for applications in Newfoundland. *Renewable Energy* 2005, 30, 835–854.
19. Amin, S. M.; Wollenberg, B. F. Toward a smart grid: Power delivery for the 21st century. *IEEE Power and Energy Magazine* 2005, 3, 34–41.
20. Baños, R.; Manzano-Agugliaro, F.; Montoya, F.G.; Gil, C.; Alcayde, A.; Gómez, J. Optimization methods applied to renewable and sustainable energy: A review. *Renewable and Sustainable Energy Reviews* 2011, 15, 1753–1766.
21. Nazari-Heris, M.; Asadi, S. Reliable energy management of residential buildings with hybrid energy systems. *Journal of Building Engineering* 2023, 71, 106531.
22. Alnejaili, T.; Labdai, S.; Chrifi-Alaoui, L. Predictive management algorithm for controlling PV-battery off-grid energy system. *Sensors* 2021, 21, 6427.
23. Roy, D.; Zhu, S.; Wang, R.; Mondal, P.; Ling-Chin, J.; Paul Roskilly, A. Techno-economic and environmental analyses of hybrid renewable energy systems for a remote location employing machine learning models. *Applied Energy* 2024, 361, 122884.
24. Tharani, K. L.; Dahiya, R. Choice of battery energy storage for a hybrid renewable energy system. *Turkish Journal of Electrical Engineering and Computer Sciences* 2018, 26, 4.

25. Jacobson, M. Z.; Delucchi, M. A.; Bauer, Z. A. F.; Goodman, S. C.; Chapman, W. E.; Cameron, M. A.; Bozonnat, C.; Chobadi, L.; Clonts, H. A.; Enevoldsen, P.; Erwin, J. R.; Fobi, S. N.; Goldstrom, O. K.; Hennessy, E. M.; Liu, J.; Lo, J.; Meyer, C. B.; Morris, S. B.; Moy, K. R.; O'Neill, P. L.; Petkov, I.; Redfern, S.; Schucker, R.; Sontag, M. A.; Wang, J.; Weiner, E.; Yachanin, A. S. 100% Clean and renewable wind, water, and sunlight all-sector energy roadmaps for 139 countries of the world. *Joule* 2017, 1, 108–121.
26. Kabeyi, M. J. B.; Olanrewaju, O. A. The use of smart grids in the energy transition. *2022 30th Southern African Universities Power Engineering Conference (SAUPEC)*, Durban, South Africa, 2022, pp. 1–8.
27. Kataray, T.; Nitesh, B.; Yarram, B.; Sinha, S.; Cuce, E.; Shaik, S.; Vigneshwaran, P.; Roy, A. Integration of smart grid with renewable energy sources: Opportunities and challenges – A comprehensive review. *Sustainable Energy Technologies and Assessments* 2023, 58, 103363.
28. Faunce, T. A.; Prest, J.; Su, D.; Hearne, S. J.; Iacopi, F. On-grid batteries for large-scale energy storage: Challenges and opportunities for policy and technology. *MRS Energy & Sustainability* 2018, 5, E11.
29. Kumar, M.; Shanmugam, K.; Pradeep, K.; Filippone, M. Grid integration and application of battery energy storage systems. *2022 IEEE International Conference on Electronics, Computing and Communication Technologies (CONECCT)*, Bangalore, India, 2022, pp. 1–6.
30. İnci, M.; Büyük, M.; Mustafa Savrun, M.; Hakan Demir, M. Design and analysis of fuel cell vehicle-to-grid (FCV2G) system with high voltage conversion interface for sustainable energy production. *Sustainable Cities and Society* 2021, 67, 102753.
31. Ahmad, T.; Zhang, D. Using the internet of things in smart energy systems and networks. *Sustainable Cities and Society* 2021, 68, 102783.
32. Elgenedy, M. A.; Massoud, A. M.; Ahmed, S. Smart grid self-healing: Functions, applications, and developments. *2015 First Workshop on Smart Grid and Renewable Energy (SGRE)*, Doha, Qatar, 2015, pp. 1–6.
33. Eom, D.; Choi, J.; Song, J.; Park, B.; Kim, S. J.; Park, S. Techno-economic analysis of grid-connected highway solar EV charging station. *International Journal of Precision Engineering and Manufacturing-Green Technology* 2024, 12, 573–590.
34. Rehman Anis ur; Khalid Haris, M.; Muyeen, S. M. Grid-integrated solutions for sustainable EV charging: A comparative study of renewable energy and battery storage systems. *Frontiers in Energy Research* 2024, 12, 1403883.
35. Güven, A.F.; Yücel, E. Sustainable energy integration and optimization in microgrids: Enhancing efficiency with electric vehicle charging solutions. *Electrical Engineering* 2024, 107, 1541–1573.
36. Amjad, M.; Farooq-i-Azam, M.; Ni, Q.; Dong, M.; Ahmad Ansari, E. Wireless charging systems for electric vehicles. *Renewable and Sustainable Energy Reviews* 2022, 167, 112730.
37. Palani, G.; Sengamalai, U.; Vishnuram, P.; Nastasi, B. Challenges and barriers of wireless charging technologies for electric vehicles. *Energies* 2023, 16, 2138.
38. Khan, S. A.; Hussain, I.; Thakur, A. K.; Yu, S.; Lau, K. T.; He, S.; Dong, K.; Chen, J.; Xiangrong, L.; Ahmad, M.; Jiyun Zhao, J. Advancements in battery thermal management system for fast charging/discharging applications. *Energy Storage Materials* 2024, 65,103144.
39. Khalid, M. S. Smart grids and renewable energy systems: Perspectives and grid integration challenges. *Energy Strategy Reviews* 2024, 51, 101299.
40. Georgiadis, A.; Collado, A.; Tentzeris, M. M. *Energy Harvesting: Technologies, Systems, and Challenges*. Cambridge University Press, 2021.

13 Power System Planning for Electric Vehicles

Jawahar Marimuthu, Bhavani Ramachandran, and Arockiaraj Sesaiya

13.1 INTRODUCTION

The integration of electric vehicles (EVs) into electricity systems creates immense challenges and opportunities to plan long term. Further penetration of EVs necessitates looking into their impact on power system operations, energy demand, as well as transmission networks [1]. Effective strategies of interdependence, response charging, and resilience of the system interdependence; effective management and control strategies are vital in the case of the incorporation of EVs and distributed generations (DG) [2]. Through their voltage-dependent characteristics, the EV loads may create instablilities and cause overload on other components of the system, demanding specific planning for the charging infrastructure. In the design of convenient and reliable charging devices, both compensation of EV consumers and power systems can be carried out by optimization techniques, including particle swarm optimization [3]. As presented in this paper, in an attempt to capture the impacts of evacuation on grid resistance during wildfire events, a power planner frame is chosen [4]. Electric vehicles can be charged smartly to promote investment in solar power capacity and minimize fossil-fuel-based capacity generation in planning power system expansion [5]. The article presents the operational background of the paper as a review of the methods of optimization in the planning of various modes of electric vehicles to reduce power losses and enhance the performance of power systems. This paper focuses on the analysis of the effects of incorporation of EV charging infrastructure on power system planning as well as operations within the Indian setting [6].

The paper screens out distributed generation, electric vehicles, and distributed generation optimization strategies to plan distribution systems with the use of electric vehicles [7]. Coupled planning of charging stations and the distribution system of electric vehicles is proposed in order to reduce costs and maximize traffic [8]. This includes the multi-purpose joint planning strategy of integrating power distribution and charging systems of electric vehicles, both to reduce associated costs and maximize traffic flow [9]. Throughout the paper, a frame to incorporate electric vehicles in the power supply system is presented, with involvement in network technology operations as well as the power market [10]. Random recharging of EVs will result in increased troubles in the daily distribution system, such as power outages, modulation of voltage, and harmonic pollution [11]. In this paper, the presence of distributed generation and optimization methods in electric vehicle planning in power

DOI: 10.1201/9781003585534-13

distribution systems is examined [12]. In the given paper, the author elaborates on the relationship between transport electrification and the planning and operation of power supply toward providing a thorough overview [13]. The article validates the methodology of planning high-performance charging infrastructure for electric vehicles [14]. Precise models of EV charging systems are necessary for consistent power system analysis as well as power system planning [15]. In this paper, the contribution of electric vehicle charging to power system planning, the necessary power generation capacity increase, and potential emissions are analyzed [16]. The paper verifies the major technologies and system settings for the best arrangement and management of energy in controlling the charging station network associated with electric vehicles [17]. The best approach toward distributed generators is the fact that the integration of electric vehicles into power systems has the ability to reduce energy losses and increase the voltage profile [18]. It also outlines how to prevent overwhelming the distribution network while designing electric cars that serve the needs of vehicle owners at the same time [19]. The present paper provides an in-depth explanation of adding electric vehicles to power systems [20]. This article, having considered the impacts of uncontrolled intelligent charging of plug-in electric vehicles, provides a sales system planning model [21]. The paper is able to verify the literature in the planning of electric vehicle infrastructure that combines transport and distribution networks [22].

13.2 CHALLENGES AND OPPORTUNITIES

13.2.1 Challenges

13.2.1.1 Increased Load on the Grid due to Charging

The rapid adoption of EVs is an important milestone in a new period of a more ecological, cleaner type of transportation. There are, however, some specific challenges facing global electrical grids, especially in light of managing the extra load that EV charging introduces. The presence of EV load is depicted as illustrated in Figure 13.1. As the number of individuals shifting to electric mobility continues to grow, their power needs will increase, and this will require modifications in terms of grid power resilience, flexibility, as well as capacity. Since charging electric vehicles causes greater loads on the grid, it is harder to find a balance between supply and demand, stabilize the grid, modernize infrastructure, and ensure reliability during peak periods.

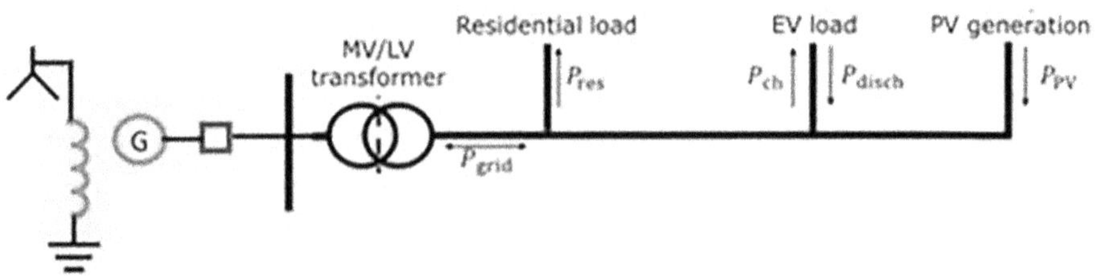

FIGURE 13.1 Single-line diagram inclusive of EV loads.

The availability of charging points for EVs is also profound because grid stability is under threat since the demand is variable. As compared to normal household appliances, the rate of power consumption of EV chargers is high, which may cause variations in voltage and compromise the quality of electricity. Regular EV charging could lead to control instabilities in the grid unless there is advanced technology in place to manage this uncertainty, and this would affect homes and businesses. There is a financial and logistical issue in enhancing the infrastructure to handle EV load. The current networks should also be strengthened with greater investments, additional transformers installed, andenhancement of grid capacity. To ensure that the grid is resilient and reliable, utility companies have to consider how much improvement can be made, which they pass on to consumers.

13.2.1.2 Peak Demand Management

Peak demand control on electrical grids is considered to be one of the most important obligatory issues of modern energy systems. Peak demand refers to the time of day or year when the energy usage is at its highest, often due to increased heating/cooling systems, appliances, or, more recently, EV charging usage. When there is an increase in demand for electricity, it becomes very tough for utility firms to ensure that they maintain low prices, ensure stability in the system, and also ensure that they are able to balance the demand and supply. These maxima are significant to control efficiently to reduce power disruptions, decrease costs, and minimize environmental impact. The major challenge that exists in the management of peak demand is maintaining a reliable supply during peak demand periods. The grid has to generate and distribute sufficient electricity to satisfy the needs of consumers during peak times. Fossil fuel power plants, as well as other current methods of power generation, have the potential to be expanded very quickly; however, they are both expensive and highcarbon dioxide–emitting. But since favorable renewable sources of energy, such as wind and solar energy, are not as reliable, it is difficult to rely on them as the primary contributors to the provision of peak demand. Grids are at risk of being unstable and experiencing blackouts unless various sources of energy are well coordinated.

The other critical issue related to peak demand is stress on the infrastructure. When high demand occurs, substations, transmission lines, and transformers become more stressed, leading to a decreased life span because the components wear faster. This burden ultimately increases maintenance costs and the need for more frequent upgrading of the infrastructure. Budgets and, ultimately, consumer energy prices end up being stretched as the utilities struggle to strike a balance between the urgency of new investments and the cost of further repairs. Besides, during peak demand, the cost of electricity to both consumers and the utilities increases.

In an attempt to satisfy demand the utilities might need to acquire or generate excess electricity during their busiest times, leading to what is known as peak pricing. This is often done by increasing bills, as the cost is passed on to the customers. In addition, it is necessary to keep reserve power sources ready, which can be switched on during busy working hours, thus increasing the total costs of energy. Another challenging factor in managing peak demand is attracting and encouraging customers to reduce their use during peak hours. Both of the following elements are needed to manage demand on the demand side, which makes people use their electric power more efficiently and during off-peak times.

13.2.1.3 Grid Stability Concerns

A stable electrical supply rests on grid stability, which ensures power supply 24/7 and the attainment of quality requirements. But there is also a great number of changing issues, and stability maintenance is more problematic. The development of renewable sources, the popularity of the use of EVs, and the fluctuation of energy consumption make it harder and harder to maintain a stable and healthy grid where the interests of grid operators are maximized. Balancing supply and demand in realtime, maintaining uninterrupted voltage and frequency, and preventing blackouts, which may lead to costly interruption of power, are all needed to keep the grid stable.

One of the threats to grid stability is renewable energy sources such as wind or solar because they are intermittent. Even though the reduction of carbon emission will demand renewable energy, this form of energy is unstable in its nature, based on the time of the day and the weather. As an illustration, solar energy production is at its peak during the daytime hours and at zero during the night, but wind energy is highly chaotic. This variability is problematic because acute variations of supply could cause problems in the grid, therefore making it hard to rely on renewable energy for grid operations. To deal with this variability (which, after all, comes unpredictably), energy storage or backup generation of power is required; neither is cheap, and both have technical challenges. Frequency and voltage differences are also a critical threat to grid stability. The grid should be able to react rapidly to changes in demand to ensure that it keeps the frequency steady, and usually this frequency is 50/60Hz; this varies according to location or area. Excessive deviation can lead to cascade failures of the entire network or damage the equipment or cause it to shut down. Similarly, changes in voltage can negatively affect sensitive systems and affect the power quality of the end users. Grid operators control these variations with reserve generation and complex control systems, but the more complex the grid is, the harder it is to control accurately. Figure 13.2 shows the phenomenon of power change in the condition of the particular state of power penetration into a grid of 80MW.

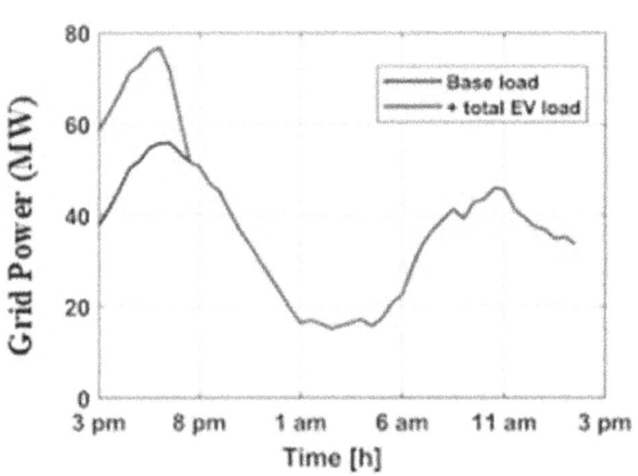

FIGURE 13.2 Power change phenomenon.

Another factor that makes grid stability a complex question is the adoption of distributed and decentralized energy resources (DERs). DERs such as small-scale storage systems, community wind farms, and household solar panels are often hooked up to the grid in an ad hoc manner. They make the grid more complex, as electricity now flows in many directions rather than in the familiar one-way flow of electricity to consumers from centralized plants. This bi-directional flow introduces new vulnerabilities and demands advanced forms of monitoring and control systems in order to be stable.

13.2.1.4 Infrastructure Development

Another important barrier is deploying charging stations. The high penetration of EVs necessitates a wide and convenient network of charging facilities, although installation of the charging points around cities and rural areas may be challenging and expensive. The extra burden is especially placed on local networks by particularly fast chargers and requires a considerable electrical capacity. Locating charging stations should be wellstrategized, synchronized with the local authorities, and financed to prevent overloading the grid and to meet demand. Moreover, the issue of equity regarding universal access to EV charging exists, as rural and underserved communities often lack the infrastructure that could support rapid EV charging.

Peak demand is another issue, as EV charging is often performed during workplace hours, i.e., in the evening when people arrive home. Such a rising demand can lead to peak load conditions that will require additional grid capacity to be installed and this could mean a higher power rate applied to all consumers unless there are other solutions, such as smart charging mechanisms or incentives to alter the timing of charging. To deal with the key points of understanding, advanced grid management technologies like demand response programs, time-of-use pricing, and smart charging infrastructure that has the capacity of deferring the load to off-peak hours are mandatory. The design of power systems to integrate renewables has impacts on EV charging process needs. In case of limitations in the power grid, there will be even more troubles with matching the unpredictability of renewable sources (solar and wind). EVs have the potential to consume great amounts of energy; hence, they require innovative fixes to meet EV demand while still ensuring a constant energy supply, including renewable energy. Other energy storage options (such as batteries) on the network can also store extra renewable energy when there is low demand and use it when there is high demand for electric vehicles. Delivery of such systems of storage is, however, costly and technically demanding.

13.2.2 Opportunities

13.2.2.1 Potential for Demand-Side Response (DSR) and Energy Storage for EVS

Integration of EVs into the electric power system enables demand-side response (DSR) and grid-scale energystorage, which can enhance grid flexibility, stability, and efficiency. These vehicles can become reliable instruments to manage peak loads, balance supply and demand, and increase the reliability of renewable energy sources as the popularity of EVs spreads. When DSR and energy storage in EVs are used,

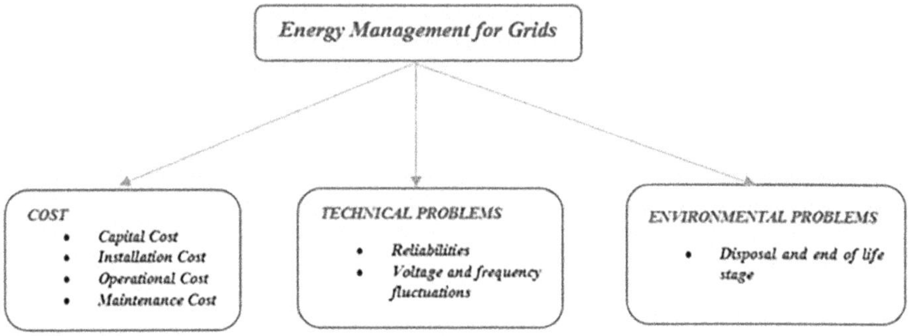

FIGURE 13.3 Practical constraints of energy management for EVs.

utility businesses and consumers can work together in order to develop a more resilient, flexible, and cost-effective electricity system. Demand-side response (DSR) is the strategy of changing or reducing power consumption during peak demand, and EVs offer an ideal platform to engage in this strategy. Smart charging technology can be used to program EVs to charge off-load when demand rates are lower and grid strain is reduced by flattening demand. In this approach, utility providers might get EV users to charge during grid-friendly times, which could help reduce the necessity of investing in costly infrastructure upgrades. DSR programs also allow financial benefits that could be offered to customers who engage in charging their EVs during off-peak hours, making EV ownership even more economical and more likely to participate in other energy efficiency initiatives. The practical limitations of energy management of EVs are plotted in Figure 13.3.

Vehicle-to-grid (V2G) technology offers a more dynamic application of DSR, on top of load shifting. With the help of V2G technology, automobiles can be used to store energy and then feed electricity back to the grid, in a way making the car a portable power station. Through V2G, the possibility of overload and blackouts may become a thing of the past, as the batteries in EVs would supply power to the grid in the event of high use. With this two-way energy flow, utility companies can dedicate fewer expensive and carbon-intense peak production generating plants since they have the capability of balancing supply and demand in a decentralized way. Owners of EVs can sell their stored energy back to the grid, thus earning monetary returns and providing an incentive to make the power system more stable. Another significant possibility is the potential of EV energy storage. The fact that they are portable batteries makes it possible to store large amounts of electricity when it is strongly produced, especially when demand is low (like on sunny or windy days). This stored energy can be utilized during periods of high demand, and this provides highly intermittent renewable energy sources with an effective source of flexibility and assistance. An example is additional solar generation in the daytime, which could be stored in EVs and used to feed houses or the grid during peak time in the evening. The large dissemination of EV storage may also provide a more distributed, high-scale way of addressing variations in renewable energy, as well as mitigate the need for stationary, more costly energy storage measures.

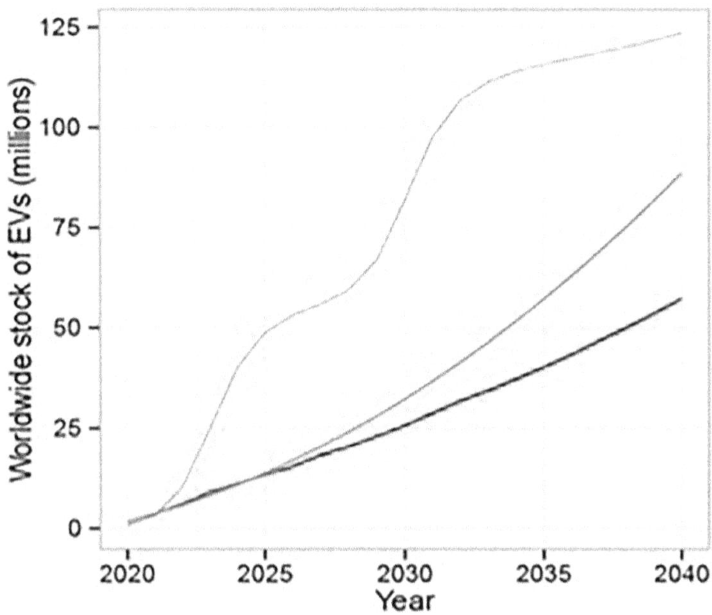

FIGURE 13.4 GEFI survey, May 2024.

13.2.2.2 Integration of Renewable Energy Sources

One of the most tempting reasons to incorporate renewable energy into EVs is the ability to significantly reduce greenhouse gas (GHG) emissions. EVs powered by renewable energy systems have extremely low carbon emissions compared to those powered by electricity that uses fossil fuels. In jurisdictions in which most of the power supply is non-fossil, EVs can operate virtually emission-free, which enhances public health and cleaner air quality. This transition to EVs that run on renewable energy goes hand in hand with broader efforts to decarbonize transportation, such as slowing down climate change, with more countries pledging to ambitious climate goals. In the month of May 2024, the Global Fuel Economy Initiative (GFEI) gave a survey display, as depicted in Figure 13.4, to estimate the stock of EVs in the world market.

The second interesting approach would be the direct connection of wind and solar power to EV charging stations. As an example, the commonly used photovoltaic (PV) cells in solar-powered EV charging stations can locally generate renewable power, decreasing operating expenses and reducing their dependence on the legacy grid. Also, wind energy would help in diversifying the energy used in charging EVs, especially in locations where wind potential is high. Besides enabling clean charging, these stations play the role of lessening strain on the grid by locally generating and storing renewable energy, especially when demanded at peak times.

13.2.2.3 Economic Benefits for the Power Sector

The increased popularity of EVs allows the power industry to have several economic benefits to enjoy, such as a larger grid intensity, a new source of revenues, and investments in and modernization of the current infrastructure. The change to a sustainable energy future, as EV adoption gathers momentum, presents a unique opportunity to utilities and power companies to revolutionize the traditional electricity market and grid efficiency and leverage the change to be the sustainable energy providers of the future. The ability to take advantage of the growth in demand for electricity that EVs cause allows the power industry to enjoy financial performance and stability.

The increase in the need for power is among the largest economic benefits of EVs to the power sector. With EVs, they consume more utilities since they do not use gas but electricity, as traditional cars would use. The demand for electricity will grow because of the growing number of people switching to EVs, enabling utilities with a stable income source to cover energy efficiency declines, as well as other drops in demand. Energy is normally less consumed during off-peak time, and because of this, such additional demand can be highly beneficial. Through time-of-use tariffs and incentives, utilities can promote charging at specific times, putting their resources to optimum use and generating more power without necessarily expanding the grid immediately. It also helps the electricity sector economically, as it allows more intensive use of the grid. Since EVs represent a controllable, variable load, utilities have the option of managing the charging of EVs strategically, in order to use the grid more efficiently. Smart charging technology effectively enhances the efficiency of grid operations since it allows utilities to shift EV charging when renewable energy is high or when demand is low. By balancing loads and averting abuses during peak periods, utilities can also reduce the need to make costly purchases of extra generation and transmission devices. Best utilization of available resources and associated savings, as well as higher ROI of power firms, are also achieved due to better grid utilization.

13.3 POWER SYSTEM PLANNING CONSIDERATIONS

The so-called conversion to EVs has disrupted the transportation industry and signals the most crucial problems and opportunities for the power system. As EVs become further incorporated into the network, planning will be quite crucial to make absolutely certain that the network is reliable, stable, and efficient. The widespread adoption of EVs provides opportunities and challenges to the power system. As the number of EVs gains a presence on the streets, it is significant that proper planning is necessary to ensure that a system of growing demand brings about justice and retains its stability. The major factors that planners of power systems should consider include:

- Load Forecasting
- Grid Infrastructure
- Energy Storage
- Demand-Side Management
- Integration of Renewable Energy

13.3.1 Load Forecasting

By taking into consideration grid stability, load forecasting is vital to facilitate resource-effective deployments and operations. At the rates of EV penetration, charging patterns, and geographic distributions, utilities would be able to make decisions regarding grid investments. With EVs, new uncertainties are added because of the behavior associated with charging, the battery level, and the availability of charging infrastructure. This article extends its focus on three components of EV load forecasting, comprising the estimation of EV penetration rate, analysis of charging patterns, and geographic distribution of EVs.

13.3.1.1 EV Penetration Rate Estimation

A penetration rate of EVs is considered an indicator of how the electric vehicle is incorporated and adopted in a specific market or region. The EV penetration rate expresses the proportion of EVs on the road vehicles in a specified region in terms of percentages.

$$\text{EV Penetration Rate} = \frac{\text{Number of EVs}}{\text{Total number of Vehicles}} \times 100$$

Where:

- EVs count both battery electric vehicles (BEVs) and plug-in hybrid electric vehicles (PHEVs) as well.
- The total number of vehicles consists of all vehicles, including internal combustion engine (ICE) vehicles.

The precise prediction of EV penetration should include data collection and predictive tools. Analysis of historical data is associated with the accumulation of historical information about vehicle sales, registered vehicles, and stock data (number of vehicles on the road). Trends work to project future penetration rates. Top-down forecasting is based on the assumption of market shares, whereas S-curve modeling occurs as an S-curve. Bottom-up forecasting entails analyzing the behavior of individuals, the rate at which regions/countries have adopted, the preparedness of local infrastructure, the state of available EV models/Brancillon, and their cost-effectiveness using microeconomic data. It will assist in making estimates of the various market segments, including passenger cars, commercial vehicles, and fleet vehicles, since they are likely to adopt EVs at varying rates in light of their unique needs and incentives.

13.3.1.2 Charging Pattern Analysis

EV charging patterns are one of the most important factors when it comes to including EVs into the existing power system and its analysis. The analysis of EV charging patterns does not only investigate the rate of how and when EVs are charged, but also the general effects of EVs on the electric power grid. It aids in forecasting charging requirements, determination of peak load periods, and developing strategies that would ensure effective implementation of charging infrastructure. EV charging tendencies are affected by factors, and time-of-use patterns have a major role in the determination of charging vehicles. Moreover, the potential effect of charging duration and charging power, which provide different processes, influences how and when EV users prefer to charge their cars. Charging behavior is also influenced by energy prices and tariffs, since most consumers will prefer to charge during off-peak periods when tariffs are lower. Finally, the charging infrastructure is important, especially when more accessible charging points may increase and more convenient charging EV owners can have. The process of analyzing EV charging patterns is based on several major methods, and the first one is data collection. This incorporates the collection of charging event logs at stations or smart chargers, which give information about the time of day, duration, power consumption, in addition to the location of the charging. Also, monitoring of on-board telematics systems data may track how much charge is on-board, driving styles and driving habits, and data collected by utilities through energy meters can be used to determine when loads will be put on a system and at what locations. After primary information is obtained, simulation and modeling solutions are also used to analyze charging behavior. Monte Carlo simulations have the ability to simulate a wide range of circumstances of consumer behavior, whereas agent-based modeling tries to simulate the act of interaction between people to determine the overall influence on the grid. In order to maximize EV charging behavior, modern technologies such as smart charging are being applied. On-demand smart chargers are able to communicate with the grid to optimize charging schedules depending on grid conditions, electricity tariffs, and availability of renewable energy. Moreover, there is a possibility of having a V2G system, wherein vehicles can receive electricity supplied by the grid but can also contribute additional energy back to the grid, thereby improving the entire grid amplification.

13.3.1.3 Geographic Distribution of EVs

The spatial location of EVs is among the salient factors in design considerations and the incorporation of electric vehicles in the power grid. Such geographic differences would require in-depth consideration so that the infrastructure (e.g., charging stations) and the power grid itself are ready to support the extra load caused by EVs without compromising reliability and efficiency of operations.

EVs are geographically dispersed by varying factors such as the structures of policies and incentives, urban and rural distributions, the availability of the charging infrastructure, and preferences and usage patterns of the vehicle types, among others. Learning about this distribution is vital to efficient power system planning, since having a large number of EVs in certain locations can place a heavy burden on the local electric grids, especially when charging schedules are not well planned out. To give an example, it may result in localized grid stress when the charging demand is at its peak, particularly in areas of high EV density. In this regard, utilities and some municipalities will need to enhance their generation and distribution grids strategically and in line with the geographic distribution of EVs to cope with the challenges. This can include reinforcement of grids in congested cities where the demand of EVs will be high, so the substations, transformers, and power lines will have to be upgraded. Furthermore, putting in place smart charging infrastructure in locations where EVs are concentrated may allow it to dynamically shift charging times and destinations, avoiding grid congestion and the necessity of extending the grid through large constructions. Moreover, the geographic presence of EVs is also quite important in the process of renewable energy integration into the power grid, either simplifying or complicating the shift toward renewable energy sources.

13.3.2 Grid Infrastructure

The electric grid is the spinal cord of the electricity distribution system, which supplies power to the power sources and ultimately to the end users. As regards the integration of EVs, the grid is to satisfy the new demand of charging points, as well as the flow of power both ways when EVs can also provide power to the power grid (vehicle-to-grid or V2G).

13.3.2.1 Capacity Expansion and Reinforcement

Perhaps the most important point of consideration in implementing EVs into the work lines of current energy infrastructure is that the electric grid should be capable of handling the increased demand without any drop in reliability. This entails enlargement and strengthening of capacity. These are some of the issues that utilities and grid operators should consider in the wake of ever-increasing demand to charge electric vehicles (EV charging points), including upgraded transmission and distribution systems, the role of DERs, and how green energy sources can be integrated. There are a number of strategies that can be deployed in order to ready the grid to support pervasive EV usage. To begin with, there needs to be proper models for prediction of future rates of EV adoption and distribution in the regions involved. By using information on how many people are registering their EVs publicly, as well as the demographic data and trends, the history may help to predict the demand and focus on upgrading the grids in a more accurate way. Enhancing distribution networks is also important; this can be done through smart substations with advanced monitoring capabilities to manage them remotely, and through battery energy storage systems (BESS) to smooth supply and demand during the peak charging time. Moreover, the improvement of voltage control and load distribution methods will help to correct grid voltage. Fast-charging infrastructure deployment must be carefully planned, and its location should be based on those regions with high EV penetration to avoid any shortage of grid capacity at electric vehicle charging points. Because they include a fast and slow charging option, utilities will be able to regulate grid stress. Moreover, other future progress, such as V2G technology, could be used to exploit the opportunity to supply energy back to the grid when demand is most challenging to serve, and eventually contribute to a more resilient and sustainable energy system.

13.3.2.2 Smart Grid Technologies for Efficient EV Management

A sophisticated power grid termed a smart grid is a high-end ICT and information grid that oversees and regulates the transportation and movement of the total electricity supply from any source of generation so as to meet various electricity consumption needs of the consumers. It integrates the renewable energy source together with the energy storage and the electric cars in the grid and maintains the stability of the grid, reliability, and efficiency. The proliferation of Electric Vehicles has both issues and opportunities in the management of the grid. The expanded number of charged cars will generate charging pressures on the network that are usually concentrated within a few hours, bringing about peak load issues that challenge the electricity system. These can be a game-changing solution because they can be applied as distributed energy resources.

Advanced Metering Infrastructure (AMI) systems are critical in offering real-time meter reads to utilities, providing the ability to see how electric car charging is affecting grid conditions. This is a critical step, which allows utilities to manage dynamic costs by applying time-of-use pricing (TOU) or demand-based pricing schemes, which allow the utilities to ensure that the electric vehicle is charged during off-peak hours to alleviate the load on the grid. Also, AMI enables real-time monitoring in that it provides detailed information about the charging times and energy use, as well as where charging occurs, to allow utilities to learn useful data about local demand partially and operate the grids accordingly. Demand response (DR) programs also augment the use of grids, allowing regulators to shift EV demand according to real-time grid conditions, attracting consumers to recharge less at peak hours or hold charging sessions during peak times. A recent potential solution is V2G technologies: in this case, the vehicles can supply energy back to the grid when needed (as distributed energy storage), in addition to the vehicles receiving energy from the grid when they need it. This interaction is improved through the use of smart charging decked with Internet of Things (IoT) capabilities that allow two-way communication with grid operators to deliver vital information to coordinate real-time EV charging loads. From a cost aspect, to enable the anticipated growth in inbound electricity demand by EVs, upgrades of distribution networks will be required, including hardening transformers and substations to larger units and upgrading smart grid technologies including AMI and demand response programs, in addition to integrating DERs to enhance grid performance and reliably serve customers.

This is the final stage in power transmission to the end user: the distribution network. Increased penetration of EVs will also affect the distribution network, and thus distribution network upgrades are necessary to support the new load of EVs, reliability, and maximum performance. Modernization and upgrading of the distribution grid will be necessary to support the expected increase in demand due to EVs. The primary electric infrastructure ought to be improved to accommodate the intense needs of EVs. Some of these important areas of focus include upgrading transformers and substations. Places with high EV penetration will also be required to replace existing transformers with transformers that can handle higher ratings to prevent overloading. In addition, the power substations in these locations may have to be expanded or reinforced, perhaps through the addition of higher-rated equipment or even with the addition of transformer banks so that they can handle the additional load. The second major aspect of this upgrade is the application of smart grid technologies that aids in load control. Advanced metering infrastructure (AMI) facilitates the ability of utilities to get real-time consumption data and easily control the loads in the grid. It is also possible for utilities to apply a DR program to avoid the possible risks that can appear due to distribution network overload during peak charging periods. Moreover, during these smart grid revampings, using DERs within the grid will be deemed profitable in enhancing efficiency and reliability of the whole energy system through tapping into rooftop solar, battery storage power, or small-scale wind power systems.

13.3.3 Energy Storage

The performance of electric vehicles is dependent on their energy storage system, which is usually the battery pack. These are the batteries that are supposed to be the primary supply of power and are the storage and users of electricity, powering the vehicle and powering utilities within the vehicle. With the increase in the number of people needing EVs, there comes the need to make the batteries more competent, effective, and durable. In this section, the author investigates the contribution of battery storage to electric vehicles, including the kinds of batteries used and how they can affect the grid and energy storage systems.

13.3.3.1 Role of Battery Storage Systems

The electric vehicle battery energy storage accommodates the electricity storage that is used to drive the motor and other auxiliary devices. In order to accommodate the increased demand for EVs, enormous renovations of the electric infrastructure are imperative. This will involve replacements of transformers and substations, especially in high EV-penetration neighborhoods where higher-capacity transformers will be required to avoid overloading. Substations in EV-heavy regions might have to be extended or strengthened, and this can be done either by using higher-rated equipment or adding more transformer banks so they can perform reliably. Also, smart grid technology will be essential to achieve good load control. AMI is deployed to enable utilities to view and take action on real-time consumption, and demand response programs give a utility a chance to ensure that the distribution network is not overloaded during times of peak charging. To meet these demands, the inclusion of DERs, including rooftop solar and battery storage systems, can increase grid capacity to meet them. EV battery storage systems are built to energize the motor and the auxiliary systems and are defined in terms of energy density, power density, cycle life, safety, and cost. At present, the most common battery technology is the Lithium-Ion (Li-ion) battery, which is energy-intensive, efficient, and has moderate values of cycle life. New types (e.g., solid-state batteries) also have the potential to achieve greater energy density and safety, since they are based on solid electrolytes, and their sodium-ion variants are cheaper since sodium is more abundant and costs less than lithium.

13.3.3.2 Economic Viability and Technical Feasibility

The rise in the use of EVs is revolutionizing the energy and transportation sectors. In order to determine the place and future of energy storage systems in mobility and grid integration, you need to evaluate the technical viability and economic viability of energy storage systems. Various fundamental factors that determine the economic viability of energy storage for EVs include costs of batteries, battery lifetime value, charging infrastructure, government incentives, and total cost of ownership (TCO). Traditionally, the costs of batteries formed a big share of EV costs, but the price of lithium-ion batteries has fallen rapidly through innovations and cost-saving technologies, economies of scale, and development of efficient production lines. Though EVs have a higher price tag as a one-off investment, their TCO can be more attractive because they are less costly to operate and have lower fuel requirements and reduced maintenance requirements. Technically, the possibility of storing energy in EVs depends on the performance of the battery, including energy capacity, charging time, life cycle, safety, and grid connection. Energy density is very important in defining the driving range of the vehicle, which is a major determinant of the acceptance rates of an EV since a higher number of miles is a major selling point.

The time to charge is still an important issue, and it can be done using a number of different charging types: Level 1 charging through standard household outlets takes between 12 and 24 hours; Level 2 charging, which is most often available on the street or in other public areas, takes 4 to 6 hours; and fast DC charging, with which many new EVs are now being equipped, offers the ability to add 80 percent charge in about 30 minutes. Both the cost and time of fast charging are further expected to be reduced by technological improvements in fast charging infrastructure or battery design. Moreover, the feasibility of the batteries depends greatly on their durability, with lithium-ion batteries lasting 8–15 years depending on the level of degradation they can sustain and the number of cycles they have gone through. To help solve the battery degradation situation, manufacturers deploy the use of advanced battery management systems (BMS) to help maintain optimal charging cycles and temperature. Finally, recycling and repurposing of EV batteries at their end-of-life can mitigate those effects and reduce costs of new batteries, and companies develop secondary uses of those batteries in fixed energy storage schemes.

13.3.3.3 Integration with Renewable Energy Sources

Battery energy storage systems (BESS), by which EVs are powered and renewable energy is stored, will be able to promote a carbon-neutral future in the shortest possible time. The integration, in addition to enabling clean movement, also increases grid stability, energy independence, and encourages the large-scale deployment of renewable sources of energy, which are intermittent. According to this chapter, the possibility of integrating EV batteries and renewable sources, the role of the V2G system, and some challenges and opportunities can be identified.

Battery systems power the EVs, and most use Li-ion batteries that can be used as a source of propulsion power as well as a platform for accessible integration of renewable energy sources. With renewable sources of energy, EV batteries, even when acting as distributed storage units, can help facilitate the inclusion of renewable energy into the power grid. V2G technology enables EVs to withdraw energy in order to charge, and also deliver energy to the grid when demanded. This will allow EVs to charge on low-cost off-peak electricity and release it during times of peak demand. This decreases reliance on fossil-fuel-powered peaking plants and aids in bringing renewable energy by delivering stored clean power when it is needed the most.

A combination of solar panels, EVs, and home energy storage, e.g., Tesla Powerwall or LG Chem battery, can create a very efficient and self-sustainable energy ecosystem on the residential scale. The solar PV systems used by homeowners can, in turn, use the solar energy produced during the day to charge the EV to an extent that their dependence on the grid will be severely limited. Also, homeowners will be able to store excess solar power in home batteries; they will be able to obtain energy stored in the battery at night or on cloudy days to use in homes or charge their Electric Vehicles. This integration does not only increase energy independence but also has many positive economic aspects. Homeowners may save a lot on energy bills, earn income through V2G programs, and will be able to create jobs in the renewable energy industry. On the whole, the scheme of solar energy, EVs, and energy storage systems is a kind of visionary solution in terms of future sustainable living and development of the economy.

13.3.4 Demands-Side Management

With the spread of EVs, the need for electricity will rise, in particular during peak hours. Demand side management (DSM) solutions, like TOU tariffs and the application of demand response programs, are the central elements of maintaining this balance with the aim of enhancing the efficiency, sustainability, and affordability of energy usage. DSM has strategies that actively control the use of electricity by consumers so as to streamline demand, lower peak loads, and enhance the reliability of the grid. This paper makes a detailed investigation of TOU tariffs and how DR grids can be harnessed to control the EV charging trend, alleviate grid load, and make the EV grid injection process easier.

13.3.4.1 Time-of-Use Tariffs (TOU)

The principal objective of TOU tariffs is to influence customers who vary his or her electricity use depending on price signals. Frequently, EV owners may charge up during off hours (late at night, etc.), and thereby they can save money by not paying as much for electricity, but at the same time, they help flatten the overall demand curve on the power grid.

The integration of EVs into the energy grid has a number of advantages when using TOU tariffs. Load shifting is one of the major benefits; these tariffs motivate holders of EVs to charge their vehicles when the demand is low so that demand peaks can be leveled and there will be less need to generate more power capacity during peak hours. Not only is the approach conducive to cost reduction for clients who can, by charging when power is cheap, considerably reduce their utility bills, it is also conducive to the encouragement of the adoption of EVs, given that adoption will be economically more appealing. Moreover, TOU tariffs will enhance the stability of the grid because they help to avoid saturation, especially during peak periods, which enables grid operators to operate energy flows better, reduce operations costs, and improve energy distribution. Also, in areas with intensive use of renewable energy sources, TOU tariffs encourage charging in the off-peak periods when renewable generation is highest, which minimizes the carbon footprint of EVs and aligns with the overall global plan to minimize the emission of greenhouse gases and achieve a more efficient and truly green grid.

13.3.4.2 Demand Response Programs

DR programs refer to the combination of efforts to get consumers to adjust their consumption of electricity in reaction to the message from the utility or the grid operator, normally during times of high demand. In the case of EVs, DR programs can be incorporated to streamline the charging processes of the vehicles and determine when to charge them optimally so that real-time management of energy use can be made depending on grid conditions. Among the merits that DR has to offer is optimized grid stability. These programs allow fine-tuning of electricity consumption in real-time and, as such, help avoid overloading the grid during peaks. In the case of EV owners, during the peak, charging can be automatically postponed to avoid any imbalance and ensure the smooth running of the grid DR also has an economic advantage at utilities. They commonly incur high expenses in order to increase generation capacity or sustain extra infrastructure when demand surges. DR programs help utilities curb peak demand, and thus they can forgo these costly investments as well as reduce the use of other expensive peaking power plants that are usually put in action only when peak demand arises. This not only literally saves money, but also helps keep electricity rates more stable for everyone.

A large number of DR programs provide financial compensation directly, which is also a significant incentive to EV owners. Such incentives aid in relieving the financial pressure of the charging infrastructure and offset part of the electricity consumption costs.

In addition, DR programs are also vital in encouraging the utilization of renewable energy. These can be coordinated in conjunction with periods when renewable energy—such as wind and solar—is producing electricity. TOU tariffs and DR programs can be significant resources in meeting the growing electricity requirement as Electric Vehicles become popular. DSM strategies may assist in shifting charging to off-peak periods to reduce the strain on the grid and bring substantial economic and environmental advantages.

13.3.4.3 Load Shifting and Peak Shaving Strategies

Increased incorporation of EVs into the power grid, together with the mounting pressure on reliable, sustainable, and affordable energy systems, has made advancing DSM strategies feasible. Load shifting and peak shaving are among this group of strategies that are most important when it comes to controlling patterns of electricity consumption, enhancing the reliability of grids, and keeping the overall costs of operations to a minimum. One of the newest technologies is load shifting, which allows balancing the energy consumption gap and decreasing expenses, particularly during the highest demand periods. It includes shifting power usage to times when demand is lower. The aim of load shifting is to straighten out the demand curve and make it easier to operate as far as utilities are concerned, and to minimize the use of peaking generation plants, which are costly to operate.

TOU pricing is one of the commonly used methods, and consumers are encouraged to adjust their electricity consumption to off-peak times with lower rates. Direct load control (DLC) is another good approach under which utilities are able to control some appliances, such as water heaters and HVAC systems, remotely. This implies that during hot days, your air conditioning will receive a slight boost during off-peak hours instead of overloading the system at a time when everyone is turning it on. It is a convenient way to allow energy to keep flowing without increasing bills and overloading the grid.

Smart meters can permit the availability of real-time data and the control of appliances and EVs, enabling consumers to adjust their energy use.

Peak shaving is the process that tries to mitigate peak demand, which transpires on the electric grid, through various approaches, including energy storage, load shedding, as well as distributed electric generation. Peak shaving lessens the need for utility companies to depend on costly and inefficient peaking power stations that are generally used during times of peak demand. Among the methods of peak shaving, one may include the introduction of battery storage systems, which may release energy to cover high-demand periods, thus actually cutting off the peak load on the grid. This practice is particularly helpful in areas that experience a high level of renewable energy penetration. There is also the possibility for electric vehicles (that have the capability of bidirectional charging, so-called V2G) to act as virtual batteries storing energy during days when people need it less and delivering power to the grid when demand is high. In addition, real-time pricing (RTP) provides a way of inducing consumers to alter their energy consumption in response to the shifting price of electricity, which changes over time during peak or off-peak periods.

13.3.4.3.1 Role of Electric Vehicles (EVs) in Load Shifting and Peak Shaving

An electric car is present in almost every abode, and there is a massive potential for load shifting and peak shaving by EVs. by incorporating EVs into DSM plans, utilities have an opportunity to stabilize the grid and save money by making use of their flexibility, which allows charging and discharging. An EV can contribute to load shifting, which is a great opportunity to optimize the use of energy. Owners of EVs need to be incentivized to charge their cars during off-peak hours, usually at night; then, the grid will not be overloaded at times of high consumption. Also, through the installation of smart charging infrastructure, it is possible to schedule charging automatically, either at times when renewable energy production is optimal, e.g., when solar power is particularly plentiful, or at times when electricity demand as a whole is low. Moreover, in comparison to peak shaving, technologies such as V2G permit EVs to send stored energy back to the grid at times of peak demand. This makes EVs a form of DERs that offer grid operators additional capacity to serve demand without having to bring peaking power plant units online. Furthermore, bidirectional charging systems in EVs have the potential to store surplus electricity during times of low energy demand and release it back into the grid when needed, which reduces peak demand and limits dependence on conventional energy sources in power generation.

13.3.5 Integration of Renewable Energy

The introduction of renewable energy into the EV ecosystem has an opportunity to decrease the carbon impact of EVs, decrease the total environmental influence, and increase the stability and effectiveness of the electrical grid.

13.3.5.1 Role of Solar Power in EV Charging Infrastructure

The most potential source of renewable energy that will enable charging of EVs is solar power. The potential of solar energy is limitless, its positioning is viable, and it is not that difficult to implement it almost anywhere, including residential premises, commercial, and public charging stations. The idea behind a charging station powered by solar energy is the concept of mounting PV arrays on the charging stations that produce electricity by the use of sunlight. This is a renewable energy source that can directly or indirectly charge electric vehicles, with energy stored in the form of batteries. Solar integration into EV charging will lead to a number of essential advantages, such as sustainability, grid independence, cost-efficiency, and assistance to the electric grid. Solar home charging systems, which use solar panels set on roofs to directly charge cars with energy obtained directly from the sun, are becoming a choice among individual EV owners and can improve the cost-effectiveness of such EVs as well as their environmental friendliness. Businesses are also in a position to install solar-powered charging points for their employees or customers, depending on the corporate sustainability agenda.

13.3.5.2 Role of Wind Power in EV Charging Infrastructure

Another important renewable fuel that can greatly contribute to the generation of power for EVs is wind power. Wind energy, similar to solar energy, is renewable, clean, and widely available, but it has different advantages and disadvantages than solar energy. Wind energy may be more ample at night or in winter months, which meets the requirements of EV charging, since cars are mainly charged during off-peak times. This is temporal complementarity between wind energy production and EV charging that can maximize the use of renewable energy. Wind-powered EV charging stations can be fitted beside EV charging stations to produce power to charge EV. These turbines may be used either individually or in pairs with solar panels, and this provides a combination of renewable energy systems.

Wind and solar hybrid systems have an amalgamation of the favorable features of the two types of energy. Whereas solar energy is more available when the sun is out, wind power is usually more powerful at night or when it is cloudy. Wind and solar energy sources can be combined to result in a more stable and consistent power supply at the charging stations, and this means that the charging stations will also rely less on the grid. Energy storage technologies, i.e., batteries, may also be easily implemented as a part of these hybrid systems to store some of the excess energy produced during periods of highest production. This energy can be used when there is low production of wind or solar energy, providing a constant source of energy to charge EVs. Moreover, the integration of wind and solar renewable energy sources to charge EVs enables such charging stations to enhance grid integration and stabilize the grid in areas where the penetration and peaks of renewable energy sources are already high but may be unreliable during periods of high load.

13.3.5.3 Grid Stability and Balancing Challenges

The interconnection that is involved in the supply of electricity is extremely complex in nature, and its model is one that has a balance of both supply and demand. The combined relationship between renewable energy and EVs adds a level of complexity because of the presence of a few interconnected factors. In the context of accelerating EV adoption, charging infrastructure is also needed and is often required during off-peak hours at night. These off-peak periods may, however, not correspond to low-demand periods as the number of EVs on the road increases, thus aggravating the electric grid with extra loads. Moreover, frequency control is important in ensuring grid stability since power systems must function at nominal frequency. When there are supply-demand imbalances, there is a possibility of fluctuations: overproduction of renewable energy can increase the frequency of the grid to meet the volatility of demand, damaging delicate equipment, or a lack of production and a sudden surge of EV charging can cause frequency drops, which may result in blackouts. EVs may also become part of frequency regulation by means of smart charging, where charging is increased or decreased depending on the conditions at that particular time on the grid. Another key factor is that of voltage stability; high renewable energy penetration and variable EV charging demand may cause voltage variability, especially in regions with high renewable generation capacity and distributed EV facilities. To tackle these grid stability and balancing issues, the following approaches (among others) can be taken: smart charging andV2G technology. Smart charging enables dynamic optimization of EV charging with regard to specific grid needs at the time of charging; EVs may be charged when renewable generation is high and when grid demand is low, flattening demand spikes. Moreover, V2G technology can also use energy stored in EVs when demand is high or when renewable sources of energy cannot meet demand, thereby contributing to frequency management and reducing the use of fossilfuels as backup power sources.

13.3.5.4 Energy Storage Solutions for Renewable Integration

This combination of RES, including solar and wind power, into the grid is a huge opportunity, but it is also a big challenge. This chapter discusses different forms of energy storage, integration of renewable energy with energy storage, and possibilities of energy storage and EVs together forming a sustainable energy ecosystem. Replacing irreversible energy with renewable energy in the grid has some challenges, mainly because the renewable sources of energy are intermittent and vary greatly because of weather and time of day, thus causing variations in power generation. This variation may mean that grid stability becomes a problem, especially in cases where there are large swings in generation or where there is a disconnect between supply and demand that may require expedited solutions. Energy storage solutions act as a buffer against these problems because they act as a buffer between energy supply and demand, so that any surplus renewable energy provided can be stored and used later, and in case of any shortage in generation, the excess energy can be supplied by stored energy.

The decentralized generation of renewable sources of power, like rooftop solar panels, also brings the problem of effective power transmission over long distances. Renewable energy is also subject to energy overproduction during windy and sunny days, thus leading to energy wastage. One of the major ways of overcoming these challenges is through energy storage systems, which help to decouple energy supply and demand, store surplus renewable energy, and make up the deficit in generation. Technology of energy storage: several types of energy storage are in development, including battery storage, based on battery energy storage systems (e.g., lithium-ion, sodium-sulfur batteries and flow batteries); pumped hydro, a form of energy storage particularly well suited to renewable energy; compressed air, storing air in large reservoirs; and thermo-chemical, storing heat in thermo-chemical systems, such as molten salt or ice storage. A conventional way of storing large-scale energy has been pumped hydro storage. It is used by utilizing surplus electricity to force water into a reservoir by pumping it uphill, to be later released to produce electricity. Although effective, its use is restricted by geographical and environmental factors. In Compressed Air Energy Storage (CAES), when there is surplus electricity, the electricity is utilized to compress the air and store it in underground caverns or vessels under pressure. When power is needed, the compressed air can be expelled and expanded in turbines. CAES systems offer large storage capacity and require certain geological formations to be implemented.

The systems involved in thermal storage are used to store energy in the form of heat or cold, and during later times this canbe reconverted into electricity or used directly. Well-known systems associated with thermal energy storage are molten salt storage, widely used in solar power plants in concentrated systems Concentrated Solar Power (CSP), where molten salt is heated to high temperatures to produce electricity in periods when there is no sun. Ice storage cooling systems are used in cooling processes, and they are important in energy demand shifting, particularly in buildings during high cooling loads. Thermal batteries store energy similarly in high-temperature materials, such as ceramics, metals, or graphite, and the thermo-stored energy is released when required to generate electricity. These energy-efficient systems differ in enriching energy efficiency and certainty in several uses.

Hybrid systems unite various forms of energy storage in an attempt to maximize performance. As an example, a hybrid system could combine both batteries (short-duration storage) and pumped hydro or CAES (long-duration storage). The systems present the advantages of all the technologies and counteract their drawbacks, providing a more affordable and dependable option for integrating renewable resources. Such energy storage systems play important roles in the areas of load shifting and peak demand management, grid frequency, reduced curtailment, increased grid resilience, and decentralized energy systems. Storage systems can be used to store extra energy when it is not needed (e.g., when energy demand is low or renewable generation is high) to release the energy when it is needed (e.g., when energy demand is high), therefore displacing peaking power plants that would otherwise be run on fossil fuels. Such systems can also be used to achieve fast response frequency regulation so that the grid frequency will remain stable under a variation in renewable energy generation. By avoiding the waste of excess renewable energy that would have gone to curtailment, it minimizes curtailment by storing the excess energy and releasing it into the grid when generation is low. With the increasing popularity of DERs in the form of rooftop solar and home battery systems, energy storage will lead to the generation of more locally resilient power systems. Such microgrids can operate independently, or with the principal grid. The adoption of modern energy storage facilities in a renewable energy environment has been found viable, as seen in some countries through the experience of successful adoption of such aspects.

13.4 CASE STUDIES AND BEST PRACTICES

13.4.1 The Netherlands: A Dense, Intelligent Charging Network

The Netherlands pioneered this as well due to its high levels of support for the adoption of EVs as well as its intelligent, dense charging infrastructure. Putting a strong focus on the creation of a simple and powerful charging infrastructure, the Dutch government has set ambitious goals to eliminate internal combustion engines by 2030. Charging is very convenient due to the presence of one of the highest concentrations of public charging stations in the country and in both urban and rural sites.

13.4.2 Archerfish Hunting Optimizer in Utrecht Part of Netherlands: A Sustainable Parking Facility with Bi-directional Charging That Can Accommodate up to 300 EVs by Alfen BV Company

The Archerfish Hunting Optimizer (AHO) is a universal method of optimization whose intention is to minimize objective functions related to real complex situations so as to obtain numerical or analytical solutions. The technique is based on the activities of archer fish during hunts of flying insects. AHO is constructed on the methodology of met aheuristics, a group of optimization methods that are nature-inspired. Such methods are centered on the notions of exploration and exploitation that are essential to the metaheuristic algorithms. Exploration (also called global optimization, or diversification) exploits a wide range of the available search space by exploring new areas. This stops the algorithm from getting trapped in local optima. Exploitation, however, can be equated to either local optimization or intensification. It pays attention to certain areas of interest to get closer and narrow the search. Metaheuristic algorithms can utilize both exploration and exploitation processes quite successfully and hence they can solve optimization problems and determine solutions as close to ideal one as possible. The research method involves the analogy of the hunting technique of archerfish. And just as archerfish utilize shooting and jumping behavior to identify their prey, so too does AHO utilize such principles to rapidly and efficiently arrive at the best solutions. This optimization approach uses what is known in the natural world to solve complicated optimization problems and ratify outcomes that are significantly close to intended objectives.The flowchart of Archerfish Hunting Optimizer is presented in Figure 13.5.

Step 1: Initialization

Initialize the input parameters current and voltage.

Step 2: Random Generation

After initialization, the input parameters are created randomly.

$$z = \begin{bmatrix} z_1 \\ z_I \\ z_N \end{bmatrix}_{N \times M} = \begin{bmatrix} z_{(1,1)} & z_{(1,j)} & z_{(1,M)} \\ z_{(I,1)} & z_{(I,k)} & z_{(I,M)} \\ z_{(N,1)} & z_{(N,k)} & z_{(N,M)} \end{bmatrix} \tag{13.1}$$

Here z represents the current and the voltage.

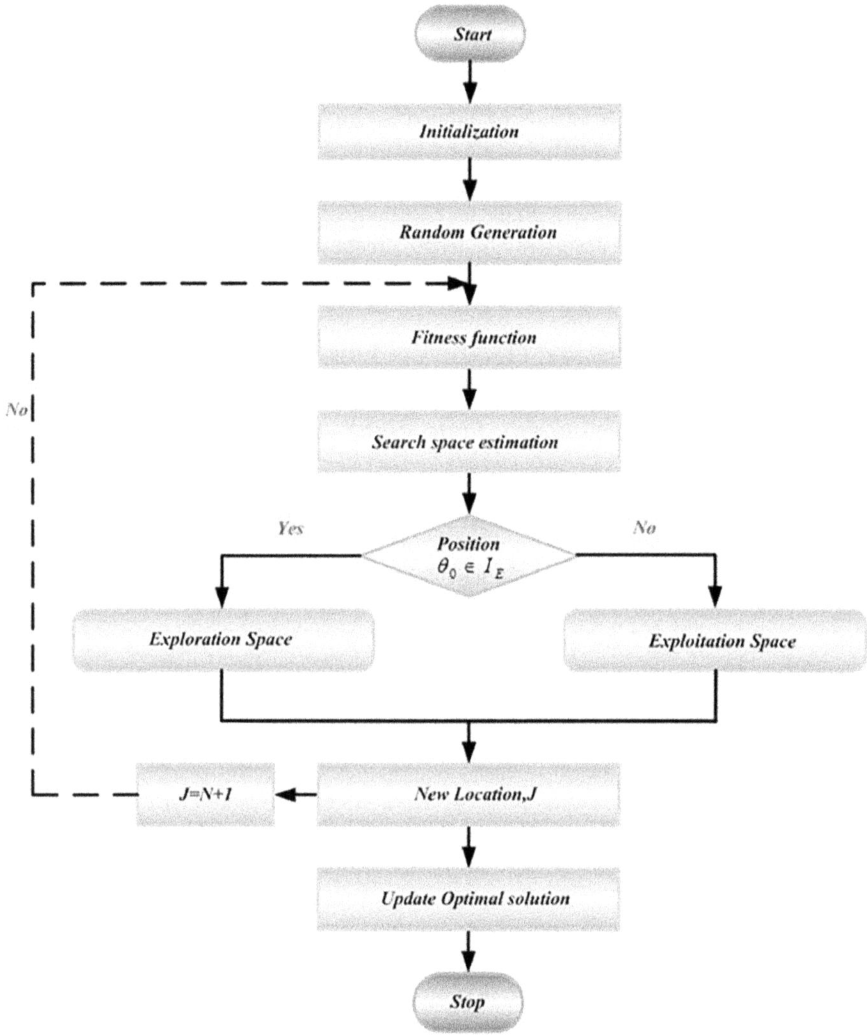

FIGURE13.5 Flowchart of Archerfish hunting optimizer.

Step 3: Fitness Evaluation

The fitness value is based on the impartial function.

$$F(t) = \text{Min}(\text{error}) \tag{13.2}$$

Step 4: Search Space Estimation

Consider an archerfish-filled search space of size d. The position of archerfish I at iteration T is illustrated below, and the flock size, or number of archerfish, is n

$$Y^{(I,T)} = (Y_1, Y_2, ..., Y_d) \tag{13.3}$$

The permitted value range for each entry in $Y^{(I,T)}$ is $Y^{(I,T)} = (Y_J) \in [Y_J^{MIN}, Y_J^{MAX}]$(where $I \in \{1,....,n\}$ and $J \in \{1,......,d\}$).

Step 5: Exploration of the Search Space

It is defined by the launch speed if air friction is small V, the perception angle θ_0, and gravity, which is accelerating. The bug will fall vertically to the water's surface K after being shot by an archerfish. Equation I is used by an archerfish to navigate to its position after hearing the vibrations of its prey.

$$Y^{(I,T+1)} = Y^{(I,T)} + E^{-\left\|Y_{PREY}^{(K,T)} - Y^{(I,T)}\right\|^2} \left(Y_{PREY}^{(K,T)} - Y^{(I,T)}\right) \tag{13.4}$$

where $Y^{(I,T+1)}$ indicates the next position of archerfish I; $Y^{(I,T)}$ denotes the recent position of archerfish I; $\|.\|$ denotes Euclidean distance, and $Y_{PREY}^{(K,T)}$ indicates the position of the prey. It is calculated with the equation E: a uniform distribution-generated vector of random numbers. At the air-water contact, it shows the effects of refraction. The place where the bug was shot is $Y^{(K,T)}$ archerfish K.

$$Y_{PREY}^{(K,T)} = Y^{(K,T)} + \left(0, ..., \frac{V^2}{2G} \times \sin 2\theta_0, ..., 0\right) + \varepsilon \tag{13.5}$$

Step 6: Exploitation of the Search Space

In order to catch its prey, the archerfish leaps at it. Similar to how gravity accelerates objects, the archerfish's speed is represented by G, V indicates launch speed, and θ_0 indicates the perceiving angle, assuming there is no air friction. An archerfish I moves in the direction of an insect when it wishes to catch one.

$$Y(I,T+1) = Y^{(I,T)} + E^{-\left\|Y_{PREY}^{(K,T)} - Y^{(I,T)}\right\|^2} \left(Y_{PREY}^{(I,T)} - Y^{(I,T)}\right) \tag{13.6}$$

$$Y_{PREY}^{(I,T)} = Y^{(K,T)} + \left(0, ..., \frac{V^2}{2G} \times \sin 2\theta_0, ..., \frac{V^2}{2G} \times \sin 2\theta_0, ..., 0\right) + \varepsilon \tag{13.7}$$

A random number between $\{1,......,d\}$ and $\frac{V^2}{2G} \times \sin 2\theta_0$ determines the place of the entry. Variable ω will take the place of fraction $\frac{V^2}{2G}$ for the sake of simplicity. It describes an archerfish's propensity to become attracted to a particular prey.

Step 7: Termination Criteria

If the stopping criteria are met and the finest outcome is obtained, the process is finished; then, it goes to step 3.

13.4.3 QNN for System Fault Identification

Quantum Neural Network (QNN): A state-of- the-art model of computations, which utilizes quantum mechanical principles to solve some form of calculations more efficiently and quickly as compared to traditional neural networks. QNNs use quantum bits, which can share many states simultaneously, and hence discuss many options simultaneously by superposition. Just as classical logic gates work, quantum gates control qubits, which are then further compounded by quantum entangling that is stronger than before in computing. QNNs are promising tools to design and address complex optimization problems and simulate quantum systems and when it comes to performing machine learning tasks in the era of quantum computing.

EV station flaw utilization of a QNN: The second possibility entails capitalizing on quantum computing to solve fault forecasting accuracy in EV stations. At first, information will be gathered and preprocessed by different sensors and state deviations and noise will be deleted. Relevant features are then derived so as to describe the performance of the EV. The peculiar nature of such a method is that it follows the quantum data encoding procedure with the use of quantum bits (qubits) to encode information-filled features. This encoded data is manipulated and analyzed through quantum states and quantum gates, and quantum parallelism and entanglement are used to improve pattern recognition. The QNN will be able to detect patterns related to station faults, and this will help create more precise detection of the faults at an efficient rate in EVs, and this will help improve the stability and safety of the power distribution systems.

In Figure 13.6, the Quantum Neural Networks (QNN) which detects faults in EV stations is a revolutionary idea of the marriage between quantum computation and

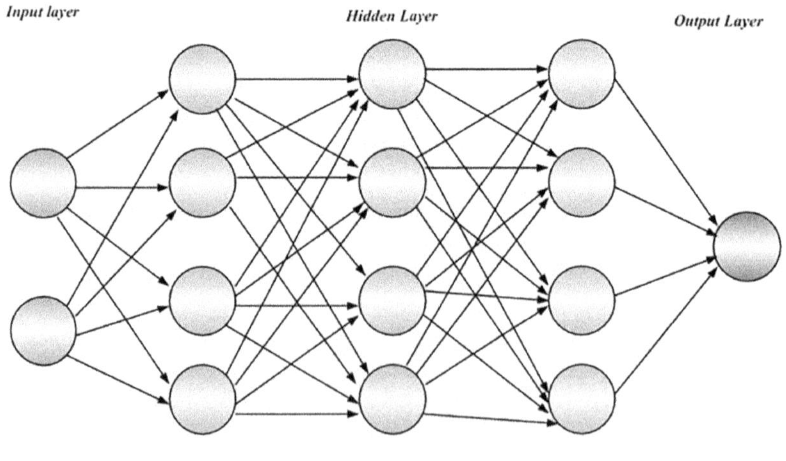

FIGURE 13.6 Structure of quantum neural networks.

machine learning. An innovative process begins with the gathering of information captured by different sensors that track the behavior and performance of the EV. Before analysis, careful preprocessing of the data is performed that includes noise removal, treatment of outliers as well as data quality checks. Afterwards, associated characteristics are scrapped out of the set, like voltage, current, and temperature values. Qubits carry and process information very differently: they are able to process in parallel and to entangle in a complex fashion. Quantum data encoding in the area of EV station fault discovery implies implementing the process of mapping the extracted features to quantum states and applying quantum gates to implement calculations. Such encoded states can be analyzed using quantum algorithms such as quantum support vector machines or quantum variational circuits. It is quantum parallelism that allows the QNN to search many potential solutions at once dramatically increasing the speed and efficiency of fault detection. The higher the number of times the QNN runs through the data training and acquires knowledge on labeled examples, the more the neural net can recognize the minute patterns which are in connection with station faults. This developing model has the potential of making power distribution systems very reliable and safe through quality and timely fault localization management. In its current stage of development, quantum computing has enormous potential in the eventual resolution of complex problems in many fields, and more specifically in power engineering, through the combination of the nature of quantum concepts and the nature of neural networks.

13.4.4 Policy and Regulatory Frameworks

13.4.4.1 Analysis of Effective Policies to Promote EV Adoption and Grid Integration

13.4.4.1.1 Development of Charging Infrastructure

13.4.4.1.1.1 Policy Overview EV adoption requires an effective and well-extended EV charging infrastructure. Governments are required to facilitate the development of charging networks, particularly in countryside,cities, and on roads.

13.4.4.1.1.2 Effective Instances

- **The Netherlands:** Through subsidies and the creation of public-private partnerships, the Dutch government initiated the large-scale deployment of public charging infrastructure, which makes EVs an accessible choice to anyone in the country. Also, they have passed legislation requiring EV charging stations to be deployed in new construction.
- **Germany:** In addition to its broader scheme to transform an energy landscape, Germany is also significantly investing in charging infrastructure. Its government aims to establish one million charging facilities by the year 2030 and gives financial incentivesfor the establishment of both publicly available charging and privately owned charging stations.

13.4.4.1.1.3 Grid Integration and Smart Charging Policies

- **Policy Overview:** Governments should establish rules that will ensure grid stability and energy consumption to meetthe raised electricity demand caused by EVs. This involves sponsoring demand response programs, deploying smart grid technologies, and integrating renewable energy resources. An example is California, where several policies have been successfully integrated to attain EV grid integration, including enabling V2G technology, the advancement of smart charging, and the provision of incentives for off-peak charging under TOU charges. To guarantee that future development is built into grid capacity, the state has also passed a bill demanding the utilities to plan for the impact of EVson the grid.
- **United Kingdom:** The Smart Systems and Flexibility Plan of the UK encourages the use of smart meters, demand-side response plans, and vehicle-to-grid centers in order to avoid the overloading of the grid resulting from EV charging. Moreover, the government is financing the creation of intelligent charging processes that can schedule EV charging to times when electricity use is minimal or when renewable power is being produced in abundance.
- **Germany:** The Energy Transition (Energiewende) plans of the country place specific emphasis on integrating the utilization of EVs and renewable energy sources. The nation has introduced incentives, as the government seeks to ensure that EV chargers are solar-powered. Advanced technology, known as the smart grid, is also being employed to manage the additional load and ensure successful distribution of energy. In Figure 13.7 below, depicting the main design of Germany 2024 environmental government policy and regulatory structures, the main thrust of this plan is seen to have policy and regulatory frameworks as its priority.

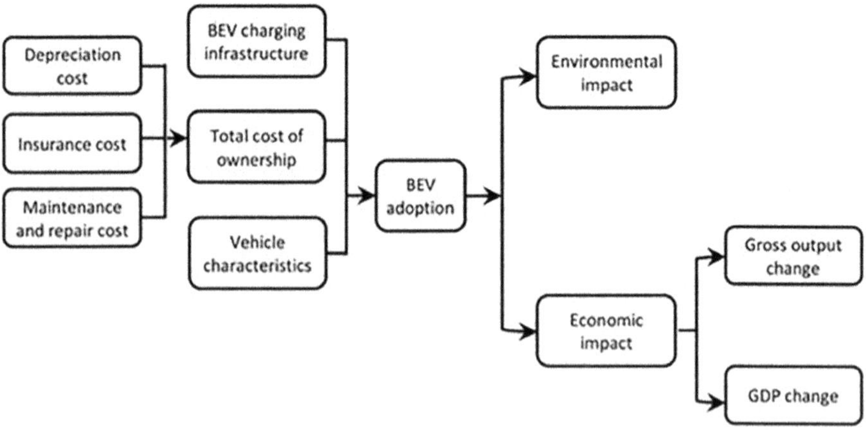

FIGURE 13.7 Primary plan of Germany's 2024 environmental government policy framework.

13.4.4.1.1.4 Carbon Pricing and Emission Standards

- **Policy Overview:** Governments have established stricter policies about vehicle pollution and carbon pricing to encourage switching to zero-emission vehicles. Such regulations make the acquisition of cars with internal combustion engines less attractive, providing businesses and customers with an economic reason to use electric cars.
- **European Union:** The EU's strict CO_2 emissions regulation has forced automakers to hasten the creation of its electric cars (EVs). To further promote the adoption of cleaner options, carbon pricing mechanisms, which penalize high-emissions cars, have also been implemented by the EU.
- **China:** The nation has also introduced a strict system of the term "dual-credit"that requires car companies to meet the required output quotas of new-energy vehicles (NEVs) and fuel economy. To meet the regulatory demands, this has compelled car companies to produce more EVs.

13.5 CONCLUSION

The incorporation into power systems of electric vehicles (EVs) proposes a compound of challenges and opportunities. Although EVsare a new technology with large environmental and financial advantages, the growing number of EVs makes it important to think about their effect on grid infrastructure, load management, and stability in general. Among the major issues, there is possible pressure on the network during peak charge. This is prone to cause grid congestion, voltage fluctuations, and even power interruptions. In order to mitigate this problem, several measures can be used such as smart charging solutions, vehicle-to-grid, and demand-side response schemes. These methods can spread the load more smoothly through time to the extent of matching charging schedules and rewarding off-peak charging. The second major factor that should be taken into account is the incorporation of renewable sources of energy into EVs. This can be used by switching to renewable sources to operate EV charging plants and limit the production of greenhouse gases and support a sustainable energy future. Nonetheless, the renewable energy sources are uneven, which means that with good forward planning and proper application of the energy storage system, there are great chances that the power supply will be clean and reliable enough. Moreover, to help EVs become predominant, it is necessary to develop charging infrastructure that will be convenient and well-developed. The latter involves setting up distribution networks as well as the installation of private and public charging places. The effective incorporation of EVs into the power systems needs a multi-pronged approach in which policymakers, utilities, and consumers will work together. The challenges and opportunities will enable us to prepare a cleaner, more sustainable, and resilient energy future.

REFERENCES

1. S. Kumar, A. K. Verma, A. Jain, P. Mathuria, and R. Bhakar, "Long-term power system planning considering increased electric vehicle penetration: A systematic approach," in *2023 International Conference for Advancement in Technology (ICONAT)*, Goa, India. IEEE, 2023, pp. 1–6. doi: 10.1109/ICONAT57137.2023.10080341.

2. Z. Yang, F. Yang, H. Min, H. Tian, W. Hu, and J. Liu, "Review on optimal planning of new power systems with distributed generations and electric vehicles," *Energy Rep.*, vol. 9, pp. 501–509, 2023, doi: 10.1016/j.egyr.2022.11.168.
3. C. H. Dharmakeerthi, N. Mithulananthan, and T. K. Saha, "Planning of electric vehicle charging infrastructure," in *2013 IEEE Power & Energy Society General Meeting*, Vancouver, BC. IEEE, 2013, pp. 1–5. doi: 10.1109/PESMG.2013.6672085.
4. D. L. Donaldson, M. S. Alvarez-Alvarado, and D. Jayaweera, "Power system resiliency during wildfires under increasing penetration of electric vehicles," in *2020 International Conference on Probabilistic Methods Applied to Power Systems (PMAPS)*, Liege, Belgium. IEEE, 2020, pp. 1–6. doi: 10.1109/PMAPS47429.2020.9183683.
5. F. Manríquez, E. Sauma, J. Aguado, S. De La Torre, and J. Contreras, "The impact of electric vehicle charging schemes in power system expansion planning," *Appl. Energy*, vol. 262, p. 114527, 2020, doi: 10.1016/j.apenergy.2020.114527.
6. B. Singh, J. P. Pandey, A. Singh, and D. K. Patel, "optimization technique for different types of electric vehicles planning: A critical survey," in *2020 International Conference on Power Electronics &IoT Applications in Renewable Energy and its Cont rol (PARC)*, Mathura, Uttar Pradesh, India. IEEE, 2020, pp. 80–85. doi: 10.1109/PARC49193.2020.236562.
7. S. Deb and S. Sachan, "Proposed power systems planning in Indian scenario for integrating EV charging infrastructure," in *Advances in Mechatronics and Mechanical Engineering*, M. S. Alam, R. K. Pillai, and N. Murugesan, Eds., IGI Global, 2022, pp. 25–37. doi: 10.4018/978-1-7998-6858-3.ch002.
8. B. Singh and P. K. Dubey, "Distributed power generation planning for distribution networks using electric vehicles: Systematic attention to challenges and opportunities," *J. Energy Storage*, vol. 48, p. 104030, 2022, doi: 10.1016/j.est.2022.104030.
9. S. Wang, Z. Y. Dong, F. Luo, K. Meng, and Y. Zhang, "Stochastic collaborative planning of electric vehicle charging stations and power distribution system," *IEEE Trans. Ind. Inform.*, vol. 14, no. 1, pp. 321–331, 2018, doi: 10.1109/TII.2017.2662711.
10. J. A. P. Lopes, F. J. Soares, and P. M. R. Almeida, "Integration of electric vehicles in the electric power system," *Proc. IEEE*, vol. 99, no. 1, pp. 168–183, 2011, doi: 10.1109/JPROC.2010.2066250.
11. B. Jayababu, R. G. Nageswara, and V. K. Kurakula, "Critical review on integration of electric vehicles into residential distribution system," *Manag. J. Power Syst. Eng.*, vol. 10, no. 4, p. 38, 2023, doi: 10.26634/jps.10.4.19215.
12. D. K. Patel, D. Singh, and B. Singh, "Impact assessment of distributed generations with electric vehicles planning: A review," *J. Energy Storage*, vol. 43, p. 103092, 2021, doi: 10.1016/j.est.2021.103092.
13. Q. Zhang, J. Yan, H. O. Gao, and F. You, "A systematic review on power systems planning and operations management with grid integration of transportation electrification at scale," *Adv. Appl. Energy*, vol. 11, p. 100147, 2023, doi: 10.1016/j.adapen.2023.100147.
14. A. Danese, B. N. Torsæter, A. Sumper, and M. Garau, "Planning of high-power charging stations for electric vehicles: A review," *Appl. Sci.*, vol. 12, no. 7, p. 3214, 2022, doi: 10.3390/app12073214.
15. A. Purvins, C. Covrig, and G. Lempidis, "Electric vehicle charging system model for accurate electricity system planning," *IET Gener. Transm. Distrib.*, vol. 12, no. 17, pp. 4053–4059, 2018, doi: 10.1049/iet-gtd.2018.5580.
16. A. Suskiet al., "Analyzing electric vehicle load impact on power systems: Modeling analysis and a case study for Maldives," *IEEE Access*, vol. 9, pp. 125640–125657, 2021, doi: 10.1109/ACCESS.2021.3111001.
17. C.-T. Ma, "System planning of grid-connected electric vehicle charging stations and key technologies: A review," *Energies*, vol. 12, no. 21, p. 4201, 2019, doi: 10.3390/en12214201.

18. K. M. Hazazi, K. Khalid Mehmood, G.-J. Cho, and C.-H. Kim, "Optimal planning of distributed generators for loss reduction and voltage profile enhancement considering the integration of electric vehicles," in *TENCON 2018 - 2018 IEEE Region 10 Conference*, Jeju, Korea (South). IEEE, 2018, pp. 0883–0888. doi: 10.1109/TENCON.2018.8650406.
19. O. Sundstroem and C. Binding, "Flexible charging optimization for electric vehicles considering distribution grid constraints," in *2012 IEEE Power and Energy Society General Meeting*, San Diego, CA. IEEE, 2012, pp. 1–1. doi: 10.1109/PESGM.2012.6344758.
20. R. Garcia-Valle and J. A. Peças Lopes, Eds., *Electric Vehicle Integration into Modern Power Networks*. New York, NY: SpringerNew York, 2013. doi: 10.1007/978-1-4614-0134-6.
21. A. S. BinHumayd, B. Lami, and K. Bhattacharya, "The effect of PEV uncontrolled and smart charging on distribution system planning," in *2016 IEEE Electrical Power and Energy Conference (EPEC)*, Ottawa, ON. IEEE, 2016, pp. 1–6. doi: 10.1109/EPEC.2016.7771783.
22. T. Unterluggauer, J. Rich, P. B. Andersen, and S. Hashemi, "Electric vehicle charging infrastructure planning for integrated transportation and power distribution networks: A review," *eTransportation*, vol. 12, p. 100163, 2022, doi: 10.1016/j.etran.2022.100163.

14 Battery Management Systems for Electric Vehicles

Malarvizhi Kandasamy, Nirmala Muthusamy, and Maithili Periyannan

14.1 INTRODUCTION

14.1.1 Objectives of Battery Management System

The main objective of this chapter is to highlight some of the challenging issues, related to the battery management system (BMS) from a control theoretic perspective and to identify selected open areas of research. The essential goals are collective amongst the entire BMS: safety and reliability, to provide complete protection of the cells or the battery from destruction, to persist the life of the battery, to sustain the battery in a state in which it can accomplish the efficient desires of the application for which it was stated, and also continuous observing and estimation of state of charge (SOC) and state of health (SOH) of the battery, to make alarms or visual signs for several standard malfunctions in the system. The active battery management system is vital to perceive battery state and check the safety of operation by cautious design concerns in charging and discharging progressions. In general, the battery management system consists of a data acquisition module, state estimation module, charge/discharge control module, cell balancing module, thermal management module, and protection circuit [1].

14.1.2 Basic Terms about Battery

Cell: It is a simple part of a battery. It consists of an anode and cathode separated by an electrolyte used to generate a voltage and current. (For Li-Ion batteries, cell voltage is 3–4 V and for lead acid battery cell voltage is 2.1 V).

Battery Module: It is a collection of battery cells with external wired connections, providing a higher voltage. (E.g., Li-ion battery module consists of six cells and provides nominal voltage of 18 V);

Battery Pack: It consists of a group of identical battery modules connected in either series or parallel combinations.

Battery cells are the tiny single electrochemical elements that produce a voltage that varies based on the selection of chemicals used to build the cell. Cells are classified into primary cells and secondary cells. The former is non-rechargeable, and the

DOI: 10.1201/9781003585534-14

latter is a rechargeable cell. The symbols for a single cell, a double-cell battery, and a four-cell battery are shown in Figure 14.1.

Nominal voltage per cell is the average output voltage of a cell when it is charged. The value of the nominal voltage is based on the reactions of the chemicals used in the battery. The operating cell voltage is either above or below the nominal cell voltage. The nominal voltage and open-circuit voltage per cell for different types of batteries are listed in Table 14.1.

The cells are either connected in series or parallel to meet the voltage and current requirements, respectively. If a number of cells are associated in series, the sum of the independent cell voltages is equal to the total voltage of the battery. Since the current flowing through the series circuit of the cells is the same, the charge capacity of the series-connected battery is the same as the charge capacity of an independent cell. For example, three 4 V, 50 Ah cells connected in series will result in a battery voltage of 8 V, a battery charge capacity of 50 Ah, and a battery energy capacity of 400 Wh. If a number of cells are connected in parallel, the total battery voltage is equal to the independent cells' voltage. Since the total battery current is the sum of all the cell currents, the charge capacity is the sum of the cells' charge capacities. For example, four 2 V, 20 Ah cells connected in parallel Result in a battery voltage of 2 V, a charge capacity of 80 Ah, and an energy capacity of 160 Wh.

The amount of stored energy per unit weight or volume of a battery determines the values of specific energy and energy density, respectively [2]. A higher specific energy cell can store more energy for a given weight, but a higher specific energy cell can store less energy for a given volume. For a given volume, a higher energy density cell can store more energy, and for a given storage capacity, a higher energy density cell stores a smaller amount of energy [3].

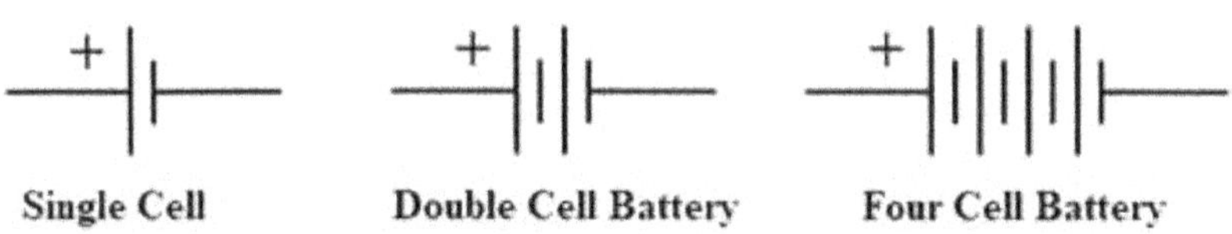

FIGURE 14.1 Symbols of different cells.

TABLE 14.1
Voltage per Cell for Different Batteries

S. No	Type of Chemical Used in Battery	Nominal Voltage/Cell (V)	Open Circuit Voltage/ Cell (V)
1	Lead-acid (PbA)	2.0	2.1
2	Nickel-Cadmium (NiCd)	1.2	1.21
3	Nickel-metal-hydride (NiMH)	1.2	1.21
4	Lithium-Ion	3.6	3.7

14.2 FUNCTIONS OF BMS

In view of the rapid growth of EVs and HEVs, the uncertainty of a battery's performance still poses a challenge to developing a comprehensive and mature BMS. A BMS should expedite functions to realize the distinct objectives by one or more of the following features:

14.2.1 Battery Cell Protection

To all BMS applications, the fundamental operation is protecting the battery from irregular operating conditions. In practice, the BMS needs to deliver full cell protection to shelter almost any contingency. The failure of the battery is predominantly caused by operating a battery outside of its identified design limits. The cost of battery swapping can be unaffordable, apart from the problems and inconveniences. For high-voltage and high-power automotive batteries, the swapping cost is mainly exact, which must function in aggressive situations and, at the same time, are subject to abuse by the user. The safety protection module provides protection against overcharge, over-discharge, and overheating [4].

In the BMS, battery cell protection is one of the primary functions and can be external to the battery. During manufacturing, safety measures can be created within the cells. The objectives of battery cell protection are:

i. to offer the essential monitoring and control to safeguard the cells from beyond operating conditions.
ii. to protect the consumer from the after-effects of battery failures.

Special care is to be taken for the protection of high-power cells, which can have terrible effects during a short circuit or physical damage because they may release an uncontrolled higher volume of energy [5]. Also, the action of protection circuits must be very fast; otherwise, heavy currents (hundreds of amps) can accumulate in a very short period (microseconds) tando damage the battery. The required degree of protection is different and based on the type of application and the type of chemistry used in the cell.

14.2.2 Charge Control

The charging control is an essential feature of a BMS. Mostly, improper charging of a battery is the main reason for the damage of the batteries. The charge/discharge control module provides charging current and voltage regulation, as well as a power electronics interface.

The battery charging approach disturbs routine function and battery life expectancy. When the battery current is increased to a large value during charging, the internal resistance of the battery also increases due to the reduced effect of chemical reactions, leading to a rise in the temperature of the battery; thereby, the core of the battery gets impaired. Conversely, if the battery current is at a very low value during charging, more time is needed for full charging, and this long charging time affects battery performance [6].

14.2.2.1 Methods of Battery Charging

Generally, based on voltage and current parameters, the following methods are commonly used:

i. Constant voltage charging (CV)
ii. Constant current charging (CC)
iii. Constant voltage constant current charging (CVCC)
iv. Pulse charging

14.2.2.2.1 Constant Voltage Charging

A constant voltage source is used to charge the battery. It is a simple arrangement and design of a control circuit. In this method, when the battery is completely charged, the battery spontaneously enters the float charging mode to offset normal battery self-discharge, and the charging current will reduce. Since the battery voltage is at a low value at the beginning of charging, the charging current is very high [7]. This increases the battery temperature and leads to a reduction in battery life. To overcome this drawback, a multi-stage voltage charging technique.

14.2.2.2.2 Constant Current Charging

Constant current is used to charge the battery. After it reaches the full charge value, the charger changes to trickle mode, which avoids overcharging. In this method, the battery is fully charged in a short time, but attention must be paid to the charging level of the battery; otherwise overcharging leads to a reduction in the battery's life.

14.2.2.2.3 Constant Voltage/Constant Current

This method has merits and demerits of a combination of constant voltage and constant current charging methods. This method decreases the charging time and avoids over charging, acting as a self-controlling current for the constant voltage charging mode. The constant current mode is used at the initial stage of charging and continues until the battery voltage reaches 80%. Once the battery voltage is set, the mode is changed to constant voltage mode. At this time, the process is called leveling or balancing mode. When the battery is completely charged, it spontaneously enters the float charging mode (Figure 14.2).

14.2.2.2.4 Pulse Charging Mode

The periodic pulse current is used to charge the battery. In pulse charging mode, the electrolyte in the battery can become more uniform and includes a stop-charging period. As a result, the full conversion of charging energy into electrical energy increases the charging efficiency.

14.2.2.2.5 Negative Pulse Charging

In this mode, a short discharge pulse with a magnitude of three times the charging current for 5 ms is applied for cell depolarization. During fast charging, these negative pulses remove any gas bubbles that have developed on the electrodes and speed up the normalization process, leading to speeding up the overall charging process

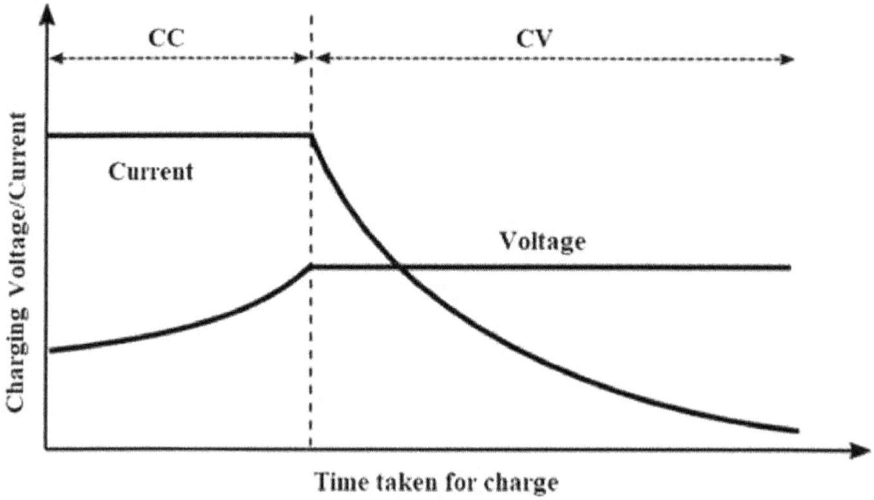

FIGURE 14.2 Charging characteristics of constant current-constant voltage method.

and improving battery lifetime. This method is called burp charging because the process of dispersal of gas bubbles is termed burping.

14.2.2.2.6 Trickle Charge

In this mode, continuous constant current charging is applied for backup utilization. This charging is mainly used to compensate for battery self-discharge and, in a few applications, once the battery gets fully charged, the charger may switch from charging to trickle charge. Based on the frequency of discharge, the charge rate fluctuates. It is not suitable for all types of batteries.

14.2.2.2.7 Float Charge

A float charge is used for emergency power backup systems during any disaster or calamity. In this method, a constant voltage is applied to the permanently parallel-connected battery and load. The applied voltage value is less than the battery's greater voltage. This method is suitable for lead-acid batteries.

14.2.2.2.8 Random Charging

Random charging is essential for automotive applications. In automotive applications, the speed of the engine is continuously changeable, and hence the energy requirement depends on this variation. In EV and HEV applications, since EVs generate large power spikes during braking, the battery must have the capability to absorb the energy generated during regenerative braking. Hence, the battery of the vehicle should tolerate this excess regenerative braking energy and provide special systems to reduce the charging current or voltage to appropriate levels.

14.2.2.3 Charging Practices

The purposes of a charger are to get charged, stabilize the charging rate, and terminate the charging operation.

14.2.2.3.1 Termination of Charging

After full charging, the dissipation of charging current causes heat and gases that are harmful to the battery. Before any hazards or damage, the principle of good charging is to be competent to identify when the regeneration of the active chemicals is extensive and to prevent the charging process. Also, ensuring cell temperature is within its reliable limits in any situation. To protect the battery life, the determination of cut-off point and termination of charging is crucial. The termination point is defined as the point at which the charger reaches a predetermined upper limit of voltage.

14.2.2.3.2 Reliable Charging

There is a chance of the battery overcharging due to any error in operation, core fault, or misuse, which causes an increase in temperature and may damage the battery. The solution for these problems is monitoring cell temperature in the battery. To avoid damage to the battery, a simple safety precaution for large-rating batteries is that a resettable fuse or measurement of temperature signals can be used to control charger operation whenever hazard indications happen.

14.2.2.3.3 Time Taken for Charging

The main challenge in batteries is the time taken for charging. The charging time is defined as the time taken for a discharged cell to be fully charged. In electric vehicle applications, the time taken for available batteries in the market is around 3–4 hours, but fossil fuel vehicles take to fill the fuel tanks just 5 minutes. During fast charging, delivery of electrical energy occurs earlier than the chemical process. It causes thermal instability, reduces cyclic life and safety of the cell, thus leads to reduced cyclic life.

14.2.2.3.4 Charging Rates

Based on the requirement, there are three different charging rates used to charge the batteries as follows:

1. **Slow Charge**: 14–16 hours to charge by 0.1C charge rate
2. **Quick Charge**: 3–6 hours to charge by 0.3C charge rate
3. **Fast Charge**: below 1 hour to charge by 1C charge rate

The **C-rate or charge rate** is rate at which a battery is charged or discharged.

14.2.2.3.5 Slow Charging

Generally, a simple type of charger is used for slow charging. After the completion of a full charge, the charger should be removed from the battery without overheating.

14.2.2.3.6 Fast and Quick Charging

In fast and quick charging, normally complex chargers are required. When the charge rate increases, the possibility of overcharging and overheating also increases. To protect the battery from overheating, the charge is terminated when it reaches full charge. Overheating is measured by detecting the end-of-charge situation using a temperature cut-off (TCO) or by using a thermal fuse while designing the battery charger.

14.2.3 Cell Balancing

The charge and discharge cycle of the batteries mainly depends on cell balancing in the battery. The minor variances between cells in a multi-cell battery chain magnify the problems in the charging and discharging cycle due to production tolerances or operating conditions. Premature failure of the battery is caused by overstressing weaker cells during the charging cycle, causing them to become even weaker. To extend the life of the battery, cell balancing is a best approach for compensating weaker cells by equalizing the charge on all the cells in the chain. Cell balancing uses passive or active state-of-charge and voltage equalization techniques. The battery operation can be balanced by using passive and active methods. Passive battery balancing performs by dissipating heat during excess capacity in the battery, but it is applicable only for low-capacity batteries. Active battery balancing methods give better efficiency, switched-capacitor boost converter (S-CBC) systemshave been developed for Valve Regulated Lithium Ion (VRLI) battery systems. The VRLI battery system consists of four batteries in a series configuration, which increases the capacity and energy of the battery system.

To meet the desired voltage, capacity, and power demand, EV batteries are generally built-up of large sequences of cells in series and parallel packs. If a battery has n cells, then the breakdown rate of the battery is the product of the number of cells and the breakdown rate of a single cell.

14.2.3.1 Self-Balancing

Self-balancing is caused in parallel chains due to an unbalanced maturation problem. The voltages across the parallel cells are equal and permit charge to move between cells even in the absence of external voltage. If any one of the parallel cells is subjected to a short circuit, then the rest of the cells get discharged through that short-circuited cell. During micro cycles, that is, quick charge/quick discharge, these problems are overemphasized in automotive applications. Long-term unbalance of cells is prevented by the process of **equalization.** Lithium batteries are less tolerant than lead-acid and NiMH batteries to cell-to-cell differences.

14.2.3.1.1 Selection of Cells

The cell equalization depends on the selection of cells. Primarily, the construction of batteries should be made up of matched cells from the same manufacturing batch. To reduce inconsistency within the cell groups, testing can be engaged to categorize and choose cells into groups. Based on storage capacity, two types of cells, large-capacity cells and small-capacity cells, are used. Many small-capacity cells are connected in parallel to form a large-capacity cell. To meet the demand for high voltage requirements, parallel-connected small cells or large cells should be connected in series.

14.2.3.1.2 Large Capacity Cells

The large cells have a minimum number of interconnections between cells, thus need less design of monitoring circuits, which reduces fabrication costs. The requirements of large cells are very small, thus increasing the cost of fabrication and wide process variability. The large cells have the capability to store large energy. But from a safety point of view, if any of the cells is subjected to short circuit, this high energy results in a blast and causes extreme pressure and fire, causing the neighboring cells to get damaged and leading to destruction of the battery.

14.2.3.1.3 Small Capacity Cells

The number of small cells is connected in parallel to get the same capacity and voltage as large-capacity cells. But it has several interconnections and needs a complex design of monitoring circuits, which in turn increases fabrication costs. In small capacity cells, entity costs are subsequently very low and have high reliability but have very little tolerance spread of the blocks. From a safety point of view, the possibility of failure is very low. If failure occurs, it releases less energy, and the failed pack can be easily eliminated.

14.2.3.1.4 Equalization

The cell equalization is another essential function of the BMS. After selection of cells, each cell's temperature must be measured, and the spreading of temperature across all cells in the battery must be confirmed to be even at all times [9]. Obviously, all the cells are assembled in one huge structure; the outer cells are cooled by ambient air. But the inner cells become hot and require proper cooling arrangements for eliminating the heat. When selecting cells, the first step of design is cell temperature equalization for the necessity of cell balancing. The BMS integrates a cell balancing method and delivers a dynamic solution to this problem to avoid overstress of individual cells. There are several methods of cell balance schemes that are generally used, and they balance time taken for charging, loss of efficiency, and component cost. The two methods are as follows:

i. Active balancing
ii. Passive balancing

14.2.3.2 Active Balancing

In the active cell balancing method, excess charge from the stronger charged cell is transmitted o the weaker charged cell. Using storage elements such as an inductor and a capacitor, active balancing can be attained. There are several methods to accomplish active cell balancing. The commonly used methods are as follows:

i. Flying capacitor charge distribution
ii. Inductive shuttle charge distribution

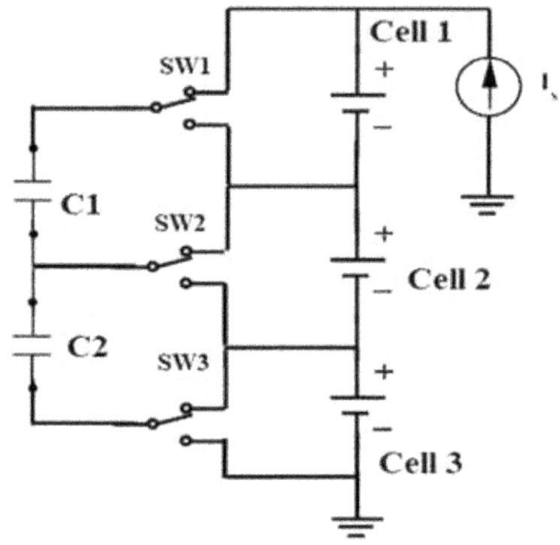

FIGURE 14.3 Charge shuttle charge distribution.

14.2.3.2.1 Flying Capacitor Charge Distribution or Charge Shuttle Charge Distribution

In a series chain of cell circuits, each cell is charged through the parallel connected capacitors as shown in Figure 14.3. A capacitor is switched by means of sequential switching. The function of a capacitor is to average the charge across the cell. The capacitor collects the charge from the high-charged cell and deposits the charge into the low-charged cell. The process speed can be enhanced using a program-based capacitor that transfers the charge quickly. But this method is complex and too costly.

14.2.3.2.2 Inductive Shuttle Charge Distribution

In this method, a transformer is used. The battery terminals are connected across the primary of the transformer, and the individual cells are connected across the secondary winding. The energy is taken from the battery and averages the voltage differences of individual cells. But balanced secondary windings are required to achieve cell balancing.

14.2.3.2.2.1 Buck-Boost Method-Based Inductive Converter Another method of active cell balancing is buck-boost method based on an inductive converter, as shown in Figure 14.4. It consists of an inductor and switching circuits. The high-charge cell supplies excess charge into the inductor, and sw1 and sw2 switches in the buck converter discharge the charge from the inductor to low-charge cells.

14.2.3.3 Passive Balancing

In this method, the excess charge from high-charge cells is determined by dissipative techniques, and a bypass resistor removes the excess energy. There are two types of passive balancing methods:

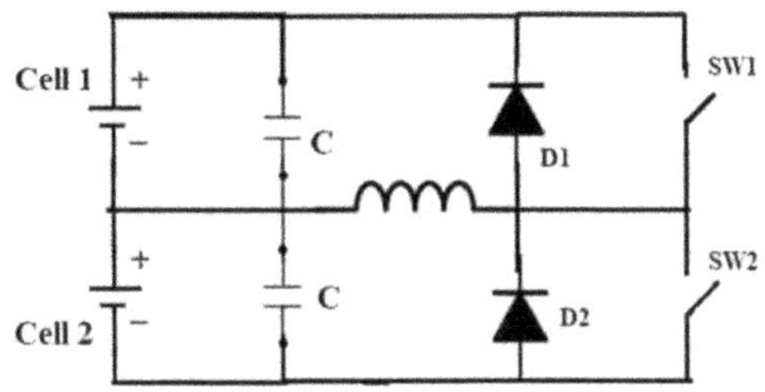

FIGURE 14.4 Inductive shuttle charge distribution.

i. Charge shunting
ii. Charge limiting

14.2.3.3.1 Charge Shunting

In a battery, a bypass resistor is connected to each cell through power electronic switches and a controller. The voltage across each cell is measured by the controller. If any one of the cell voltages is above the rated voltage, then the controller activates the power electronic switch of that high-charge cell and starts to discharge the excess energy through the resistor. This method is fast but costly.

14.2.3.3.2 Charge Limiting

In the charge-limiting method, the voltage of each cell is supervised continuously. During the charging period, if the cell voltage reaches the full-charge voltage, then the charging of the entire battery is stopped, and the charging of other cells in semi-charging is also stopped.

Likewise, even if one cell discharges to the minimum cut-off voltage during the discharging period, the battery pack is disconnected from the load until the pack is charged again. But it is an inefficient method of charging.

14.3 STATE OF CHARGE (SOC) DETERMINATION

An efficient BMS ensures the optimum use of the battery and accurately monitors the SOC and SOH of a battery. The state-of-charge determination requires temperature and current measurements. Many applications require knowledge of the SOC of the battery or of the individual cells in the battery chain [10]. To confirm optimal control of the charging process, knowledge about the state of charge may be provided to the user as a warning of the capacity remaining in the battery, or it could be required in a control circuit. Precise assessment of the battery's state of health (SOH) and SOC is a vital function that an advanced battery management system must effectively carry out [12,13].

In order to achieve the most efficient management system, numerous methods have been suggested to estimate the battery SOC. The estimated battery SOC data from various methods are analyzed, and their optimistic and undesirable aspects are deliberated. As a result, the progressive and proficient methods are emphasized, and endorsements are articulated.

14.3.1 What Is SOC?

The state of the battery, such as the voltage of individual cells, total voltage, and minimum or maximum voltage, is monitored by the BMS. It also shows the charge level of the battery by monitoring SOC or depth of discharge (DOD) level of the battery [11]. The current into or out of the battery and the amount of power (SOP) available for a defined time interval, given the current power usage and temperature, are also monitored by BMS.

To know the balance of battery energy compared to the energy it had once the battery is fully charged, the battery can operate for how many hours before recharging again. It shows the quantity of the quick-span ability of the battery [12,13]. An ideal SOC indication must be the graded capability of a brand-new unit rather than the present cell amount. The cell age decreases by cell capacity. Once the battery is fully charged, the SOC will show only 80% of its graded capacity. When the battery cell is going to end its span, it may have an SOC of 100%; nevertheless, its actual capability is 80%, and the change factors should be relevant to the expected volume rather than its actual capacity.

Monitoring the battery's SOC is one of the BMS's functions. The SOC might infer the consumer and regulate the charging and discharging procedure [14]. Several bids involve an understanding of the specific units of the SOC of the battery-operated. The SOC can be determined using three different methods:

a. **Direct Measurement:** The battery voltage decreases linearly during the discharge cycle; SOC may be measured directly using a voltmeter.
b. **Coulomb Counting:** To provide the virtual amount of charge, it is incorporated with the current value going into the battery or flowing out of the battery.
c. **Combination of the Two Techniques:** The two approaches might also be merged. As the real charge approaches either end, the voltmeter can be used to calibrate the SOC and track the battery voltage. In the meantime, the battery current can remain combined to establish the comparative charge going into and flowing out of the battery. The SOC refers to the remaining capacity (Q_{rem}) as a percentage of the maximum available capacity.

$$\mathrm{SOC} = Q_{\mathrm{rem}} / Q_{\mathrm{max}} \tag{14.1}$$

Depth of Discharge: The depth of discharge can be defined as the percentage of the battery that has been discharged as a percentage of maximum capacity of the battery. The ratio of 80% DOD clearance is shown as a "deep discharge." An efficient BMS ensures the optimum use of the battery and accurately monitors the SOC and SOH of a battery. The state-of-charge determination requires temperature and current measurements.

The empirical data collected in this part offer a thorough examination of how they perform, protection, and the effectiveness of the BMS in electric cars. The following are the main conclusions.

14.3.2 Battery Efficiency and Performance

In comparison to traditional unmanaged battery packs, an optimized BMS increases battery efficiency by 10%–15%, according to experimental data from real-world EV applications.

By employing sophisticated SoC estimating techniques like Kalman filters and neural networks, the accuracy of SOC estimation has increased from ±5% to ±1%.

14.3.3 Thermal Management and Safety

According to temperature monitoring of lithium-ion batteries, active thermal management prevents possible thermal runaway accidents by reducing battery overheating by 30%.

Based on empirical testing, a well-designed BMS extends battery life by 95% by lowering the danger of overcharging and deep draining.

14.3.4 Power Density and Energy Density

By effectively controlling power supply, BMS-assisted power optimization increases vehicle acceleration by up to 12%, according to EV acceleration testing. Tests of battery capacity retention reveal that, even after 1000 charge-discharge cycles, cells under BMS management retain 80% of their initial capacity, while unmanaged cells deteriorate more quickly.

14.3.5 Accuracy Requirements of SOC

A knowledge of the SOC is particularly important for large lithium batteries. Lithium is highly chemically responsive than other batteries, and to operate the battery within a safe and sound operating area, it needs an automated BMS. So that it ensures a lengthy cycle span. The main purpose of the BMS is to regulate the SOC. Moreover, locomotive applications, major uses of significant lithium batteries, require extremely accurate command of the SOC for effective and reliable management of the power streams.

- The range is found by using SOC in EV applications. It would present an entire quantity based on the capability of the battery during the 20% error in the recent capacity, and it will cause 20% or more because of battery aging.
- The SOC governs the on/off position of the engine in HEV applications. Errors in SOC greater than 5% might really change the system fuel effectiveness. An accuracy significantly better than 5% is therefore desirable.
- Motorized fuel gauges are disreputably rough in SOC accuracy of 5%; uncertainty in accomplishing this could be suitable for such claims.

14.3.6 State of Charge (SOC) Determination

A solid electrolyte contact forms at the negative electrode of aging batteries.

A battery's lifespan is influenced by several factors, including cell design, battery operation, and environmental conditions. The primary deterioration variables influencing battery status are depicted in Figure 14.5.

Battery state estimation includes four key states, which are shown in Figure 14.6: SOC, SOH, internal temperature, and joint state estimation. In this chapter, the details about the SOC are discussed.

14.3.6.1 Estimation of SOC

Direct estimation and model-based estimation are the two most used commonly methods for SOC estimation.

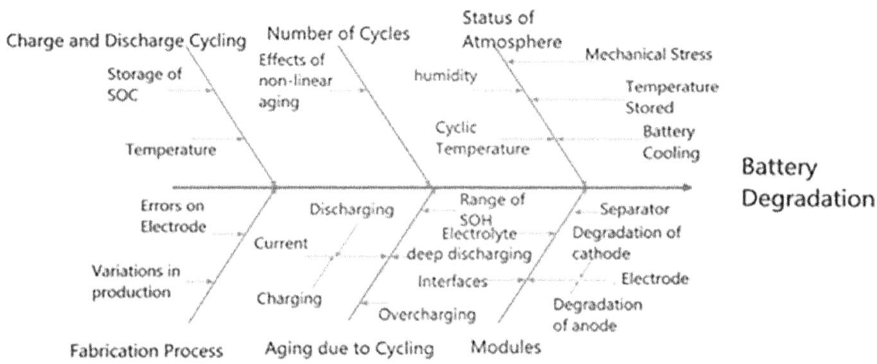

FIGURE 14.5 Diagrammatic representation of battery degradation and aging.

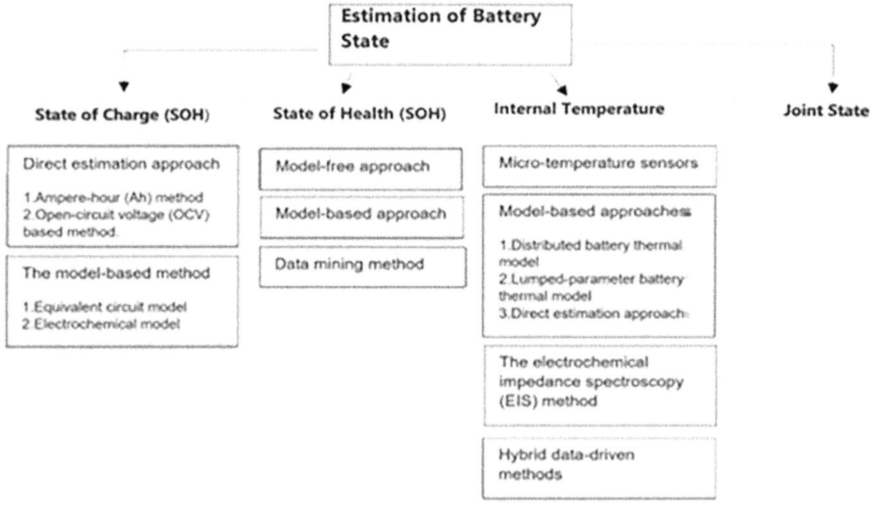

FIGURE 14.6 Estimation of battery in four states.

14.3.6.1.1 Direct Estimation Approach

This method is based on the direct measurement of electric battery factors (voltage and current). This would be a simple provision; the battery may release at a steady state level. The current flow and time multiplications are equal to the charge in a battery. The discharging current does not remain persistent in all batteries; practically, when the battery diminishes, it follows a non-linear manner. One measurement apparatus is capable of incorporating the current period. Then, this approach hinges on releasing the energy of the battery to see how much charge it includes.

The estimation techniques are the ampere-hour (Ah) and open-circuit voltage (OCV) approaches.

The **Ah technique** measures the charging or discharging current simply. The following formula may be used to calculate the SOC:

$$\mathrm{SOC}(n) = \mathrm{SOC}(n_0) + \int_{n_0}^{n} \eta' \, I(t)\,dt / C_N \tag{14.2}$$

SOC(n_0): Represents the initial known state of charge.
η: Denotes the efficiency of the battery during charging and discharging.
$I(t)$: Indicates the current flowing during the charging or discharging process.
Cn: Refers to the nominal capacity of the battery.

Though establishing the preliminary SOC and measurement precision can be a perplexing procedure while adjusting the Ah technique for the SOC evaluation algorithm.

14.3.6.1.2 OCV-Based SOC Estimation

The excellent precision of this approach in determining the SOC is well known. OCV and SOC have a nonlinear connection, which it depends on. The battery must, however, be unplugged from chargers and loads for a considerable amount of time in order to achieve balance for accurate measurement. This resting period is a major disadvantage that reduces the method's applicability for EV applications. In low-temperature situations, it can surpass two hours. Real-time OCV computations have been implemented to overcome this constraint, allowing the SOC estimate while the vehicle is moving.

14.3.6.2 SOC from Specific Gravity (SG) Measurements

It is the traditional method of regulating the charge condition of lead-acid batteries. The weight of the effective chemical changes is measured by this technique. It diminishes the solution's specific gravity in immediate percentage to the state of charge. The real SG of the electrolyte is treated as a sign of the state of charge of the battery. SG quantities have usually been measured using a pressure-type hydrometer, which is slow and inconvinient. Currently, automated sensors that provide a digital measurement of the SG of the electrolyte can be integrated into the cells to provide a constant readout of the battery situation/status.

14.3.6.2.1 *Current-Based SOC Estimation*

The energy included in an electric charge is computed in Coulombs and is equivalent to the integral over current flow time which provides the charge. The enduring capability in a cell can be computed by determining the current entering (charging) or leaving (discharging) the cells and accumulating this over time. There are three current sensing approaches as follows:

- **Current Shunt**: The easiest way of establishing the current is by utilizing the voltage drop over a low-ohmic-value, high-accuracy, string, sense resistor trapped in the middle of the battery, and the output capacity is known as a current shunt. The battery warms up and leads to incorrect results in the system. It also causes power losses in the battery.
- **Hall effect** transducers avoid this problem, but they are more expensive. Unfortunately, these sensors cannot tolerate high currents and are susceptible to noise.
- **GMR** magneto-resistive sensors have better performance and deliver a superior indicator point, and they are more expensive than others. They have better temperature stability than Hall effect devices.

14.3.6.3 SOC Estimation from Internal Impedance Measurements

Through the cell charge-discharge cycles, the structure of the active elements in the cell fluctuates as the chemicals are transformed between the charged and discharged states, and this is reflected in changes to the cell impedance. Consequently, the magnitude of the cell's interior impedance can be used to verify SOC, though these methods are not widely utilized owing to complexities in evaluating the impedance. Although the cell is effective, there are difficulties in inferring the information; meanwhile, the impedance is also temperature dependent.

14.3.6.3.1 *Calculating the SOC of Lithium-Ion Batteries*

There are theoretical and practical methods of Li-ion battery.

14.3.6.3.2 *Theoretical Method*

If sufficient data is available, the State of Charge (SOC) may be inferred using measured cell characteristics and operating circumstances.

14.3.6.3.3 *Practical Method*

Alternatively, the performance physical characteristics of a normal model cell (or cells) can be evaluated, and the outcomes used as a pattern to characterize the functioning of the remainder of the inhabitants. Compared to theoretical computations, the estimation procedure is made simpler by using lookup tables generated from real cell data. The majority, if not all, of the elements affecting SOC are naturally taken into consideration by this method, producing ratings that are more accurate and trustworthy.

14.3.6.3.4 Algorithms for Estimation of Charge

Numerous approaches, including fuzzy logic, Kalman filtering, neural networks, and recursive, self-learning algorithms, have been used to increase the accuracy of both the state of health (SOH) assessment and the SOC estimation [15].

- **Fuzzy Logic:** A straightforward method for deriving clear conclusions from imprecise, ambiguous, or unclear data is fuzzy logic. It looks like human judgment in conjunction with its capability to operate on estimated data to discover accurate explanations.
- **Kalman Filter:** Kalman filtering focuses on an age-old problem: How do you make precise data out of imprecise information? How do you revise a "best" assessment for the status of a scheme as modern? An HEV motorized application is an illustration of this condition.
- **Neural Network:** An artificial intelligence model that mimics the human brain's information processing, memory, and learning functions is called a neural network. It mimics the brain's capacity to see trends, adjust by trial and error, and uncover hidden connections in data.

In two important ways, neural networks are similar to the human brain:

1. Through a process of learning, they acquire knowledge.
2. The connections between neurons, which are represented by synaptic weights, preserve their learned experience.

The actual energy and benefit of neural networks remains in their capability to characterize equally direct and indirect (non-linear) connections and in the skill to understand these interactions immediately from the data that is being developed.

Neural network methods help assess battery performance, which differs in computing the impact of various constraints on the whole, well-defined by numerical accuracy.

14.3.6.4 Two-Parameter SOC Estimation and Accuracy Improvement

Then, by merging appropriately weighted data from the voltage-based and current-based methods into a single number, the overall accuracy of the SOC calculation may be increased.

14.3.6.4.1 Consumer Battery Condition Indicators

Tiny main cells are currently obtainable along with an on-cell equivalent SOC gauge known as battery testers or fuel gauges. Analogous to a thermometer on the side of the cell, a printed strip is included to show the approximate balance of the battery capacity.

14.4 STATE OF HEALTH (SOH) DETERMINATION

SOH is a degree of a battery's or a battery pack's proficiency to provide its specified output compared to its ideal conditions. SOH is a sign of the battery's health over its age. This is vital for assessing the promptness of backup power apparatus and is an indicator of whether maintenance actions are needed.

14.4.1 SOH Determination

- The SOH is a portion of a battery's ability to produce its required production.
- The condition of alternative power equipment serves as an indicator of whether maintenance actions are necessary.

14.4.2 What Is the SOH?

The SOH is a "measurement" that indicates the typical situation of a battery and its capability to provide the required performance associated with a brand-new battery. It considers factors such as charge recognition, inner resistance, voltage, and self-discharge. It is an amount of the prolonged period capacity of the battery and provides an "indication," not a pure measurement, of the "lifetime energy throughput" of the battery, how much energy has been expended and how much is left [16]. The SOH is a symptom of the moment which is achieved in the life phase of the battery, and a quantity of its situation relative to a brand-new battery.

14.4.3 What Is the SOH Used for?

Its objective is to deliver a signal of the performance that can be anticipated from the battery in its present condition or to deliver a warning of how large amount of the beneficial lifespan of the battery has been destroyed and how much continues prior to it must be substituted.

But the documented practice of the past of the battery is applied to establish the SOH, as in the Log Book Function further down. This is especially helpful for reviewing the provision of high-ranking value EV and HEV batteries.

14.4.4 How Is the SOH Determined?

In a Lithium-ion battery, the cell capacity weakens relatively linearly with age or cycle life; the terminated, or residual cycle life time, depending on the characterization utilized, is frequently employed as a rough portion of the SOH [8]. A battery's State of Health (SOH) indicates both its general health in relation to its ideal state and its ability to store and deliver energy. The SOH of a brand-new battery is normally 100%; however, it steadily drops with time [17]. But sometimes, if a battery's performance isn't up to par, its initial SOH might be less than 100%. The following is a mathematical representation of a battery's SOH:

$$\mathrm{SOH}(t) = \mathrm{SOH}(t_0) + \int_{\tau=t_0}^{\tau} \delta_{\mathrm{func}}(x)\, d\tau \tag{14.3}$$

SOH(t_0): Represents the initial state of health of the battery.
δ_func(x): Denotes the aging rate function.
x: Includes various influencing factors such as current, temperature, and SOC that contribute to battery aging.

This method estimates the current SOH without making any changes to the battery cells by using external measurement equipment. For precise SOH measurement, specialized proprietary equipment is available.

The model-free method approximates the battery SOH by utilizing factors called the ancient capability (Caged) or the improved interior resistance (Rinc). The pattern is introduced merely by the equations beneath, where Cn and Rn are the nominal capacities and internal resistance for a brand-new idle battery:

$$\text{SOH} = \frac{C_{\text{aged}}}{C_n} \cdot 100\% \tag{14.4}$$

$$\text{SOH} = \frac{R_{\text{inc}}}{R_n} \cdot 100\% \tag{14.5}$$

The direct measurement method uses a traditional capability test drive or a vibration current experiment to calculate the battery's aged capacity (Caged).

Data-mining SOH estimation methodologies have been developed and are commonly used for detecting and forecasting the cell condition of Li-ion batteries. Frequently utilized data-powered analytical algorithms for detecting the complex aging of Li-ion batteries apply artificial neural networks.

14.5 THE LOG-BOOK FUNCTION

14.5.1 History (Log-Book Function)

Another potential function of the BMS is monitoring and storing the battery's antiquity. The past data are needed to estimate the SOH of the battery, but also to determine whether it has been subject to misuse. Parameters such as number of cycles, maximum/minimum voltages, maximum charging/discharging currents, and temperature can be recorded for subsequent evaluation. This can be an important tool in assessing warranty claims. An alternate technique of stipulating the SOH supports the assessment based on the treatment record of the battery rather than a few quantified factors [18]. The amount of charge-discharge phases completed by the battery is an evident quantity.

14.5.1.1 Authentication and Identification

An authentication is frequently used to avoid the use of unapproved batteries in critical battery applications. It might be used to prevent compatibility difficulties in conjunction with system security and energy supervision programs, or through the relevant software change, or it may be to prevent destruction to the status of the manufactured goods and the brand if inferior or unreliable cells are used.

14.5.2 BMS (Battery Management System)

Authentication by the BMS operates by integrating into the battery pack a distinctive method of pinpointing the battery, such as a code printed onto a memory gadget.

Authentication can be employed to limit the customers of the battery application to capable people or it may be planned to set boundaries on the upper limit current that can be taken from the battery depending on the customer's authority.

14.5.3 RFID (Radio Frequency Identification)

It is also feasible to build parallel documentation and verification programs using RFID techniques [19].

- **Holograms:** Holograms support an inexpensive alternative technique for detecting batteries, but they are less appropriate for computerization. Even with computer scanning, they are difficult to reproduce yet easy to read. They provide consumers with a trustworthy way to verify a product's legitimacy. They do not completely stop counterfeiting, even though they can aid in the detection of fake goods. In order to preserve manufacturing data and provide traceability in the case of quality problems, holograms can also be added to labels.
- **Barcode Labels:** Using barcode labels is the easiest and lowest-cost technique for delivering data such as battery type, date of manufacture, and serial number with the battery. They are helpful for proof of identity, traceability, and supply management.

14.6 BATTERY COMMUNICATIONS INTERFACES

14.6.1 Communications

Nearly all BMS systems integrate a few kinds of communications between the battery and the charger or test drive equipment [20]. It is termed an intelligent battery. Some have connections to other systems integrated through the battery for examining its status or its record. The communications interface is a data link used for performance monitoring, data logging, diagnostics, smart battery modules (SBMs), or for setting system parameters, or it may be a communications channel carrying system control signals [20].

Apart from the SMBus (System Management), which is specific to batteries, all the communications interfaces and protocols mentioned below were designed for general-purpose or battery applications.

14.6.2 Data Buses

RS232 Connection: A pattern for serial broadcast of information between two peripheral devices through cable distances up to 50 feet. Discrete broadcast and accept cables deliver maximum duplex communication. This is utilized for information speeds up to 20,000 bits per second.

EIA-485 (Formerly RS485) Connection: A pattern for serial broadcast of information among numerous devices with cable distances up to 4000 feet (Possible 32 channels). Suitable for data rates up to 100K bits per second or 10M bits per second, depending on the length of the cables.

14.6.3 Inter-Integrated Circuit (I2C) Bus

The low-level speed bus is the I2C bus. A pattern for serial broadcast of information between interior components inside a system instead of peripheral communications.

This is also termed a double-cord synchronous bus. It runs with data rates up to 3.4 Mbits and is suitable for master-slave applications.

14.6.4 USB Universal Serial Bus

The USB standard is intended to restore the many patented serial and parallel information networks between computers and secondary devices and makes use of a star topology. The data rate specified in the original version was 1.5 Mbits/sec, but this has increased with subsequent versions up to 3 Gigabits/sec. Containing hubs, 127 devices are attached to the bus. Cable lengths are limited to 5 metres (16 feet).

The USB connection is employed for examining the battery condition or establishing command boundaries.

14.6.5 CAN Bus

The manufacturing model for on-panel vehicle communications is the CAN (Controller Area Network) bus. It was originally designed in the 1980s by Bosch in Germany with the purpose of eliminating hundreds of wires from the wiring harness, reducing the size and weight. The CAN bus was termed as a double-cable, serialized telecommunications bus intended for interacting with knowledgeable sensors and actuators in a concentrated multiplexing structure.

The CAN bus allows 12 volts. Vehicle functions such as temperatures, alarms, or the positions of switches are monitored by sensors connected to the bus. To commence the proper reaction, the controller is able to supervise the actuators as needed.

The communication protocol specifies the following aspects:

- The method for addressing devices on the bus
- The format of the data word (message structure)
- Priority assignment for messages
- The sequence of data transmission
- Control signal mechanisms
- Techniques for detecting errors
- Strategies for error handling or correction
- The speed at which data is transmitted

Automotive BMS consequently employs the CAN bus as its leading communications channel.

14.6.6 LIN Bus

The LIN bus is another automotive communications standard, initiated in 1998, like the CAN bus.

It is a one-wire local communication network operating at a data rate of 20 kbps. It consumes additional electronics than the CAN bus, although it is more adaptable and uses a smaller amount of wiring.

14.6.7 FlexRay Bus

Drive-by-wire, steer-by-wire, and brake-by-wire are examples of more complicated control applications that the newly created FlexRay bus was created to satisfy. Unlike the CAN bus, which supports simpler sensors and actuators, the FlexRay bus can enable rapid, dynamic control systems with a data payload per frame 20 times bigger than the CAN bus.

14.6.8 SM Bus

The SMBus (System Management Bus) is a double-wire, 100 kHz, serial bus intended for usage in conjunction with small power Smart Battery Systems (SBS) with the restricted purposes of communicating smart batteries with built-in intelligence, with their combined chargers. Sometimes found in simple vehicle applications, it does not have the range of capabilities for controlling devices connected to the power lines that the CAN and LIN buses have.

14.7 CONCLUSION

Since they provide information on the battery's short- and long-term conditions, estimating the State of Charge (SOC) and State of Health (SOH) is essential when creating a battery management system. In order to help develop a sophisticated BMS for electric car applications, this chapter attempts to give a basic overview of different algorithms, emphasizing their benefits and drawbacks. The review comes to the conclusion that there is no one-size-fits-all approach to battery state estimation, and the choice should be based on the particular application and the complexity of the system.

REFERENCES

1. M. S. H. Lipu, et al., "A review of state of health and remaining useful life estimation methods for lithium-ion battery in electric vehicles: Challenges and recommendations," *J. Clean. Prod.*, vol. 205, pp. 115–133, 2018.
2. A. Augeard, et al., "Contribution to the study of electric arcs in lithium-ion batteries," *IEEE Trans. Components, Packag. Manuf. Technol.*, vol. 6, no. 7, pp. 1066–1076, 2016.
3. M. A. Hannan and S. Member, "State-of-the-art and energy management system of lithium-ion batteries in electric vehicle applications: Issues and recommendations," *IEEE Access*, vol. 6, pp. 19362–19378, 2018.
4. B. Pattipati, et al., "Open circuit voltage characterization of lithium-ion batteries," *J. Power Sources*, vol. 269, pp. 317–333, 2014.
5. H. Dong, et al., "Lithium-ion battery state of health monitoring and remaining useful life prediction based on support vector regression-particle filter," *J. Power Sources*, vol. 271, pp. 114–123, 2014.

6. Moghaddam, A.F.; 7nati, M.; Sun, H.; Van den Bossche, A. Electric vehicles charging concepts for lithium based batteries. In *Proceedings of the 2018 7th International Conference on Renewable Energy Research and Applications (ICRERA)*, Paris, France, 14–17 October 2018; pp. 397–401.
7. Shang, Y.; Xia, B.; Zhang, C.; Cui, N.; Yang, J.; Mi, C. An automatic battery equalizer based on forward and flyback conversion for series-connected battery strings. In *Proceedings of the 2017 IEEE Applied Power Electronics Conference and Exposition (APEC)*, Tampa, FL, 26–30 March 2017, pp. 3218–3222.
8. Lievre, A.; Sari, A.; Venet, P.; Hijazi, A.; Ouattara-Brigaudet, M.; Pelissier, S. "Practical online estimation of lithium-ion battery apparent series resistance for mild hybrid vehicles," *IEEE Trans. Veh. Technol.* vol. 65, pp. 4505–4511, 2016.
9. Lu, L.; Han, X.; Li, J.; Hua, J.; Ouyang, M. "A review on the key issues for lithium-ion battery management in electric vehicles," *J. Power Sources*, vol. 226, pp. 272–288, 2013.
10. Daowd, M.; Omar, N.; Van Den Bossche, P.; Van Mierlo, J. "A review of passive and active battery balancing based on MATLAB/Simulink," *J. Int. Rev. Electr. Eng.*, vol. 6, no. 7, 2011.
11. Saqli, K.; Bouchareb, H.; Oudghiri, M.; M'sirdi, N. An overview of State of Charge (SOC) and State of Health (SOH) estimation methods of Li-ion batteries. In *IMAACA 2019*, Lisbon, Portugal, 2019.
12. Vaideeswaran, V.; Bhuvanesh, S.; Devasena, M. Battery management systems for electric vehicles using lithium ion batteries. *2019 Innovations in Power and Advanced Computing Technologies (i-PACT)*, vol. 1, pp. 1–9, 2019.
13. Piller, S.; Perrin, M.; Jossen, A. "Methods for state-of-charge determination and their applications," *J. Power Sources*, vol. 96, no. 1, pp. 113–120, 2001.
14. Wang, S.L.; Shang, L.P.; Li, Z.F.; Xie, W.; Yuan, H.F. "A novel online SOC estimation method for the power lithium battery pack based on the unscented Kalman filter," *DEStech Trans. Environ. Energy Earth Sci.*, pp. 98–105, 2017.
15. Harting, N.; Schenkendorf, R.; Wolff, N.; Krewer, U. "State-of-health identification of lithium-ion batteries based on nonlinear frequency response analysis: First steps with machine learning," *Appl. Sci.*, vol. 8, no. 5, p. 821, 2018.
16. Omar, N.; Monem, M.A.; Firouz, Y.; Salminen, J.; Smekens, J.; Hegazy, O.; ... Van Mierlo, J. "Lithium iron phosphate based battery–Assessment of the aging parameters and development of cycle life model," *Appl. Energy*, vol. 113, pp. 1575–1585, 2014.
17. You, G.W.; Park, S.; Oh, D. "Real-time state-of-health estimation for electric vehicle batteries: A data-driven approach," *Appl. Energy*, vol. 176, pp. 92–103, 2016.
18. Rieback, M.R.; Crispo, B.; Tanenbaum, A.S. RFID Guardian: A battery-powered mobile device for RFID privacy management. In *Australasian Conference on Information Security and Privacy*, 2005, pp. 184–194. Springer.
19. Garche, J.; Jossen, A. Battery management systems (BMS) for increasing battery life-time. In *TELESCON 2000. Third International Telecommunications Energy Special Conference (IEEE Cat. No. 00EX424)*, 2000, pp. 81–88. IEEE.
20. Chatzakis, J.; Kalaitzakis, K.; Voulgaris, N.C.; Manias, S.N. "Designing a new generalized battery management system," *IEEE Trans. Ind. Electron.*, vol. 50, no. 5, pp. 990–999, 2003.

15 Green Energy-Driven Transportation Infrastructure for Fuel Cell Hydrogen Electric Vehicles

Muthukaruppasamy Subburaj,
Arun Sampaul Thomas Gnanaraj,
Nandha Gopal Jothiperumal,
and Parimalasundar Ezhilvannan

15.1 INTRODUCTION

The transport sector is liable for the release of hazardous pollutants that have a notable effect on the health of the public. Green hydrogen has transformed transportation by providing a pollution-free alternative to traditional gasoline or diesel engines. It is a sensitive component of a completely clean system of energy and is regarded as the lead in transforming the energy zone's framework. There is increasing curiosity in the analysis of green hydrogen to evaluate its potential function in the advancement of transportation. It is an ecologically supportive fuel that is advantageous for nearly all businesses. It is a feasible replacement for traditional carbon-based fuel and can substantially reduce the emissions of hazardous substances in buses, aircraft, vehicles, and ships. Green hydrogen will be advantageous to all nations that are still experiencing environmental pollution and transportation emissions from a global perspective. The negative effects of global warming will be mitigated by the substantial reduction of hazardous exhaust emissions that will result from the use of hydrogen-based fuel cell vehicles [1]. The use of hydrogen fuel cells has the potential to significantly cut emissions from sturdy vehicles, which are major sources of nitrogen oxide emissions and transportation-related pollutants. The transportation sector, which relies on fossil fuels, is a significant contributor to the emission of deleterious greenhouse gases. When renewable hydrogen is employed as a fuel source, no greenhouse gases are released. The combustion of hydrogen generates only water vapors that are entirely innocuous. There is a broad scope of applications for the use of renewable hydrogen in fuel cells. This process produces water and power by reacting the hydrogen gas within the vehicle with oxygen taken from

DOI: 10.1201/9781003585534-15

the environment. The vehicle is propelled by a motor that is powered by the electric energy that is used to power a battery. Since this technique only produces water as an emission, it is a clean and ecologically friendly mode of transport [2]. Developments in green hydrogen have the ability to boost the economy and encourage carbon reduction, especially when used in the transportation sector. The economy is being fortified, and new employment is being created as a result of the substantial investment being made in the advancement of infrastructure for the storage and distribution of green hydrogen. The aviation sector can also be decarbonized through aircraft propulsion that is based on renewable hydrogen. An airplane operates on green hydrogen by burning energy in hydrogen combustion engines or electrochemical conversion using fuel cells, yielding no harmful emissions [3,4]. Nevertheless, it is imperative to finalize innovative aircraft design concepts, as hydrogen propulsion necessitates hydrogen gas storage. Aviation experts have shown that propellant systems centered around hydrogen may be scaled up or down depending on the power and energy density requirements of medium-range and long-distance aircraft. On the other hand, shorter-range aircraft sectors could benefit more from designs powered by hydrogen fuel cells. This innovative energy source promotes sustainability in the freight business, in addition to the aviation sector, automobiles, and trucks [5]. Green hydrogen has many other benefits. The transportation sector at sea is responsible for around 7%–8% of the world's greenhouse gas (GHG) emissions [6,7]. One potential alternative energy source that might help to reduce pollution in this industry is hydrogen. In order to mitigate carbon emissions, numerous passenger ships are establishing frameworks for the implementation of green hydrogen-based propulsion. A concept for ships entails the thermal stimulant dissolution of methane to produce hydrogen gas and solid carbon [8]. The hazardous waste is transformed into solid carbon, which is employed for alternative needs. In this manner, renewable hydrogen is energizing various sectors of the transportation industry and establishing the foundation for a future that is both sustainable and emission-free.

15.1.1 Types of Hydrogen

Hydrogen is a fuel that is both pure and versatile. Hydrogen is employed in many sectors, including the generation of electricity and the operation of vehicles. Hydrogen is a versatile energy source that is accessible in a diversity of patterns, each with its own different settings and applications. The primary distinguishing factor among the five distinct forms of hydrogen is their manufacturing processes. In this section, we look at the various forms of hydrogen and their applications.

15.1.1.1 Grey hydrogen

It is through a method called steam methane reforming (SMR). Natural gas is heated with vapor in this method. Grey hydrogen is a notable contributor to emissions due to the subsequent atmospheric release of carbon dioxide.

15.1.1.2 Blue Hydrogen

The same SMR method that is employed to produce grey hydrogen is also employed to produce blue hydrogen, but with the inclusion of carbon capture and storage (CCS) technology. By extracting carbon dioxide generated during the SMR process and depositing it underground, this innovation mitigates greenhouse gas emissions. Blue hydrogen is a transitional fuel to reduce emissions while allowing for the continued use of existing infrastructure. Its principal application is in industrial applications, despite the growing interest in its use for power generation and transportation.

15.1.1.3 Hydrogen That Is Environmentally Friendly

Electrolysis, which entails the utilization of electricity to free the water (water splits into hydrogen and oxygen), is the process by which green hydrogen is generated. Green hydrogen is an entirely emissions-free fuel, as the electricity utilized in this process can be derived from components of wind and solar. The cost of renewable hydrogen is still relatively high; however, it is anticipated that this will decrease as renewable energy becomes more prevalent. It has the potential to be employed in a diverse array of applications.

15.1.2 What Is the Environmental Impact of Grey Hydrogen?

Some 70 million tonnes of hydrogen are presently consumed globally, as indicated by the International Energy Agency (IEA) [9]. The issue is that the majority of this hydrogen is manufactured from coal and natural gas, which are the most environmentally harmful methods. Consequently, only 0.1% of the hydrogen we consume is environmentally friendly. To provide context, the production of this grey hydrogen generates 900 million tonnes of carbon dioxide yearly. The annual CO_2 emissions of the economies of the United Kingdom and Indonesia are combined with tonnes of CO_2 [10]. In other words, the production of this hydrogen is equivalent to the pollution produced by two countries with a combined population of over 300 million. It is imperative to comprehend hydrogen's source in light of this alarming data. Green hydrogen is an unparalleled instrument for the replacement of production models that contribute to climate change and the decarbonization of energy consumption that is more resistant to electrification, such as heavy industry or long-haul transport.

15.1.3 What Is Green or Renewable Hydrogen, and How Does It Operate?

For this reason, green hydrogen is a pure, sustainable fuel with a zero pollution index that can serve as both a basic material and an energy vector. It is produced using renewable energies. It is an important contribution to the battle against climate change and is derived from its ability to substitute fossil fuels in industry and uses that are hard to reduce CO_2, as well as its potential as an energy storage system. A method of electrolysis that is powered by renewable energies, such as solar or wind, is employed to manufacture renewable hydrogen. Electrolysis entails the use of electrodes to split water molecules into hydrogen and oxygen through electrical current. Hydrogen stored in specific containers is led to a fuel cell when it is required for energy conversion. There, it re-establishes its bond with oxygen from the atmosphere, yielding electricity. Obviously, the secondary product is water, which leads to an imperishable pure process that generates energy without emitting any CO_2.

15.1.4 What Is an Electrolyzer and Why Is It Essential for Renewable Hydrogen?

The electrolyzer uses electricity to separate the hydrogen and oxygen molecules that comprise water through a chemical process known as electrolysis. The production of hydrogen in a sustainable manner, which does not result in the emission of carbon dioxide into the atmosphere, can serve as the foundation for a decarbonized economy. The key to attaining renewable hydrogen is the impact of this process, which enables the breakdown of molecules using electricity, specifically water molecules. An electrolyzer is a device that is capable of separating water molecules into their individual oxygen and hydrogen atoms [11]. An appliance that is able to split molecules of water into their component oxygen and hydrogen atoms is referred to as an electrolyzer. In order to promote the fracture of the bonds that are very robust between them, a procedure that is recognized as electrolysis requires the use of electricity. The creation of electrolyzers that are not only effective but also efficient will be a prerequisite for the broad use of hydrogen in an industrial way and the implementation of hydrogen fuel cells. It was in the year 1800 when electrolysis was first discovered. In that exact year that Alessandro Volta invented the electric battery, other scientists tried to establish a connection between their electrodes in a vessel filled with water [12]. Hydrogen and oxygen were found to split at the electrodes, and they discovered that the current passed through the water. An electrode array is placed within an electrolyzer and subjected to a high voltage and current via a membrane. Because of this, an electric current is generated in the water, which breaks it down into its component parts—oxygen and hydrogen. When two processes generate oxygen simultaneously, the former may be discharged into the air or stored for potential use as a therapeutic or industrial gas in the future. Industries and transportation vehicles like ships, planes, and trains may all benefit from hydrogen fuel cells, which store the gas in a compressed or liquefied form.

15.1.4.1 Different Types – Electrolyzers

Currently, electrolyzers are classified according to their dimensions and functionality. The most frequently employed are:

i. **Alkaline Electrolyzer**: They use a mixture of water and a liquid electrolyte, such as sodium hydroxide or potassium hydroxide, for example. Hydrogen is produced by a cell's membrane, anode, and cathode. The cells are usually set up in a sequence so that they can produce more oxygen and hydrogen all at once. As current flows through the electrolytic cell framework, hydrogen ions move across the electrolyte from each cell's cathode to its anode. Gas particles, one containing hydrogen and one containing oxygen, are produced during this process. The cathode end of the electrolyzer is affected by this process. They do not need noble metals as a catalyst and have been running for more than a century [13]. However, they are inconvenient pieces of machinery that only generate hydrogen with a medium concentration and are not very versatile.

ii. **Electrolyzer with a Proton Exchange Membrane (PEM)**: Electrolyzers that use proton exchange membranes typically use solid polymer electrolytes. Applying electricity to a battery causes water to split into hydrogen and oxygen. Hydrogen gas is created on the cathode side when the hydrogen protons traverse the membrane. Their simple cooling and capacity to produce hydrogen of the greatest purity make them the most popular. Their small size and ability to produce very pure hydrogen make them ideal for use in environments where sources of clean energy are intermittent. On the other hand, using valuable metals as catalysts makes them a little more expensive [14].

iii. **Solid Oxide Electrolysis Cell (SOEC)**: Since they function at higher temperatures, SOECs may prove to be much more effective. High-temperature electrolysis (HTE) uses a solid ceramic as the electrolyte. The reaction between water in the cathode and electrons from the external circuit produces hydrogen gas and ions with negative electrical charges. Afterwards, oxygen passes through the movable ceramic membrane and reacts at the anode, creating gaseous oxygen and electrons for the external circuit. Compared to the previous generation, they are technologically behind [15]. While the mentioned electrolyzers are the most effective and economical options at the moment, there are more kinds with great growth potential. Photo electrolysis is one such process that uses sunlight to split water molecules instead of electricity. However, the semiconductors that are required by this device are not yet completely mature. The following are the positive aspects along with the drawbacks of electrolyzer technologies (Table 15.1):

15.1.5 Function and Types of Fuel Cells

The electrochemical elements known as fuel cells use the chemical power of fuel to generate electricity, with heat and water becoming the only by-products. When it comes to producing clean, sustainable energy, they are considered an exciting technology [16,17]. After that, all the different types of fuel cells are described in detail. Anode-to-cathode proton transfer in proton exchange membrane fuel cells (PEMFCs) is accomplished by means of a polymer electrolyte membrane. The low

TABLE 15.1
Positive Aspects Along with Drawbacks of Electrolyzer Technologies

Electrolyzer Type	Advantages	Disadvantages
Alkaline	Low cost, reliable	Unable to respond quickly to changing power, large footprint
PEM	Responds quickly to changing power, small footprint	High cost
Solid Oxide	High efficiency	Unable to respond quickly to changing power. Not at an industrial scale.

temperature of operation and high power density of PEMFCs make them ideal for portable and transportation-related uses. An impending innovation in PEMFCs is the effect of new materials, membrane electrode assemblies, and catalyst developments on PEMFC performance. One kind of fuel cell is the solid oxide fuel cell (SOFC), which moves oxygen ions between the cathode and anode using a solid oxide electrolyte. SOFCs are ideal for stationary power production applications due to their high efficiency and ability to operate at extreme temperatures. Innovations in SOFC technology, including new materials, cell designs, and system construction, and how these factors affect SOFC operation in recent years. An alkaline electrolyte is used by an AFC to move hydroxide ions from the cathode to the anode, making it a kind of fuel cell [18]. Being very efficient and compatible with a wide variety of fuels, AFCs are ideal for use in space and submarines. New materials, electrode architectures, and catalysts are all part of AFCs. A proton exchange membrane serves as the electrolyte in direct methanol fuel cells (DMFCs). Because of the abundance of energy and low operating temperature, DMFCs are ideal for mobility and portable uses. A special kind of fuel cell, known as a microbial fuel cell (MFC), converts organic materials into electricity by the use of microbes. Regardless of their diversity of fuels and poor power density, MFCs show promise for bio-energy generation and effluent treatment. Modern progress in MFCs, including new electrode designs, materials, and system integration, has an effect on their efficiency [20]. This review of the literature places an emphasis on the recent and promising developments in fuel cell technology and potential uses in the energy generation industry. To make them more accessible for commercial uses, we need further research to improve their effectiveness and reduce their prices.

15.1.6 The Importance of PEMFC

It is the proton-exchange membrane fuel cell (PEMFC), sometimes called the polymer electrolyte membrane fuel cell (PEMFC), that is mainly under development for fuel-cell use in transportation, stationary power sources, and portable devices [19]. In contrast to PEM electrolysis, which relies on electricity, PEMFCs produce electricity by operating on an asymmetrical concept. For fuel cell vehicles (FCVs), PEMFCs are the way to go because of how efficient they are and how well they work in vehicles. PEMFCs operate at a lower temperature than other types of fuel cells, which results in enhanced responsiveness and faster vehicle launch periods. Fuel cells of various types are also employed in fuel-cell automobiles, in addition to PEMFCs. Carbon monoxide, hydrogen, and methane are capable of directly generating electricity, necessitating higher temperatures for SOFCs. AFCs, which are highly efficient, were implemented in the initial prototypes of fuel cell cars. DMFCs and DEFCs eliminate the need for hydrogen storage by employing methanol and ethanol as fuels, respectively.

The article's concept focuses on the following objectives:

i. This article aims to do an extensive review of studies on viable, healthy, and productive transportation systems.

ii. This article aims to establish a distinctive framework for transportation for fuel cell electric vehicles and a bidirectional electrical grid link utilizing sustainable energy sources such as wind and solar.
iii. This article aims to power a water electrolysis station with renewable energy to produce green hydrogen for sophisticated fuel cell hydrogen electric vehicles.

Section 15.2 provides the origination of green energy-based hydrogen production systems. Section 15.3 provides a detailed infrastructure of green energy-based hydrogen production systems. Section 15.4 provides the current plot and future position of the renewable hydrogen and fuel cells market. Section 15.5 provides the economic challenges of renewable energy-based hydrogen production systems.

Section 15.6 provides scientific advancements to confront the challenges. Finally, the conclusion specifies the need to solve ecological threats through an amalgamated green energy-based hydrogen framework to enable a sustainable, low-carbon future.

15.2 ORIGINATION OF GREEN ENERGY-BASED HYDROGEN PRODUCTION SYSTEMS (GEHPS)

The need for clean and renewable energy sources is increasing as the globe progresses toward a green future. Hydrogen and other sources have become significant contributors to the shift toward a zero-carbon future. All three of these energy sources underscore their potential and possibilities as follows.

i. **Green Hydrogen**: Green hydrogen is generated through water electrolysis to obtain hydrogen and oxygen. The power used for electrolysis may originate from renewable resources (solar and wind), making green hydrogen a clean and sustainable energy carrier. The International Energy Agency observed that hydrogen requires governmental assistance similar to that which solar and wind energy received during their initial phases, since these sectors have significantly developed and expanded.
ii. **Solar Energy**: Solar energy is obtained by transforming sunlight into electrical energy using either PV panels or concentrated solar power systems (CSP) [20]. Solar energy has seen exponential expansion in recent decades and constitutes a substantial share of worldwide renewable energy generation. The declining prices of solar panel production and installation have made solar energy more economical, therefore enhancing its appeal for both residential and commercial uses.
iii. **Wind Energy**: Wind power is produced through capturing the physical energy of wind by turbines, which transform the motion of the wind into electricity. Wind energy has seen substantial expansion in recent decades, and its costs have significantly reduced, enhancing its competitiveness with other forms of energy. The generation of hydrogen by wind energy may emerge as a crucial element in a worldwide zero-carbon future, with the capacity for an increase in hydrogen production from existing levels.

Although hydrogen, solar, and wind energy are all renewable resources, they vary in their uses and prospects for future development. Hydrogen, as an energy carrier, has the benefit of facile storage and transmission, which is advantageous for extensive industrial applications and long-range energy conveyance. The creation of green hydrogen may use surplus electricity produced from solar and wind energy, thus serving as a supplementary technique to these renewable sources. Conversely, solar and wind energy serve as direct power generators and are more appropriate for decentralized and household uses. Solar energy is particularly appealing in countries with enough sunshine, while wind energy is more appropriate for areas with stable wind patterns.

Hydrogen, solar, and wind energy have distinct benefits and are essential for the worldwide shift toward sustainability. Green hydrogen serves as an energy carrier, enhancing solar and wind energy via storage and transit solutions. Concurrently, solar and wind energy continue to expand as principal sources of renewable power production. As technology and infrastructure advance, these three clean energy sources will assume a progressively significant role in reducing global carbon emissions and fostering a cleaner, healthier society.

a. GEHPS 1: Photovoltaic cell (PV) + Electrolyzer (Ely) + H2 tank ++ Fuel cell (FC)

   As shown in Figure 15.1, hydrogen serves as another resource that can be utilized during periods of unpredictable weather. In addition to using sunlight as its main clean energy source, the electrical system employs hydrogen generated by an electrolyzer and utilized by the fuel cell to generate power as a backup secondary substance for use during times of erratic weather and alternating daylight and darkness.

b. GEHPS 2: Wind Turbine (WT) + Electrolyzer (Ely) + H2tank + Fuel cell (FC)

   The speed of the wind is the principal source of renewable energy that is used by the energy network as depicted in Figure 15.2. In addition to using the velocity of the wind as its main renewable energy resource, the energy network employs hydrogen generated by an electrolyzer and utilized by the fuel cell to generate power as a backup secondary resource for use during times of erratic weather.

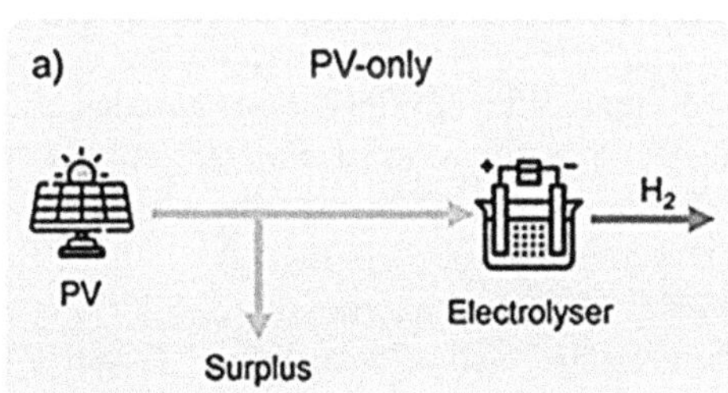

FIGURE 15.1 PV with electrolyzer.

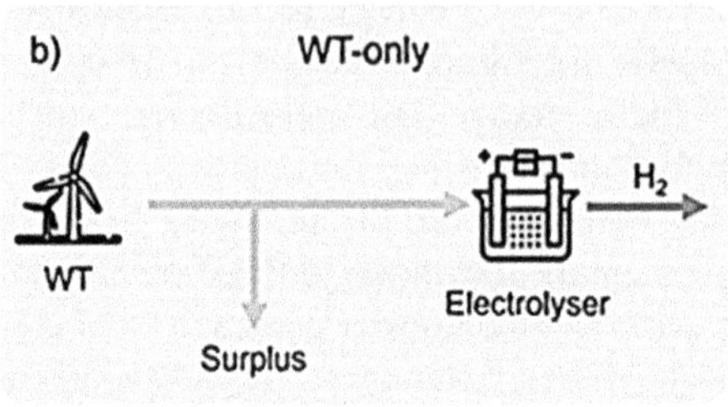

FIGURE 15.2 WT with electrolyzer.

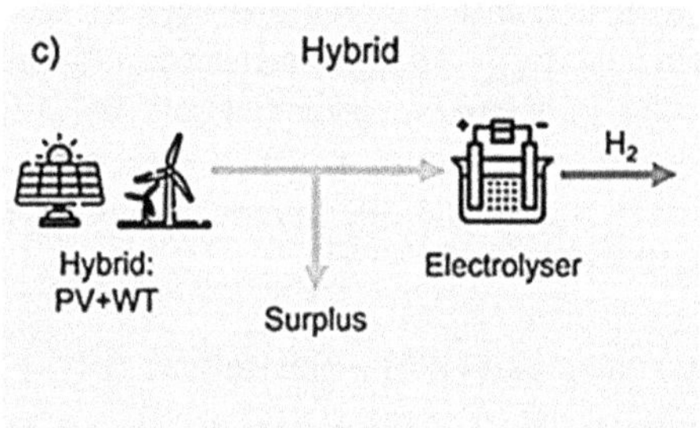

FIGURE 15.3 PV + WT with electrolyzer.

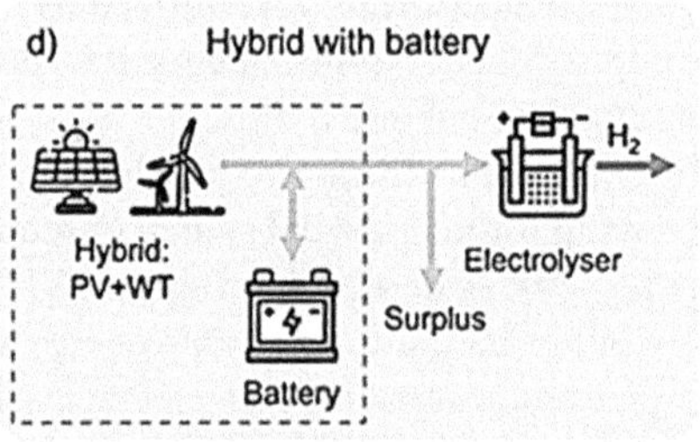

FIGURE 15.4 PV + WT + battery with electrolyzer.

c. GEHPS 3: Photovoltaic cell (PV) + Wind Turbine (WT) + Electrolyzer (Ely) + H2tank + Fuel cell (FC)

   As depicted in Figure 15.3, rays of the sun coupled with the force of the wind provide the primary and most sustainable source of power production for the energy system. Additionally, it makes use of hydrogen that is produced by an electrolyzer and then transferred it into the fuel cell in order to create energy during periods of weather that are not very common.

d. GEHPS 4: Photovoltaic cell (PV) + Wind Turbine (WT) + Battery + Electrolyzer (Ely) + H2tank + Fuel cell (FC)

   Solar energy, when paired with gusts of wind, constitutes the primary renewable energy source for power generation within the system. As per Figure 15.4, batteries, on the other hand, fulfil the role of a backup supply in the event that the weather is unpredictable. In order to create energy,

TABLE 15.2
Performances of GEHPS

Solar Hydrogen Production Systems Performances	Wind Hydrogen Systems of Production Performances	Solar/Wind Hydrogen Systems Performances
• The most popular technique for green hydrogen creation. • Less upkeep. • Expanding the hydrogen manufacturing. • Decreases the manufacturing cost.	• AC/DC converter is needed to charge the electrolysis. • High wind-velocity setup.	• High efficiency. • H_2 manufacturing is greater than wind/H_2 and PV/H_2 production.

hydrogen is first produced locally with the use of an electrolyzer, and then it is consumed by the fuel cell. Also, the performance of GEHPS is pointed out in Table 15.2.

15.2.1 Fundamental Model PEM Structure of Fuel Cell Vehicles Utilizing Green Hydrogen

A greater demand for environmentally friendly, long-term energy options with lower carbon footprints than conventional industrial chemical processes has arisen as a result of the fast development of hydrogen proton-exchange membrane fuel cells. It is of great technological importance to identify suitable alternative propulsion systems to substitute powertrains that depend on energy from fossil fuels. The utilization of hydrogen obtained from sporadic renewable sources, such as solar and wind power, and oxygen from the environment is expected to result in the release of greenhouse gases approaching zero in the green hydrogen economy, where fuel cells in electric vehicles are essential for low-carbon transportation [21]. Fuel cell transportation is deemed essential regarding the advancement of green hydrogen electric vehicles, recognized as a premier solution for contemporary transportation due to their benefits, including rapid refueling, robustness, less vibration, and pollution-free transport. Polymer electrolyte membrane fuel cell (PEMFC) has garnered interest due to benefits for automobile propulsion in innovative e-vehicles and its significant technological advancements over the decade [22]. Battery electric transportation may serve as an intermediary strategy leading to the eventual adoption of automotive hydrogen technology, or the market must focus solely on pure battery electric vehicles. Batteries function as an energy-saving component that accumulate electrical energy generated through various energy resources, while fuel cells act as energy transformation components that primarily utilize renewable hydrogen. Hydrogen has benefits over batteries due to its superior energy density and shorter refueling duration compared to lithium-ion systems [23,24]. Conventional electric hybrid car prototypes need the use of Li-ion batteries, acknowledging its limitations

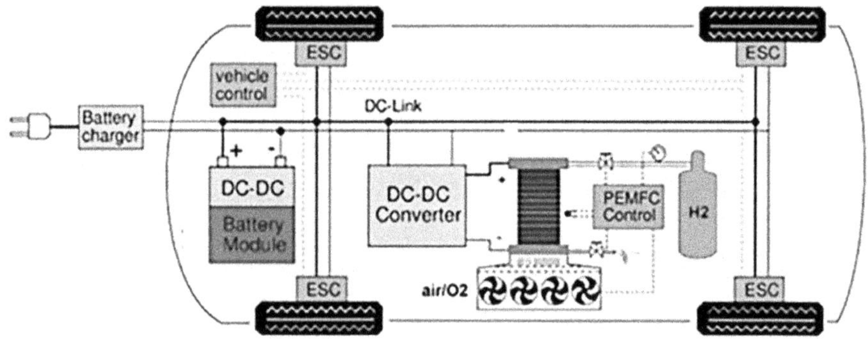

FIGURE 15.5 Elementary model of PEMFC vehicle.

regarding weight, cost, energy inefficiencies, finite lifespan, and environmental considerations. Electric transportation, integrating motors with hydrogen PEMFCs, is illustrated in Figure 15.5. The PEMFC is an uncontrolled energy reliant on several conditions, including temperature, humidity, and fuel requirement. Consequently, a converter unit (PCU) is required to interface with the loads. A DC-DC converter utilizes power from the source to optimize the task point of the fuel cell, ensuring the optimum voltage meets the requirements of the load under standard circumstances and transients.

15.3 DETAILED INFRASTRUCTURE OF GREEN ENERGY BASED HYDROGEN PRODUCTION SYSTEMS (GEHPS)

Wind/solar power serves as renewable resources for energy generation to facilitate electrolysis. Solar energy is utilized in photovoltaic structures, concentrated solar energy systems, and wind power to operate an electrolysis device. Converters are necessary for integrating the power of clean sources with the input of electrolysis. Instead of the substantial costs associated with energy transmission, non-conventional energy sources are utilized to deliver energy in distant areas [25]. The surplus power produced through renewable sources is used for electrolysis for hydrogen production. Electrolysis is energized through green energy sources. During periods of minimal sunlight, the battery is considered an energy-saving one. The PV with H2 method has numerous benefits compared to clean energy sources. The CSP/H2 approach utilizes thermal energy to operate electrolysis to convert water to steam. Thermal content at CSP/H2 facilitates uninterrupted hydrogen production. An investigation was done to assess the robustness under uniform working circumstances. Additionally, wind power is used to energize the electrolysis unit by delivering electrical energy via an AC/DC converter. Wind energy is accessible for the entire day, unlike solar energy, which is limited to daylight; nonetheless, it is an inconsistent energy source due to its inherent characteristics [26].

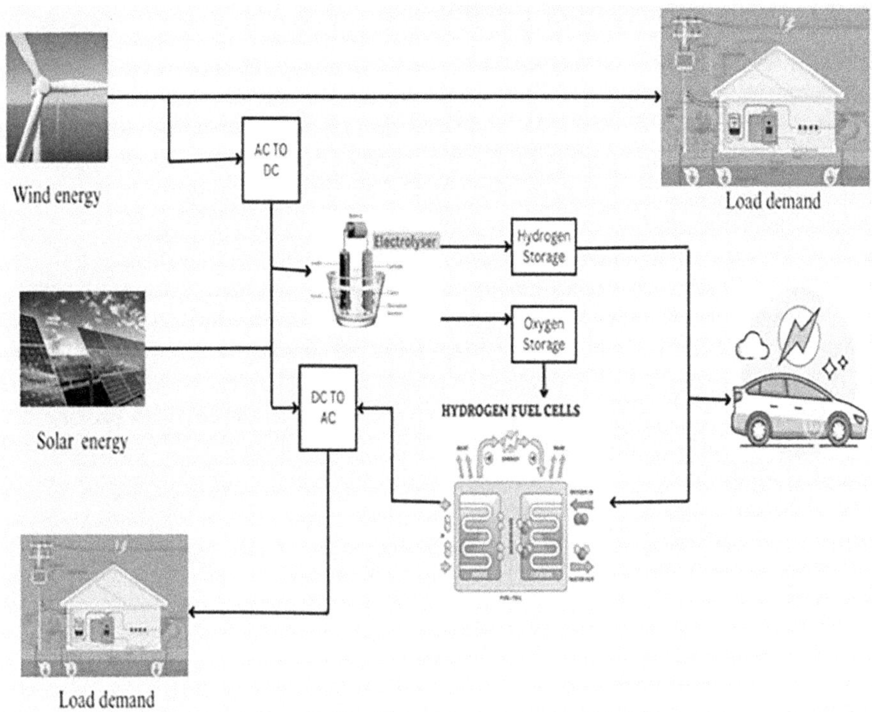

FIGURE 15.6 Detailed model of GEHPS.

Photovoltaic (PV) systems are particularly advantageous in distant locations for energy generation in the absence of a power grid. Furthermore, generating hydrogen from solar and wind energy to supply electrolysis approach is compelling, especially during periods of surplus the energy when hydrogen may be sold, stored, or transformed back into electrical energy by fuel cells. The PV/H2 network and the wind/H2 network are the components that make up the hybrid system, as shown in Figure 15.6. As a result of an increase in the amount of electricity that is supplied to the electrolysis section, the efficiency of the solar paired with wind and hydrogen method is improving. The following findings are shown as examples:

- Refueling automobiles at hydrogen power stations where hydrogen output is considered in good quantity.
- Incorporating batteries into the solar-wind/hydrogen approach facilitates uninterrupted operation and improves system efficacy.
- Hybrid system utility surpasses individual systems.
- Supplying electrical energy through fuel cells from the hydrogen production approach to residential properties.
- The hydrogen production system of solar and wind energy is beneficial for generating H2 as well as for electricity, cooling, heating, and desalination.

15.3.1 Application Cases of GEHPS

A hybrid system is shown in Figure 15.7, which consists of loads from both the home and the electric vehicle together with supplies of energy. The goal of a non-conventional energy system, which is comprised of a large number of household electric vehicles, is to achieve the optimal configuration of a variety of devices at the lowest possible cost while maintaining their performance. The devices, including PV, wind power, and H_2 fuel cells, are used to meet requirements. The objective is to meet the load demand through various renewable sources such as photovoltaic arrays, wind, and hydrogen fuel cells; any excess energy after satisfying the load is used for battery charging and for selling back to the grid. The storage battery linkage with the grid is used to fulfil the need when green resources are inadequate to meet aggregate needs for residential, along with electric vehicle loads. The grid will then cover the load requirement if the renewables cum battery reduce the strength to provide the required electricity at any stage. In the event of an emergency caused by natural disasters or power outages from all sources, the diesel generator will be activated to

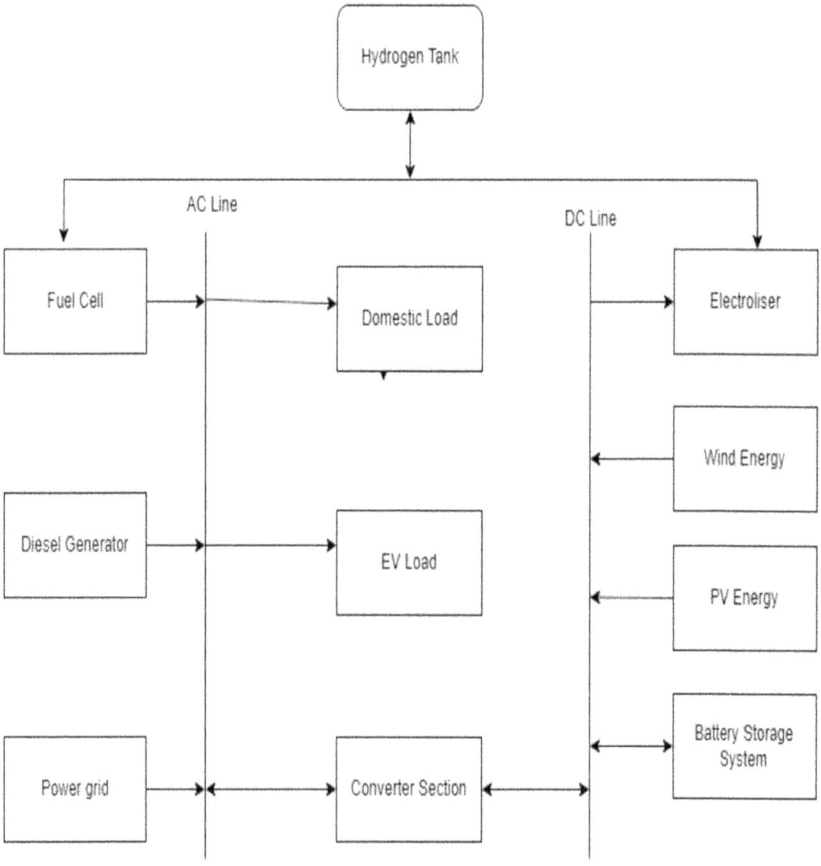

FIGURE 15.7 Application cases of GEHPS.

meet customer demand during such emergencies. The distribution of energy through various sources to the customer is regulated by the cycle charging dispatch strategy to maximize performance. In a dispatch strategy, a grid or generator is used to provide sufficient energy to meet the needs of loads, therefore protecting customers from the outage of power. Promising the security of customers in hybrid renewable energy entails guaranteeing dependable and effective energy consumption when mitigating threats and possibilities. In a diligent setup and renewable energy system, the reliability of customers and providers is enhanced through the amalgamation of technology, legislation, and best practices that guarantee dependable and sustainable energy generation and distribution.

Six distinct scenarios are given to understand the setup of the system and choose the best price-effective approach with minimum discharge. The following representations depict these scenarios:

i. **Scenario 1 (Baseline Scenario of a System Linked to Grid)**: This scenario pertains to grid connection, with energy supply relying entirely on the conventional power system.
ii. **Scenario 2 (Comprehensive Non-conventional Energy Integration with Hydrogen)**: In this scenario, all non-conventional sources, including sun and air, are integrated into the framework, with hydrogen incorporation as an auxiliary medium for storing energy. This arrangement enables the storage of surplus renewable energy as hydrogen for use during times of diminished energy output.
iii. **Scenario 3 (Renewable Connection Without Hydrogen)**: Analogous to scenario 2, renewable sources are included in the system, but hydrogen storage is omitted. The system depends exclusively on renewable sources for its power supply.
iv. **Scenario 4**: This situation resembles scenario 2, but lacks grid connection. The system functions autonomously, using renewable resources and hydrogen storage.
v. **Scenario 5**: This segment mirrors scenario 4, although it employs distinct operational tactics or setup to enhance the system's effect in the absence of grid connection.
vi. **Scenario 6 (Baseline Situation of Off-Grid Setup)**: This section resembles scenario 5, although with additional modifications to improve the system's effectiveness and functionality.

Escalating energy demand resulting from heightened computerization and less labor in the contemporary period necessitates the investigation of novel and alternate resources of energy, propelled by high technological progress. Wind power, hydrogen, solar power, and hybrid systems have developed as viable alternatives for electricity generation, especially in isolated areas, sometimes termed hinterlands. These systems provide significant promise and are anticipated to have extensive applications. A major part of the emphasis is on the generation and use of H_2 as a sustainable power supply. Hydrogen contains the capability of energy storage for the next generation's use and can be employed in many areas. Substituting traditional energy sources with this fuel enables the reduction of pollutants and the mitigation of the environmental effects linked to energy production and use. The incorporation of green sources results in robust energy expenses, employment generation, and financial benefits. Research professionals are urged to concentrate on the techno-economic dimensions of Green Energy-Based Infrastructure using sophisticated approaches for enhanced research. Also, solar/wind-based green hydrogen fuel cell systems are progressively emerging as high technologies, efficient and mitigating strain for the welfare of the environment. As their deployment escalates, prospective advantages transcend environmental sustainability, fostering comprehensive socioeconomic growth.

15.3.2 Comparison of Green-Hydrogen Energy System and Hybrid Renewable Energy System

The Green-Hydrogen Energy System and the Hybrid Renewable Energy System (Solar + Wind) are both environmentally friendly ways to generate energy, but they are used for different purposes and have different benefits. The two systems are compared below based on a number of factors.

Factors	Green Hydrogen Energy System	Hybrid Renewable Energy System
Energy Generation and Storage	i. Electrolyzers that separate water into hydrogen and oxygen are powered by renewable energy sources, namely wind and solar electricity. ii. The hydrogen is separated and stored for future use in fuel cells, turbine combustion, or other industrial processes. iii. Provides a means of storing energy for the future, which is helpful for use throughout different seasons.	i. Creates power by harnessing the energy of the solar and wind turbines. ii. When there is more power than people need, it is usually either sent to the grid or stored in batteries (such as flow batteries or lithium-ion ones). iii. Conveniently caters to domestic, business, and industrial needs.
Grid Dependence	i. Capable of delivering off-grid energy solutions, particularly for industrial applications and isolated areas. ii. Can facilitate grid stabilization by storing surplus renewable energy when supply surpasses demand.	i. It may function as either grid-connected or off-grid. ii. Grid-connected systems may need grid stability solutions, such as battery storage or hybrid integration with other sources.
Environmental Impact	i. Utilizes water as a feedstock, necessitating appropriate water management practices. ii. The storage and transportation of hydrogen may result in energy losses. iii. Hydrogen generation from renewable sources eliminates carbon emissions, in contrast to gray or blue hydrogen.	i. Minimal environmental effect during operating. ii. Manufacturing solar panels and wind turbines needs the extraction of resources and energy use. iii. The disposal of batteries and the supply of rare-earth materials present environmental challenges.

(Continued)

(*Continued*)

Factors	Green Hydrogen Energy System	Hybrid Renewable Energy System
Scalability and Future Prospects	i. Anticipated to expand as green hydrogen becomes more economical and efficient. ii. A crucial facilitator for profound decarbonization in industrial sectors and transportation. iii. Necessitates policy endorsement and financial investments for extensive implementation.	i. Already extensively implemented and capable of scaling. ii. Advancements in battery technology and smart grids will augment its efficiency.
Applications	i. Appropriate for sectors like transportation (trucks, ships, aircraft) and extensive energy storage. ii. Can substitute fossil fuels in electricity generation and industrial heating.	Optimal for direct electrical use.

Summary:

The Green Hydrogen Energy System offers superior long-term energy storage and decarbonization options in comparison to hybrid renewable energy systems.

In a sustainable future, the Green Hydrogen Energy System enables energy storage and the decarbonization of industry, while the Hybrid Renewable Energy System just powers the grid.

15.4 RENEWABLE HYDROGEN AND FUEL CELLS MARKET CURRENT PLOT AND FUTURE POSITION

15.4.1 By Type

The Air-Cooled Type was the dominant type in the Hydrogen Fuel Cells Market in 2024, commanding more than 56.4% of the market share, as shown in Figure 15.8. Air-cooled fuel cells are preferred because of their ease of use and reduced maintenance requirements, which make them suitable for situations where space and weight limitations are of the utmost importance. Providing cost-effective solutions for an extensive scope of industries, which includes transportation, these fuel cells make use of air as the oxidizer and are cooled by the air that is present in the surrounding environment.

In contrast, the Water-Cooled Type is another notable area within the Hydrogen Fuel Cells Market. Despite possessing a diminished market share in 2023, water-cooled fuel cells are recognized for their superior efficiency and enhanced performance under challenging operating circumstances. These fuel cells use water as a coolant, facilitating elevated power densities and extended operating lifespans, rendering them optimal for applications necessitating continuous and dependable power supply, such as backup power systems and grid-scale energy storage.

Both Air-Cooled and Water-Cooled fuel cells are integral to the Hydrogen Fuel Cells Market, addressing varied customer demands and application specifications [27,28]. Although Air-Cooled fuel cells prevail in the market owing to their simplicity

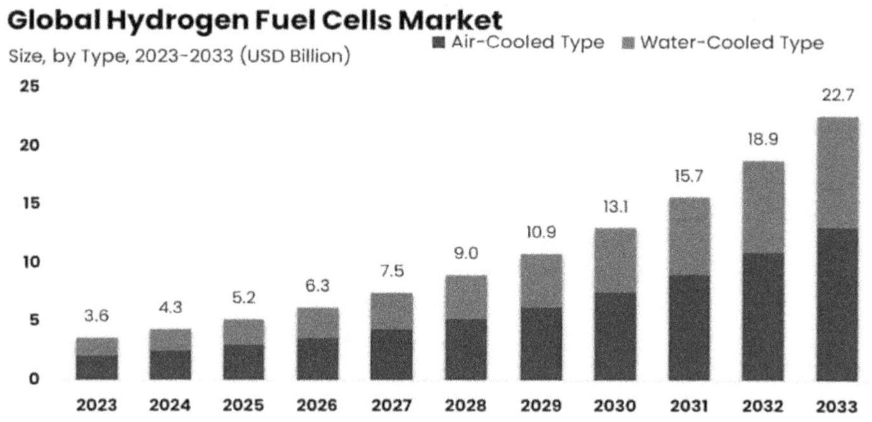

FIGURE 15.8 Hydrogen fuel cell market.

and cost-effectiveness, Water-Cooled fuel cells provide enhanced performance and efficiency, facilitating their utilization in essential applications where dependability and performance are crucial.

15.4.2 By Products

In 2024, proton exchange membrane (PEM) fuel cells led the hydrogen fuel cells market, with a market share of 45.6%. PEM fuel cells are distinguished by their high efficiency, fast start-up times, and adaptability, making them suitable for diverse applications such as automotive, portable electronics, and stationary power production [29]. Their capacity to function at low temperatures and provide clean energy with minimal emissions facilitates their extensive adoption and commercial supremacy.

Phosphoric acid fuel cells (PAFCs) constituted a substantial portion of the hydrogen fuel cells market. PAFCs provide exceptional efficiency and dependability, especially in stationary power production applications such as cogeneration systems and distributed energy networks [30]. Their advanced technology and proven efficacy provide them with ideal options for extensive power production projects, enhancing their significant market position.

Additionally, solid oxide fuel cells (SOFCs) have developed as a significant component within the hydrogen fuel cells market [31]. SOFCs function at elevated temperatures and may efficiently convert various fuels directly into electricity. They are used in stationary power production, integrated heat and power, and secondary power units for transport, demonstrating adaptability and promise for sustainable energy solutions.

Moreover, molten carbonate fuel cells (MCFCs) maintained a significant role in the hydrogen fuel cells market [32]. MCFCs function at high temperatures and may use fuels such as natural gas and biogas to generate electricity effectively. Their capacity to harness waste heat for combined heat and power applications improves their total energy efficiency, making them appropriate for grid-scale power generation and industrial uses.

15.4.3 By Application Analysis

In 2024, stationary applications prevailed in the hydrogen fuel cells market, capturing more than 45.3% of the market share [33]. Stationary fuel cells are widely used throughout residential, commercial, and industrial domains to provide dependable and clean energy for static power production, reserve power systems, and distribution of power. Their capacity for continuous and efficient operation, together with a low emissions profile, makes them ideal for energizing buildings, telecommunications towers, and essential infrastructure.

Transportation has emerged as a prominent sector in the hydrogen fuel cells market. Hydrogen fuel cell cars (FCVs) generate zero-emission mobility, with enhanced driving settings and less refilling time. FCVs are increasingly being used globally, particularly in heavy-duty sectors of buses, trucks, and trains, where electrification poses difficulties related to range and payload specifications.

Furthermore, portable applications have retained a significant foothold in the hydrogen fuel cells market. Portable fuel cells are tiny, lightweight, and provide on-demand power for many applications, including consumer electronics, military equipment, and off-grid power solutions. Their mobility and rapid refilling characteristics make them appropriate for isolated areas, outdoor pursuits, and crises when traditional power sources are inaccessible or unfeasible.

The hydrogen fuel cells market is propelled by several applications, each catering to distinct energy requirements and environmental issues. Although stationary fuel cells dominate the market due to their extensive use in stationary power production, transportation and portable applications provide substantial prospects for fuel cell technology to transform sustainable energy solutions in mobility and off-grid power sectors.

15.4.4 By End-User Examination

In 2024, FCVs became the predominant end-use sector in the hydrogen fuel cells market, securing more than a 43.3% market share. FCVs use hydrogen fuel cells, providing zero-emission transportation with extended driving ranges and reduced refueling durations relative to conventional engine cars. The heightened focus on diminishing carbon emissions and alleviating air pollution has led to the advent of fuel cell vehicles, especially in areas favoring sustainable transportation alternatives. Utilities constituted a substantial component of the hydrogen fuel cells market. Utilities use fuel cells for stationary power production, delivering dependable and efficient energy to residential, commercial, and industrial users. Fuel cell power plants provide grid stability, energy security, and low emissions, making them appealing alternatives for utilities aiming to incorporate renewable energy sources and diminish reliance on fossil fuels.

Furthermore, Defence maintained a significant presence in the Hydrogen Fuel Cells Market. Defence applications include military vehicles, remote power systems, and portable energy solutions, whereby fuel cells provide quiet operation, less heat signatures, and extended durability relative to conventional power sources. Fuel cell technology empowers military forces to improve operational efficiency while reducing logistical difficulties and environmental consequences in crucial missions.

TABLE 15.3
Parameters of Renewable Hydrogen Fuel Cell Market

S. No	Parameters of Renewable Hydrogen Fuel Cell Market
1.	**Market Trends** Several automakers were developing and introducing hydrogen fuel cell vehicles (FCVs). Unlike other battery-electric cars, these vehicles only produce water vapor and have greater driving distances.
2.	**Market Drivers** Many home sources of hydrogen allow one to lessen reliance on imported fossil fuels. This promotes nations' diversification of their energy sources and improves energy security.
3.	**Market Opportunities** To make an alternative to fossil fuels and lower carbon emissions, hydrogen finds use in industrial operations like chemical manufacture and steel fabrication. This offers opportunities to make energy-intensive businesses more environmentally friendly.
4.	**Market Challenges** Low energy density by volume of hydrogen makes storage and transportation difficult. Liquefying or compressing hydrogen calls both energy and specific infrastructure. Hydrogen may also seep through containment materials, which presents transportation and safety issues [35,36].
5.	**Market Restraints** Many areas lack the built infrastructure needed for hydrogen generation, delivery, and refueling. Widespread acceptance of this infrastructure is challenging as expanding it takes time and money.

Further applications of hydrogen fuel cells span several industries, including telecommunications, maritime, aviation, and off-grid electricity [34]. These applications use fuel cells for auxiliary power, remote monitoring systems, and alternative energy solutions, enhancing the adaptability and potential of fuel cell technology in meeting energy demands across many sectors (Table 15.3).

15.5 ECONOMIC CHALLENGES

Production of green hydrogen is costly and might persist without government backing and intervention. This is true as, considering distance and demand, reduced hydrogen production costs are seen in emerging nations with abundant natural resources for electricity generation. Even in places with plenty of renewable energy sources, electricity accounts for a major portion of generation costs; electrolyzers and other charges have little bearing. Authorities have to improve their energy budget to promote green hydrogen. For instance, renewable hydrogen fuel might be funded using the savings from lowering fossil fuel subsidies. However, rising production would need the development of hydrogen infrastructure, which is a big undertaking that requires a sound strategy and political support [30]. Thus, the government should cooperate with reputable companies in building green hydrogen infrastructure in order to design a strategic plan for the success of such infrastructure in the market and its expansion.

15.6 SCIENTIFIC ADVANCEMENT TO CONFRONT THE CHALLENGES

The industrialization of hydrogen production procedures that are fueled by sources of clean energy, along with the creation of markets and infrastructure, are all areas that need more development and research. There is a number of scientific advancements that have been established for future study in order to hasten the implementation of hydrogen and to construct systems that are integrated with hydrogen. Scientific advancement is required in the following factors.

i. **Advancement in Electrolyzer Development:** The following areas deserve to be the focus on prospective electrolyzer research: increasing the capacity of proton electrode membrane electrolyzer and alkaline electrolyzer, investigating catalysts and electrodes that do not contain platinum-group metals, creating hydrogen with high cleanliness from water and methanol, solving problems with safety in electrolysis, enhancing electrode/cell design and technologies for separating membranes, studying the principles underlying performance and resilience, and assessing the ecological effects and efficiency of electrolyzers [37,38].
ii. **Advancement in Restoring and Distributing Hydrogen:** Future investigations should focus on coming up with new ideas for large-scale above- and underground hydrogen buffer storage space to allow industrial processes to keep running, studying and choosing polymer components for cryogenic storage facilities, looking into novel substances for hydrogen storage, making liquid hydrogen tanks specifically for aircraft use, and using a variety of approaches for building the infrastructure needed to move hydrogen from places where it could be produced to places where it is needed [39–41].
iii. **Development of Fuel Cell:** Fuel cell technology is one of the most important areas of hydrogen technology. Researchers need to concentrate on finding the best materials and stack technologies for electrolysis and fuel cells, with the goal of making the systems work better and cost less overall. To deal with problems that are unique to this industry, such as power output and longevity, new fuel cell ideas that are made for heavy-duty cars are required. For long-term success, it is important to make fuel cell parts last longer and be easier to find, along with developing improved hydrogen cleaning technologies [42,43].

15.7 CONCLUSION

This work comprehensively addresses the development of combining green sources with green hydrogen framework for effective generation, storage, and consumption. It emphasizes technical advances such as better energy storage and electrolyzers. As the need to solve ecological threats increase, the requirement for an amalgamated green energy and green hydrogen framework to enable a sustainable, low-carbon future becomes evident. Even while there are still challenges to overcome, ongoing technological advancements, as well as increased governmental and societal support for green energy solutions, indicate a promising path for innovative techniques. The success of merging renewable sources (solar/wind energy) with green hydrogen infrastructure will eventually rely on collaborative efforts among various areas, including technical innovation, regulatory assistance, and public participation. As research advances and prices fall, amalgamated green energy systems have the strength to play a critical role in the worldwide energy evolution, notably continuing to the reduction of carbon emissions across diverse precincts and the development of a more tenable and irrepressible energy environment.

REFERENCES

1. Weger, L. B.; Leitao, J.; Lawrence, M. G. Expected impacts on greenhouse gas and air pollutant emissions due to a possible transition towards a hydrogen economy in German road transport. *International Journal of Hydrogen Energy* 2021, 46(7), pp. 5875–5890.
2. Herrmann, F.; Rothfuss, F. Introduction to hybrid electric vehicles, battery electric vehicles, and off-road electric vehicles. In *Advances in Battery Technologies for Electric Vehicles*, Bruno Scrosati; Jürgen Garche; and Werner Tillmetz (Eds.), Woodhead Publishing, 2015, pp. 3–16.
3. Yusaf, T.; Mahamude, A. S. F.; Kadirgama, K.; Ramasamy, D.; Farhana, K.; Dhahad, H. A.; Talib, A. R. A. Sustainable hydrogen energy in aviation–A narrative review. *International Journal of Hydrogen Energy* 2024, 52, pp. 1026–1045.
4. Farook, S.; Thrinath, B. S.; Muthukaruppasamy, S.; Ponmurugan, P.; Kannan, V. K.; Chand, J. P. An grid-integrated electric vehicles with hybrid energy storage for optimal power management. In *IEEE International Conference on Information Technology, Electronics and Intelligent Communication Systems (ICITEICS)*, Bangalore, India, IEEE, 2024, pp. 1–6.
5. Mrazova, M. Sustainable development-the key for green aviation. *INCAS Bulletin* 2014, 6(1), 109.
6. Galimova, T.; Fasihi, M.; Bogdanov, D.; Breyer, C. Impact of international transportation chains on cost of green e-hydrogen: Global cost of hydrogen and consequences for Germany and Finland. *Applied Energy* 2023, 347, 121369.
7. Nandha Gopal, J.; Muthukaruppasamy, S.; Rajagopal, V.; Thomas, G. A. S.; Parimalarsundar, E. Artificial *intelligence governance and comprehensibility in renewable energy systems.* In *Explainable AI (XAI) for Sustainable Development*, Chapman and Hall/CRC, 2023.
8. Parkinson, B.; Tabatabaei, M.; Upham, D. C.; Ballinger, B.; Greig, C.; Smart, S.; McFarland, E. Hydrogen production using methane: Techno-economics of decarbonizing fuels and chemicals. *International Journal of Hydrogen Energy* 2018, 43(5), pp. 2540–2555.
9. Noyan, O. F.; Hasan, M. M.; Pala, N. A global review of the hydrogen energy eco-system. *Energies* 2023, 16(3), 1484.
10. IEA. *The Future of Hydrogen*, Report Prepared by the IEA for the G20: Seizing Today's Opportunities, Japan, 2019
11. Wang, S.; Lu, A.; Zhong, C. J. Hydrogen production from water electrolysis: Role of catalysts. *Nano Convergence* 2021, 8(1), 4.
12. Brauns, J.; Turek, T. Alkaline water electrolysis powered by renewable energy: A review. *Processes* 2020, 8(2), 248.

13. Pathak, S.; Upadhyayula, S. A review on the development of supported non-noble metal catalysts for the endothermic high temperature sulfuric acid decomposition step in the Iodine–Sulfur cycle for hydrogen production. *International Journal of Hydrogen Energy* 2022, 47(31), 14186–14210.
14. Salvi, B. L.; Subramanian, K. A. Sustainable development of road transportation sector using hydrogen energy system. *Renewable and Sustainable Energy Reviews* 2015, 51, 1132–1155.
15. Karakaya, C.; Kee, R. J. Progress in the direct catalytic conversion of methane to fuels and chemicals. *Progress in Energy and Combustion Science* 2016, 55, 60–97.
16. Jaiswal, K. K.; Chowdhury, C. R.; Yadav, D.; Verma, R.; Dutta, S.; Jaiswal, K. S.; Karuppasamy, K. S. K. Renewable and sustainable clean energy development and impact on social, economic, and environmental health. *Energy Nexus* 2022, 7, 100118.
17. Muthukaruppasamy, S.; Arun Sampaul Thomas, G. Artificial intelligence and block-chain technologies for revolutionizing electric vehicles and charging infrastructure. In *A Sustainable Future with E-Mobility: Concepts, Challenges, and Implementations*, Lakshmi.D; Neelu Nagpal; Neelam Kassarwani; Vishnu Varthanan G. and Pierluigi Siano (Eds.), IGI Global, 2024, 130–149.
18. Kaur, M.; Pal, K. Review on hydrogen storage materials and methods from an electro-chemical viewpoint. *Journal of Energy Storage* 2019, 23, 234–249.
19. Islam, A.; Islam, T.; Mahmud, H.; Raihan, O.; Islam, M. S.; Marwani, H. M.; Awual, M. R. Accelerating the green hydrogen revolution: A comprehensive analysis of technological advancements and policy interventions. *International Journal of Hydrogen Energy* 2024, 67, 458–486.
20. da Costa, G. P.; Garcia, D. M.; Van Nguyen, T. H.; Lacharmoise, P.; Simão, C. D. Advancements in printed components for proton exchange membrane fuel cells: A comprehensive review. *International Journal of Hydrogen Energy* 2024, 69, 710–728.
21. Osman, A. I.; Mehta, N.; Elgarahy, A. M.; Hefny, M.; Al-Hinai, A.; Al-Muhtaseb, A. A. H.; Rooney, D. W. Hydrogen production, storage, utilisation and environmental impacts: A review. *Environmental Chemistry Letters* 2022, 20, 153–188.
22. Chakraborty, S.; Dash, S. K.; Elavarasan, R. M.; Kaur, A.; Elangovan, D.; Meraj, S. T.; Said, Z. Hydrogen energy as future of sustainable mobility. *Frontiers in Energy Research* 2022, 10, 893475.
23. Togun, H.; Aljibori, H. S. S.; Abed, A. M.; Biswas, N.; Alshamkhani, M. T.; Niyas, H.; Paul, D. A review on recent advances on improving fuel economy and performance of a fuel cell hybrid electric vehicle. *International Journal of Hydrogen Energy* 2024, 89, 22–47.
24. Arun Sampaul Thomas, G.; Muthukaruppasamy, S.; Saravanan, K.; Muleta, N. Revolutionizing smart grids with big data analytics: A case study on integrating renewable energy and predicting faults. In *Data Analytics for Smart Grids Applications-A Key to Smart City Development*, Devendra Kumar Sharma; Rohit Sharma; Gwanggil Jeon and Raghvendra Kumar (Eds.), Springer Nature Switzerland, 2023, pp. 179–198.
25. Onalaja, O.; Abdullahi, Y.; Taofeek, L. A. *Review of Hydrogen Fuel Cell Technology and Its Application*. Authorea Preprints, 2024.
26. Porté-Agel, F.; Bastankhah, M.; Shamsoddin, S. Wind-turbine and wind-farm flows: A review. *Boundary-layer Meteorology* 2020, 174(1), 1–59.
27. Qasem, N. A.; Abdulrahman, G. A. Recent comprehensive review of fuel cells: History, types, and applications. *International Journal of Energy Research* 2024, 2024(1), 7271748.
28. Horizon Data Book. *Fuel Cell Market Size & Trends*. https://www.grandviewresearch.com/industry-analysis/fuel-cell-market, 2024.

29. Ferguson, K.; Dubois, A.; Albrecht, K.; Braun, R. J. High performance protonic ceramic fuel cell systems for distributed power generation. *Energy Conversion and Management* 2021, 248, 114763.
30. Fragiacomo, P.; Piraino, F.; Genovese, M.; Corigliano, O.; De Lorenzo, G. Experimental activities on a hydrogen-powered solid oxide fuel cell system and guidelines for its implementation in aviation and maritime sectors. *Energies* 2023, 16(15), 5671.
31. McPhail, S. J.; Aarva, A.; Devianto, H.; Bove, R.; Moreno, A. SOFC and MCFC: Commonalities and opportunities for integrated research. *International Journal of Hydrogen Energy*, 2011, 36(16), 10337–10345.
32. Arun Sampaul Thomas, G.; Muthukaruppasamy, S.; Kumar, S. S.; Krishnan, S. A. Review of green IT, computing-enabled, sustainability-based cities: Nurturing living standards. *Green Computing for Sustainable Smart Cities,* Neha Sharma; Jai Prakash Verma; Sunil Gautam; Valentina Emilia Balas and Saravanan Krishnan (Eds.), CRC Press, 2024, 20–37.
33. Maia, B. *Dr Conor O'Brien Stationary. Fuel Cell Markets 2025–2035: Technologies, Players & Forecasts.* https://www.idtechex.com/en/research-report/stationary-fuel-cell-markets/1037, 2024.
34. Yue, M.; Lambert, H.; Pahon, E.; Roche, R.; Jemei, S.; Hissel, D. Hydrogen energy systems: A critical review of technologies, applications, trends and challenges. *Renewable and Sustainable Energy Reviews* 2021, 146, 111180.
35. Muthukaruppasamy, S.; Thomas, G. A. S.; Gopal, J. N.; Parimalasundar, E.; Duraipandy, P. Applications of machine learning/deep learning for power grid: Reliability and resilience of smart grid systems. In *Deep Learning in Engineering, Energy and Finance,* Vivek S. Sharma; Shubham Mahajan; Anand Nayyar and Amit Kant Pandit (Eds.), CRC Press, 2024, pp. 250–281.
36. Taibi, E.; Miranda, R.; Vanhoudt, W.; Winkel, T.; Lanoix, J. C.; Barth, F. *Hydrogen from Renewable Power: Technology outlook for the energy transition*, International Renewable Energy Agency (IRENA) Report- Abu Dhabi, UAE, p. 52, 2018.
37. Zhang, S.; Liu, Y.; Zhang, M.; Ma, Y.; Hu, J.; Qu, Y. Sustainable production of hydrogen with high purity from methanol and water at low temperatures. *Nature Communications* 2022, 13(1), 5527.
38. Zhang, J.; Zhang, L.; Liu, J.; Zhong, C.; Tu, Y.; Li, P.; Cui, Z. OH spectator at intermetallic narrowing activity gap between alkaline and acidic hydrogen evolution reaction. *Nature Communications* 2022, 13(1), 5497.
39. Barth, R. R.; Simmons, K. L.; San Marchi, C. W. *Polymers for Hydrogen Infrastructure and Vehicle Fuel Systems* (No. SAND2013–8904). Sandia National Lab. (SNL-CA), Livermore, CA; Pacific Northwest National Laboratory, Richland, WA, 2013.
40. Azadnia, A. H.; McDaid, C.; Andwari, A. M.; Hosseini, S. E. Green hydrogen supply chain risk analysis: A European hard-to-abate sectors perspective. *Renewable and Sustainable Energy Reviews* 2023, 182, 113371.
41. Deivendran, P.; Ilaiyaraja, S.; Selvakanmani, S.; Raghuram, K. S. Scalability and security requirements for the Internet of Things architecture. In *Artificial Intelligence for Internet of Things*, Thillaiarasu, N; Suman Lata Tripathi and V. Dhinakaran (Eds.), CRC Press, 2022, pp. 109–147.
42. Xin, H. Catalyst design with machine learning. *Nature Energy* 2022, 7(9), 790–791.
43. Zhai, S.; Xie, H.; Cui, P.; Guan, D.; Wang, J.; Zhao, S.; Ni, M. A combined ionic Lewis acid descriptor and machine-learning approach to prediction of efficient oxygen reduction electrodes for ceramic fuel cells. *Nature Energy* 2022, 7(9), 866–875.

16 IIoT and Digital Twin Technologies to Improve Energy Efficiency and Lower Carbon Footprint

Thirunavukarasu Suryalakshmi and Gurumoorthy Yohanandhan Rajaa Vikhram

16.1 INTRODUCTION

The tire industry's carbon footprint depends upon the method of production, raw materials, the method of transportation, and the tire disposal method when their life comes to an end [1]. Production of energy, transportation, and industrial production are the most significant areas of the globe contributing to the carbon footprint [2]. The industrial sector is responsible for about 21% of total carbon emissions, with processes like manufacturing, energy production, and transportation relying heavily on fossil fuels. This makes reducing emissions in these areas essential for lessening the environmental impact and tackling climate change. The main sources of CO_2 in industry include electricity, natural gas, water gas, and oil, as shown in Figure 16.1. Since electricity generation is a significant contributor to carbon emissions, efforts are being focused here. There is increasing emphasis on sustainability,

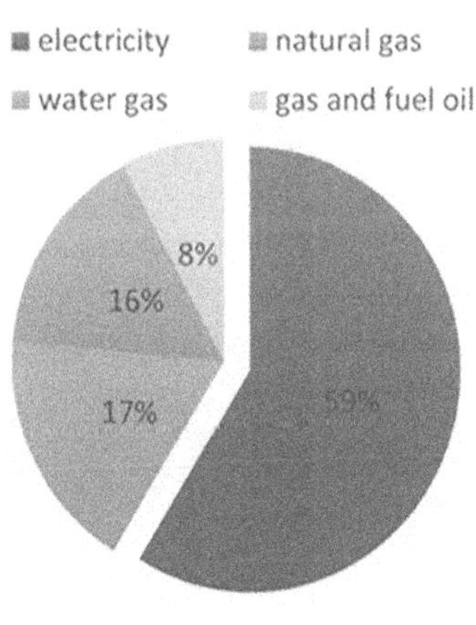

FIGURE 16.1 Carbon footprint percentage in industry.

DOI: 10.1201/9781003585534-16

and sustainable manufacturing practices are continuously evolving [3]. Through such a carbon-reduction approach, improved energy production can be obtained, and the 2050 target of becoming a carbon-neutral nation can be aided, decreasing the total amount of emitted carbon (Table 16.1) [4].

End-to-end management of equipment for energy consumption (EECM) forms the backbone of cutting down energy usage and boosting energy efficiency [5]. A number of technologies and techniques have been put forward to reduce carbon dioxide emissions via carbon capture, utilization, and storage (CCUS) [6]. Implementing energy management through the use of virtual models to mimic and simulate energy systems in real time helps to optimize, forecast performance, and detect inefficiencies in industrial manufacturing systems. This facilitates better energy conservation and reduced costs [7]. By combining data and conducting analytics, these techniques enhance sustainability as well as resilience in the industrial setting. Digital twin (DT) technology drives the process through the use of intelligent sensing, co-simulation, data analysis, and intelligent control technologies. Digital twins align virtual and physical product offerings through the use of data such as physical models, sensor data, and a history of operations [8]. The concept of a virtual digital twin of a real product, referred to as the DT, originated during the 2003 University of Michigan Executive Course on Product

TABLE 16.1
List of Acronyms

Definition	Acronyms
Artificial Intelligence	AI
Amazon Web Services	AWS
Carbon Capture, Utilize, and Storage	CCUS
Cubic Feet Per Minute	CFM
Comma Separated Values	CSV
Digital Twin	DT
Digital Twin Shop Floor	DTS
Equipment Energy Consumption	EECM
Function Structure Behavior Control Intelligence Performance	FSBCIP
Greenhouse Gas	GHG
Heating, Ventilation, and Air Conditioning	HVAC
Industrial Internet of Things	IIoT
kilowatt hour	kWh
Liter Per Minute	LPM
Long Short-Term Memory	LSTM
Normal Cubic Meter	NM^3
Node Java Script	NODEJS
Programming Logic Control	PLC
Product Lifecycle Management	PLM
Smart Manufacturing System	SMS
Sparse Autoencoder–Support Vector Machine	SSAE-SVM
Damadics Actuator Benchmark Library	DABLib

Lifecycle Management (PLM). In DT-based intelligent factories, the technology finds a strong use case when it comes to addressing problems such as the influence of equipment factors such as lifecycle management as well as data transmission delays on the energy usage of the physical world [9]. DTs are capable of conducting semi-physical simulation techniques, which help save money and a lot of time through the minimization of the need for physical reconfiguration as well as commissioning. This can be attributed to the fact that DTs enable the early identification of problems, design flaws, and defects in the smart manufacturing system (SMS). Researchers have put forward a framework referred to as Function-Structure-Behavior-Control-Intelligence-Performance (FSB-CIP) to measure how DT technologies are interconnected and to support the advancement of SMS design [10]. Also, the technology can be applied to the medical sector [11,12], as it generates virtual replicas of physical systems, assets, or processes. In the medical sector, DT technology helps specialists simulate events, process data, and improve operations, such as enhanced treatment, better patient care, and improved industry efficiency.

16.1.1 Advantages of Using DTs in Smart Manufacturing

Enhanced Productivity and Efficiency: Through simulating manufacturing processes and pinpointing potential bottlenecks and inefficiencies, manufacturers can optimize their operations, thereby increasing productivity and minimizing waste as shown in Figure 16.2.

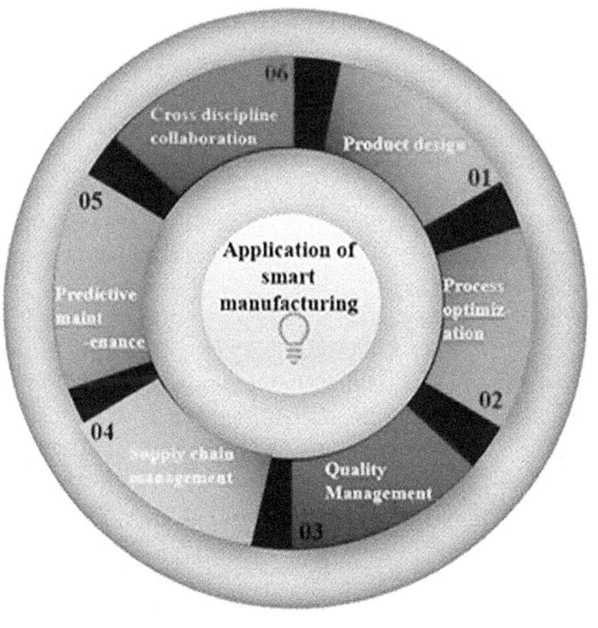

FIGURE 16.2 Smart manufacturing – Benefits.

Predictive Maintenance: Digital twins help keep track of how machines are working and spot problems early, before they cause breakdowns or expensive repairs. This allows maintenance to be planned ahead of time.

Improved Quality Control: Real-time monitoring of machines improves productivity and quality control.

Cost Reduction: Manufacturers can save on materials, energy, and labor costs by identifying optimization opportunities and reducing waste.

By 2025, DT are expected to be integrated into 89% of all Internet of Things (IoT) platforms. By 2027, they are likely to become a standard feature in IoT systems. Approximately 36% of executives across various industries recognize the benefits of DT, with nearly half planning to adopt them by 2028. The development of DT involves creating a virtual representation of devices and leveraging IoT capabilities by using sensors to collect data from connected physical devices [13]. DT is a key concept in the Fourth Industrial Revolution, gaining importance through technological advancements, particularly the Industrial Internet of Things (IIoT), which enables real-time data exchange. The progression of industrial revolutions [14] spans several decades. The First Industrial Revolution began around 1850, using water and steam power to mechanize production. The Second Industrial Revolution utilized electrical power for mass production. The Third Industrial Revolution integrated electronics and information technologies to automate production. The term Fourth Industrial Revolution was introduced by Klaus Schwab in 2015, characterized by advanced technologies like DT and IIoT.

The Fourth Industrial Revolution brings together different technologies, merging the physical, digital, and biological worlds. It focuses on using the IIoT and DT strategies to promote environmental sustainability in manufacturing. To improve energy efficiency in manufacturing systems [15], it is necessary to measure and analyze energy use in real time. This helps identify useful patterns and insights for each part of the manufacturing process. The rise of IIoT and DT has greatly improved monitoring systems, making them key parts of Industry 4.0 [16]. FlexSim provides a 3D DT model, as shown in Figure 16.3, allowing users to visually simulate different systems [17], such as manufacturing, supply chains, warehouses, and healthcare. It includes a library of ready-made objects and components, helping users easily build and customize their simulations.

The IIoT involves connecting devices and sensors to the internet within industrial processes to improve efficiency and productivity [18]. IIoT allows machines to communicate and share data in real time, enabling monitoring, analysis, and control of industrial operations. As digital transformation impacts all sectors, manufacturing is also evolving with IIoT [19], playing a key role in moving traditional industries toward Industry 4.0. The widespread use of IoT devices [20] generates large amounts of unstructured data. To make the most of this data, it is essential to integrate it with artificial intelligence (AI) techniques, such as machine learning, to support informed decision-making. In the future, organizations could also apply IoT, AI, and DT for smart education management [21]. Currently, the manufacturing industry is in the early stages of digital transformation, mainly using isolated solutions rather than full-scale digital integration. This limits the benefits of DT technology, which can

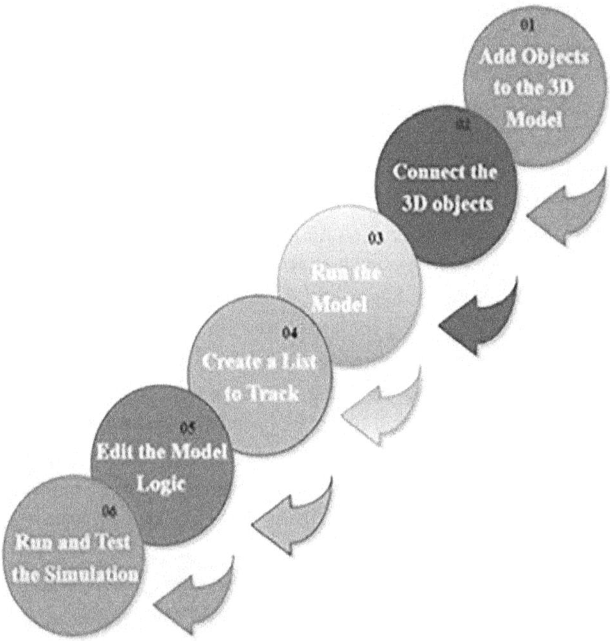

FIGURE 16.3 Procedures for constructing a FlexSim model.

improve maintenance, boost efficiency, and enhance decision-making [22–24]. The sustainability [25] of IoT platforms with multiple devices can be explored by studying sensing strategies across devices with different settings.

In many industries, IIoT is being used in compressed air systems for real-time monitoring, predictive maintenance, and optimization. Compressed air is widely used due to its convenience, safety, and clean nature, but it is also costly only 10%–30% of the energy consumed actually reaches the end use. Energy is lost through heat, leaks, pressure drops, and improper use. Energy-saving measures, such as lowering compressor pressure, reducing air inlet temperature, increasing storage capacity, recovering heat from compressors, and minimizing leaks, can save 20%–60% of energy. These strategies typically have a payback period of less than two years. This study focuses on pneumatic cylinders and air motors [26], which are vital in robotics and factory automation. They are popular due to their cost-effectiveness, simple maintenance, high force output, and easy installation.

With compressed air readily available in most industrial environments, pneumatic cylinders and valves are widely used in material handling, robotics, industrial processes, and food processing. Therefore, it is crucial to study simulation and control models for these actuators. Advancing techniques [26] to analyze the dynamic behavior of pneumatic actuators remains a priority. Rotary vane air motors utilize vanes to convert compressed air into mechanical energy, driving rotary motion. These compact and durable motors are commonly used in various industrial applications due to their simplicity and reliability. However, potential leakages in seals or vanes can impact efficiency, requiring regular maintenance for optimal performance. One major downside of air motors is their lower efficiency compared to electric motors. This inefficiency stems from the need for an electric motor to power the air compressor, which then supplies the air to drive the air motor. Throughout this process, energy is lost due to factors like air pressure drop in pipes, elbows, and fittings. Consequently, the overall system efficiency can be reduced to below 20% of that achieved by a direct-drive electric motor.

Air leakages [28] in pneumatic systems occur when compressed air unintentionally escapes due to issues such as worn seals, faulty connections, damaged parts, or poor maintenance. As shown in Figure 16.4, these leaks lead to higher energy use, reduced system performance, and increased operational costs. Identifying and fixing these leaks is crucial to maintaining the efficiency of pneumatic systems used in industrial and manufacturing settings. To assess and calculate compressed air losses, ultrasonic detectors are used to collect measurement data [29]. Air leaks can be caused by factors such as continuous vibrations from tools and compressors or thermal expansion due to seasonal temperature changes. Small maintenance investments to address these issues can significantly reduce energy consumption in pneumatic systems.

Energy losses within pneumatic systems can also result from leaks in the compressed air distribution network, incorrect settings in humidification systems, and inefficiencies in the movement of conditioned air within facilities. These energy losses [30] are often linked to poor equipment conditions, improper operational settings, and air leaks. Machine learning algorithms are used for fault diagnosis by analyzing data to detect abnormalities, helping to identify and resolve issues efficiently [29]. Unsupervised machine learning explores hidden patterns in unlabeled data, unlike supervised learning, which relies on labeled datasets. It organizes data based on natural patterns and similarities, using large amounts of publicly available unlabeled data. This method helps reduce data complexity, making it easier to perform tasks such as labeling. Unsupervised learning [31] provides initial insights, which experts can use to make informed decisions. DT technology is also applied for

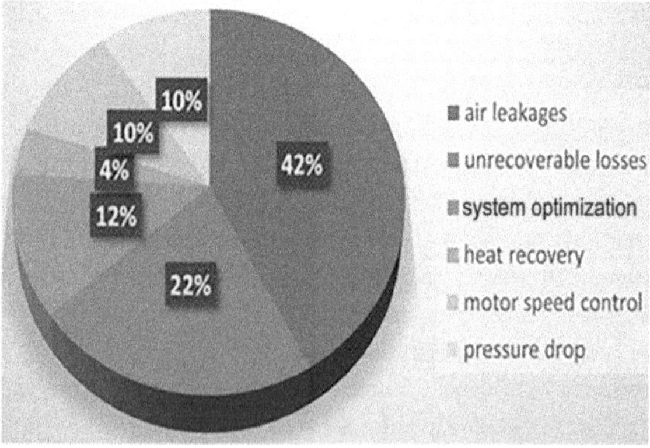

FIGURE 16.4 Factors affecting energy efficiency in pneumatic system.

fault detection and health management [32] across various devices, enhancing system reliability and efficiency. Machine learning has gained widespread success in various fields today, including fault diagnosis systems for reciprocating compressors in the petroleum industry. It enhances DT by enabling real-time data analysis, predictive modeling, and anomaly detection [24]. This improves system performance, predicts future behavior, and supports data-driven decision-making. By integrating machine learning, DT can adapt to real-world changes, increasing efficiency, reliability, and operational insights.

Several studies have explored fault diagnosis methods, including:

a. **Early leak detection** [33] to reduce energy waste in compressed air systems, supporting sustainable manufacturing.
b. **Rolling bearing fault diagnosis** using convolutional neural networks and transformers [34].
c. **SSAE-SVM model** for diesel engines operating in challenging conditions [35].
d. **Pneumatic system diagnosis** through extensive investigations [36].
e. Pneumatic actuators using adaptive multi-kernel multi-classification relevance vector machines with DABLib software [37].
f. **Bidirectional LSTM neural network** for diagnosing faults in pneumatic circuits [38].
g. **Leakage fault detection** in circuits with two parallel pneumatic cylinders using unsupervised machine learning techniques, such as Gaussian process classifiers, support vector machines, and k-nearest neighbor algorithms [39].

According to the relevant study works, the work at hand seeks to study air leakages while carrying out a belt cutter process that uses eight pneumatic cylinders and two air motors. Industry data are processed through the application of k-means clustering with the use of anomaly detection. Through the programming of the FlexSim software, a DT model of the pneumatic system is formulated to aid prediction and system behavior understanding. Through the use of machine learning and DT technology, the work at hand suggests an integrated framework to effectively regulate and optimize the use of pneumatic systems, thereby improving reliability and the performance of the belt cutter process.

16.2 PROPOSED WORK

The study seeks to marry IIoT and DT technologies in the tire manufacturing sector to improve energy efficiency and lower the tire manufacturing sector's carbon footprint. Sensors and devices that are interconnected will be utilized across the manufacturing process to allow real-time data capture. DT will be utilized to simulate and optimize operational conditions, guaranteeing effective resource usage. Semantic data [40] will further be utilized to create DT, facilitating the improvement of prognostic ability, which is a necessary capability for IIoT usage. Such a strategy forms an invaluable part of Industry 4.0 strategies, facilitating predictive maintenance as well as optimization of operations.

The main contributions of the work at hand are the following:

a. **Actual Data Analysis**: In the current study, energy usage as well as the minimization of the carbon footprint have been analyzed through IIoT technology.
b. **Machine Learning Model**: A k-means clustering algorithm with anomaly detection is put forward for the determination of faults in pneumatic systems.
c. **DT Model Development**: A DT model for the belt cutting process of tire manufacturing is developed and modeled through the implementation of FlexSim software.
d. **Usage and Impact**: The products of fault diagnosis and DT models are utilized during the manufacturing process of belt cutters, which enhances energy efficiency while reducing the carbon footprint.

16.2.1 Process Model

The block diagram of the process model is shown as Figure 16.5 wherein the flow meter senses the air uptake of the process, and the actuation of the pneumatic system is controlled by a PLC. Data gathering occurs through an IIoT gateway, which facilitates subsequent analysis and optimization of the process.

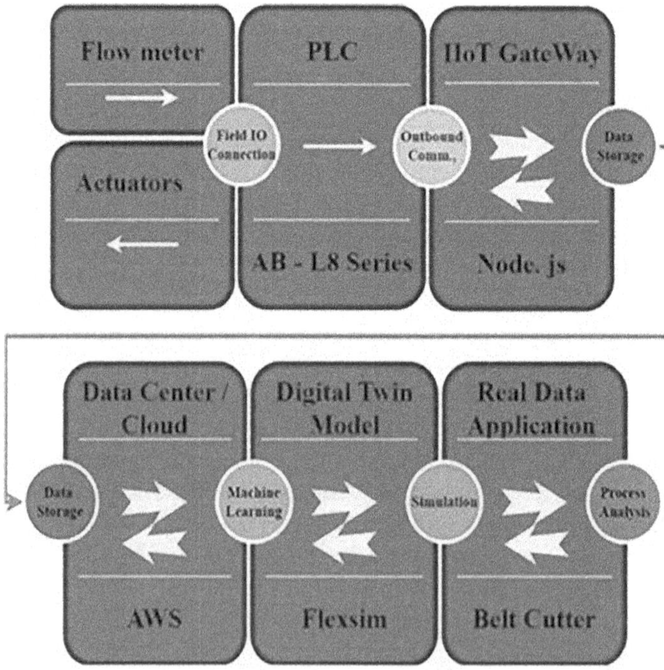

FIGURE 16.5 Block diagram of process model.

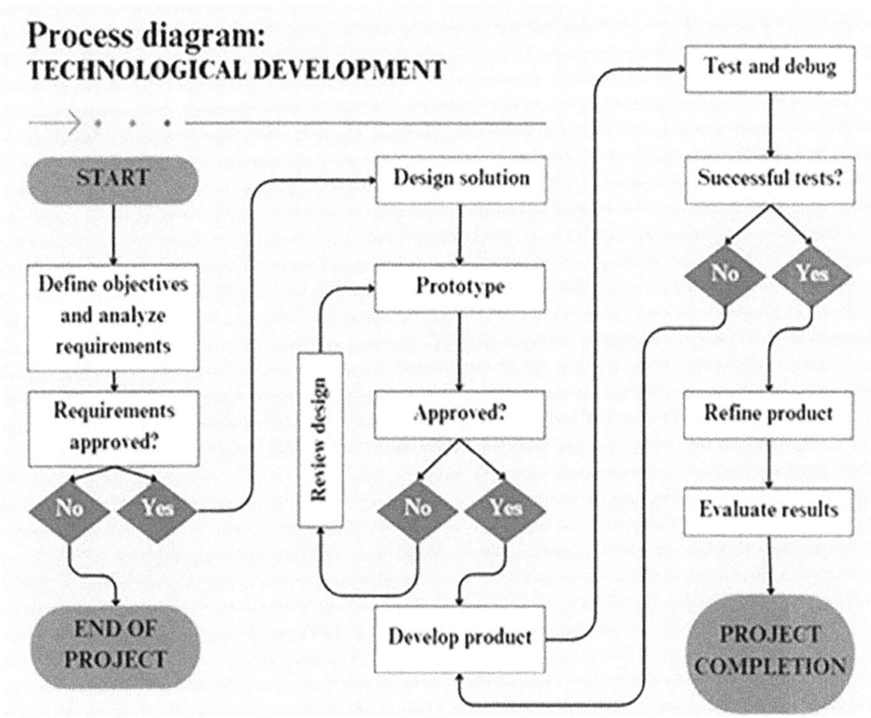

FIGURE 16.6 Flowchart for data analysis.

High-power computing DT services such as structural analysis, deep learning, multi-objective optimization, and higher-level decision-making encompass extensive computing. Such services are offered as Docker containers [18] on cloud servers. The data analysis flowchart, shown as Figure 16.6, exemplifies the integration of IIoT with machine learning algorithms. DT utilizes the data thus analyzed by such algorithms to improve knowledge, predict behavior, and optimize activities. Through such a continuous, repetitive process, the process facilitates continuous improvement as well as efficient decision-making in various industries.

16.2.2 Manufacturing Process in Tire Industry

Tire production process comprises several critical stages, as shown in Figure 16.7. Primarily, it starts with raw material preparation, which includes rubber, cloth, and chemicals. The raw materials are mixed and compounded into a mixture applied to make the tire tread, sidewalls, and inner lining. The tire itself is made by building up the plies, which are utilized for a variety of applications. After assembling the tire, it enters a mould for a curing process to achieve the finished shape. Subsequent to the curing process, the tire undergoes inspection and testing for quality and lifespan and then is prepared for shipment. Such a comprehensive process generates a superior-quality tire, which can accommodate a broad spectrum of vehicles.

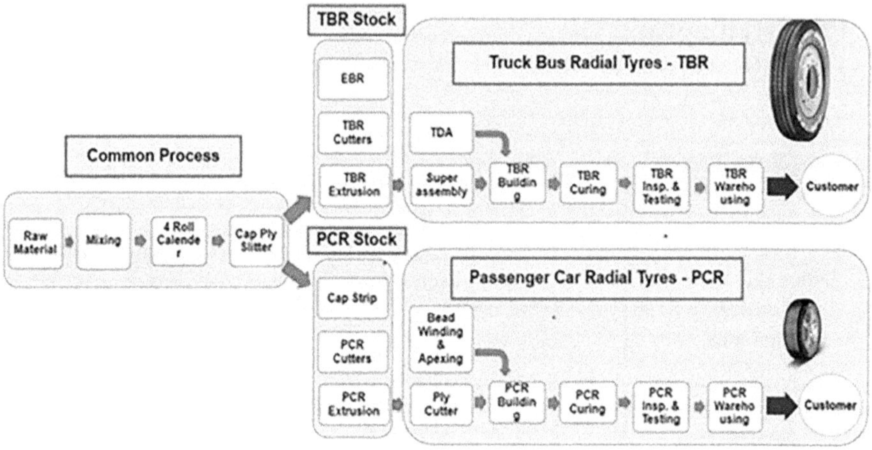

FIGURE 16.7 Steps in manufacturing process.

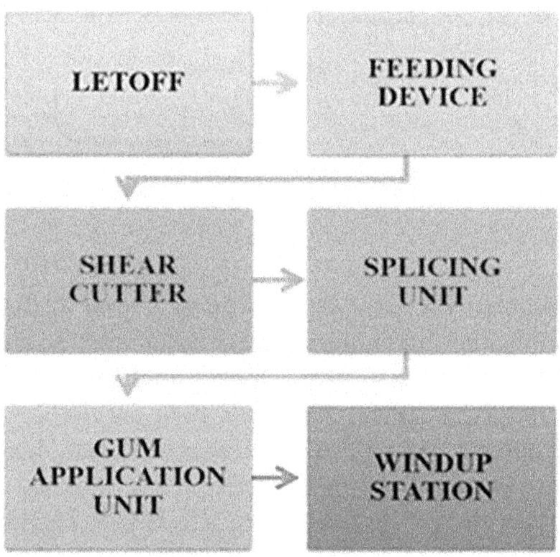

FIGURE 16.8 Belt cutter process flow.

16.2.3 Belt Cutter Process

In tire production, belt cutting is a major process, specifically for the manufacturing of steel-belted radial tires. Belt cutting involves cutting steel belts, which are used to reinforce the tire, to the desired shape and width. Figure 16.8 shows the process flow of belt cutting, while Figure 16.9 shows the tire manufacturing plant layout. Advanced machines and automation are used to cut rubber-coated cloth belts to the desired sizes accurately. It assists in attaining accurate layering, which improves the

FIGURE 16.9 Tire manufacturing plant.

TABLE 16.2
Vital Components Involved for Study

Si.No	Assembly	Sub Assembly	Capacity	Nos
1	Winding station 1 & 2	Pull in/pull out	16	2
2	Poly wind station	Material leading	9	1
3	Splicing unit	Right/left	6	1
4	Winding 1 & 2	Material leading	8	2
5	Pick and place unit	Material feeding	30	1
6	Two air motors	Poly film removal	200	2

strength, endurance, and performance of the tire. The belt cutting process uses eight cylinders and two air motors, with all the components playing major roles in enabling the system to work effectively. They are used to offer precision, efficiency, and reliability during cutting, as shown in Table 16.2.

16.3 DATA COLLECTION VIA IIoT

In an industrial setting, data gathering through IIoT involves the use of Node.js programming to deploy devices and sensors directly onto machines as well as processes. Such devices receive real-time data, such as pressure, temperature, and the efficiency of a machine. Such information is relayed through a network to a central system. Edge computing can process and filter the data onsite and then send it to a cloud platform like AWS for storage and analysis. It is important to use secure communication methods and maintain data integrity. Analytics tools then process this data to provide useful insights, helping with predictive maintenance, process optimization, and improving overall operational efficiency. Central control devices analyze the collected data to identify the most vulnerable parts of the system, focusing on key factors such as reliability, security, productivity, and specific functional requirements [27].

16.3.1 Electricity and Carbon Footprint Calculation

The measurement of carbon emissions related to electricity consumption is determined by the emission factor, which is calculated as0.85 kg of CO_2 per kilowatt-hour (kWh). The reference data is taken from CO_2 Emission sources, Factor Database, specifically version06, which is issued by the Central Electricity Authority (CEA) of the Government of India [46].

Figure 16.10 shows the steps for calculating the carbon footprint. To find the total carbon emissions (in kilograms of carbon dioxide), the amount of energy used (in kilowatt-hours per year) is multiplied by 0.85, which is the emission factor in India. Figures 16.11 and 16.12 display the total energy used and production levels at the tire manufacturing plant from 2020 to 2023. Table 16.3 shows how daily energy use is calculated based on the plant's daily production.

To account for a 20% leakage rate in the process, calculations are made to estimate the related energy consumption and carbon emissions. Through process optimization, it is possible to save 1.764 kWh of energy per day and reduce the carbon footprint by 13.137 metric tons annually, as shown in Table 16.3. By measuring energy losses and the resulting carbon emissions from leaks, organizations can identify inefficiencies and adopt strategies to minimize environmental impact and enhance resource efficiency.

Applying machine learning to pneumatic systems helps detect and predict air leakages. By analyzing data from sensors and actuators, machine learning models can identify patterns that indicate leaks, enabling proactive maintenance [41,44].

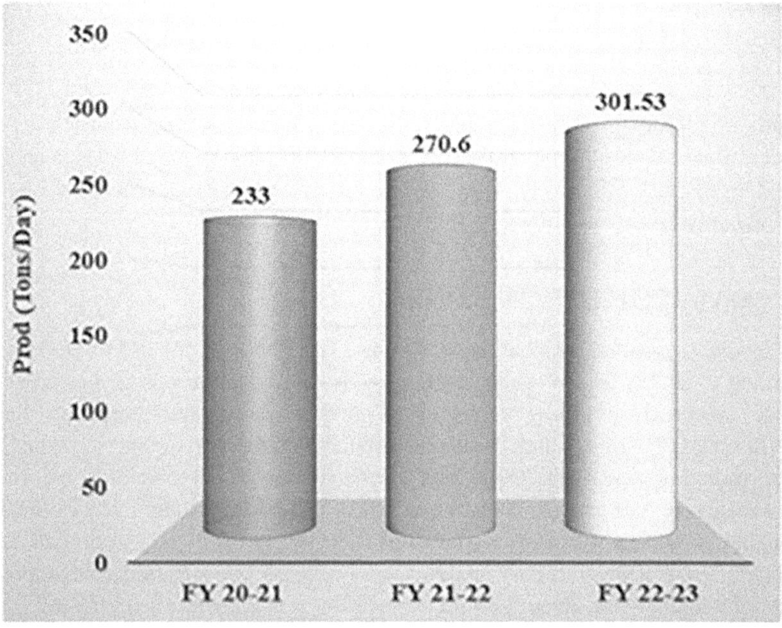

FIGURE 16.10 Plant production (in tons per day).

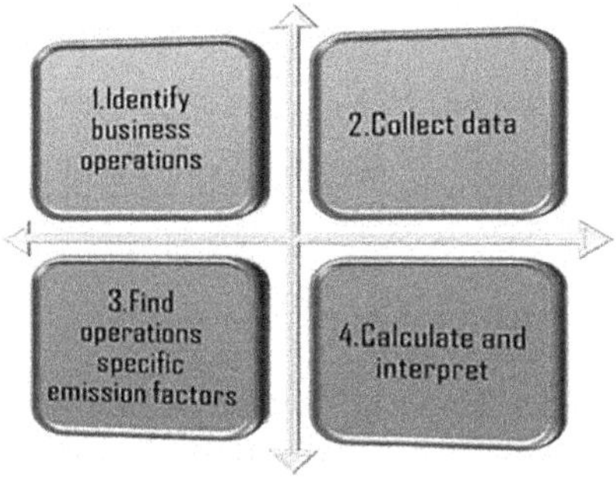

FIGURE 16.11 Steps for CO_2 emission calculation.

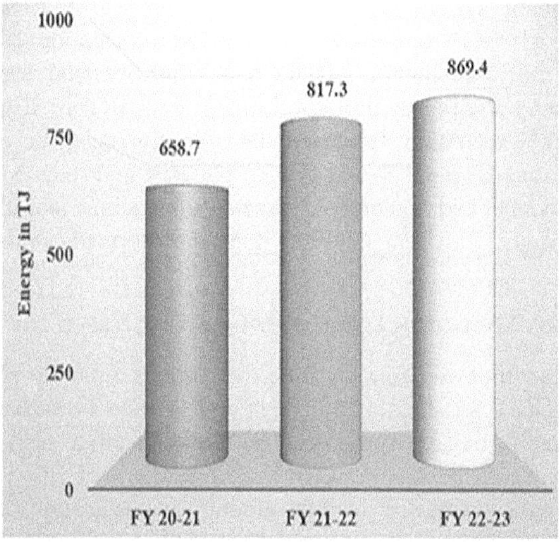

FIGURE 16.12 Plant absolute energy consumption (in TJ).

TABLE 16.3
Carbon Footprint Calculation

Si. No	Assembly	Sub Assembly	Capacity
1	Daily energy consumption	8.01 GJ/ton * 8 * 2.7778 kWh	2216 kWh
2	Plant production	301.53 tons per day * 2216 kWh	667,016 kWh
3	Daily carbon emissions	66,7016 kWh * 0.85 kg/kWh	566,963.6 kg

≈ 566 tons of carbon

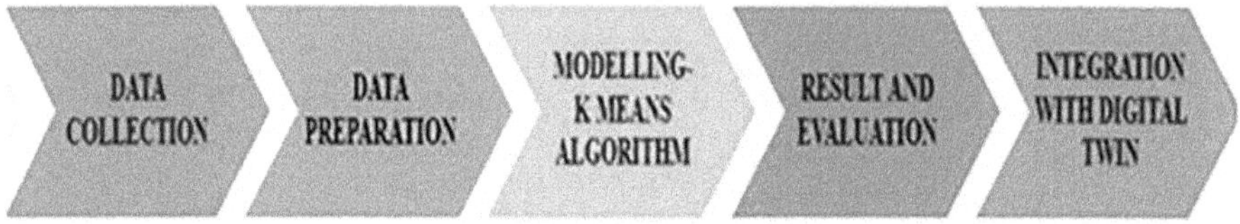

FIGURE 16.13 Stages in unsupervised learning.

At the same time, a DT, which is a virtual model of the pneumatic system, is created as shown in Figure 16.13. This digital model simulates real-time behavior, allowing for simulation-based analysis. Combining machine learning with the DT improves system efficiency by identifying potential leaks, optimizing energy consumption, and supporting predictive maintenance strategies, resulting in a more reliable and sustainable pneumatic system. Unsupervised machine learning uses algorithms to discover hidden patterns in datasets without labeled information. Unlike supervised learning, which relies on human-labeled data, unsupervised learning organizes unstructured data by finding natural similarities and patterns. It works with large amounts of publicly available unlabeled data. Analyzing these datasets through exploratory data analysis helps reveal hidden patterns and reduce data complexity, making it easier to label and organize. Unsupervised machine learning [31] approach provides initial insights into data, enabling domain experts to use them effectively.

16.3.2 Anomaly Detection Using K-Means Clustering

Clustering is a branch of machine learning that groups data into similar categories based on shared characteristics [42,45]. It is widely used in marketing to segment customers and identify behavior patterns. The main objectives of clustering are:

- Maximizing the separation distance among various groups.
- Minimizing the distance among members of the same group.

Clustering analysis becomes a powerful data exploration method used in all engineering domains, such as design, manufacturing, production scheduling, modeling, monitoring, quality control, and process optimization. Clustering analysis assists in information extraction about prices, identifying data patterns from the data given by a time series, and boosting the forecasting process. Anomaly detection, another name used to address the process of analyzing outliers, assists in the detection of unusual data points that differ from others. Anomaly detection is used fully through machine learning in various applications, such as unusual data set pattern detection. K-means clustering is a simple yet effective unsupervised learning method used for the process of detecting anomalies. It is a highly used clustering algorithm, partitioning the data into distinct groups through the categorization of samples into clusters with similar characteristics. It is used with unlabeled data sets and assumes a centroid for a group of clusters. Data is categorized into 'k' clusters, with 'k' representing the number of sets identified through the work defined by the user. The K-means algorithm undergoes an iterative process, clustering data into sets with common characteristics. 'K' is used to indicate the number of sets, e.g., K = 5 indicates set number 5, while set number 10 is indicated thrbyough K = 10. Clustering inertia is the principal target of the K-means algorithm, ensuring that data points within a set remain as close as possible to each other.

16.4 RESULTS AND DISCUSSION

The sequential activation of individual cylinders served as a diagnostic aid in identifying air leakage, which can be illustrated in Figure 16.14. By plotting the activation pattern, it assists in identifying any irregularities or deviations in the performance of the cylinders.

Let $D^{(i)(t)}$ denote the digital signals from the ith pneumatic actuator at time t, and $F(t)$ represent the flow meter readings in response to cylinder actuation. The recorded data is given by the equation:

$$\text{Recorded Data} = \Sigma\left[i = 1 \text{ to } M\right] D^{(i)(t)}, F(t) \text{ for } t = 1 \text{ to } T$$

A correlating plot of flow meter readings with time facilitates a comparative analysis of airflow across all eight cylinders and air motors, allowing for the identification of potential air leakage by comparing the actual flow values with their theoretical counterparts as depicted in Figure 16.15. It provides a visual insight into the performance of each cylinder over time, aiding in pinpointing discrepancies between observed and expected airflow. This analytical approach proves invaluable in understanding and diagnosing issues related to air leakage within the system, ensuring optimal efficiency and performance in the monitored cylinders.

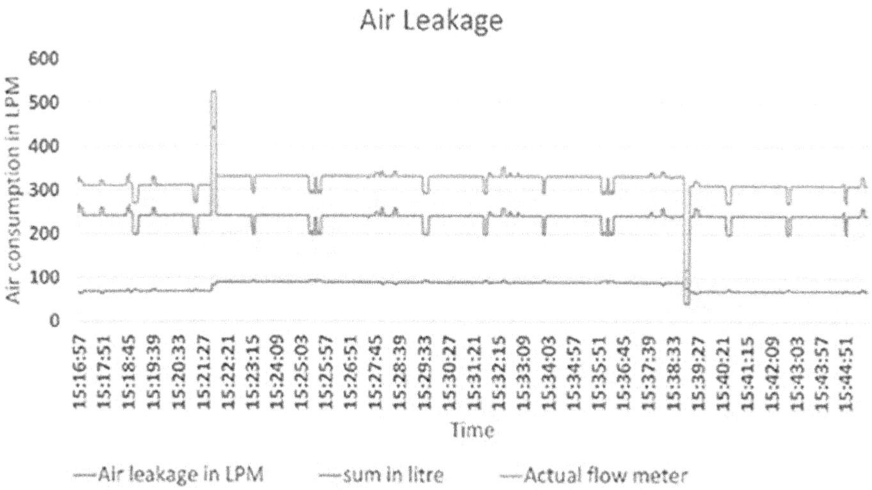

FIGURE 16.14 Process study with theoretical calculation.

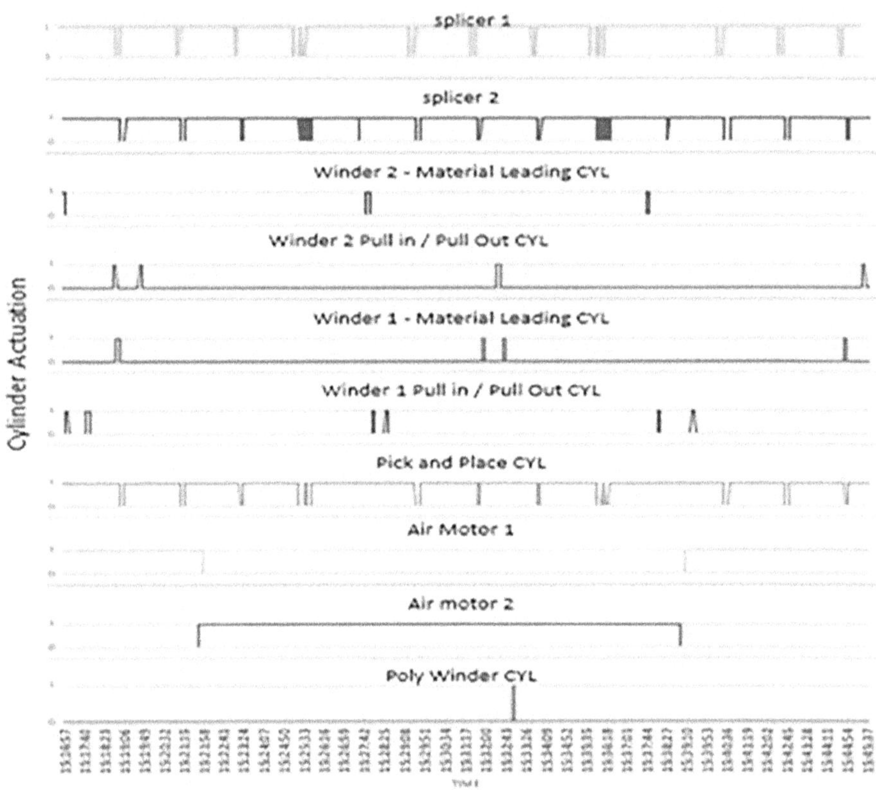

FIGURE 16.15 Data map of digital actuation of cylinder integrated with IIoT.

16.4.1 Fault Diagnosis Using Anomaly Detection Method

Unsupervised learning is applied to detect dissimilar data using the k-means clustering method. Jupyter Notebook software [43] is utilized for anomaly detection, leveraging Python code and interactive visualizations to analyze data and identify unusual patterns or outliers. It combines machine learning algorithms such as isolation forests or autoencoders to detect anomalies, extracting information about likely irregularities for the purposes of investigation. The procedure involves importing necessary libraries, e.g., k-means through scikit-learn for clustering, NumPy for numerical computation, and Matplotlib for data visualization. Visualization involves the use of pyplot. Data are separated into three clusters as 0, 1, and 2, which are represented as blue, yellow, and green colors, as depicted in Figure 16.16.

The anomalies along with the clusters are depicted as shown in Figure 16.17 through a scatter plot. Each of the clusters is labeled using different marker styles, whereas the anomalies are indicated with bold markers. The number of the clusters (n clusters) used while executing the algorithm of k-means is 3. An anomaly threshold is set (in this case, 2.5) to identify outliers.

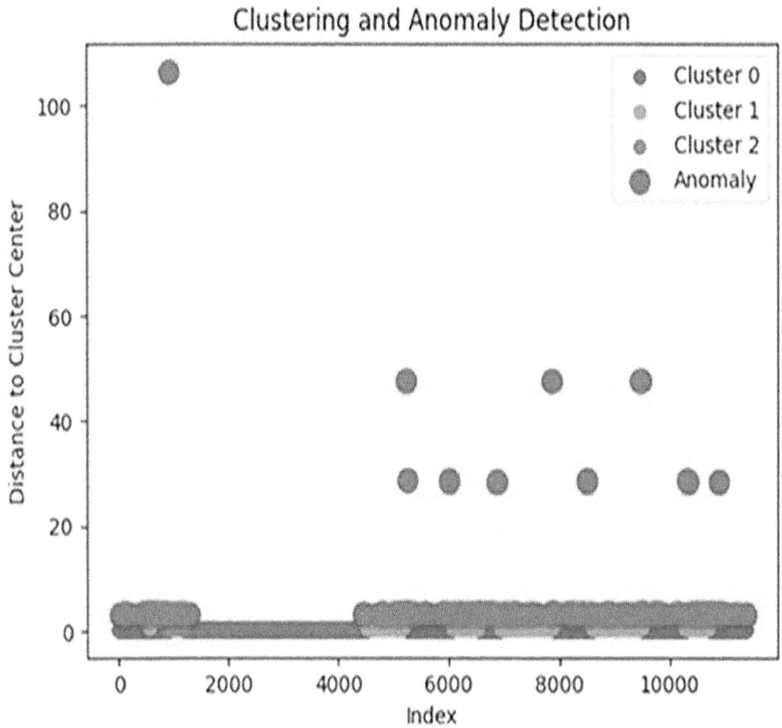

FIGURE 16.16 Scatter plot exhibiting three clusters.

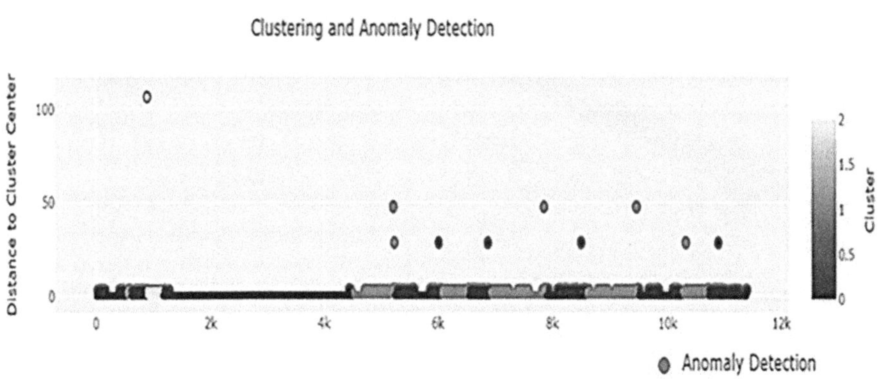

FIGURE 16.17 Visualization of clusters and anomalies in scatter plot.

A scatter plot is used to show clusters and anomalies, where each cluster is shown with points in different colors, and anomalies are marked in red. This helps to easily see the grouped data points and any outliers in the dataset. The red points clearly highlight unusual patterns, making it easier to understand the data structure and spot any irregularities. In this analysis, anomalies were identified based on their distance from the clusters. It was found that cylinder winder 1 pull-in and pull-out in the belt cutter process had more leakages, indicating that it was faulty. By detecting this issue and either fixing or replacing the faulty cylinder, air consumption was reduced. This improvement is demonstrated using a DT model and simulation results.

16.4.2 DT Model

Simulation modeling for process optimization and system performance prediction has made DT an essential technology in the manufacturing industry. In tire manufacturing, a DT model is created for the belt cutter process with pneumatic cylinders and air motors to check for air leakage and improve energy efficiency. The DT model for the belt cutter process is developed using FlexSim software, as shown in Figure 16.18.

The DT model includes a feeding device where the material is discharged from the station, followed by a splicing and gum application unit. After that, the material moves to the poly winding station. This model uses eight cylinders with different capacities and functions, as shown in Figure 16.19. The analysis focuses on air leakages in these cylinders by measuring total air consumption using a flowmeter. Each cylinder is operated one by one, and the air consumption is recorded until the end of the process. The actual total air consumption for the entire process, shown in

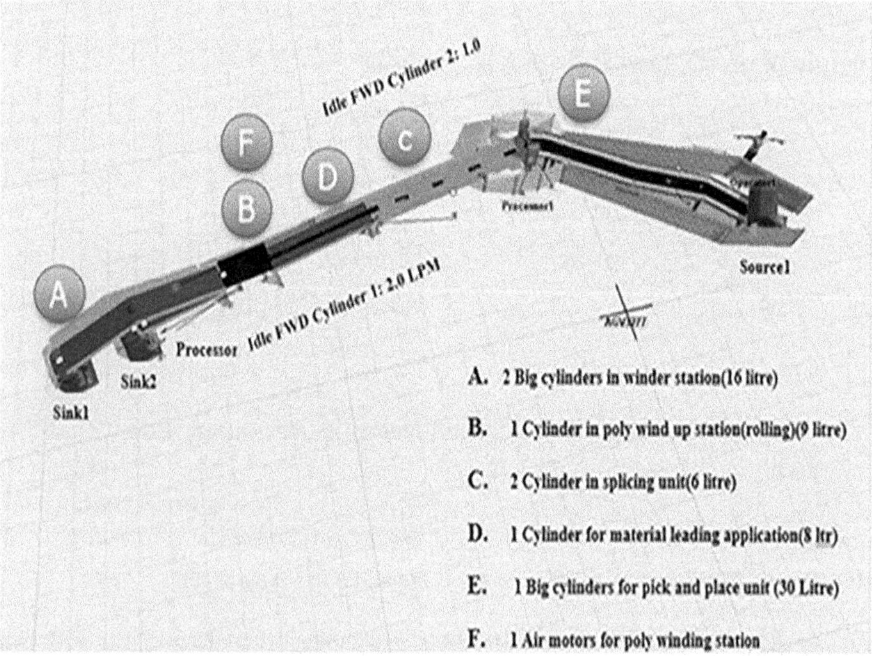

FIGURE 16.18 Utilizing digital twin (FLEXSIM) simulation for process optimization in belt cutter process.

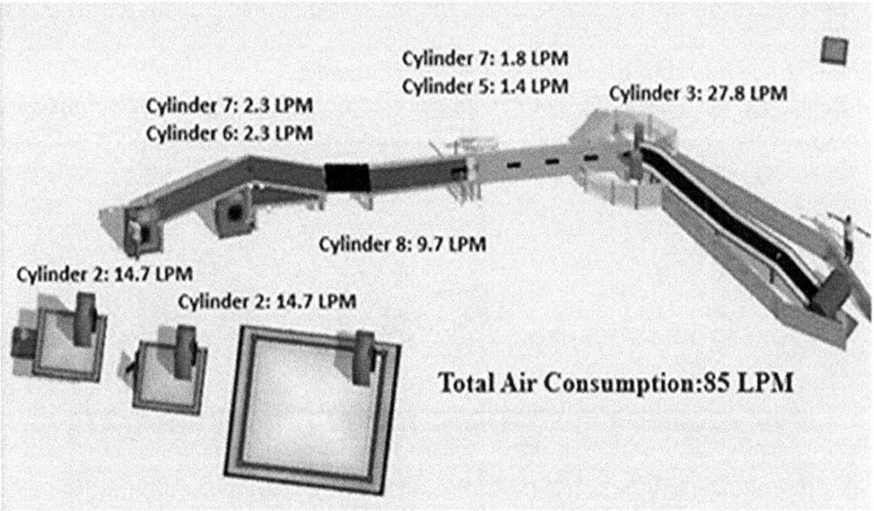

FIGURE 16.19 Position of cylinder and air motors in the digital twin model.

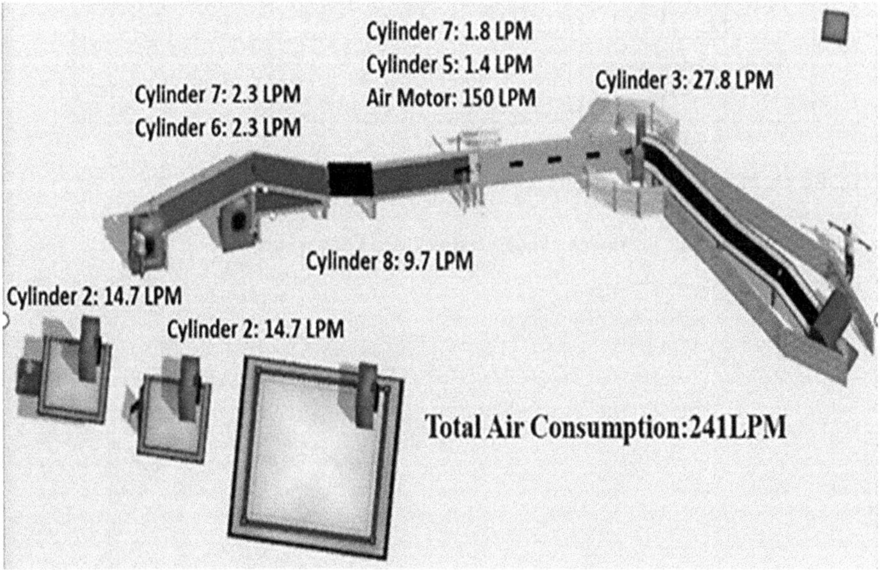

FIGURE 16.20 Process overview with total air consumption in simulation.

Figure 16.20, is compared with theoretical consumption, revealing and calculating air leakages. Through this simulation, it was found that cylinder winder 1 pull-in and pull-out in the belt cutter process had significant leakage. The analysis also examines two air motors used in the process as possible sources of air leakage.

By simulating the air motor alone, the air consumption is analyzed to check for leaks. Additionally, the model explores replacing these air motors with electric motors to see how this affects the system's performance.

Replacing the leaky air motor with an electric motor significantly improves energy savings, as shown in Figure 16.21. A comparison of air consumption with and without IIoT and DT is presented in Figures 16.22 and 16.23, showing a reduction

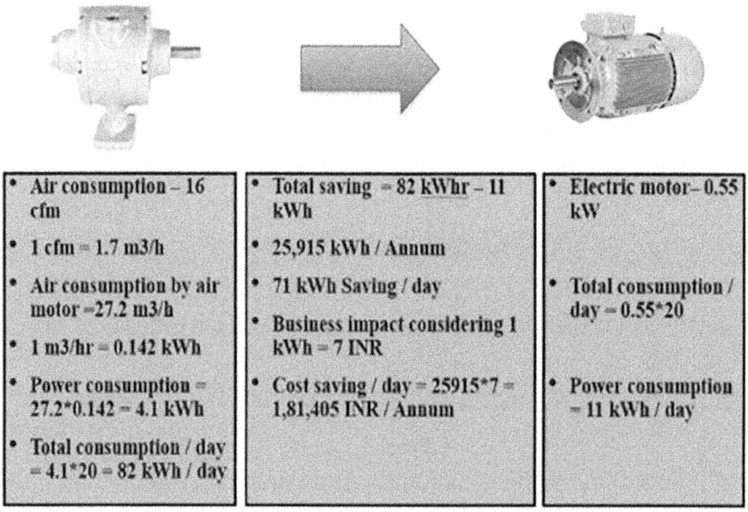

FIGURE 16.21 Energy saving by replacing electric motor.

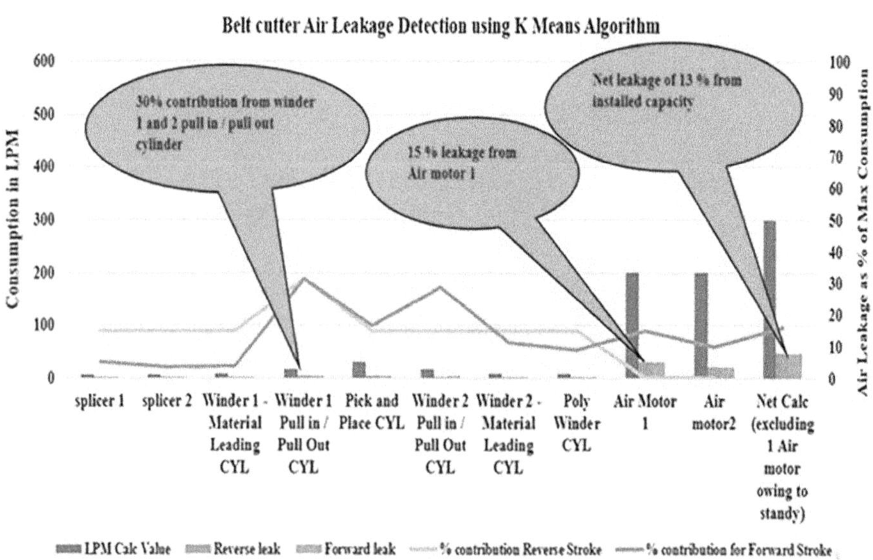

FIGURE 16.22 Air leakage estimation in the belt cutter process without IIoT and DT.

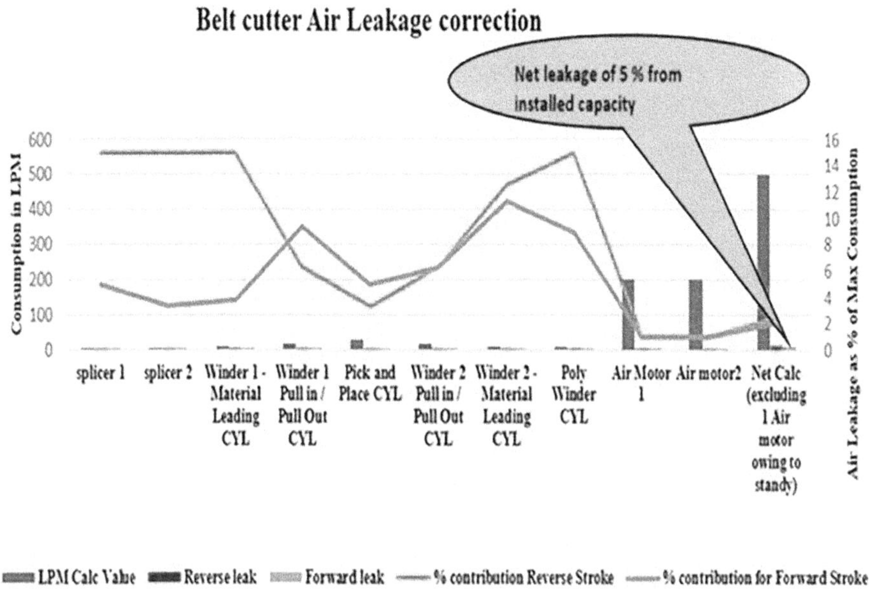

FIGURE 16.23 Air leakage estimation in the belt cutter process with IIoT and DT.

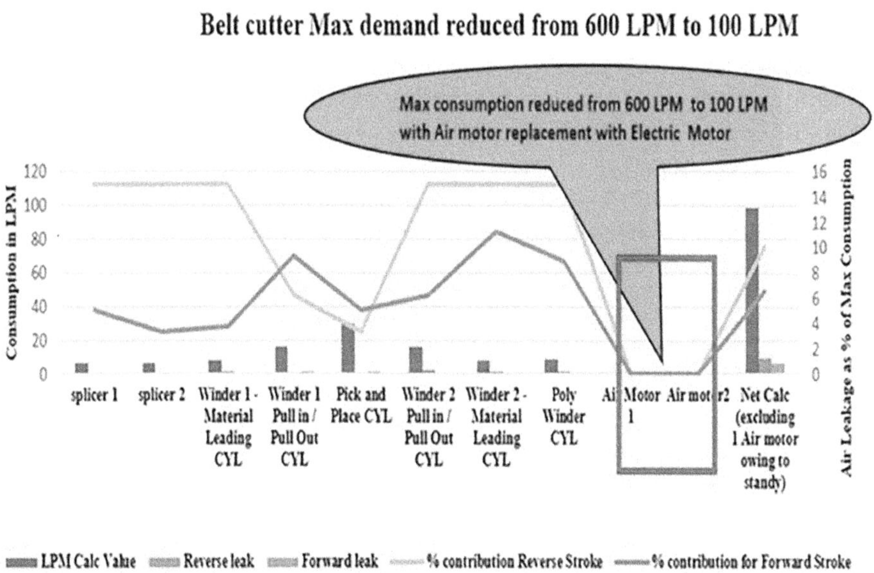

FIGURE 16.24 Reduction in air consumption using IIoT and DT (proposed work).

TABLE 16.4
Energy Efficiency and Carbon Footprint Reduction

Parameter	With Existing Work	With Proposed Work by Simulation
Total Energy Consumption Reduced in kWh	8%	67%
Calculated Air Saving in LPM with Max. Demand – 300 LPM (Table 16.2)	24	201
Reduction in Air Consumption for a Day in Nm^3/h	1.44	12.06
Reduction in Air Consumption for a Day in Nm^3/Day	34.56	289.44
Saving in kWh@0.18 Compressor Efficiency/ Day	6.22	52.09
kWh to CO2 Equivalent/Day	5.28	44.28
Reduction in CO2 per Annum in MT of CO2	1.93	16.16
Considering Only 70% Saving in MT of CO2 in Present and Proposed Work	1.35	11.31

in air leakage from 15% to 8%. This approach was implemented in the industry by replacing the faulty cylinder, resulting in noticeable energy savings.

The proposed idea of replacing the air motor with an electric motor could further reduce air consumption from 600 to 100 LPM, as shown in Figure 16.24, and is currently being considered by the plant. By analyzing air leakages and exploring the benefits of using electric motors, the model provides valuable insights for optimizing energy savings. This detailed analysis and simulation offer a useful framework for decision-making on system upgrades and enhancing energy efficiency. As shown in Table 16.1, the air savings in LPM for a system with a maximum demand of 300 LPM for air motors and pneumatic cylinders are 24 LPM for the implemented solution and 201 LPM for the proposed solution.

As shown in Table 16.4, reducing air leakage by 8% leads to a carbon footprint saving of 1.93 metric tons per year. Additionally, using DT modeling to replace air motors with electric motors can reduce average air consumption by 67%, resulting in a saving of 16.63 metric tons per year. Together, these measures save a total of 18.56 metric tons of carbon emissions annually, lowering the carbon footprint of the belt cutter from 24.13 metric tons to 8.04 metric tons per year. To ensure continued progress, it is important to track these savings yearly. Therefore, 70% of the total savings, equal to 12.66 metric tons per year, has been accounted for. By adopting the proposed method, the belt cutter process has achieved a notable reduction in carbon emissions, as shown in Figure 16.25.

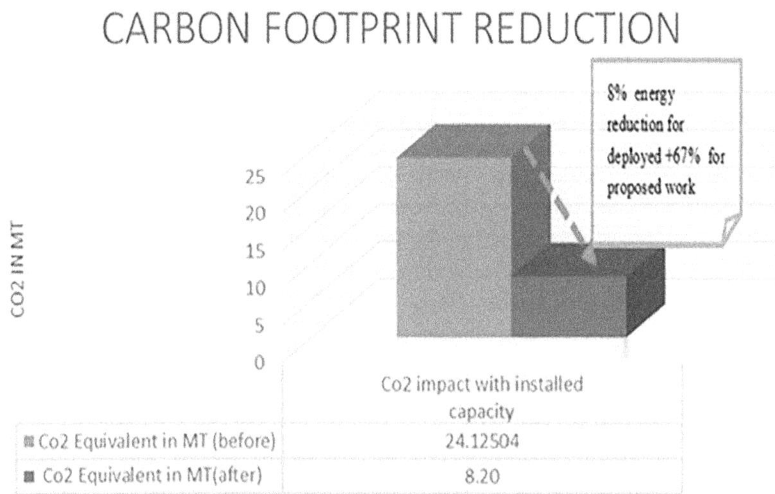

FIGURE 16.25 Reduction of CO_2 for present (existing) and proposed work.

Through detailed calculations, including air leakage adjustments, there has been an 8% decrease in average air consumption. In the future, integrating a DT could enable further design optimizations, targeting a 67% reduction in air consumption. This approach not only significantly reduces environmental impact but also demonstrates potential for ongoing improvements and sustainability in the process. The formulated solutions for the current work are thoroughly discussed, justifying the adoption of unsupervised learning via the k-means clustering algorithm for the detection of air leakage in pneumatic systems. Also, the adoption of a Digital Twin (DT) model for assessing cylinder actuation and replacing air motors with electric ones reveals a futuristic approach toward the enhancement of energy efficiency as well as the reduction of the amount of emitted carbon.

16.5 LIMITATIONS AND FUTURE WORK

Despite the significant promise the proposed solution demonstrates, various challenges and limitations must be addressed to enhance its practical relevance and workability in industrial practice. Among the limitations are data quality issues, accuracy of the model, limitations of implementation, as well as the absence of verification in the real world. A short description of these challenges is as follows:

16.5.1 Data Quality and Availability

The model, as proposed, depends on sample datasets, which were artificially created through a controlled setup. Data quality limitations, e.g., inconsistency, noise, and the lack of adequate historical data, could influence the accuracy and generalizability of the model. Also, data preprocessing, noise elimination, and real-time updating of the DT model, specifically with long-term or annually changing data, are areas yet to be investigated. The work on the existing solution, as yet, is confined to the process of validating it, and work remains to alleviate these issues.

16.5.2 Model Performance and Scalability

Although the k-means algorithm effectively separates faulty and healthy cylinders, the robustness of the model for varied working conditions (e.g., variation in pressure, variation in temperature, or variation in other external parameters) has been insufficiently tested. Additionally, the scalability of the model to other manufacturing lines or other pneumatic systems has not been explored. The solution's generalizability to different industrial applications requires further exploration.

16.5.3 Implementation Constraints

The integration of IIoT sensor devices with current systems, as well as data transfer guaranteeing security, were revealed as significant practical challenges that were not fully explored. Also, while the air motor versus electric motor substitution was explored thoroughly, probable system compatibility, dynamic changes, and system operation adjustment needs following motor substitution were not sufficiently explored through this research. Additional studies could elaborate on these limitations of deployment further.

16.5.4 Real-World Validation and Reliability

Results obtained here are primarily via simulation tests conducted within controlled conditions. Though the simulation yields considerable information, experimental verification under various operating conditions enhances the results' robustness and credibility. Moreover, the ability of the DT model to effectively emulate the inherent intricate behaviors of the pneumatic systems, e.g., system dynamics in real time, hasn't been thoroughly evaluated. Future work should involve real-case tests as well as DT model optimization to enable the model to effectively perform during industrial operations.

16.6 CONCLUSION

In the current study, an air leakage detection system for the classification of a group of cylinders using publicly accessible datasets uses an unsupervised learning *k*-means algorithm. Distinguishing and differentiating cylinders with air leaks and well-functioning cylinders, the model runs effectively. Apart from improving operational efficiency, the method facilitates the effective use of proactive maintenance, which increases the lifespan as well as the reliability of industrial cylinders. In the current work, air leakage through the use of anomaly detection as a fault diagnosis aid occurs in the belt cutter process's pneumatic system. As a result, faulty cylinders were repaired, while others were replaced. Several benefits were thus documented, such as 6.22 kWh daily energy conservation as well as a 1.35 metric ton carbon reduction. Through the use of a belt cutter process-based digital twin (DT) model, the analysis of the air motor operation as well as the movement of cylinders was facilitated. An alternative model was also created through the use of electric motors instead of the air motor. Through simulation, an energy-saving amount of 52.09 kWh was achieved, thereby decreasing the number of carbon emissions to 11 metric tons. This shows the project's relevance in contributing to cost-saving benefits as well as environmental benefits. Such concepts instituted through the current work can extend to use across the industrial sector as well as in fields such as the health sector, robotics, cities, the oil and gas sector, and schools. Digital transformation across these fields remains a stimulant for economic growth as well as societal advancement.

ACKNOWLEDGEMENT

This research work is supported by JK Tyres & Industries. I extend my sincere thanks to Mr. Bharathithasan, Electrical and Instrumentation Head at JK Tyre and Industries, for his invaluable guidance and support. His expertise and leadership have greatly contributed to my learning experience during my time at the organization. I am extremely grateful to all the staff who worked alongside me at JK Tyre and Industries for their patience, openness, and friendly working environment provided to me during the project.

REFERENCES

1. J. González-Gonzâlez, S. García-Méndez, F. De Arriba- Pérez, F. J. Gonzâlez-Castaño, and Barba-Seara, "Explainable automatic industrial carbon footprint estimation from bank transaction classification using natural language processing," *IEEE Access*, vol. 10, 2022, pp. 126 326–126 338.
2. P. Wadke, V. Gonal, D. Watwani, P. Chavan, and S. Sahu, "Carbon footprint: Causes, impacts and sector-wise survey," in *2023 5th Biennial International Conference on Nascent Technologies in Engineering (ICNTE)*, Navi Mumbai, India 2023, pp. 1–6.
3. I. Stepanovskaya, "Reduction of carbon footprint: Digital management strategies," in *2022 15th International Conference Management of Large-Scale System Development (MLSD)*, Moscow, Russian Federation 2022, pp. 1–5.
4. H. Han, J. Park, S. Kim, S. Lee, M.-I. Choi, and S. Park, "Carbon reduction method for intelligent energy transformation based on energy data analysis," in *2022 IEEE 5th Student Conference on Electric Machines and Systems (SCEMS)*, Busan, Republic of Korea, November 2022, pp. 1–6.
5. F. Lima, L. K. B. Y. Nonogaki, J. Chang, and A. A. Massote, "Estimation of energy consumption in manufacturing lines using machine learning into industry 4.0 context," *2022 Portland International Conference on Management of Engineering and Technology (PICMET)*, Portland, OR, USA, 2022, pp. 1–7.
6. H. Han, S. Kim, S. Choi, J. Park, T. Lee, S. Park, and S. Park, "Optimal energy use ratio adjustment simulation to reduce carbon emissions," *2022 IEEE/IAS 58th Industrial and Commercial Power Systems Technical Conference (I&CPS)*, Las Vegas, NV, USA, 2022, pp. 1–5.
7. W. Yu, P. Patros, B. Young, E. Klinac, and T. G. Walmsley, "Energy digital twin technology for industrial energy management: Classification, challenges and future," *Renewable and Sustainable Energy Reviews*, vol. 161, 2022, p. 112407.
8. J. Liu, X. Cao, H. Zhou, L. Li, X. Liu, P. Zhao, and J. Dong, "A digital twin-driven approach towards trace- ability and dynamic control for processing quality," *Advanced Engineering Informatics*, vol. 50, 2021, p. 101395.

9. M. Soori, B. Arezoo, and R. Dastres, "Digital twin for smart manufacturing, a review," *Sustainable Manufacturing and Service Economics*, vol.2, 2023, p. 100017.
10. J. Leng, D. Wang, W. Shen, X. Li, Q. Liu, and X. Chen, "Digital twins-based smart manufacturing system de- sign in industry 4.0: A review," *Journal of manufacturing Systems*, vol. 60, 2021, pp. 119–137.
11. G. Poletti, L. Antonini, L. Mandelli, P. Tsompou, G. S. Karanasiou, M. I. Papafaklis, L. K. Michalis, D. I. Fotiadis, L. Petrini, and G. Pennati, "Towards a digital twin of coronary stenting: A suitable and validated image-based approach for mimicking patient-specific coronary arteries," *Electronics*, vol. 11, no. 3, 2022, p. 502.
12. G. Coorey, G. A. Figtree, D. F. Fletcher, V. J. Snelson,S. T. Vernon, D. Winlaw, S. M. Grieve, A. McEwan, J. Y. H. Yang, P. Qian et al., "The health digital twin to tackle cardiovascular disease—a review of an emerging interdisciplinary field," *NPJ Digital Medicine*, vol. 5, no. 1, 2022, p. 126.
13. C. Metallidou, K. E. Psannis, D. D. Vergados, and M. Dossis, "Digital twin and industrial internet of things architecture to reduce carbon emissions," *2022 4th International Conference on Computer Communication and the Internet (ICCCI)*, Chiba, Japan, 2022, pp. 185–189.
14. S. Khan, T. Arslan, and T. Ratnarajah, "Digital twin perspective of fourth industrial and healthcare revolution," *IEEE Access*, vol. 10, 2022, pp. 25732–25754.
15. F. Pires, B. Ahmad, A. P. Moreira, and P. Leitão, "Digital twin based what-if simulation for energy management," *2021 4th IEEE International Conference on Industrial Cyber-Physical Systems (ICPS)*, Victoria, BC, Canada, 2021, pp. 309–314.
16. J. F. D. Santos, B. K. Tshoombe, L. H. B. Santos, R. C. F. Araújo, A. R. A. Manito, W. S. Fonseca, and M. O. Silva, "Digital twin-based monitoring system of induction motors using iot sensors and thermo- magnetic finite element analysis," *IEEE Access*, vol. 11, 2023, pp. 1682–1693.
17. J. Li, S. Wang, J. Yang, H. Zhang, and H. Zhao, "A digital twin-based state monitoring method of gear test bench," *Applied Sciences*, vol. 13, no. 5, 2023, p. 3291.
18. Y. Gao, H. Li, G. Xiong, and H. Song, "Aiot-informed digital twin communication for bridge maintenance," *Automation in Construction*, vol. 150, 2023, p. 104835.
19. J. Wu, Y. Dang, H. Jia, X. Liu, and Z. Lv, "Prediction of energy consumption in digital twins of intelligent factory by artificial intelligence," in *2021 International Conference on Technology and Policy in Energy and Electric Power (ICT-PEP)*, Jakarta, Indonesia, 2021, pp. 354–359.
20. B. Babayigit and M. Abubaker, "Industrial internet of things: A review of improvements over traditional scada systems for industrial automation," *IEEE Systems Journal*, vol. 18, 2023, pp. 1–14.
21. T. Pexyean, K. Saraubon, and P. Nilsook, "Iot, ai and digital twin for smart campus," *2022 Research, Invention, and Innovation Congress: Innovative Electricals and Electronics (RI2C)*, 2022, Bangkok, Thailand pp. 160–164.
22. Y. Qamsane, J. R. Phillips, C. Savaglio, D. Warner, S. C. James, and K. Barton, "Open process automation- and digital twin-based performance monitoring of a process manufacturing system," *IEEE Access*, vol. 10, 2022, pp. 60823–60835.
23. D. Menon, B. Anand, and C. L. Chowdhary, "Digital twin: Exploring the intersection of virtual and physical worlds," *IEEE Access*, vol. 11, 2023, pp. 75152–75172.
24. Arva Arsiwala, Faris Elghaish and Mohammed Zoher, "Digital twin with Machine learning for predictive monitoring of CO2 equivalent from existing buildings," *Energy and Buildings*, vol. 284, 2023, p. 112851.
25. C. Wang, Z. Cai, and Y. Li, "Sustainable blockchain- based digital twin management architecture for iot de- vices," *IEEE Internet of Things Journal*, vol. 10, no. 8, 2023, pp. 6535–6548.

26. A. A. Nikitin, E. A. Sorokin, T. N. Nikitina, D. A. Sokolov, and I. V. Andreychikov, "Mathematical model of a single-acting pneumatic cylinder actuator," *Journal of Physics: Conference Series*, vol. 1889, no. 4, 2021, p. 042079.
27. M. Jiménez, E. Kurmyshev, and C. Castaneda, "Experimental study of double-acting pneumatic cylinder," *Experimental Techniques*, vol. 44, 2020, pp. 355–367.
28. K. Abela, P. Refalo, and E. Francalanza, "Analysis of pneumatic parameters to identify leakages and faults on the demand side of a compressed air system," *Cleaner Engineering and Technology*, vol. 6, 2022, p. 100355.
29. E. Soylu, H. B˙ilg˙ in, S. Cagman, and U. Unver, "An ultrasonic application to evaluate energy costs of air leakages at the compressed air systems," *2021 6th International Conference on Smart and Sustainable Technologies (SpliTech)*, Bol and Split, Croatia, 2021, pp. 1–3.
30. R. K. Pongiannan, M. Krishnamoorthy, S. Srinivasan, M. Chandrasekaran, C. Bharatiraja, B. Praveen Kumar, S. K. Cherukuri, T. S. Babu, and H. H. Alhelou, "Cost-effective energy conservation techniques for textile spinning mills," *IEEE Access*, vol. 10, 2022, p. 49839–49852.
31. C. S. Wickramasinghe, K. Amarasinghe, D. L. Marino, C. Rieger, and M. Manic, "Explainable unsupervised machine learning for cyber-physical systems," *IEEE Access*, vol. 9, 2021, pp. 131824–131843.
32. B. He and K.-J. Bai, "Digital twin-based sustainable intelligent manufacturing: A review," *Advances in Manufacturing*, vol. 9, no. 1, 2021, pp. 1–21.
33. M. Rajashekarappa, J. Lené, E. T. Bekar, A. Skoogh, and A. Karlsson, "A data-driven approach to air leakage detection in pneumatic systems," *2021 Global Reliability and Prognostics and Health Management (PHM-Nanjing)*, Nanjing, China, 2021, pp. 1–7.
34. W. Liu, Z. Zhang, J. Zhang, H. Huang, G. Zhang, and M. Peng, "A novel fault diagnosis method of rolling bearings combining convolutional neural network and transformer," *Electronics*, vol. 12, no. 8, 2023, p. 1838.
35. H. Bai, X. Zhan, H. Yan, L. Wen, Y. Yan, and X. Jia, "Research on diesel engine fault diagnosis method based on stacked sparse autoencoder and support vector machine," *Electronics*, vol. 11, no. 14, 2022, p. 2249.
36. M. Borg, P. Refalo, and E. Francalanza, "Failure detection techniques on the demand side of smart and sustainable compressed air systems: A systematic review," *Energies*, vol. 16, no. 7, 2023, p. 3188.
37. Z. Feng and J. Yang, "Research on fault diagnosis algorithm of self-validating pneumatic actuator," *Chinese Journal of Sensors and Actuators*, vol. 35, 2022, pp. 785–791.
38. K. Leu van, Y. Sun, L. Chen, M. Fang, and M. Zhang, "Research on fault diagnosis method of pneumatic circuit used bidirectional lstm neural network," *Computers and Digital Engineering*, vol. 50, 2022, pp. 367–372.
39. H. Zhu, Z. Wang, H. Wang, Z. Zhao, and W. Xiong, "Leakage fault diagnosis of two parallel cylinders in pneumatic system with a minimal number of sensors," *Electronics*, vol. 12, no. 15, 2023.
40. G. Bhatti, H. Mohan, and R. R. Singh, "Towards the future of smart electric vehicles: Digital twin technology," *Renewable and Sustainable Energy Reviews*, vol. 141, 2021, p. 110801.
41. F. Slama, I. Ismail, and L. Latrach, "Classify project programming languages (plp) in github: Clustering in machine learning," in *2021 20th International Conference on Ubiquitous Computing and Communications (IUCC/CIT/DSCI/SmartCNS)*, London, United Kingdom, 2021, pp. 184–189.
42. G. Bulivou, K. G. Reddy, and M. G. M. Khan, "A novel method of clustering using a stochastic approach," *IEEE Access*, vol. 10, 2022, pp. 117925–117943.

43. I. Trawinski, H. Wimmer, and J. Kim, "Anomaly detection in intrusion detection system using amazon sage-maker," in *2023 IEEE/ACIS 21st International Conference on Software Engineering Research, Management and Applications (SERA)*, Orlando, USA, 2023, pp. 210–217.
44. M. Klymash, O. Hordiichuk-Bublivska, M. Kyryk, L. Fabri, and H. Kopets, "Big data analysis in iiot systems using the federated machine learning method," in *2022 IEEE 16th International Conference on Advanced Trends in Radioelectronics, Telecommunications and Computer Engineering (TCSET)*, Lviv-Slavske, Ukraine 2022, pp. 248–252.
45. P. Bavithra Matharasi, "A review on categorizing the covid19 dataset using k-means clustering algorithm," in *2021 Innovations in Power and Advanced Computing Technologies (i-PACT)*, Kuala Lumpur, Malaysia 2021, pp. 1–5.
46. https://cea.nic.in/wp-content/uploads/baseline/2023/01/Approved_report_emission__2021_22.pdf

17 Power System Planning for Electric Vehicles

Design, Analysis, and Testing of Electric Powertrain for Electric Vehicles

Subramani Nagaraj, Vadivel Ayyakannu, Aadithya Nithyanandan, Sharan Rabinson Mohanraj, and Gowtham Ayyalusamy

17.1 INTRODUCTION

The team Phoenix Racing was started in the year 2012. It designs, fabricates, manufactures, analyzes, and tests an electric ATV. The team participates with the vehicle in various ATV racing events like E-Baja conducted by SAE India, Mega ATV, and FMAE (Fraternity of Mechanical and Automotive Engineers) BAJA. The team comprises four main sub-wings, namely Powertrain, Suspension & Steering, Roll Cage, Brakes & Unsprung. The team starts its work by defining objectives. The team collaborates with all the sub-wings and designs the buggy with these objectives. The primary objective of the powertrain department is improved traction and control by implementing 4WD. Improved traction is achieved through the construction of the powertrain. The 4WD vehicle has the capability of splitting equal torque to all wheels.

The team collaborates with all the sub-wings and designs the buggy with some objectives. The primary objective to be achieved by the powertrain department is improved traction and control by implementing 4WD. Improved traction is attained through the construction of the powertrain. The 4WD vehicle has the capability of splitting equal torque to all wheels. The performance targets set are improved acceleration, better gradeability, maximum speed, and efficient torque. The design of components and selection of OEM products starts with powertrain calculations. The components are selected based on their performance, safety, standards, cost, and ease of availability. The electrical components and systems are fitted as per the rulebook guidelines. Prime consideration is given to safety.

The subsystem is developed with Design Validation Plan (DVP) and Design/Performance Failure Mode and Effects Analysis (D/P FMEA) plans, and all the

DOI: 10.1201/9781003585534-17

components and system were verified using software as well as real-world testing. The developed powertrain architecture is integrated, and the complete vehicle is tested. Various problems were identified during the fabrication and testing phase and were solved using the best and most appropriate solutions. The developed vehicle also participated in the BAJA SAE India 2023 event and achieved remarkable positions.

17.2 LITERATURE SURVEY

The BAJA SAEINDIA Rulebook 2023 (BAJA SAE India Committee, 2022) provides a unified framework for both mBAJA (mechanical, fossil fuel-based engine-driven) and eBAJA (electric) categories, streamlining the design and competition guidelines. For teams developing electric ATVs, the rulebook provides critical insights into safety protocols, component specifications, and performance benchmarks. Adherence to these guidelines ensures compliance and competitiveness in BAJA SAEINDIA events. The team considered this book a primary reference and implemented every aspect detailed in the rulebook and designed the vehicle and its components accordingly.

The research book *Automotive Transmissions – Fundamentals, Selection, Design and Application* (Naunheimer et al., 2011) offers an in-depth exploration of automotive transmission systems, encompassing manual, automated manual, automatic, continuously variable, and hybrid drives for both passenger and commercial vehicles. It systematically addresses the entire development process—from initial product planning and conceptual design to detailed engineering, manufacturing technologies, and reliability testing. The team gained in-depth design knowledge about the power transmission systems from this book.

The technical paper "All-Wheel and Four-Wheel-Drive Vehicle Systems" (Dick et al., 1995) presents a comprehensive overview of All-Wheel Drive (AWD) and Four-Wheel Drive (4WD) systems, focusing on their design evolution, performance benefits, and application in both on-road and off-road conditions. Dick emphasizes the fundamental differences in drivetrain architectures, such as part-time and full-time 4WD systems, torque distribution mechanisms, and the role of center differentials and locking strategies. The team gained knowledge from this paper, including the advantages of equal torque distribution across all wheels for enhanced traction on rough terrain, the mechanical requirements for drivetrain synchronization to prevent wheel slip, and the importance of differential selection and gear-train efficiency in maintaining off-road stability and performance.

The research work "Progress in Automotive Transmission Technology" (Xu et al., 2018) provides a comprehensive review of the evolution and current trends in automotive transmission systems, including manual, automatic, continuously variable transmissions (CVT), and electrified drivetrains. Particular emphasis is placed on the integration of CVT systems with electric drives to improve torque control and driving smoothness. The team gained some insights from this work, such as the selection of a CVT, the associated tuning efforts performed using MATLAB®/Simulink, and modular design and integrated control as essential factors for adapting electric powertrains to varied terrain and performance conditions.

The research paper "Design, Analysis, and Simulation of a Four-Wheel-Drive Transmission for an All-Terrain Vehicle" (Chidambaram et al., 2022) presents a comprehensive approach to designing and simulating a 4WD transmission system tailored for ATVs. The authors detail the integration of mechanical and electrical subsystems to achieve optimal torque distribution and vehicle stability across varied terrains. The study emphasizes the importance of balancing power delivery with vehicle dynamics to enhance off-road capabilities. The team gained insights into how to design the 4WD transmission according to real-time use cases.

The paper "Design and Fabrication of BAJA SAE India All-Terrain Vehicle" (Tomar et al., 2012) presents a comprehensive overview of the design and fabrication process of a BAJA SAE India ATV. The authors detail the engineering methodologies employed in developing a robust and efficient ATV suitable for off-road conditions. The study emphasizes the importance of material selection, weight optimization, and adherence to safety standards. Finite Element Analysis (FEA) was utilized to assess the structural integrity of critical components, ensuring the vehicle's performance and durability. By referencing this paper, the team gained valuable insights into the foundational design principles and practical challenges encountered during the development of an all-terrain vehicle.

The study "Powertrain Modelling and Range Analysis for All-Terrain Electric Vehicles" (Parmar et al., 2020) introduces a novel Reverse Engineered Powertrain (REP) modeling approach to estimate the range of all-terrain electric vehicles (EVs) under dynamic conditions. The Reverse Engineered Powertrain (REP) model simulates real-world driving scenarios by incorporating variable velocities and terrain slopes, enabling precise calculation of energy consumption throughout a race track. By analyzing the forces acting on the vehicle at each moment, the model determines the electrical energy drawn from the battery, facilitating accurate range predictions. For electric 4WD ATV projects, this paper offers valuable insights into integrating terrain-specific dynamics into powertrain simulations and optimizing component selection with energy management strategies.

The study "Electric Vehicle Energy Consumption Modelling and Estimation—A Case Study" (Miri et al., 2020) presents a comprehensive methodology for modeling and estimating the energy consumption of electric vehicles (EVs), using the BMW i3 as a case study. The authors developed a forward-facing simulation model in MATLAB/Simulink, incorporating detailed representations of the vehicle's powertrain components, including the electric motor, transmission, and battery system. The model also accounts for auxiliary power consumption and implements a regenerative braking strategy to reflect real-world driving conditions. For electric 4WD ATV projects, this paper offers valuable insights into constructing accurate energy consumption models that consider various driving scenarios and vehicle dynamics.

The design report "Baja SAE UTEP 2014 Design Report – Team Jaabaz, VIT University" (Nagubandi, 2015) presents a detailed structured approach taken for subsystem development including frame, drivetrain, suspension, and braking systems. Notably, the team prioritized durability, cost-efficiency, and ease of maintenance, which are essential factors for off-road vehicle design. A manual transaxle was selected over a continuously variable transmission (CVT) to reduce power losses and ensure reliable acceleration, a decision supported by torque and speed calculations.

Although the powertrain in this study was internal combustion-based, the insights into mechanical packaging, component selection, and safety considerations are directly relevant to electric ATV projects. The team gained a good understanding of ATV design.

The paper "Design and Development of Electrical Powertrain of ATV" (Kumar et al., 2020) presents a comprehensive study on the design and development of an electric powertrain for an ATV, tailored for the eBAJA SAE India competition. The research focuses on creating an energy-efficient and robust electric drivetrain capable of navigating challenging off-road conditions. The authors detail the selection and integration of key components, including a brushless DC (BLDC) motor, pulse width modulation (PWM) controller, and DC-DC converters, to ensure optimal performance and reliability. The study emphasizes the importance of tractive effort calculations, terrain resistance analysis, and system optimization to achieve a balance between performance and energy consumption. This paper provides valuable insights into the practical challenges and solutions associated with electric powertrain development for ATVs, particularly in the context of student-led engineering projects.

The textbook *Theory and Performance of Electrical Machines* (Gupta, 2015) is a comprehensive resource that delves into the fundamental principles and operational characteristics of various electrical machines. The book covers a wide range of topics, including the construction, working, and performance analysis of DC machines, transformers, synchronous machines, and induction motors. The team gained a clear view of different types of electrical machines and applied the knowledge in selecting the most efficient machine for the vehicle's electric powertrain.

Design Data Book of Engineers (Professors of PSG College of Technology, 1968) is a comprehensive reference that provides essential engineering data for mechanical design and analysis. The book encompasses a wide array of topics, including material properties, heat treatment processes, fits and tolerances, bearing calculations, gear design, and standard mechanical components. In the context of developing an electric 4WD all-terrain vehicle (ATV), this data book served as a critical resource for selecting appropriate materials and components that meet the required strength and durability criteria. For instance, the detailed information on gear materials and bearing specifications aids in designing a reliable drivetrain system capable of withstanding off-road stresses.

Electrical Safety, Fire Safety Engineering and Safety Management (Sunil et al., 1975) is a comprehensive resource that addresses the critical aspects of safety in electrical and fire engineering domains. Designed for engineering students, safety professionals, and personnel involved in the operation and maintenance of electrical systems, this book offers in-depth coverage of safety protocols, risk assessment, and management strategies essential for preventing accidents and ensuring compliance with safety standards. From this book, the team gained insights into operating and handling the electrical equipment used in the vehicle.

17.3 COMPONENTS OF E–POWERTRAIN

- Battery pack
- DC–DC converter
- Electric motor
- Motor controller
- Battery management system
- Main control unit panel
- Grounded Low Voltage (GLV) ancillary systems (lighting, alarms)

17.4 MATHEMATICAL MODELING

The powertrain department starts designing with mathematical modeling. Basic achievable parameters such as acceleration, gradeability angle, mass of the vehicle, etc., are predetermined, and consecutive calculations such as acceleration, maximum tractive effort, gear ratios, and gear profiles are calculated. The high voltage (HV) components like motor and battery pack are selected based on the powertrain calculations.

17.4.1 Total Tractive Effort

Gross mass of the vehicle = 310 kg
Gross weight = 3,041 N
Tire static radius = 279.4 mm
Assuming transmission system efficiency as 0.71,
Frontal area = Maximum width * Maximum height = 1.6 m^2

Coefficient of friction:

For concrete = 1
For Dry sand = 0.65

Coefficient of rolling resistance:

For concrete = 0.014
For dry sand = 0.05

$$\text{Rolling resistance} = \text{Gross weight} * \text{Co-efficient of rolling resistance}(\text{Dry sand})$$

$$= 3{,}041 * 0.05 = 152.05\,\text{N}$$

$$\text{Aerodynamic resistance} = (1/2) * \text{Air density} * (\text{velocity})^2 * \text{Frontal area}$$

$$* \text{Aerodynamic drag co-efficient}$$

$$= (1/2) * 1.12 * (16.66)^2 * (1.6) * (0.44) = 13.136\,\text{N}$$

$$\text{Grade resistance} = \text{Gross weight} * \sin\alpha$$

$$= 3041 * \sin(26.56) = 1359.7 \text{ N}$$

$$\text{Total Tractive effort} = \text{Rolling Resistance} + \text{Aerodynamic resistance}$$

$$+ \text{Grade Resistance}$$

$$= 152.05 \text{ N} + 13.136 \text{ N} + 1359.7 \text{ N} = 1524.85 \text{ N}$$

Considering efficiency, (P_{out}/P_{in}),

$$\eta_{\text{tran}}s = \eta_{\text{CVT}} + \eta_{\text{Gearbox}} + \eta_{\text{Propeller shaft}} + \eta_{\text{Differential}} + \eta_{\text{Drive shaft}}$$

$$\rightarrow V = 8 \text{ km/hr.} @ 50\% \text{ gradeability}$$

$$\text{Max Tractive effort at } 50\% \text{ gradeability} = T_m$$

$$= \text{Co-efficient of friction of dry sand} * \text{Gross weight of the vehicle} * \cos\alpha$$

$$= 0.65 * 3041 \text{ N} * \cos(26.56) = 1768.047 \text{ N.}$$

Max Tractive effort at 50% gradeability is less than the T_m.

Hence, no slip occurs.

$$\text{Torque produced at wheels} = \text{T.T.E.} * \text{Tire static radius (in meter)}$$

$$= 1524.85 \text{ N} * 0.2695 \text{ m} = 410.947 \text{ N.}$$

17.4.2 Acceleration

A car's ability to accelerate swiftly from a stop or the time it takes to do so, is also referred to in the context of vehicles. Low gear is always used for acceleration and negotiating hills since the loss of speed is greater there and the pulling force at the wheels is largest. The torque required is assumed to be 600 Nm.

$$T_m = T(\text{design torque}) / r(\text{tire static radius}) = 600/0.2794$$

$$= 2147.45 \text{ N}$$

where T_m = Max. tractive effort for design torque

$$\text{Traction force} = T_m - \text{ Total Tractive Effort (T.T.E)}$$

$$= 2147.45 \text{ N} - 1524.85 \text{ N} = 622.6 \text{ N}$$

$$\text{Acceleration}, a = \text{Friction force/Mass of the vehicle}$$

$$= 2.008 \text{ m/s}^2$$

Thus, the powertrain calculations are completed, and the final points are summarized as shown in the table 17.1 below in order to pursue OEM main powertrain components.

17.4.3 Machine and Selection

$$\text{Energy required for 1 second} = \text{T.T.E.} * v = 3000 \text{ Watts}(\text{W})$$

$$\text{W.K.T. Motor rated power}, P = 5 \text{ kW} = 5000 \text{ W}$$

Thus, a machine with about 5 kW is selected as the electric motor for the vehicle.

The team mainly selected a Permanent Magnet Synchronous Motor (PMSM) since it offers better torque—speed characteristics compared to other motor types. Also, it offers higher efficiency compared to other types. In addition, its better control capability makes it a dynamically controllable machine in all scenarios such as low speed—high torque for initial acceleration, driving over a heavy gradient, driving at high speed, sudden braking, etc.

17.4.4 Battery Pack Selection

Since the battery voltage and capacity should not exceed 60 V and 120 Ah, the battery pack is selected accordingly. The pack also complies with the IEC (International Electrotechnical Commission) 60529 IP67 (Ingress Protection) standard (ability to withstand water and dust), as per the rulebook (C.3.1). The pack is also selected such that it has a built-in BMS for enhanced protection. We selected the NMC (Nickel Manganese Cobalt)-based type since it offers high energy density, which is suitable for the ATV's high torque requirement.

17.5 ELECTRIC POWERTRAIN ARCHITECTURE

The electric powertrain architecture is derived from the complete powertrain architecture. Various components are interconnected in a sequential manner in order to form a complete drivetrain. The following flowchart explains this in a better way (Figures 17.1 and 17.2; Table 17.1).

17.6 HIGH VOLTAGE TRACTIVE SYSTEM

It deals with the HV battery and electric motor, two parts of a high-voltage system. Every component that is electrically connected to the motor(s) and tractive system accumulator(s) is referred to as a part of the vehicle's tractive system. The voltage between any two terminals of the vehicle's tractive system must never, under any circumstance, exceed 60V DC. The maximum tractive system voltage must be rated for all tractive system components. The contactor, fuse, and battery management systems/solutions corresponding to each battery cell that stores the electrical energy used by the tractive system are referred to as the tractive system accumulator(s).

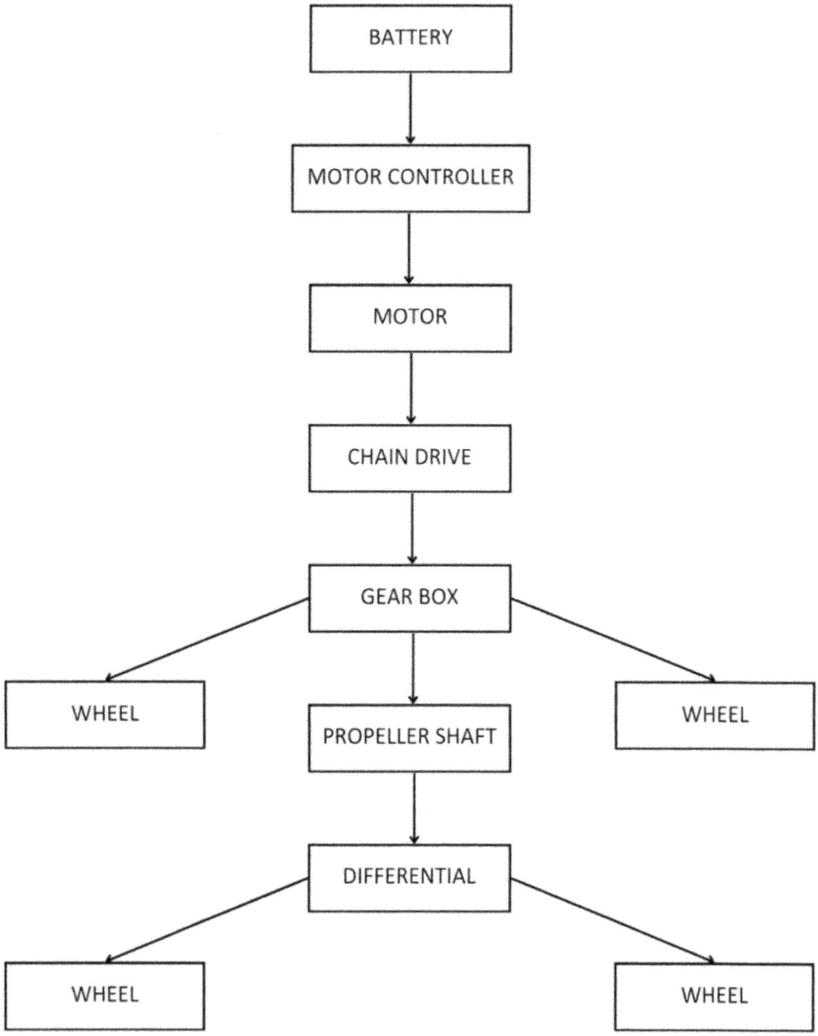

FIGURE 17.1 Flowchart of power flow of the E–Powertrain.

Through a motor controller (traction controller), the tractive system motor(s) must be connected to the accumulator(s). It is strictly forbidden to bypass the control system and attach the tractive batteries directly to the motor(s) (Figures 17.3 and 17.4).

17.7 BATTERY PACK

The energy required to power the vehicle is stored in the battery pack, which is made up of many cells. Direct current output is provided by a battery pack. The bulk of portable consumer devices currently employ lithium-ion batteries due to their high energy density per mass compared to other electrical energy storage solutions. They

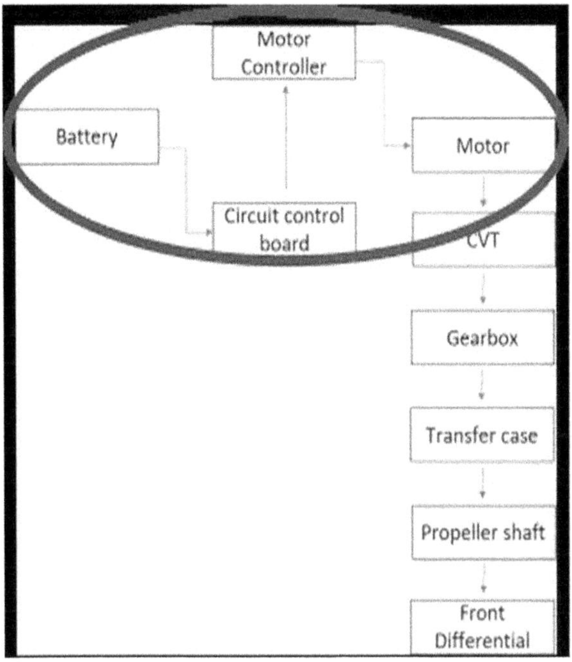

FIGURE 17.2 Power flow and position of the electrical powertrain in the flow.

TABLE 17.1
Powertrain Components and System Parameters Specifications

Transmission Unit data		
Net Transmission Ratio	Max	Min
	33.15	7.65
CVT Model	Gaged Gx9	
CVT Ratio	Low	High
	3.9:1	0.9:1
Engagement Speed	900 rpm	
Shift Speed	3800 rpm	
Gearbox	2 Stage Red	
Gearbox Ratio	8:1	
Drive Shaft	Tripod Joint	
Half Shaft Material	AISI 1050	
OD	19 mm	
Plunge Length	40 mm	
Articulation Angle	32 degrees	
Radius of Gyration	0.2 m	

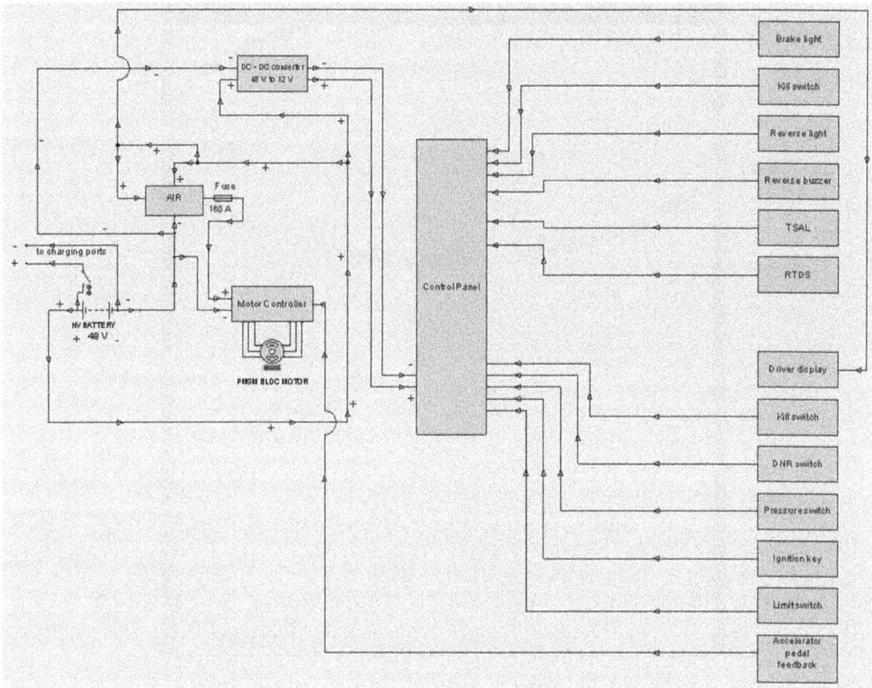

FIGURE 17.3 Electric powertrain architecture circuit diagram.

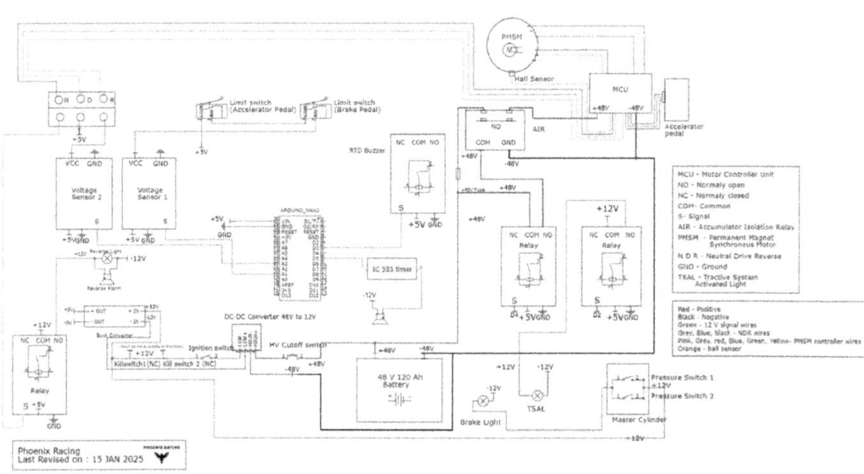

FIGURE 17.4 Schematic electric powertrain architecture and new schematic to be implemented.

also work well at high temperatures, have a high power-to-weight ratio, are highly energy-efficient, and discharge themselves slowly. The battery pack is bought as original equipment manufacturer (OEM) from Eifer Megacorp. It is of NMC chemistry type and has 48 V 120 Ah capacity. Every time, for design and fabrication of the roll cage (Chassis), a battery pack prototype is made using cardboard in order to check the clearance as per rulebook. The roll cage to accumulator clearance is considered for the swapping process and designed according to that (Figures 17.5 and 17.6; Table 17.2).

FIGURE 17.5 Accumulator pictures and specifications.

FIGURE 17.6 Battery prototype made using cardboard with accurate dimensions for clearance checking by inserting it in the roll cage.

TABLE 17.2
Accumulator Configuration

Accumulator	
Accumulator Series Configuration	13
Accumulator Parallel Configuration	24
Maximum Voltage (V)	54.6
Nominal Voltage (V)	46.8
Cut-off Voltage (V)	35.75
Nominal Capacity (Ah)	120
Approximate Weight (kg)	40
Dimensional (Length x Width x Height in mm)	428*428*178
Max Continuous Discharge Current (A)	180
Max Instantaneous Discharge Current (A)	200
Charge Voltage (V)	54.6
Operating Temperature Range (°C)	(-20) -(55)
Battery Cooling Type	Nature Cooling
IP Rating	67
Vendor Details	Eifer Megacorp Pvt Ltd
Charger	48 V 30 A
Configuration	13 s 24 p

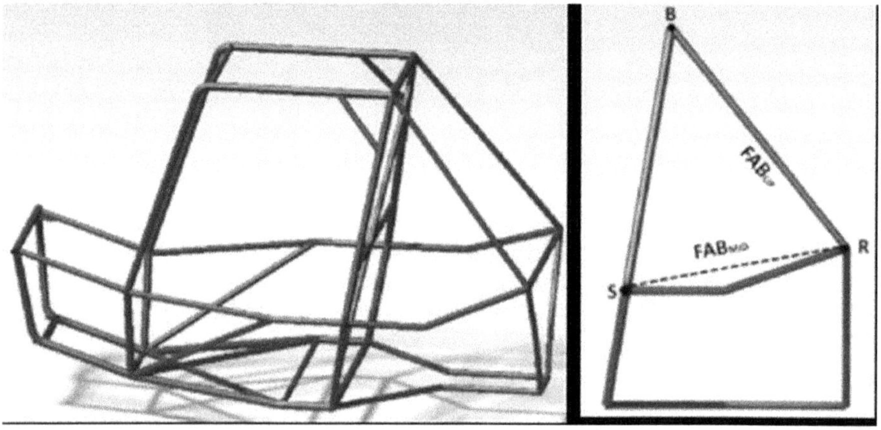

FIGURE 17.7 Position of the battery pack on the roll cage.

17.7.1 Mounting of Battery Pack

The battery pack is mounted in the area above the S-R plane such that the path of battery removal during battery swapping should be easier and accessible both in the horizontal and vertical directions of movement (Figure 17.7). The battery and electrical component mounting members are analyzed with a significant load impact to test their rigidity and safety, such that the electricals are safely mounted. The battery pack is bought as OEM from Eifer Megacorp of India as per the rulebook and design requirements. The laboratory test for checking standards is done. One battery pack has a nominal voltage of 48 V, and another one is 52 V. Certain analyses such as battery mounting bracket load analysis, roll-over impact analysis, etc., are done in Ansys Workbench to identify how the impact force is affecting the battery-mounted area. The battery mounting point is selected such that it is more safe during any impact and it won't change the center of gravity point of the entire vehicle (Figures 17.8–17.13; Table 17.3).

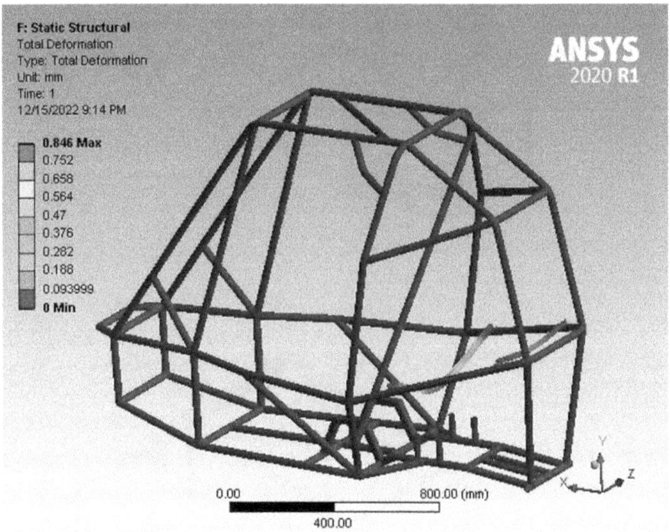

FIGURE 17.8 Battery mounting "L" brackets load analysis.

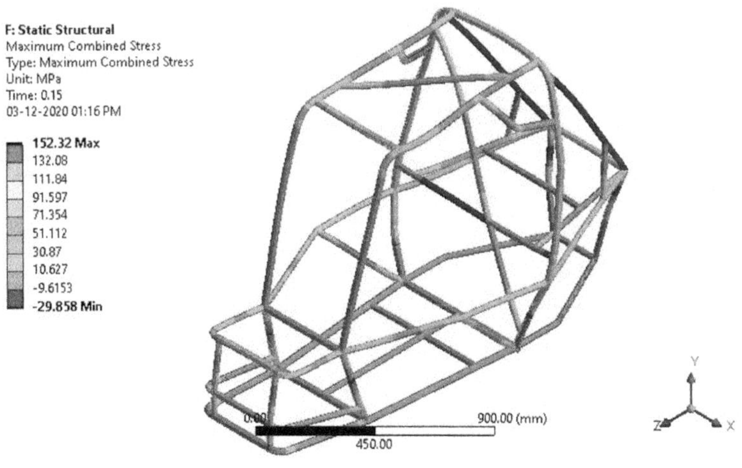

FIGURE 17.9 Roll over impact analysis.

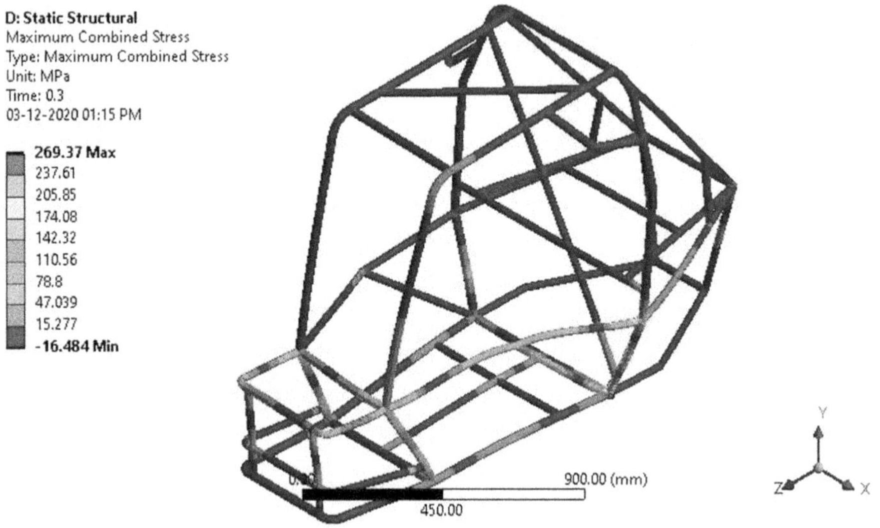

FIGURE 17.10 Cabin force impact analysis.

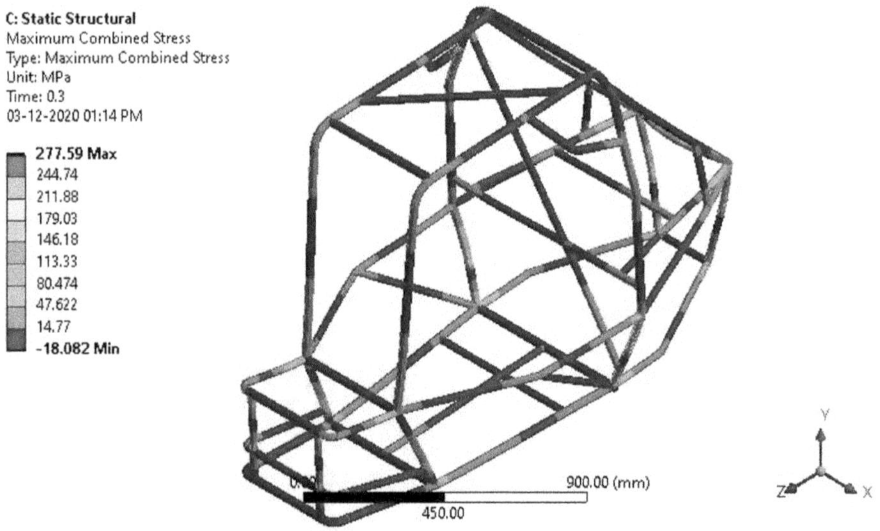

FIGURE 17.11 Rear motor impact analysis.

STC LAB **STANDARD TESTING & COMPLIANCE (STC LAB)**

Tel : 0129-4321765 | Email : Info@stclab.in | website : www.stclab.in

TC-9629

TEST REPORT

TEST REPORT AS PER: IS/IEC 60529:2001

Page No.: 1of 4

ULR No.:-	TC962922000000059F	Group/ Discipline:	ELECTRICAL/ENVIRONMENTAL TEST FACILITY
Test Report No.	STC/Test/20220228001		Date of Issue: 15/03/2022

Name & Address of Customer:	EIFER MEGACORP PRIVATED LIMITED CHINTALAPUDI; 11-112, SARDA THEATRE NEAR OLD BUS STAND, West Godavari, Andhra Pradesh, 534460	
Name & Address of Manufacturer (If required):	--	
Customer Ref. & Date:	Nil	W.O. No.: 20220228001
Date of Sample Receipt: 28/02/2022	Start of Test Date: 28/02/2022	End of Test Date: 03/03/2022

PART A - PARTICULARS OF THE SAMPLE SUBMITTED

Sample description	Metal Box(419*418*173)
Grade/ variety/ type/ class/ size etc.	48V, 120Ah (13s 24p)
Declared values, if any	IP67
Code no., BIS seal and IO's sign. if any	Nil
Batch no., date of manufacture and Brand Name	Manufacture: "EIFER MEGACORP PRIVATED LIMITED"
Model	--
Quantity	01 Nos.
Condition of the sample	OK
Reference specification (s)	IS/IEC 60529:2001
Environmental conditions	Temperature (25±10)°C & Relative Humidity(45-75)%

PART-B: SUPPLEMENTARY INFORMATION

1. If an Item is tested, acknowledging deviations from specified conditions as requested by customers, the results may be affected due to this deviation.
2. Details of the drawings, graphs, tables, sketches or photographs as referred in the test report, if any:
3. The standards/ Instruments used are maintained in accordance with ISO / IEC 17025 and are traceable to National and International Standards

Notes:

1. This report is not to be reproduced except in full/partial without approval of the laboratory in writing.
2. This report refers only to the particular sample detailed above.
3. The results reported in this Test report are valid at the time of and under the stipulated conditions of measurement.

Tested by	Approved by
(signature)	(signature, stamp: Standard Testing And Compliance, STC LAB)
(Esha Tyagi/Testing Engineer)	(Shivendra Kumar/Technical Manager)

Format No. - STCLAB/F/EL/06

Address : #14/7, Parmeshwari Colony, Mathura Road, Faridabad (Haryana) - 121008, India

FIGURE 17.12 Battery test report.

FIGURE 17.13 Battery cells arrangement.

TABLE 17.3
Battery Cells Configurations

Cells	
Cell Make/Model/Style	**HLY/Lithium/26650**
Cell Chemistry (LFP, NCA, LTO etc.)	NMC
Cell Nominal Capacity (mAh)	5000
Maximum Voltage (V)	4.2
Nominal Voltage (V)	3.6
Minimum Voltage (V)	2.75
Maximum Peak Discharge Current (30s) (A)	15
Maximum Continuous Discharge Current (A)	10
Maximum Peak Charge Current (<80% SoC) (30s) (A)	15
Maximum Continuous Charge Current	2.5
Maximum Cell Cut-Off Temperature (discharging) (°C)	65
Maximum Cell Cut-Off Temperature (charging) (°C)	60
Vendor Details (if any)	Eifer Megacorp Pvt Ltd

17.7.2 Battery Management System (BMS)

The Daly BMS (Battery Management System) is fixed as OEM and is fitted inside the battery pack enclosure itself. The BMS satisfies all the standards specified in rulebook clause C.4.6. The battery BMS data is analysed using the SmartBMS app, which communicates with the BMS using Bluetooth communication. The battery parameters are measured using the BMS tag and SmartBMS app on a mobile phone using Bluetooth communication (Figures 17.14–17.16; Table 17.4).

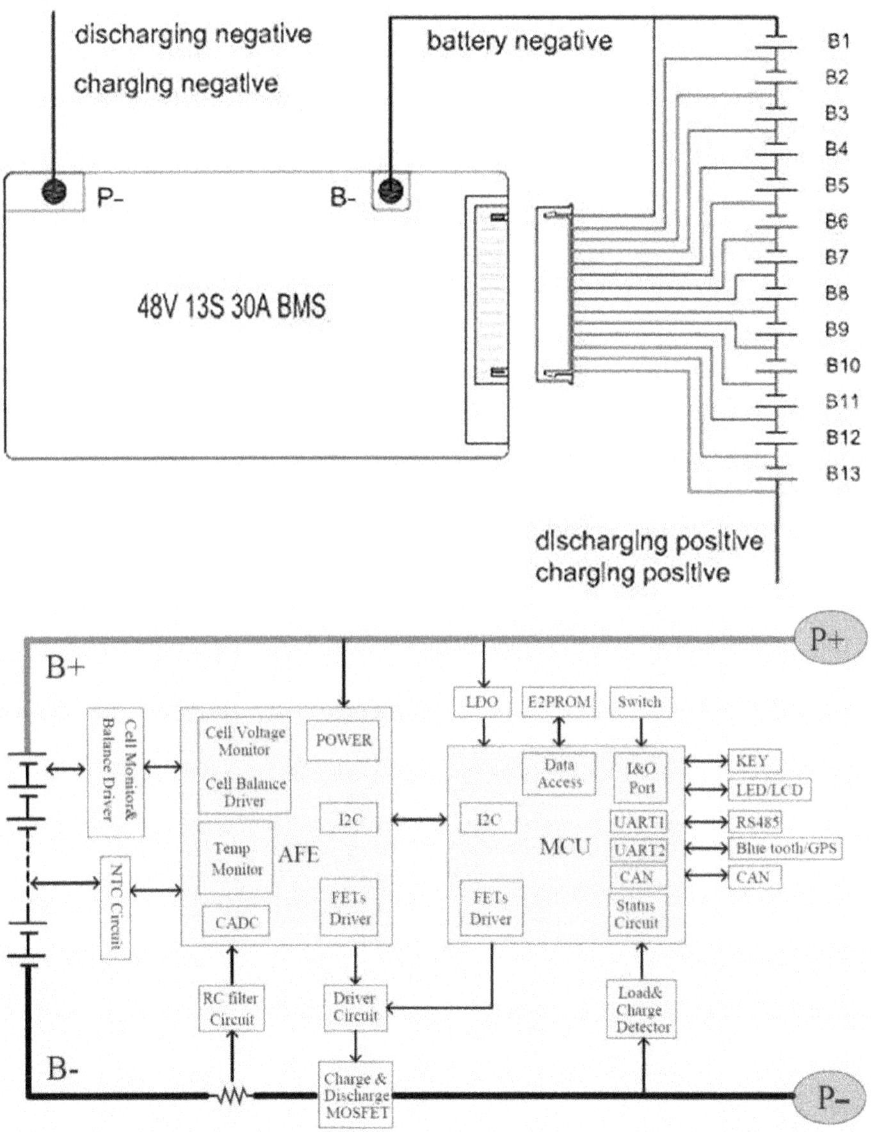

FIGURE 17.14 BMS circuit diagram.

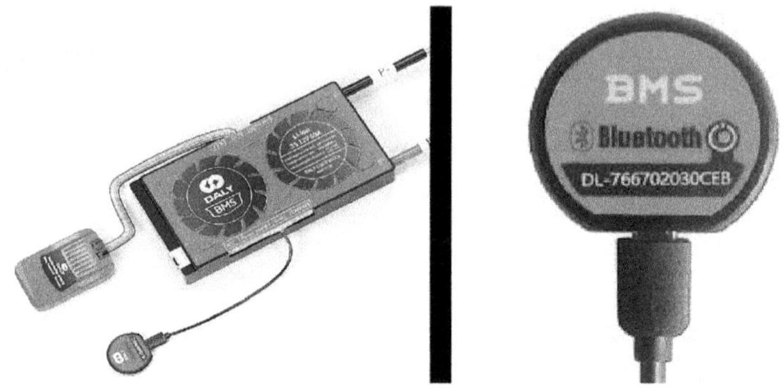

FIGURE 17.15 Daly BMS and Daly BMS tag used for transmitting BMS data to a mobile phone using Bluetooth.

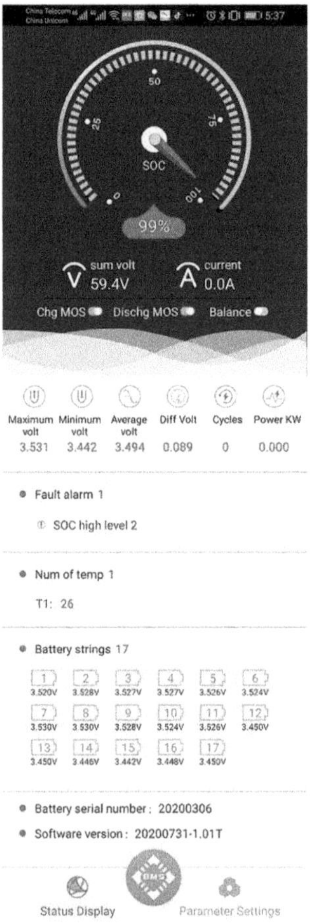

FIGURE 17.16 SmartBMS app.

TABLE 17.4
Battery Management System Details

BMS		
Discharge	**Details**	**Specifications**
	Maximal continuous discharging current (A)	180
	Peak discharging current (A)	200
	Over current protection current (A)	200
Charge	Charging voltage (V)	54.6
	Charging current (A)	100
Overcharge Protection	Overcharge detection voltage (V)	4.25
	Overcharge detection delay time (seconds)	1
	Overcharge release voltage (V)	4.15
	Cell balancing detection voltage (V)	3.8
	Cell balancing release voltage (V)	4.25
Cell Balancing	Cell balancing current (A)	2
	Over-discharge detection voltage (V)	2.7
Over-Discharge Protection	Over-discharge detection delay time (seconds)	1
	Over discharge release voltage (V)	2.8
	Over current detection voltage (V)	2.75
Over Current Protection	Over current detection delay time (seconds)	1
	Over current protection condition	Off load
	Detection condition	Off load
Short Protection	Detection delay time (seconds)	5000 μs
	Release condition	Off load
Over - Temperature Protection Cut-Off (°C)		60
Vendor Details		Eifer Megacorp Pvt Ltd

17.8 CHARGER

The charger is also bought as OEM from Eifer Megacorp. Charger specifications are given below. A unique charging station area is allotted for charging the battery as per standards. The charging environment at the garage is implemented as per charging safety standards mentioned in rulebook C.4.7 & C.4.8 (Figure 17.17).

17.9 ELECTRIC MOTOR

The motor is chosen as a Permanent Magnet Synchronous Motor (PMSM) for better control and traction on any type of surface. It is brought from Datai Corporation. It has six phases and the coils are energized according to the Hall sensor feedback signal and throttle input signal (Figure 17.18; Table 17.5).

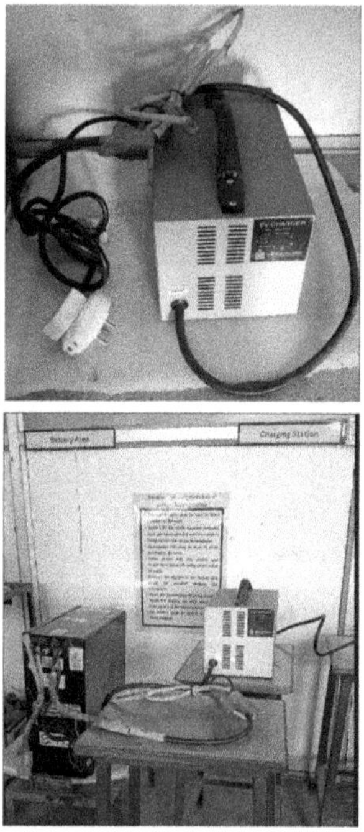

Parameter	Range	
Model	ES-4835	ES-8325
Battery Pack Type	Li-Ion LiFePO4 or NCM 13 to 16 Cell Series Lead Acid – 3/4 Series Flooded or VRLA	Li-Ion LiFePO4 or NCM 19 to 23 Cell Series Lead Acid – 5 Series Flooded or VRLA
CHARGING VOLTAGE	48V to 65V ± 0.5V (Configurable in 0.5V steps)	68V to 84V ± 0.5V (Configurable in 0.5V steps)
CHARGING CURRENT	15A - 35A ± 1A (Configurable in 1A steps)	10A - 25A ± 1A (Configurable in 1A steps)
CHARGE PROFILE	Li-Ion: Three Stages – CC, CV, Charge Termination with Current Threshold and Timer. Lead Acid: Four Stage – Bulk CC, Absorption CC, Absorption CV and Float charge	
MAINS OPREATING RANGE	120VAC±10V -280VAC±10V, 40 Hz TO 60 Hz	
INPUT POWER FACTOR	Active PF correction, >0.95	
EFFICIENCY	89% Typical at nominal input	90% Typical at nominal input
LCD Display Parameters	SCREEN1 – Charger Bar Graph, Elapsed Time – Instantaneous Charging Current and Battery Voltage	
	SCREEN2 – Selected Battery Type, Set CC Current, Set CV Voltage	
	SCREEN3 – Mains Voltage and Frequency, Ambient Temperature (optional)	
LED INDICATIONS	3 LEDs – Green – Charging On, Yellow – Charging Stopped, Red - Fault	
AUDIO INDICATIONS (Buzzer)	Power On, Charging Start or Stop – Short Beeps Battery Charge Completion – Interminent beeps upto 20minutes. Mains Failure or any Faults - Long Beeps	
REVERSE BATTERY PROTECTION	Electronically Protected – Charger will not start unless battery is connected in correct polarity.	
MAINS VOLTAGE PROTECTION	Withstand up to 320V AC RMS with Surge Protection.	
MAINS OVER CURRENT PROTECTION	By FUSE, in case of abnormal condition	
COOLING SYSTEM	Forced Air Cooled	
THERMAL PROTECTION	Electronically protected with internal temperature sensor.	
CHARGER OPERATING TEMPRATURE	0°C TO 45°C	
Battery Cable	3mts 6sqmm with Anderson Type 50A Connector. 50A Terminal Block.	
Mains Cable	1.8mts 1.5sqmm 3Core Cable with 16A molded Plug.	
HUMIDITY	95% RH Non-Condensing	
ENCLOSURE	Powder coated sheet metal cabinet	
NET WEIGHT	6.0 kgs	

FIGURE 17.17 Charger specifications and charger picture & Battery charging setup at garage.

17.10 DC–DC CONVERTER

A direct current (DC) to direct current (DC) converter is a circuit or electrical device that alters the voltage level of a DC source. It is a type of electromagnetic power converter. Tiny batteries have very little power, whereas high-voltage power transfer has a lot. DC-to-DC converter circuitry is often used to adjust the output voltage.

One exception is high-efficiency light emitting diode (LED) power sources, which are a form of DC-to-DC converter that manages the current through the LEDs. Another option is to employ simple charge pumps, which can raise the output voltage by two or three times (Figure 17.19).

17.11 MOTOR CONTROLLER

A motor controller has the power to govern how an electric motor functions in accordance with predetermined regulations. A motor controller enables automatic or manual starting and stopping, forward or reverse rotation, speed selection and regulation, torque restriction or control, overload protection, and electrical failure prevention. To control a motor's direction and speed, motor controllers can use electromechanical switching or power electronics (Figure 17.20; Table 17.6).

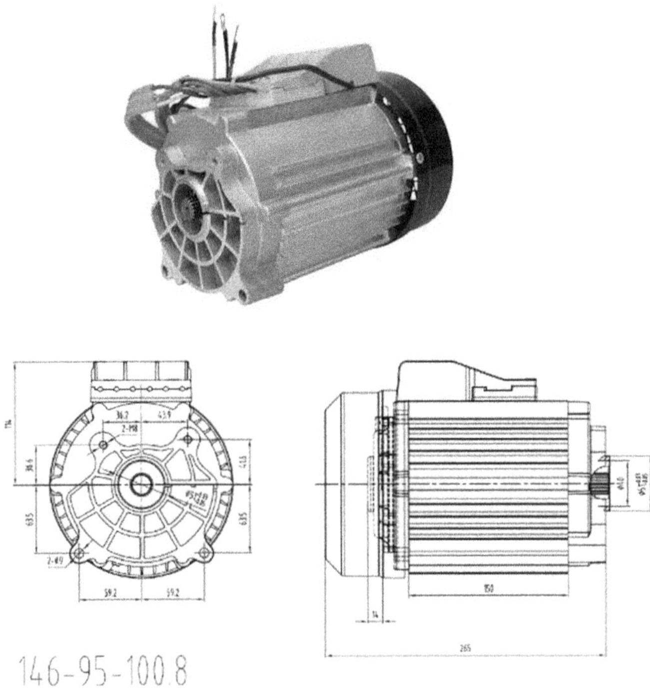

PMSM CONFIGURATION	
Voltage	48 V
Max Torque	75 @300 RPM
Rated / Peak Power	5 / 7.2 KW
Rated / Peak Speed	3900 / 4500rpm
Rated / Peak Torque	25 / 75 Nm
Water Resistant	IP67

FIGURE 17.18 Picture and engineering diagram of motor.

17.12 MAIN CONTROL UNIT

The main control unit provides logical control and implementation of all the starting and security sequences according to the rulebook (C.5.5). The drive mode operates whenever the brake pedal is pressed, Neutral mode is applied, and there

TABLE 17.5
Motor Configuration

Motor			
Make/Model		**DATAI**	
Operating Voltage (V)		48 V	
Cont. Rated Power (kW) @ RPM	5 kW	@	3600 RPM
Max Peak Power (kW) @ RPM	7.2 kW	@	2700 RPM
Rated Speed (RPM)		3900 RPM	
Maximum Speed (RPM)		4500 RPM	
Cont. Rated Torque (Nm) @ RPM	25 Nm	@	2700 RPM
Max Peak Torque (Nm) @ RPM	75 Nm	@	300 RPM
Cooling Type		Air cooling	
Cooling system is powered of		Connected with motor shaft	
Age of the Motor if reused (in Years)		0	
Operating Temperature range (°C)		0–50	
Vendor details		Eifer Megacorp Pvt Ltd	

is no accelerator press. The control panel is implemented using NO (Normally Open)/NC (Normally Closed) relays and a Darlington pair type Integrated Circuit (IC). The HV wire is connected to the control panel using an HV terminal block (Figures 17.21–17.24).

17.12.1 Parts of Main Control Unit

17.12.1.1 Relay Module

This module has a significant number of relays that get different types of signals from various parts of the vehicle, such as accelerator pedal push, brake pedal press, D-N-R (Drive-Neutral-Reverse) switch position, and kill switch position. According to the rulebook, the vehicle must start if and only if there is a brake press, no accelerator push, and neutral mode. The decision of actuating the respective relay in order to achieve the rule conditions is made by building the circuit accordingly. HL JQC 12–24 V relay and ULN2903 Darlington pair ICs are used for implementing the control logic (Figures 17.25–17.27).

17.12.1.2 Wire Connectors

The signal-carrying wires from various parts of the vehicle are given to the wire connector plugs, which make a connection with the relay logic circuit. The unwanted slots are filled with filler blocks, and the wires are labelled with plastic ring labels (Figure 17.28).

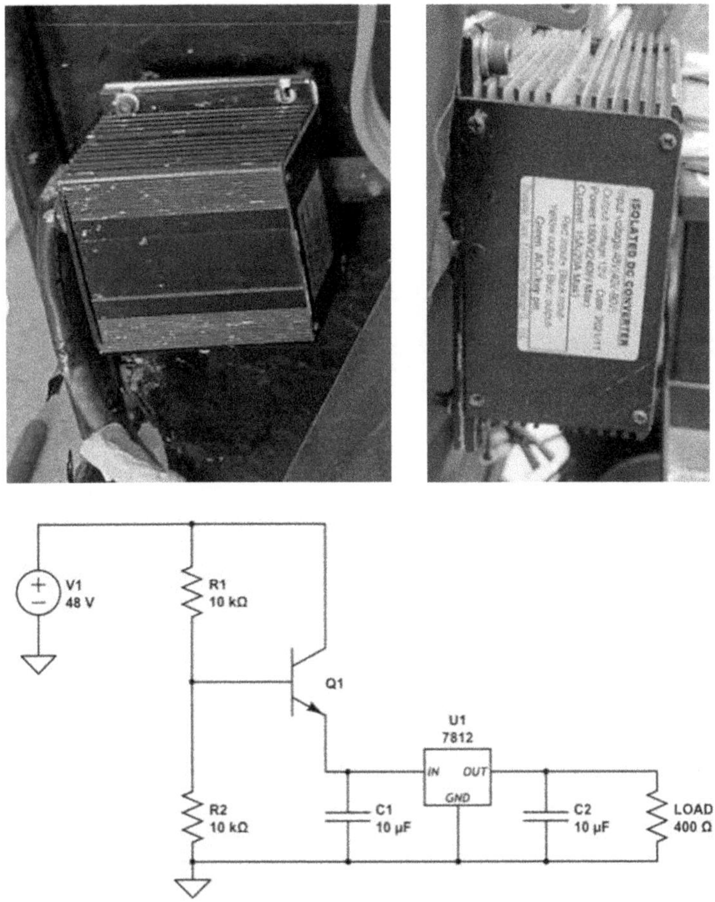

DC-DC Converter	
Type	Isolated type
Input rated voltage (V)	48
Input voltege renge (V)	40-90
Effleleney (%)	90
Output voltage (V)	12
Output current (A)	15
Output rated power (W)	180
Output peak power(W)	240
Voltage regulation (%)	±10%
Load regulation(%)	±10%
Ripple (full load test)	300mv
No-load eurrent (A)	0.01A
Working Temperature (C)	.- 25°C +85°C
Ip Rating	IP65
Protections (Over-volt, Under-volt etc.,)	1. Temperature protection 95°C 2. Overcurrent protection 3. Short-circuit protection 4. Overvoltage protection
Fure rating(A)	15A

FIGURE 17.19 Picture, circuit diagram, specifications of DC–DC converter.

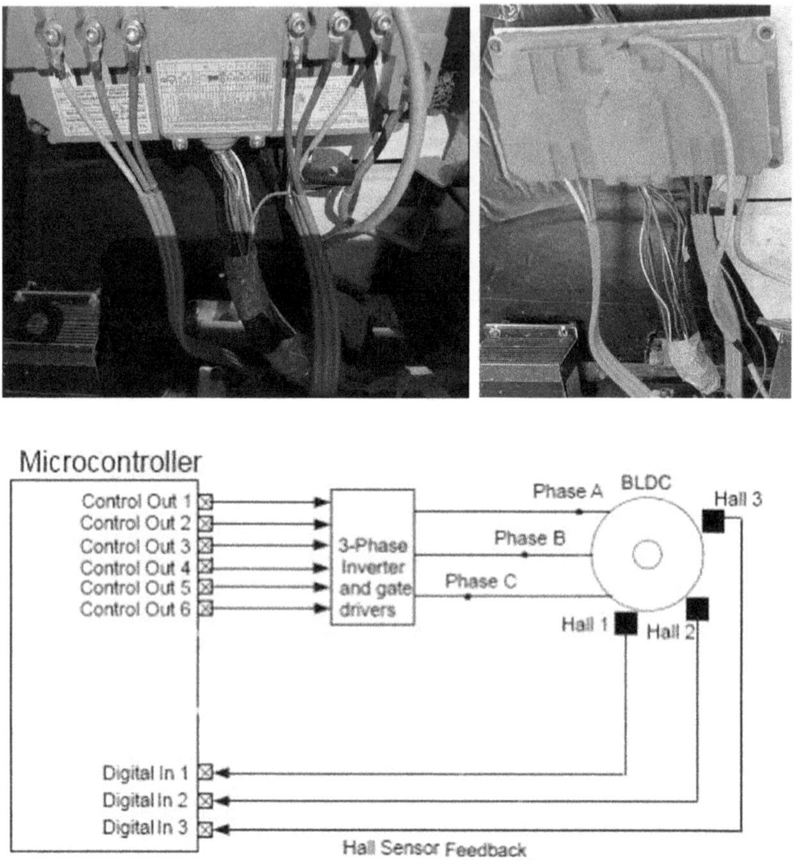

FIGURE 17.20 Internal circuit diagram and specifications of motor controller.

TABLE 17.6
Motor Controller Details

Electric Motor Controller	
Make/Model	**DATAI**
Available Battery Voltage (V)	48
Available Motor Power (kW)	5
Peak Phase Current (A)	25
Rated DC Current (A)	120
Maximum DC Current (A)	150
Communication Protocol	Not applicable
Throttle Input Type	Pedal feedback
Operating Temperature Range	0°C – 50°C
Vendor details	Eifer Megacorp Pvt Ltd

FIGURE 17.21 Real implementation of main control circuit with wire harness and main components.

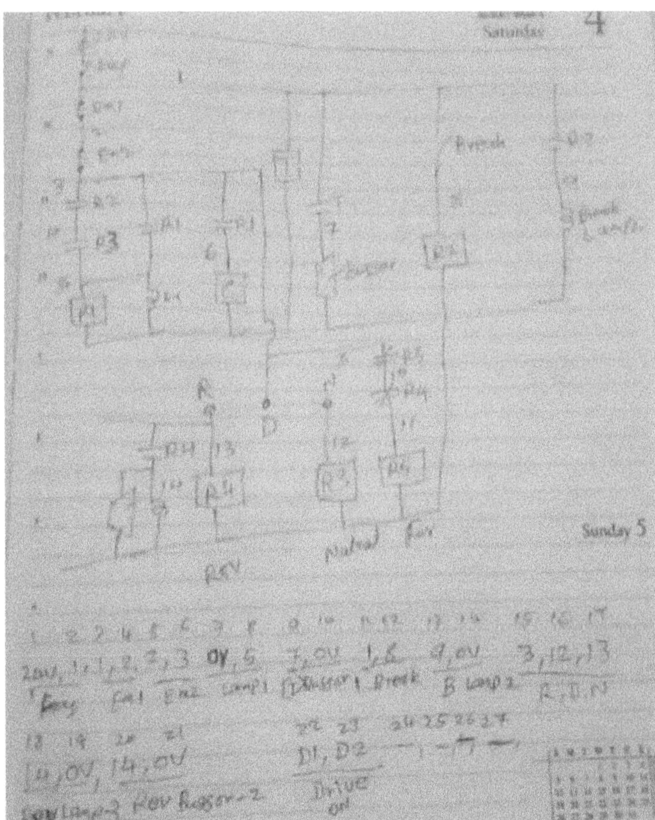

FIGURE 17.22 Circuit diagram of main control unit.

FIGURE 17.23 Control unit with spade pin terminal blocks.

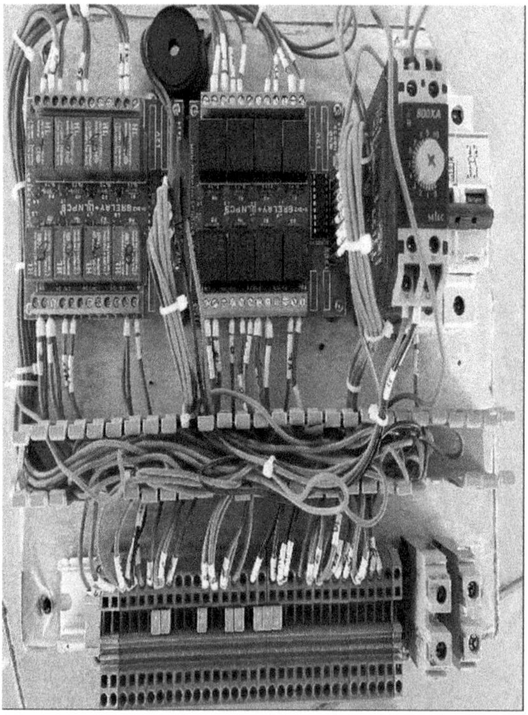

FIGURE 17.24 Control unit with bolt terminal blocks.

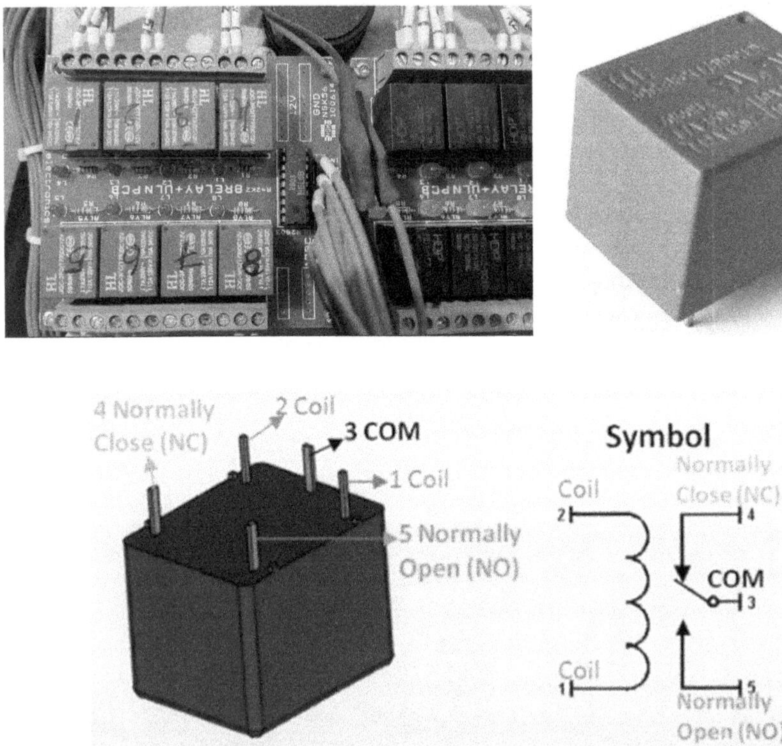

FIGURE 17.25 HL JQC 12-24 V relay picture and relay pinout.

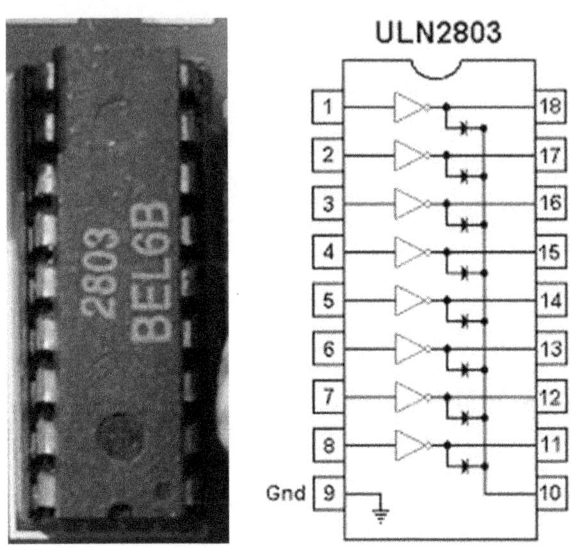

FIGURE 17.26 ULN2903 BEL6B Darlington pair ICs.

Trigger Voltage (Voltage across coil)	5V DC
Trigger Current (Nominal current)	70mA
Maximum AC load current	10A @ 250/125V AC
Maximum DC load current	10A @ 30/28V DC
Package configuration	Compact 5-pin configuration with plastic moulding
Operating time	10msec
Release time	5msec
Maximum switching	300 operating/minute (mechanically)

FIGURE 17.27 Relay electrical data.

FIGURE 17.28 Wire connector slots.

17.12.1.2.1 Open Slot Wiring Cable Raceway

The wires are routed with cable raceways to ensure a neat and clear wire harness design and easy troubleshooting and connection purposes. This raceway also ensures a non-shaky environment for wires in case of heavy vibration and shaky movement, thus preventing loose connections and short circuits (Figure 17.29).

FIGURE 17.29 Wires arranged in a cable raceway.

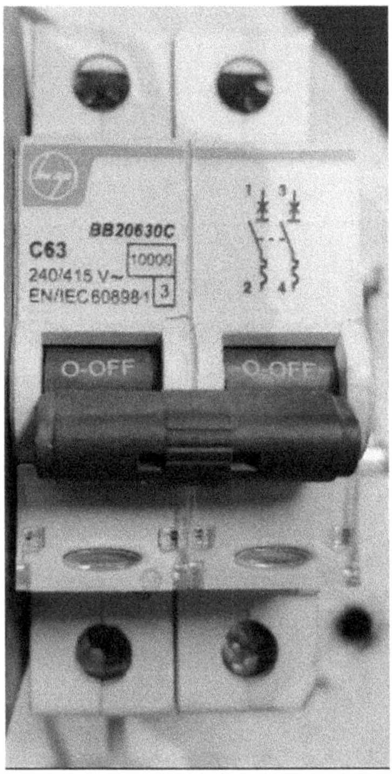

FIGURE 17.30 L&T 10000 BB20630C MCB.

17.13 PROTECTIVE MCB FOR CHARGING

A protective Miniature Circuit Breaker (MCB) is fitted as a protective device to safeguard the control unit from sudden current surges and short circuits during charging. The MCB is turned ON during onboard charging (if needed) by isolating the main circuit. L&T 10000 BB20630C is used as the charging protective MCB (Figure 17.30).

17.14 TIMER FOR RTDS (READY TO DRIVE SOUND)

According to rulebook 2023 C.5.7, the Selec 800 XA 12-level industrial timer module is used for providing a timer signal for making the ready to drive sound (RTDS). The RTDS is made for 3 seconds whenever the driving sequence starts to indicate

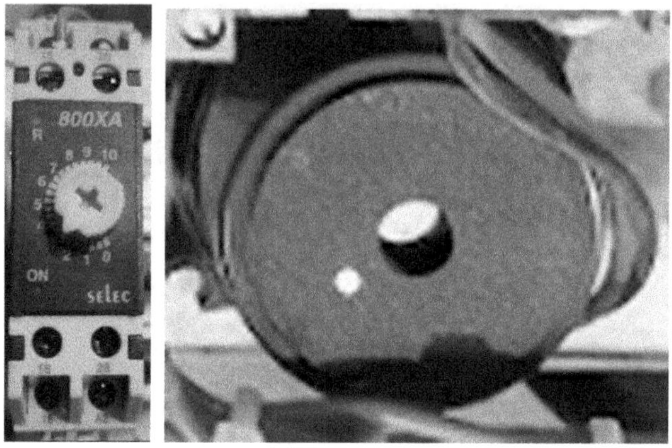

FIGURE 17.31 Selec 800X timer and buzzer.

that the vehicle is ready for driving. A sound buzzer of 80dB is used to deliver the sound for 4 seconds (Figure 17.31).

17.15 TERMINAL BLOCKS

At first, Connectwell CTS16U pin-type terminal blocks were used. Later, it was found that the wires inserted in the pin-type blocks were easily loose connected because of vibration. To avoid this, ring connector-type bolted terminal blocks were selected. The selection of these blocks mainly depends upon input current limit, vibration factor, and ease of implementation (Figures 17.32 and 17.33).

17.15.1 Electro–Thermal Analysis of Electric Connector Plate of the Terminal Block

The electric metal plate of the terminal block is analyzed in Ansys Workbench with a given voltage of 48,000 mV and 100,000 mA to simulate the electro-thermal properties. For that, a Computer-Aided Design (CAD) model is designed in SolidWorks (Figures 17.34–17.37).

By the above simulation studies and with professional assistance, it is found that the bolted terminal block is safe and efficient and has open-air circulation. Hence, the bolt-type terminal blocks are used for connecting the HV wires.

17.15.2 Mounting Plate

The whole control unit is mounted on the firewall plane by attaching a plate to the RRH (Rear Roll Hoop) plane's LDB (Lateral Diagonal Bracing) member (Figure 17.38).

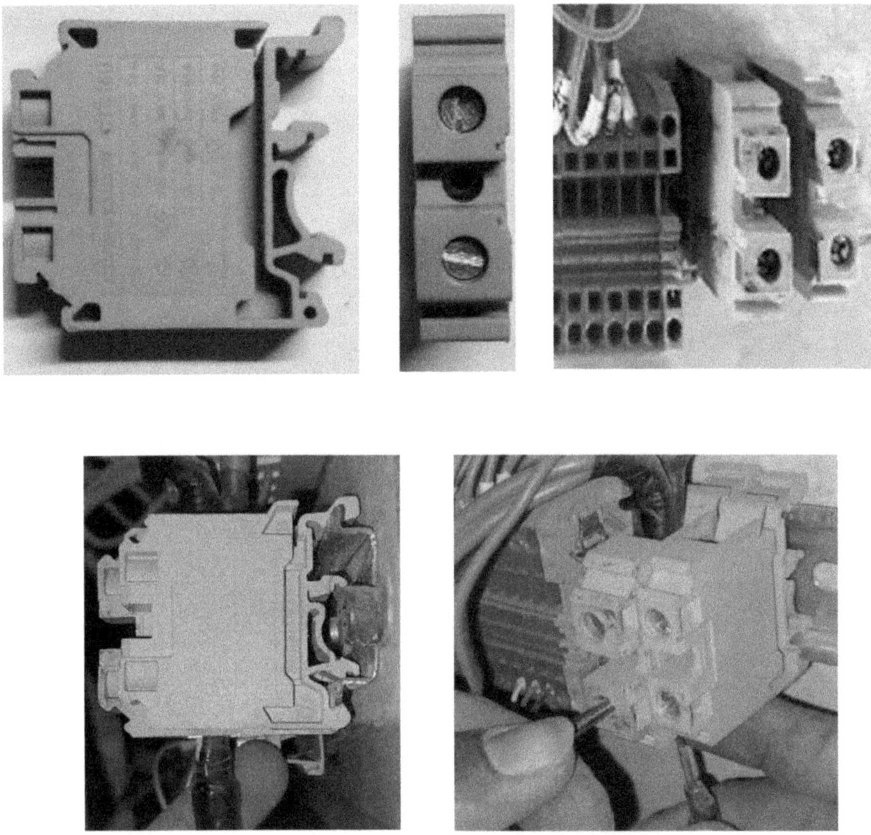

FIGURE 17.32 Old Connectwell CTS16U pin-type terminal blocks.

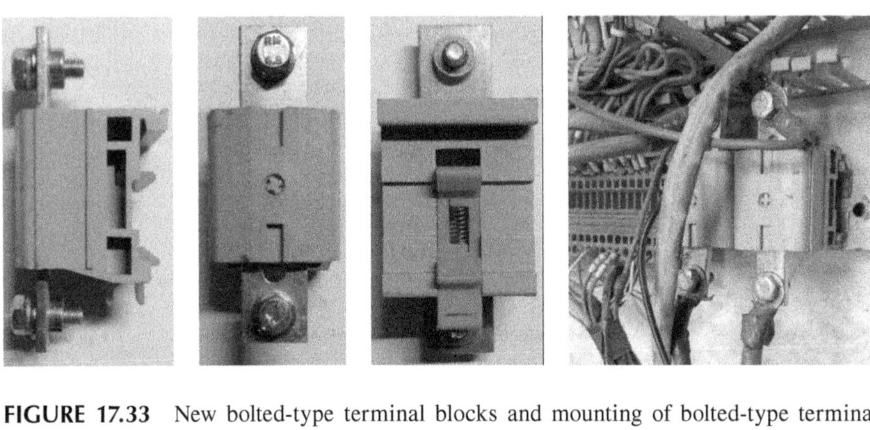

FIGURE 17.33 New bolted-type terminal blocks and mounting of bolted-type terminal blocks.

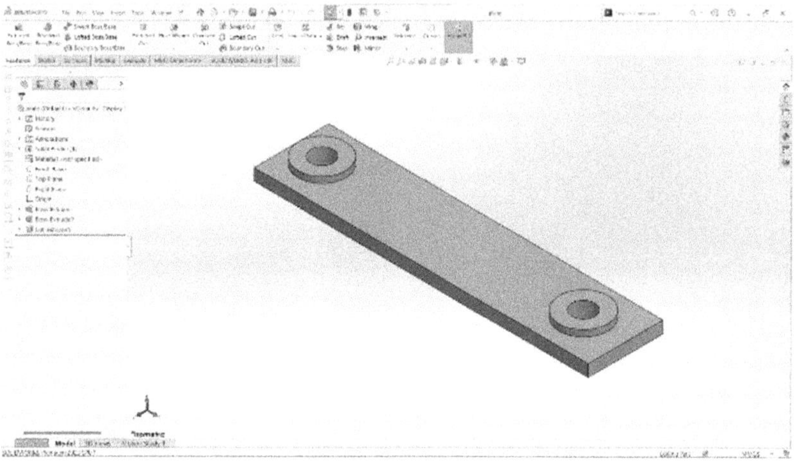

FIGURE 17.34 CAD model of an electric metal part of a terminal block.

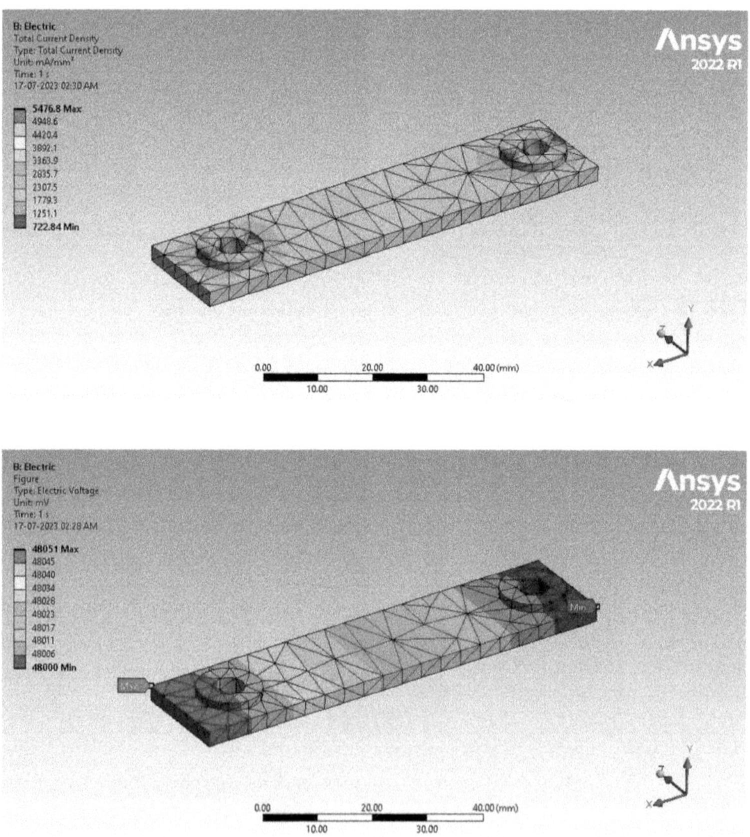

FIGURE 17.35 Total current density (mA/mm^2) and transient electric voltage changes for short duration (mV/1 s).

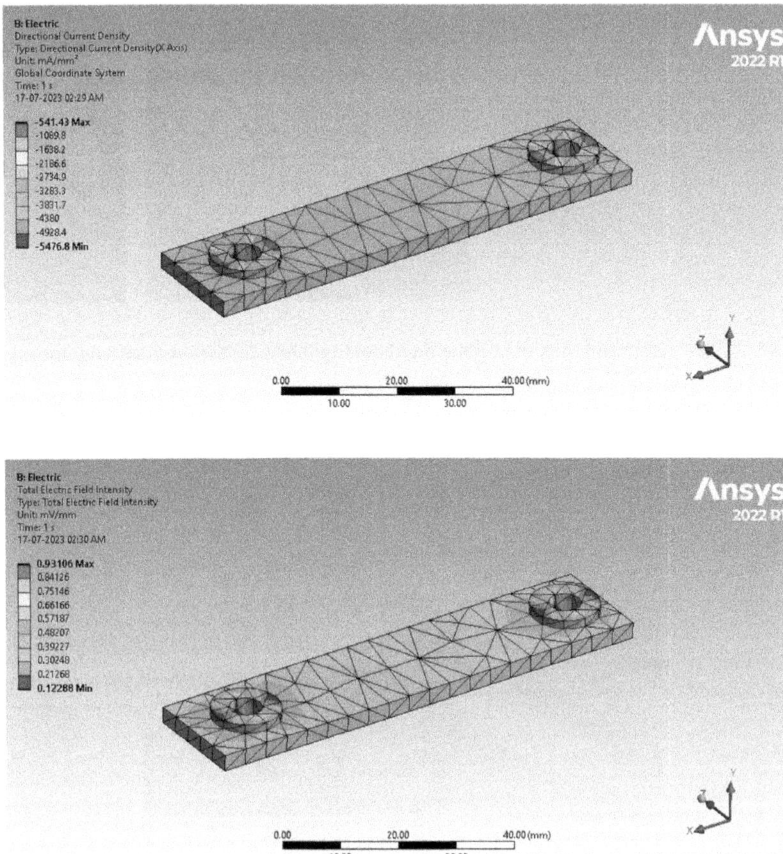

FIGURE 17.36 X-axis based directional current density (mA/mm²) and electric field intensity (mV/mm).

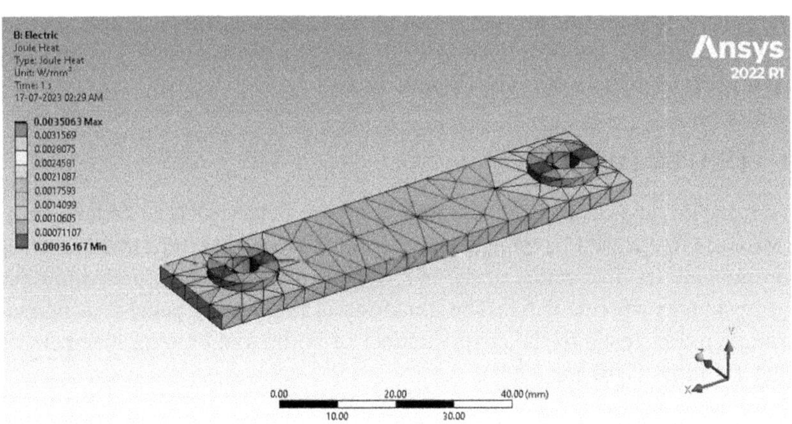

FIGURE 17.37 Joule's heating (W/mm²).

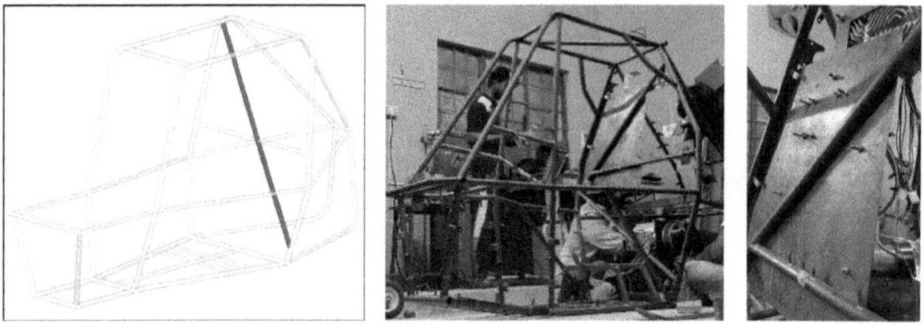

FIGURE 17.38 Mounting position, plate fixed during fabrication phase, and mounting position and plate fixed during fabrication phase.

FIGURE 17.39 Image of AIR.

17.16 ACCUMULATOR ISOLATION RELAY (AIR)

The AIR is a protective HV equipment used to act like a switch between the battery and the main circuit. It latches only when the conditions are safe. Whenever the kill switch is pressed, it immediately latches off and cuts off the circuit from the battery. The AIR is of normally open type (Figure 17.39).

17.17 DRIVER DISPLAY

A driver display is used to indicate important parameters such as battery temperature, State of Charge (SoC), voltage, current, etc. Two bright lamps (red and blue) are used to indicate the live status of the HV tractive system and grounded low (GLV) voltage system as per rule C.5.8. The mounting of the display panel is designed and 3D printed and fastened to the cross member of both front bracing members of the roll cage (Figures 17.40 and 17.41).

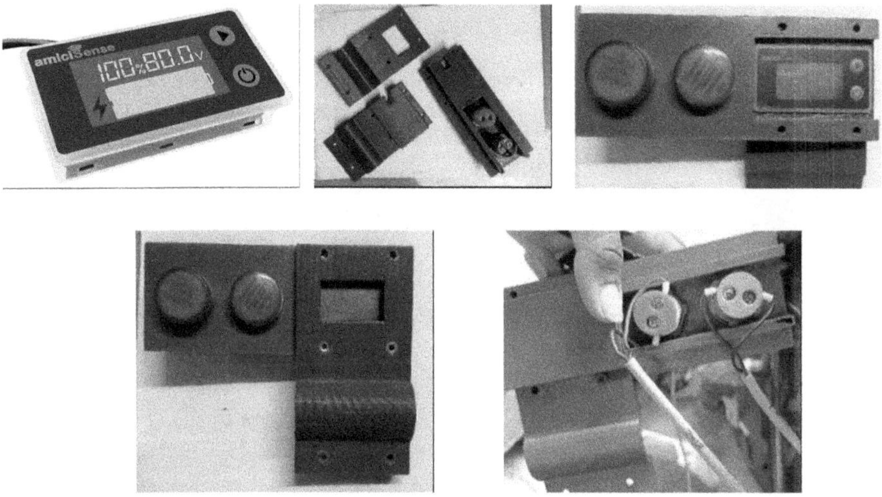

FIGURE 17.40 Driver displays with red and blue panel lights fitted inside a 3D printed holder case.

FIGURE 17.41 Mounting of driver display in the vehicle.

17.18 FUSE

The isolation relays' maximum peak current must be less than the rating of the fuse used to protect the circuit for the accumulator(s) tractive system.

A fuse with a rating higher than the electrical system's current rating is provided to protect each of the electrical systems (both the GLV and the tractive system). A fuse's current rating cannot exceed the peak current cut-off rating. The fuse has an instant-blow type and a 500 ms time delay. It is of automotive quality (Figure 17.42).

Since the fuses withstand very high current cycles during running conditions and produce enormous heat, a fuse holder made of good insulating material is mounted, on which the fuse is fixed. The TosunLux FBA01 fuse holder is imported from Minnesota, US, and is used as the fuse holder (Figure 17.43).

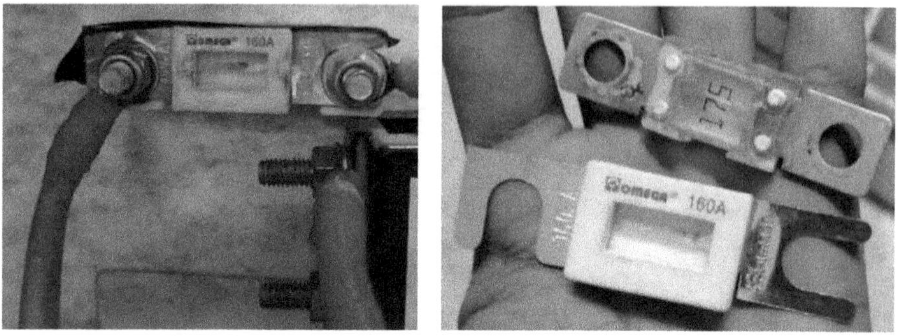

FIGURE 17.42 Images of HV protection fuses.

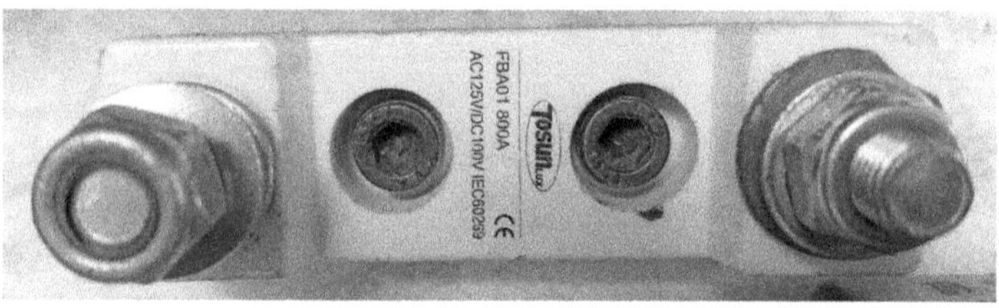

FIGURE 17.43 TosunLux FBA01 CE certified IEC grade fuse holder.

17.19 HV CUT-OFF SWITCH

An HV-rated cut-off switch is connected between the tractive system accumulator(s) and the DC-DC converter. The DC-DC converter is de-energized (no voltage detected at the input terminals) when the HV-rated cut-off switch is off. When the HV-rated cut-off switch and GLV ignition/master switch are on, it continues to power the safety appliances and auxiliaries irrespective of the kill switch position. The Milltec c10 240V model is used as an HV cut-off switch (Figure 17.44).

17.20 GROUNDED LOW VOLTAGE SYSTEMS (GLV)

The GLV system consists of lights, alarms, and other ancillaries. The Tractive System Active Light (TSAL) indicates the operation of the high voltage traction system. The reverse light and reverse alarm indicate the working of the reverse operation. The brake light operates whenever the brake is applied. The driver indication lamps indicate the GLV and HV status to the driver. The ignition key is used to start the vehicle. The FNR (forward-neutral-reverse) switch is used to toggle between forward drive, reverse mode, and neutral condition. The kill switches are used to open the complete circuit of the vehicle whenever an emergency occurs. The ready to drive sound operates whenever the vehicle is started for 3 seconds (Figure 17.45).

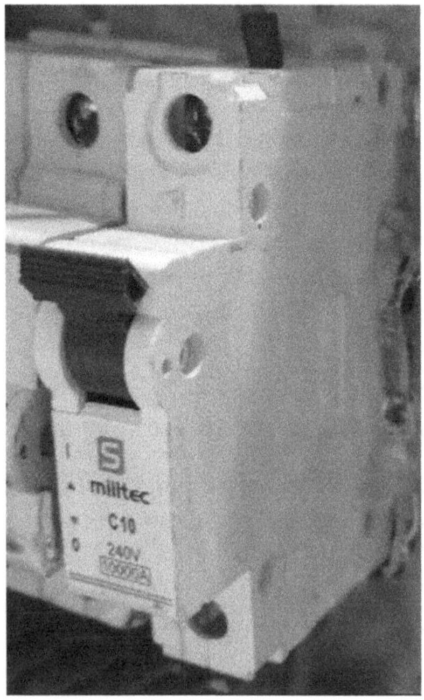

FIGURE 17.44 Image of HV cut-off switch.

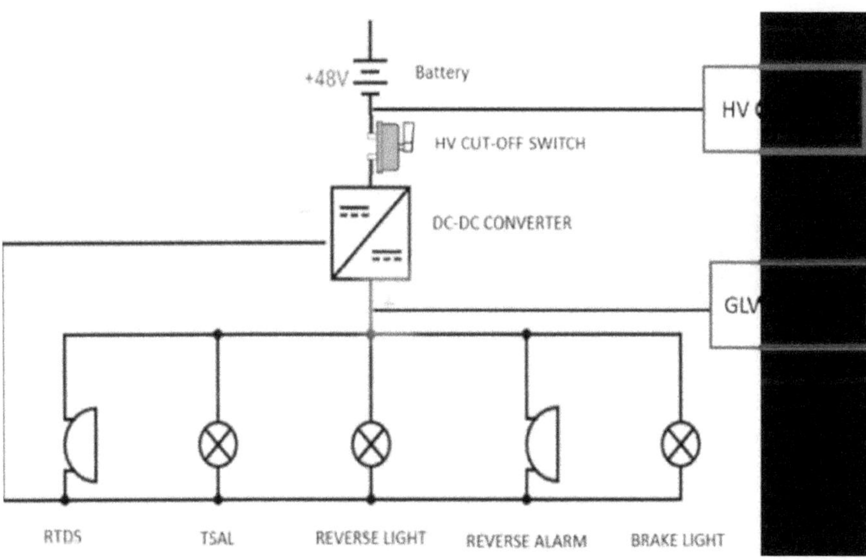

FIGURE 17.45 GLV circuit diagram of the vehicle circuitry.

17.20.1 Brake Light

The brake light is purchased as OEM, satisfying the rulebook constraints. An Automotive Indian Standard (AIS) homologation-marked brake light is purchased from an authorized dealer. The light is energized whenever the brake is applied. The brake light energizing circuit consists of two automotive pressure switches, a brake light, a relay, and a power source (DC-DC converter here). The light is in adherence to BAJA SAE J586 and J759 standards. Hydraulic pressure switches are mounted on the ancillary output ports of master cylinder. They are activated by brake fluid pressure during braking conditions and act as a switch, closing the brake light circuit (Figures 17.46 and 17.47).

17.20.2 Reverse Light and Reverse Alarm

The reverse light and reverse alarm are bought as OEM with AIS the as specified in rulebook. The reverse light and reverse alarm are energized whenever the D-N-R switch is put in R (reverse) mode. The light has an “R” engraving and satisfies SAE J759 standard. The reverse alarm has J17941 standard. The reverse alarm and reverse light are positioned as per the rulebook (27.6 inches above the ground) (Figure 17.48).

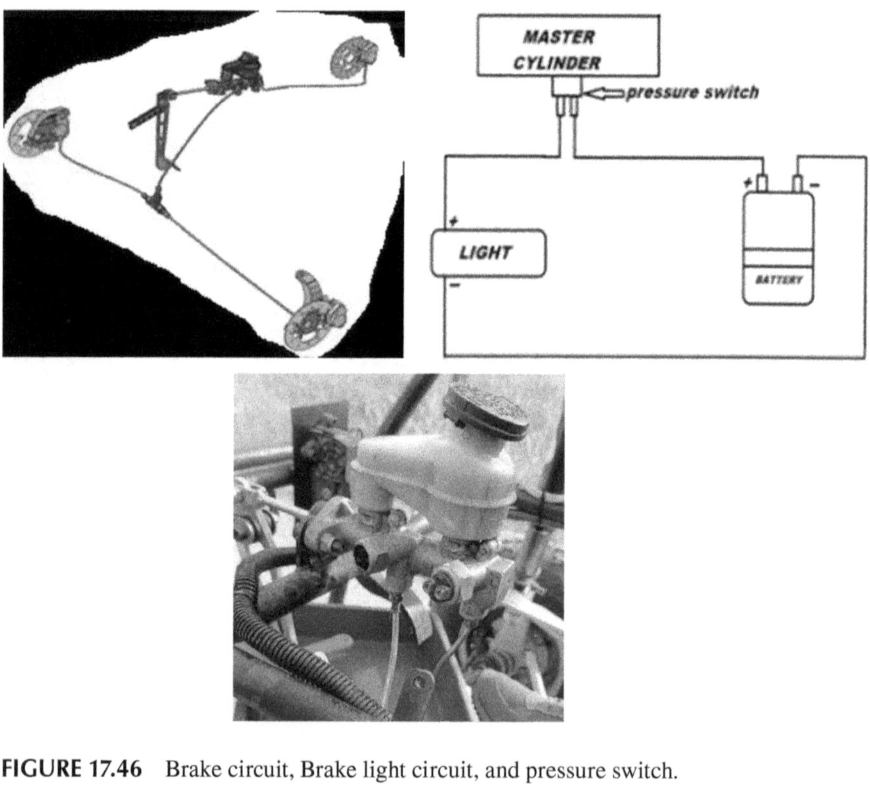

FIGURE 17.46 Brake circuit, Brake light circuit, and pressure switch.

FIGURE 17.47 Brake light, pressure switch, and arrangement of pressure switches on the master cylinder.

FIGURE 17.48 Reverse light and reverse alarm, their mounted in the vehicle.

17.20.3 Kill Switch

The kill switches are used to completely cut off the powertrain circuit from the power source in case of emergenciesy and casualties. They de-energize all the components except the brake lights. The kill switch is of the E-stop, push-to-kill, and pull-to-rotate type. Two kill switches are mounted on the vehicle. One is at the driver's cockpit side along the SIM (Side Impact Member), and another one is at the lateral side of the vehicle. The rear kill switch is positioned on the right side of the vehicle, aft of the plane of the RRH, and forward of the right FABUP for easy access for track workers or volunteers (Figure 17.49).

The mounting brackets for both the kill switches are designed in AutoCAD software and manufactured using laser cutting. The manufactured brackets are fixed using Tungsten inert gas (TIG) welding. Utmost consideration is given to kill switch mounting since it is more prone to rollover and collision (Figures 17.50 and 17.51).

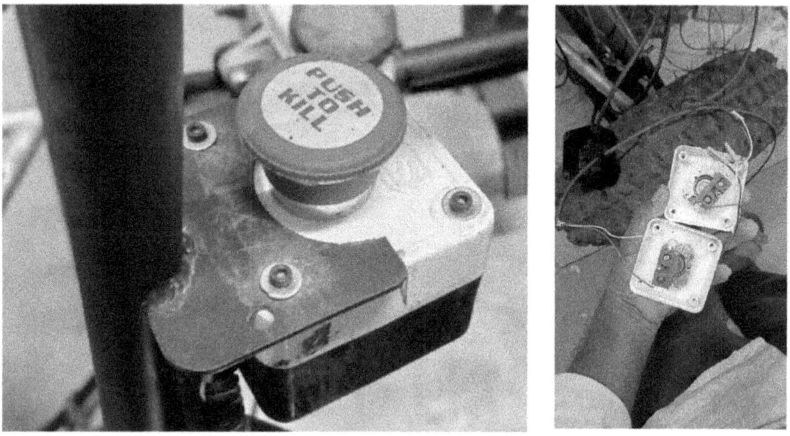

FIGURE 17.49 Kill switch – Original image and opened view of two connected Kill switches.

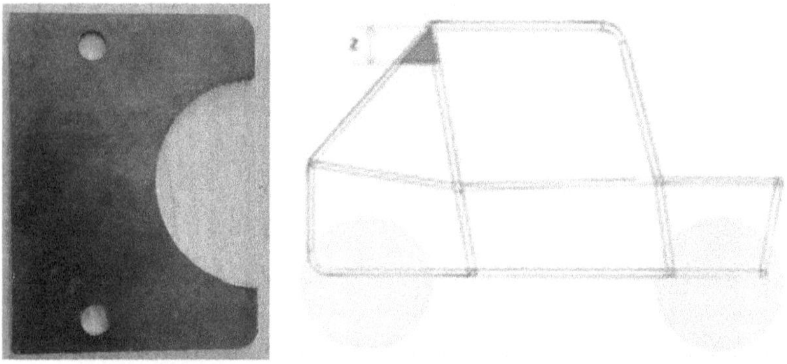

FIGURE 17.50 Mounting bracket of kill switch and rear kill switch mounting position as given in the rulebook.

FIGURE 17.51 Mounting of kill switches on the vehicle.

17.20.4 TSAL Light

TSAL (tractive system activated light) is used to indicate the tractive system ON status. The TSAL is IP65 rated and has >350 lumens of brightness. It is mounted inside a cage fitted on the centre of BLC (overhead lateral cross member) (Figure 17.52).

17.20.5 Ignition Key

The ignition key completes the circuit flow when turned. The AIR and tractive system de-energize when it is turned off (Figure 17.53).

17.20.6 F-N-R Switch

The F-N-R (Forward-Neutral-Reverse) switch is used to switch over between forward driving, neutral condition, and reverse mode. The signal from this switch is given to both the motor controller as well as the main control unit for motor operation and vehicle operating sequence as per the rulebook (Figure 17.54; Table 17.7).

FIGURE 17.52 TSAL fitted inside a cage and the infographic given in the rulebook.

FIGURE 17.53 Ignition key and the position of the ignition key on the vehicle.

FIGURE 17.54 Ignition key and the position of the ignition key on the vehicle.

TABLE 17.7
GLV System Specifications

GLV Components Configuration	
1. TSAL Specifications	
Make/Model	Cali – International
Color	Amber
Flash RRate	5
Operating Voltage (V)	12 V
IP Rating	65
2. Ready to Drive Sound	
Make/Model	INEBIZ
Control Voltage (V)	14
Sound Intensity (dB) at 2m	81
3. Fuse	
Current Rating (A)	160 A/150 A
Voltage Rating (V)	54.6 V/48 V
Type (Instant Blow/Delay blow)	Instant blow
4. AIR	
Make/Model	Main Contact HV Contactor
Contact Current (A)	200 A
Contact Voltage (V)	48 V DC
Type (Normally Closed/Normally Open)	Open

(Continued)

TABLE 17.7 (*Continued*)
GLV System Specifications

GLV Components Configuration	
5. Firewall	
Insulating Layer Thickness (mm)	0.8 mm
Insulating Material Make/Model	UL94O
6. Kill Switches	
Type	Push to kill
Number of Kill Switches	2
Working Voltage (V)	12
7. Ignition Switch	
Type (Switch, Key, etc.)	Key
Working Voltage (V)	12
8. Lights	
Name of Light	Brake light
Make/OEM	MINDA
Color	Red
Name of Light	Reverse light
Make/OEM	REECE
Color	White

17.21 WIRINGS AND INSULATION

The HV wiring is insulated with sleeve insulation. The HV wiring is connected using ring connectors to ensure a safe and better connection. The firewall and the inner walls of the control panels are insulated using UL94-V0/FAR25 Nomex material ("H" class), which is rated as "H" class and can withstand high temperatures and high voltages. To test its reliability, a small flame is induced, and the material's heat-spread stopping property is checked. The connection joints are insulated. The HV components are marked with high voltage warning stickers. The wires are insulated with insulative orange sleeves to protect against thermal runaway. The + and – terminals are given proper color codes for polarity identification (Figures 17.55–17.58).

The connectors used in previous years caused loose connections and unreliable effects such as heating. So, ring connectors (RC) and insulated twin-end lugs were used. The wires with lugs and connectors were crimped using a crimping tool for better grip (Figure 17.59).

The wire conduits or hoses were used to protect the wires from external damage and for neat wire routing. The wiring at the cockpit and the wiring at the motor vicinity is given more priority since it involves more safety in crash and rotating components. The zip ties were mostly used to tie up the wire routes with the chassis to protect from vibrations and vehicle movement. Duct tapes and insulation tapes were used for protecting the terminals and wires from electrical damage, short circuits (Figures 17.60 and 17.61).

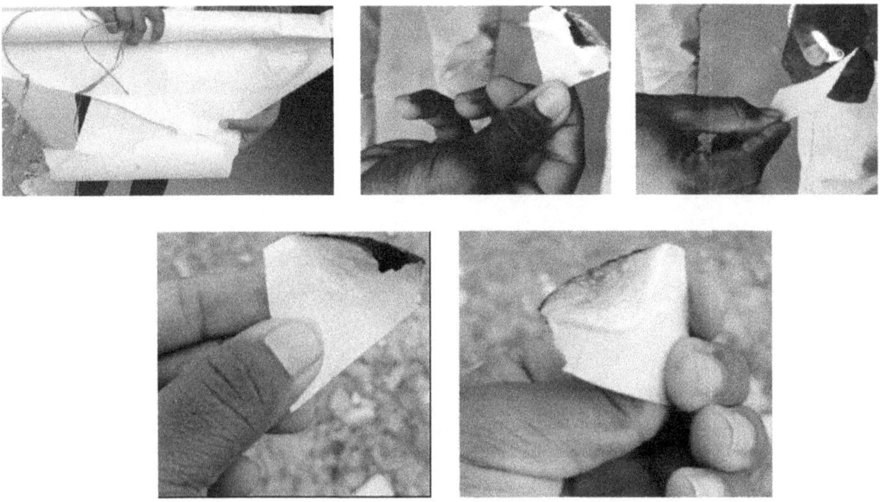

FIGURE 17.55 H class Nomex material and flame testing.

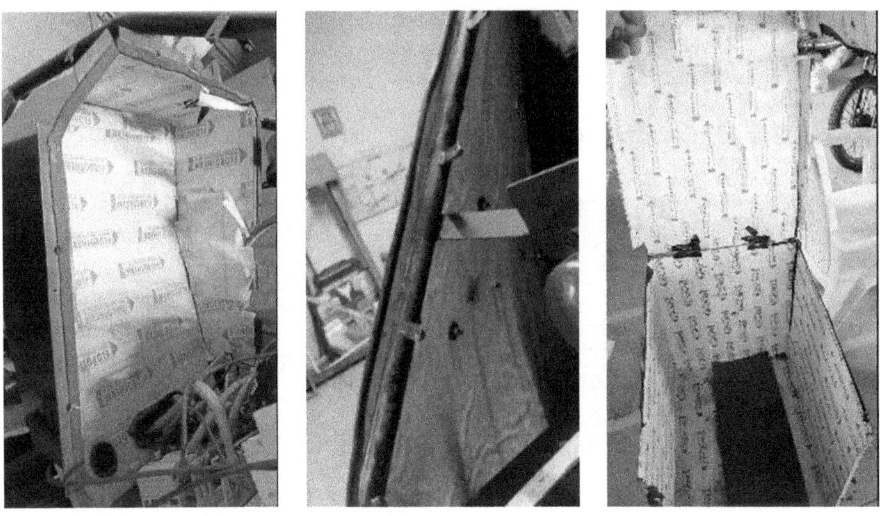

FIGURE 17.56 HV warning stickering on motor, UL 94 Nomex sheeting on control panel casing and side protection panels.

Boundary conditions were set as (Figure 17.67),

- MATLAB Road Length = 30 m
- Torque = 560 Nm

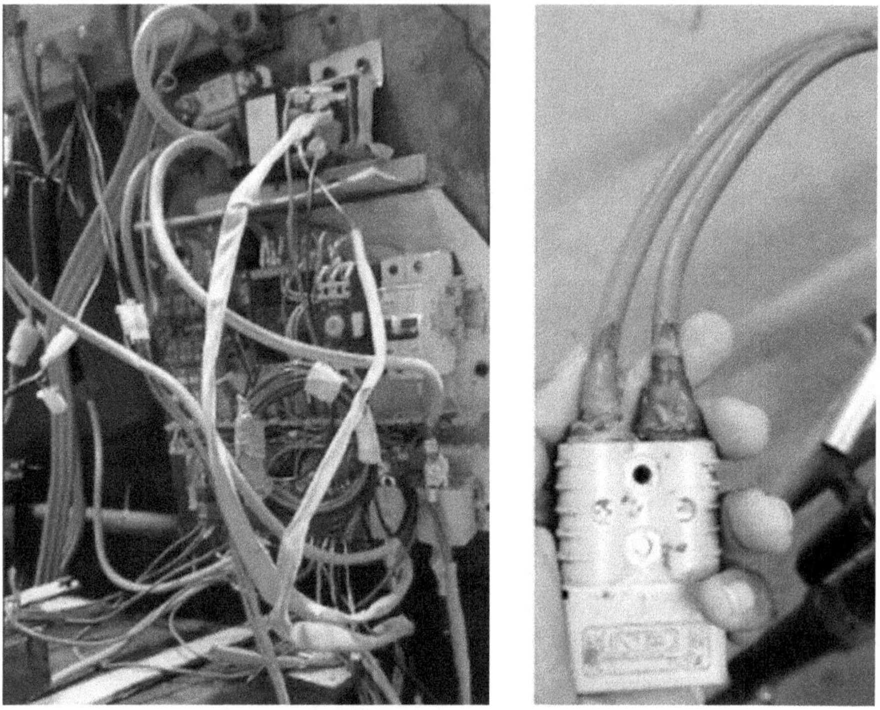

FIGURE 17.57 Orange insulative sleeve protection and colour coding of terminals.

FIGURE 17.58 High voltage warning in motor, DC-DC converter, battery pack.

17.22 CVT TUNING USING MATLAB SIMULINK

The tuning of the GAGED GX9 CVT is simulated in MATLAB®/Simulink, and the purpose is to engage the buggy @500 Revolution Per Minute (RPM) by changing the flyweights and spring stiffness in the primary pulley, and tuning the shift speed in the secondary pulley is in progess. The graph ultimately illustrates the engagement of the buggy when the centrifugal force exceeds the opposing force exerted by the spring (Figure 17.62).

FIGURE 17.59 RC for HV and GLV systems, lugs, RC open cable ring lug.

FIGURE 17.60 Member-tied non-exposing small conduit wire routing.

17.22.1 Starting Sequence

The starting sequence of the vehicle is followed according to the rulebook. The HV system activates if and only if there is no accelerator push and a mandatory brake push from the driver and the mode of the drive system is turned to Neutral mode (Figure 17.63).

FIGURE 17.61 Main conduit from cockpit to main control and motor controller, motor HV wiring route inside a conduit.

The limit switch placed behind the accelerator pedal is used to detect whether there is a throttle push or not. The neutral position is identified by using the signal from the F-N-R switch. The brake press is identified by pressure switch circuit connected to the master cylinder and the main control unit (Figure 17.64).

17.22.2 Electrical Control Panel Protecting Case

The electrical control panel case is designed in SolidWorks, and Computer-Aided Engineering (CAE) analysis is done in Ansys. Aluminium is selected as the material of construction. An internal temperature of 35 °C is fixed, and ambient temperature analysis was performed. The case protects the main control circuit from water or mud splashes, external disturbances, dust, etc. (Figures 17.65 and 17.66).

17.22.3 Acceleration Test Analysis Using MATLAB/Simulink

The main objective is to find the acceleration of the vehicle with two different tires of dimension 22 × 7 inch and 23 × 7 inch. The acceleration test was carried out for a 30 – meter straight road with different tires of dimensions 22 × 7 inch and 23 × 7 inch.

The result was found to be that the 22 × 7-inch tire had better acceleration than the 23 × 7-tire.

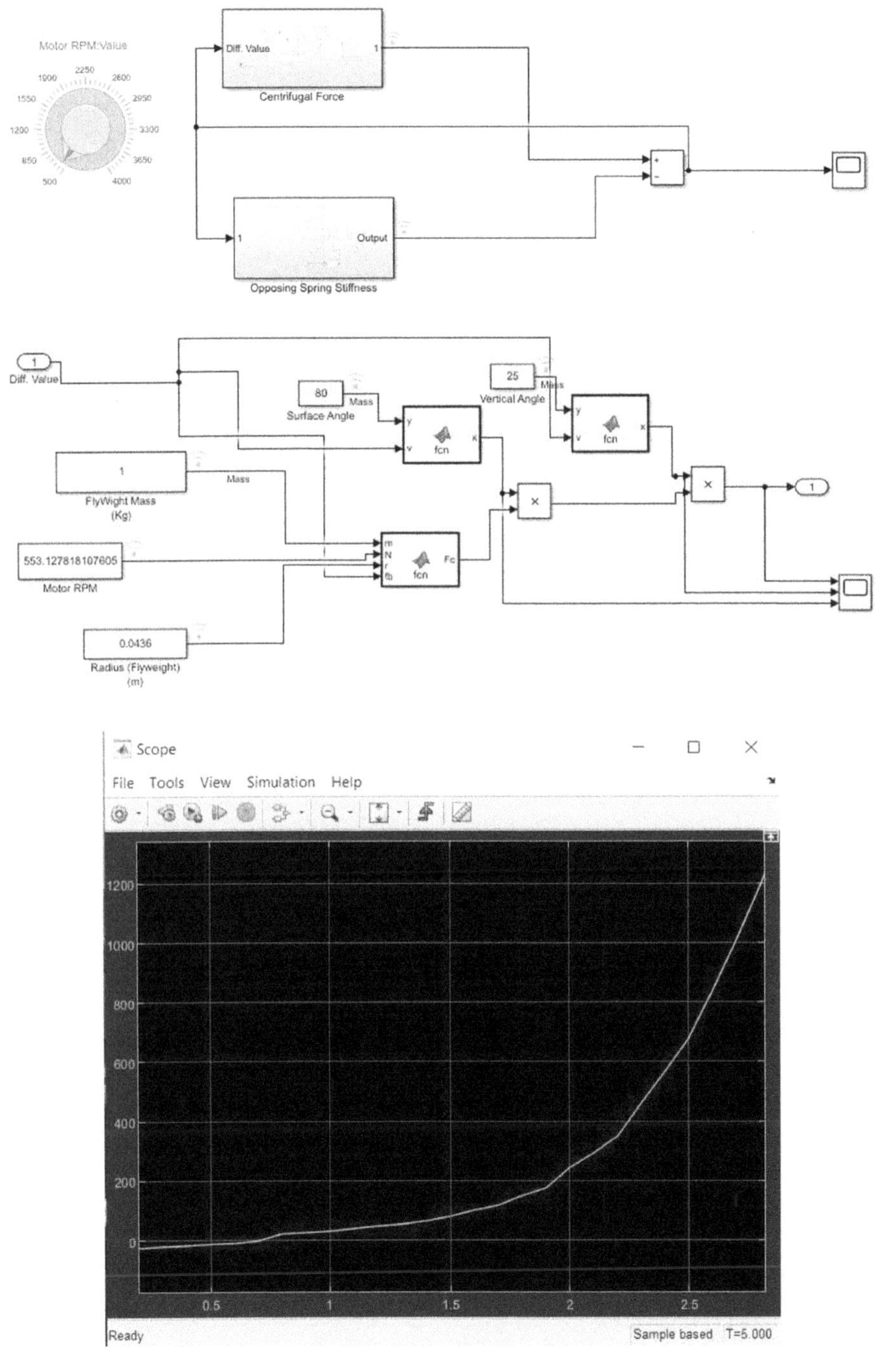

FIGURE 17.62 MATLAB Simulink simulation of CVT setup for finding relation between CVT springs and engagement RPM.

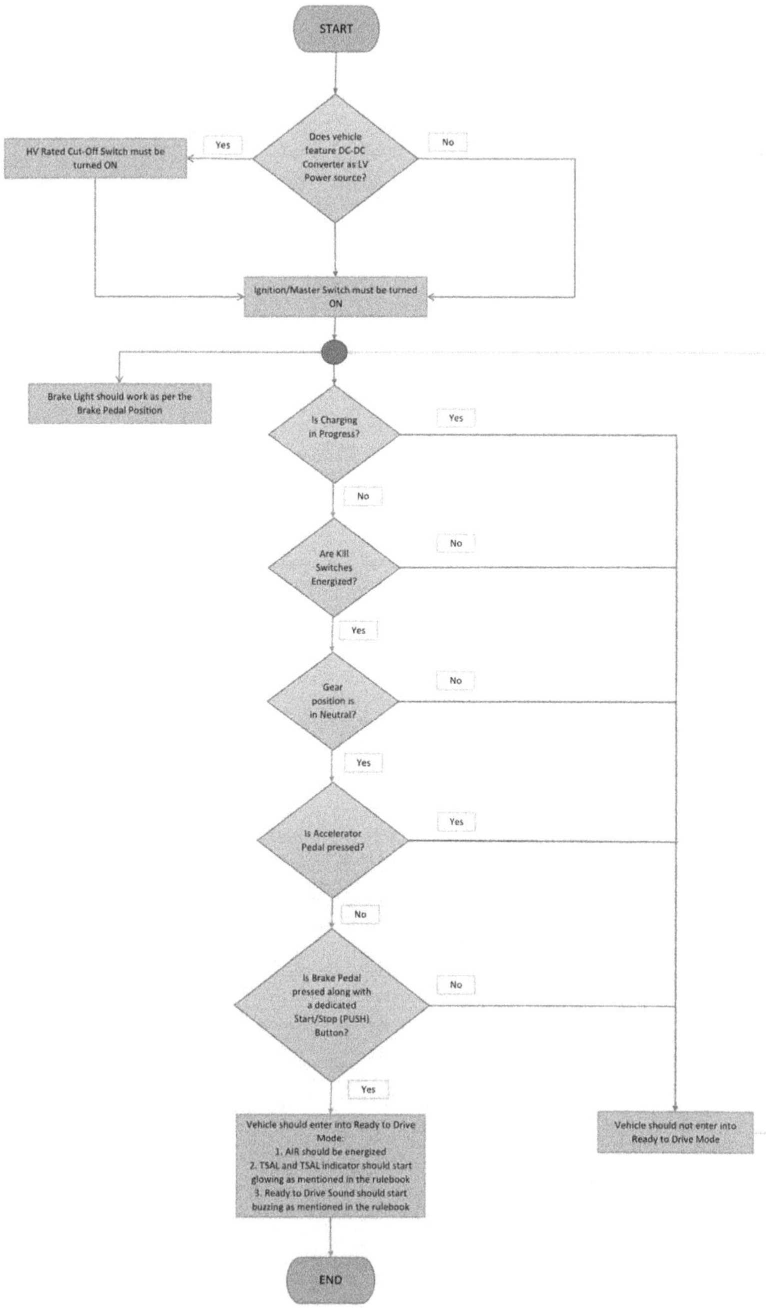

FIGURE 17.63 Drive sequence flowchart given by BAJA SAE India.

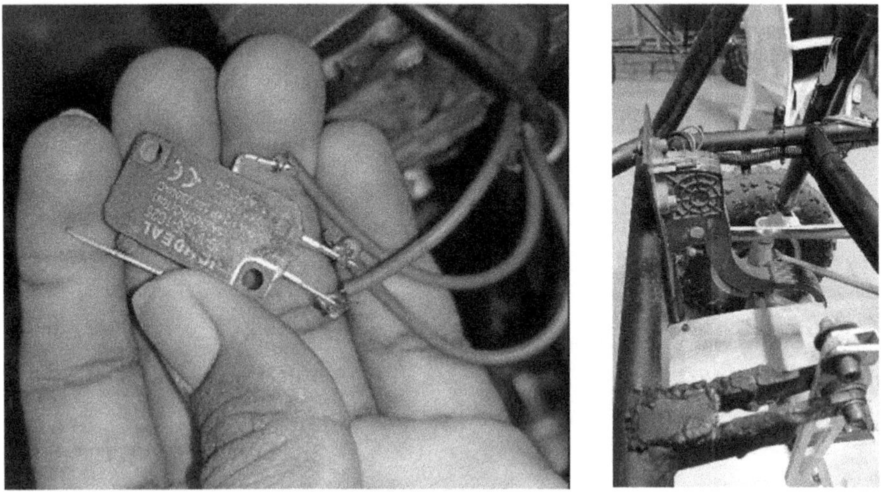

FIGURE 17.64 Limit switch and placement of limit switch behind the accelerator pedal.

Heat flux (in W/m^2)	Min: 65.397
	Average: 102.13
	Max: 108.96

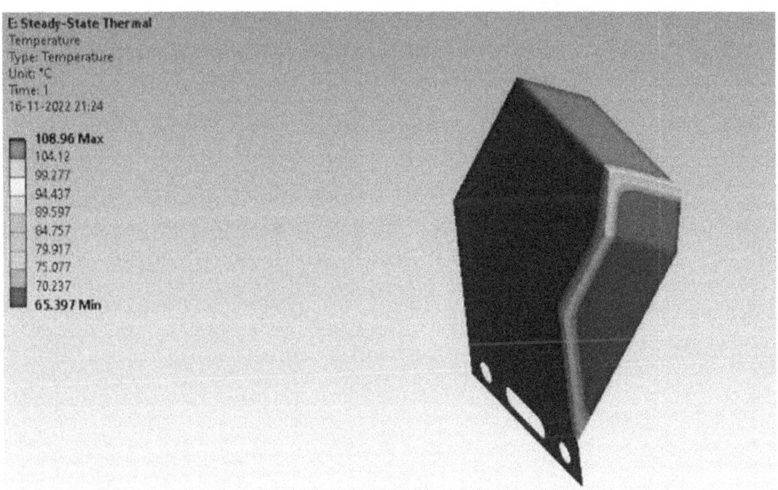

FIGURE 17.65 CAE analysis of electrical control unit casing.

FIGURE 17.66 Implementation of main control circuit protection case.

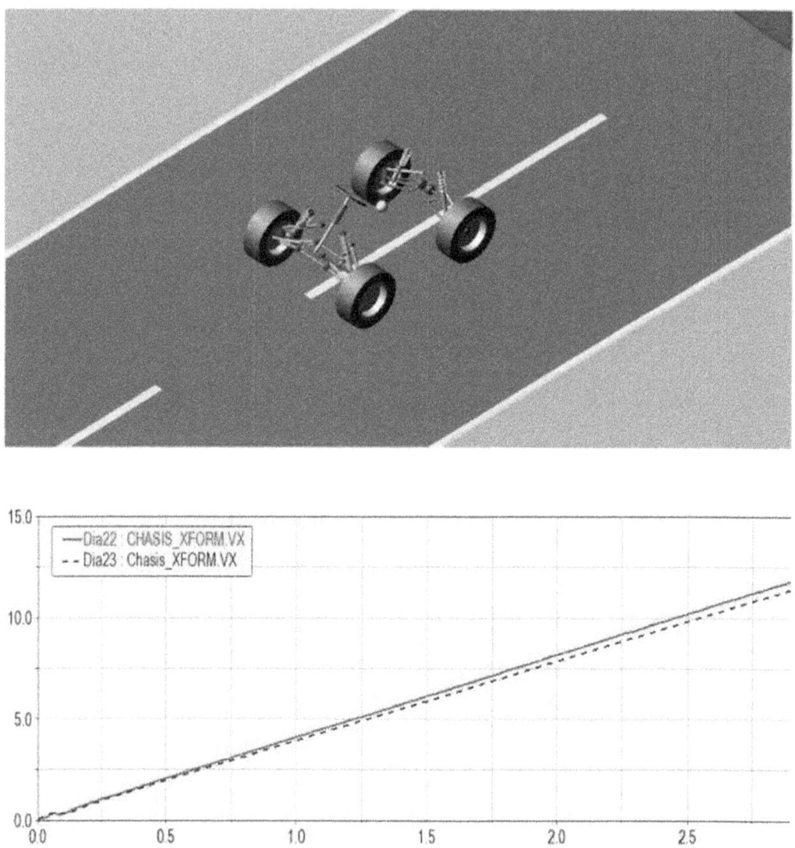

FIGURE 17.67 MATLAB Simulink analysis with GUI of vehicle speed with relation to wheel size and road length to find maximum acceleration.

FIGURE 17.68 CAD of CVT casing and CVT casing with wire mesh – implementation.

17.22.4 Design of Drivetrain Casing

From the studies, CVT is prone to slippage at higher temperatures. To minimize the operating temperature, forced convection through well-placed air vents in the CVT casing is the best possible solution. To understand the flow of air at operating temperature, Computational Fluid Dynamics (CFD) iterations were performed with various designs of the CVT casing. Ansys is used to find the heat transfer and maximum temperature of the CVT within the given boundary conditions (Figure 17.68).

17.23 DESIGN OF MOTOR MOUNT PLATE

The motor mount plate is designed such that the motor is placed with no hindrance and in correct alignment in the powertrain transmission. The measurements with accurate tolerance play a great role in the successful manufacturing of this part (Figures 17.69 and 17.70).

17.24 POSITIONING OF MOTOR CONTROLLER

The motor controller is positioned such that the heat is easily dissipated from the heat sink (Figures 17.71).

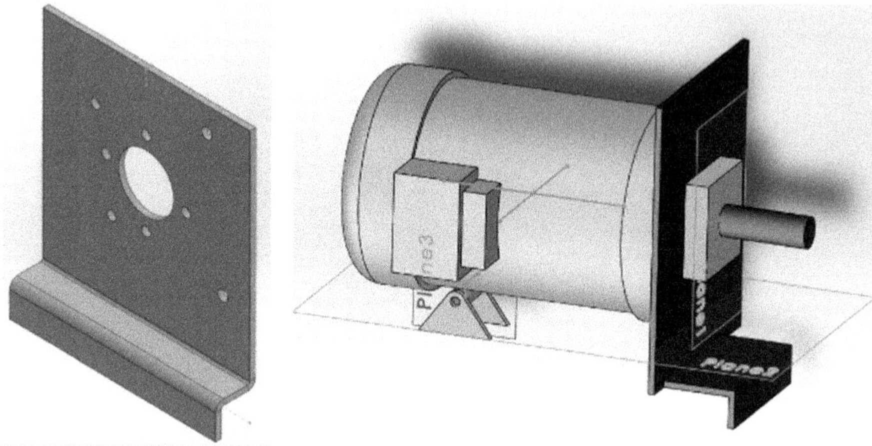

FIGURE 17.69 CAD of motor mount plate.

FIGURE 17.70 Assembly of motor mount.

FIGURE 17.71 Mounting of motor controller with ventilation to heat sink.

17.25 MOUNTING BRACKETS OF MOTOR CONTROLLER AND DC-DC CONVERTER

The motor controller and DC-DC converter were mounted using a platelike Nomex insulated bracket on the backside of the RRH plane (Figure 17.72).

17.25.1 Test Run Incidents

17.25.1.1 Wire Loosening and Melting – Abrupt Heat Formation

During the test runs, the electrical side made small problems like wiring point loosening due to vibration, wire excess heating, uncontrolled load, etc. All problems

FIGURE 17.72 Mounting bracket.

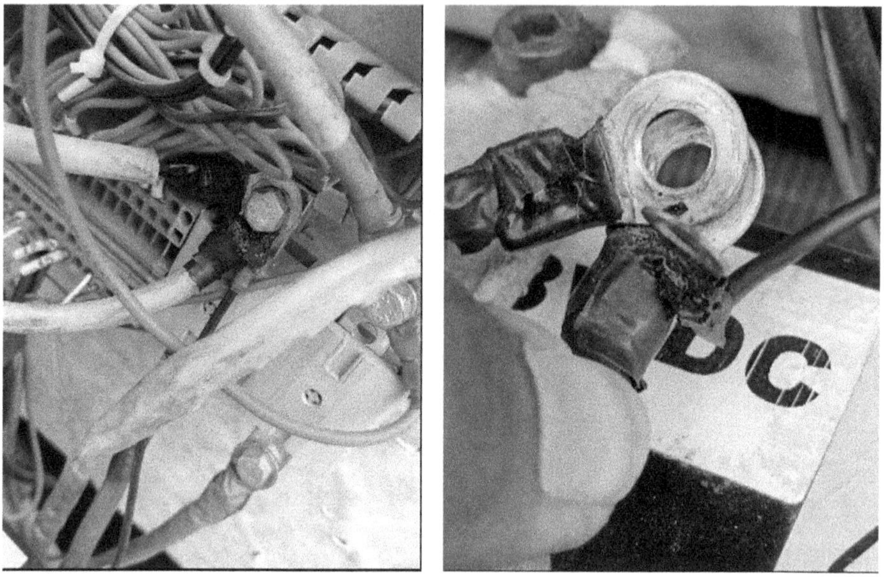

FIGURE 17.73 Wiring junction point melted because of vibration-induced fastener loosening.

were immediately addressed with the author's work and also with skilled persons assistance, and solutions were implemented by keeping safety as the first priority.

Wire excess heating was identified as vibration. Because of vibration, the improperly fastened lugs got slight movement, and over the course of time, it creates enormous heat at the fastened point, which spreads to nearby wiring insulation, causing them to melt (Figure 17.73).

17.25.1.2 Uncontrolled Power Management

The driving cycle is unpredictable and the torque-speed requirements vary for the ATV. Because of high load requirements, the wire harness is easily damaged. To avoid this, the load current and maximum instantaneous discharge current were inferred using the SmartBMS app.

Since the app can provide data only within Bluetooth short communication range, the mobile phone is given to the driver, and the results were screen recorded. Accordingly, the fuse and wiring thickness, along with ventilation, are arranged for the next test run (Figure 17.74).

17.25.1.3 Breakage of Pressure Switches

The pressure switches are not able to handle enormous brake pressure, and they easily deform during test runs. So, the dimensions and OEM brand are changed, and a better switch is implemented (Figure 17.75).

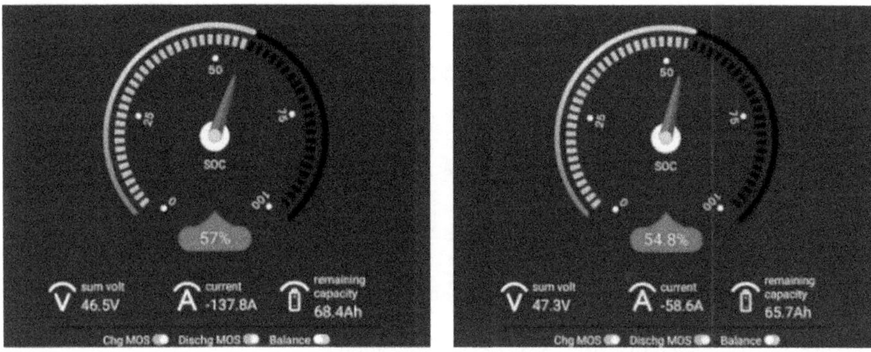

FIGURE 17.74 Varied current discharge ranging from >140 and <20 A.

FIGURE 17.75 Deformed pressure switch.

17.25.1.4 Fuse Failure

During test runs, the 160 A fuse got blown, and it was later found that the continuous excess current demand caused the fuse to blow. Then, the driving cycle pattern is adjusted, and the driver is told to follow the same. This led to smooth performance with normal heating of the fuse (Figure 17.76).

FIGURE 17.76 Blown fuse after heavy load application.

17.26 DESIGN VALIDATION PLAN CHART

Design validation plan (DVP) for the whole project is created to identify project flow obstacles in accordance with the main objectives. Each vehicle subsystem's part/aggregate system is considered for the DVP. Certain benchmark performance criteria/target are considered, and accordingly, the validation test/method for validating those performance targets is also defined with appropriate test resources or equipment (Table 17.8).

17.27 DFMEA AND PFMEA ANALYSIS OF THE PROJECT

The design/process failure mode effect analysis creates an in-depth view of solving problems which arises throughout the development of the project. The problems are identified based on their severity, frequency, and ease of detection, and the solving approach is formulated based on the priority level (Table 17.9).

17.27.1 Safety Standards

All the components, systems, subsystems, assemblies, and designs are made according to industrial standards such as AIS, IEC, IEEE (Institute of Electrical and Electronics Engineers), and AISI (American Iron and Steel Institute). To make sure that there is high-grade output with increased safety. At all places such as the work area, charging station, and swapping area, standards are followed for correct and efficient results (Figure 17.77).

17.27.2 Graphs

The graphs re given here for reference. Some graphs are provided by the OEM equipment suppliers (1, 2, 3), and some are plotted by the team (4, 5, 6) by testing the systems and components (Graphs 17.1–17.6).

TABLE 17.8
Design Validation Plan Chart

Sl. No.	Level: Vehicle, Aggregate, or Part	Performance Target	Acceptance Criteria/Target	Validation Test and Method	Test Resource/ Equipments Used
1.	Weld test (Part)	The weld should withstand the load conditions.	No welding defects must exist.	(Liquid penetrant test) Weld shall be cleaned. The penetrant and developer are applied to the test surface and checked for surface defects.	Cleaner, Penetrant, Developer
2.	Crash test (Vehicle)	The vehicle should not disintegrate.	The vehicle should resist physical deformation.	By using Hypermesh software, the crash test is simulated.	Hypermesh software
3.	Egression test (Aggregate)	The driver must get out of the buggy within 4 seconds	Time taken to eject must not exceed 5 seconds	The driver is buckled up with all safety equipment, and the ejecting time of the driver is noted.	Stopwatch
4.	Tire wear test (Part)	The buggy must run without any change in static camber angle	The tire must wear as per the given camber angle without any offset wear.	Buggy is set up as per the designed camera, and tire wear is noted after each test.	Vernier Calliper, Bevel Protractor
5.	Brake test (Aggregate)	Stopping distance should be within 7.2 m	All the wheels must be locked during braking, and the vehicle must stop within 8 m.	The buggy is accelerated to 40 kmph on a straight track, and the brakes are applied. Different road conditions are considered, and a brake test is done.	Test track
6.	Straight line stability test (Vehicle)	The vehicle must maintain its steering wheel angle at 0°	Steering wheel angle during dynamic condition must be within ± 2°.	The buggy is accelerated to a certain speed and the hands are taken off the steering wheel.	Test Track
7.	Turn circle radius (Aggregate)	Turn circle radius must be 1.833 metre	It must turn within the specified radius.	The buggy is made to steer into the figure of O, with its radius measured.	Measuring tape

(Continued)

TABLE 17.8 (*Continued*)
Design Validation Plan Chart

Sl. No.	Level: Vehicle, Aggregate, or Part	Performance Target	Acceptance Criteria/Target	Validation Test and Method	Test Resource/ Equipments Used
8.	Hill climb test (Aggregate)	Gradeability should be 50%	The buggy must be able to climb a 25^0 slope	The buggy is made to climb a 25^0 slope for a distance of 30 meters.	Ramp inclined at 25^0 from the ground
9.	Acceleration test (Aggregate)	The buggy must accelerate up to 1.42 m/s^2	The buggy must cover 30 meters in 10 seconds (1.42 m/s^2).	Buggy is accelerated on a 30-meter straight track, and the time taken is noted.	Test track, stopwatch
10.	Top speed test (Aggregate)	The buggy must be able to achieve a top speed of 53 kmph	Top speed should be at least 50 kmph.	Buggy is made to run at its maximum speed, and the vehicle speed is measured.	Speed sensor
11.	Gear profile test (Part)	The gear profile must be involute.	The gears must mesh with each other.	The gear is measured by the shadowgraph technique.	Profile projector
12.	Gear life cycle test (Part)	Life span of the gears must be 200 hours	The gears must withstand the wear for the given life cycle.	The gearbox is connected to a motor and made to run periodically for 10 hours under load conditions. Gear wear is checked at regular intervals.	Test rig
13.	Drop test (Vehicle)	The vehicle must withstand its structural integrity when dropped from a height of 5 feet.	No breakage of components must occur	The buggy is dropped from a 5-foot height. The drop is video-graphed and checked for body motions.	camera

TABLE 17.9
Design/Process Failure Mode Effect analysis

Item/Function	Potential Failure Mode	Potential Effect(s) of Failures	Sev	Potential Cause(s)/ Mechanism(s) of Failures	Occ	Current Design Control	Det	RPN	Recommended Action(s)
Roll Cage Fixtures	Leads to intense assembly errors and geometrical inaccuracies	Assembly Problems	5	Poor welding fixtures that do not arrest the intended DOF	8	Wooden fixtures	4	160	Accurate dimensioning
Throttle Pedal	Disengagement from fore and aft bracing (low), the signal sent by the throttle pedal to the motor is hindered	Reduce performance in some cases, not operable at all	8	Improper welding, excessive pedal effort from the driver. The wiring between the pedal and motor controller is halted.	6	Accurate fabrication	8	384	Correct design with assembly Monitoring and accurate final fabrication
BMS	Failure of BMS due to thermal runaway and overpower	Uncontrolled accumulator parameters that affect other equipment and may cause hazards	5	Seldom monitoring of battery parameters, Improper tolerances	8	Battery checking through the app	4	160	Frequent monitoring of accumulator parameters
Machine (Hubs and Uprights)	Improper dimensions	May result in manufacturing malfunctions	6	Improper tolerances	4	Manual filing	2	48	Use of alternate fixtures with high accuracy

(Continued)

TABLE 17.9 (*Continued*)
Design/Process Failure Mode Effect analysis

Item/Function	Potential Failure Mode	Potential Effect(s) of Failures	Sev	Potential Cause(s)/ Mechanism(s) of Failures	Occ	Current Design Control	Det	RPN	Recommended Action(s)
Design of Mounting Brackets	Unpaired brackets leading to assembly problems	Brackets tend to bend under shock loads	7	Improper welding/ positioning, bracket thickness	6	Manual marking of mounting points on the roll cage	3	126	Improvised design with proper thickness (3mm)
Welding of Suspension Mounting Brackets	Rupture of brackets onto the rollcage, bracket displacement	Unbalanced Geometry may result in poor steering and damping	6	Angular variation (brackets)	8	Rework	6	288	Ensure the brackets are parallel to each other
Steering Arm	Rupture or bending of steering arm bolt	Can cause considerable damage to the steering arm	7	Usage of 8.8 grade bolts	6	Frequent change of bolts	4	168	Usage of bolts with high strength
Gearbox casing	Gear teeth failure	Results in uneven meshing	8	Improper heat treatment	3	Frequent oil refill	2	48	Proper assembly of the gearbox
Tie Rod Ends	Limits the steer angle and the motion of the tie rod	May result in poor steering or failure of rod ends	8	Fail to attain the required articulation	4	Frequent change of rod ends	2	64	Increased articulation of rod ends

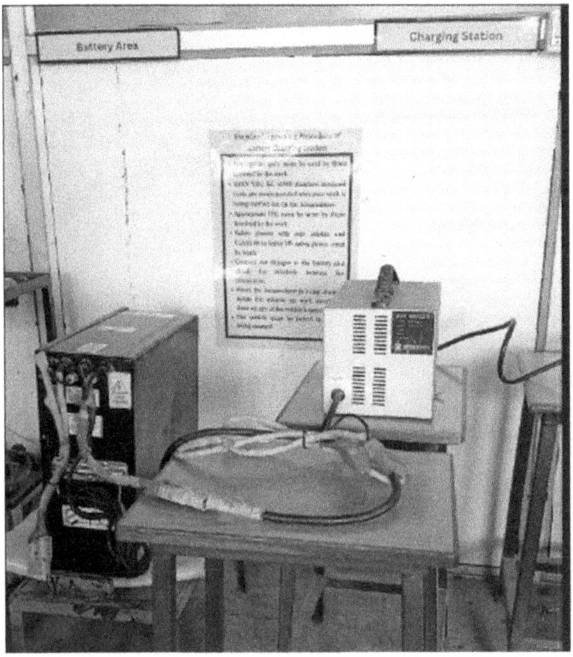

FIGURE 17.77 Charging station and battery pack (with implementation of given standards).

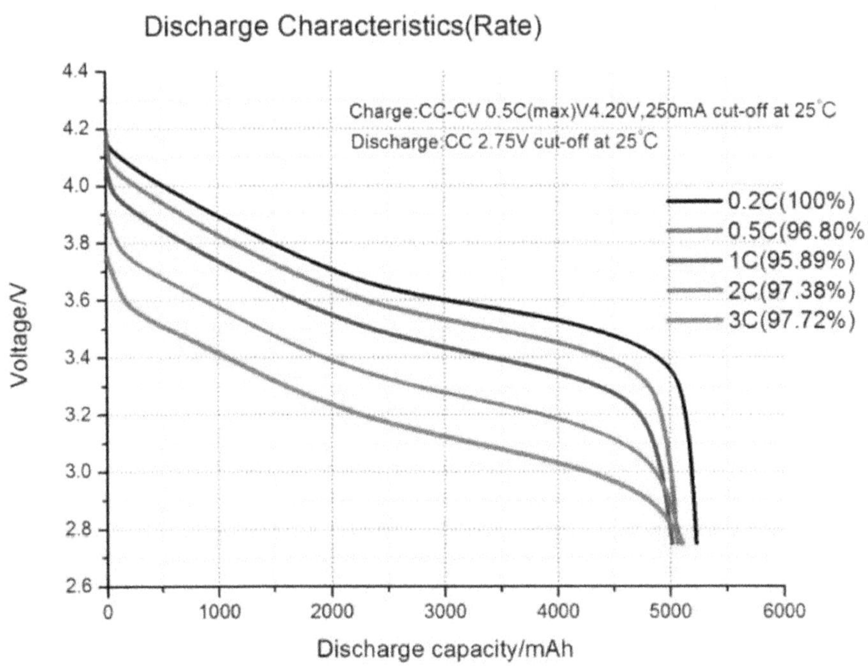

GRAPH 17.1 Charge and discharge characteristics curve.

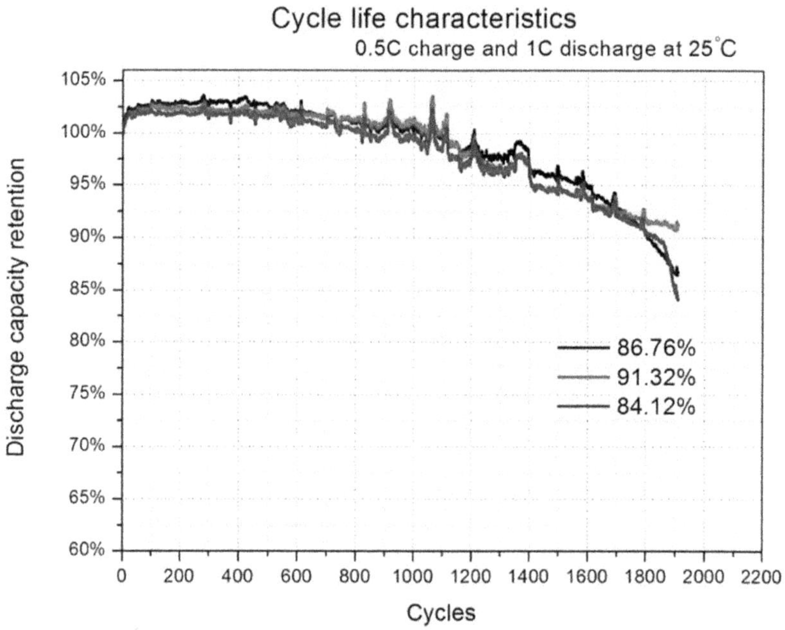

GRAPH 17.2 Cycle life characteristics curve.

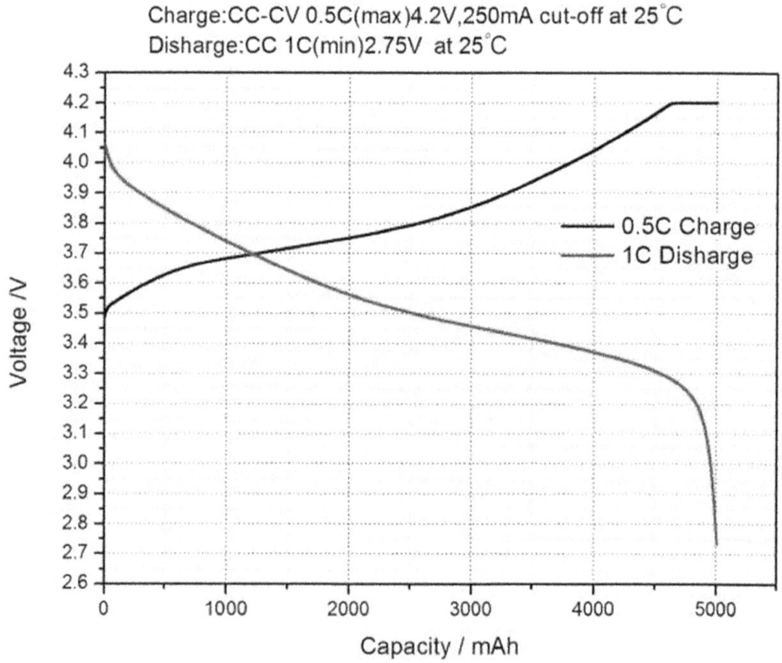

GRAPH 17.3 Rate discharge characteristics curve.

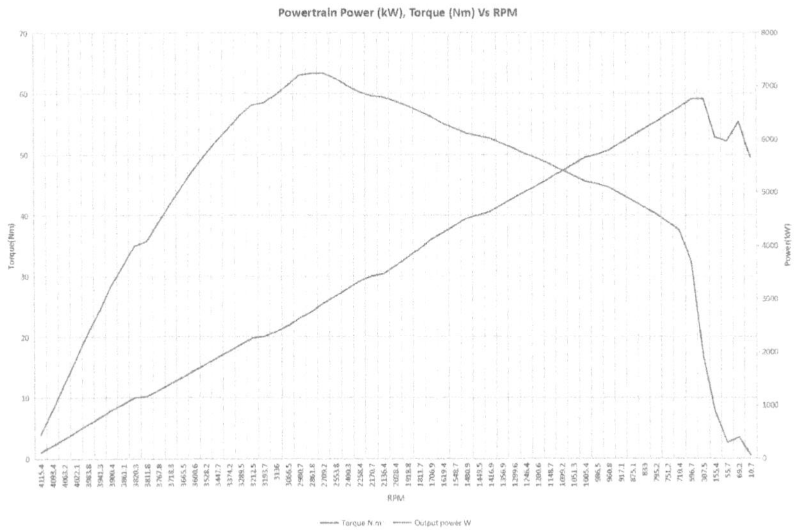

GRAPH 17.4 Powertrain power (kW), torque (Nm) Vs RPM.

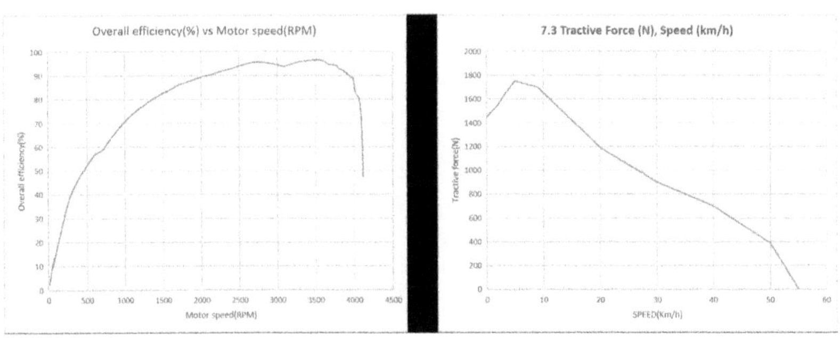

GRAPH 17.5 Overall efficiency (%) vs motor speed (RPM) and tractive force (N), speed (km/h).

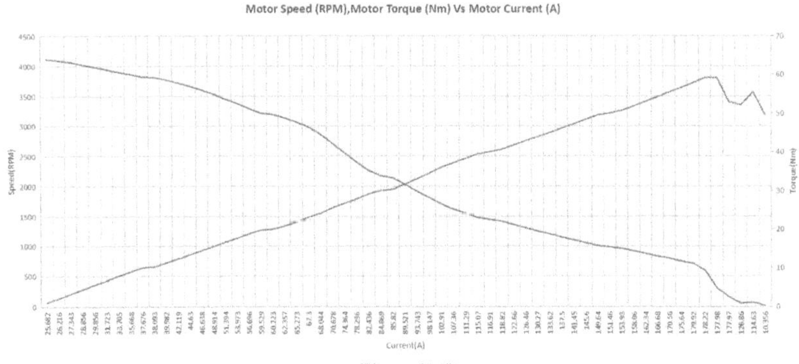

GRAPH 17.6 Motor speed (RPM), motor torque (Nm) vs motor current (A).

17.27.3 Lab Test Reports

The electric powertrain components were thoroughly certified in an NABL accredited lab for product reliability. The lab reports were documented for future reference, such as Technical Inspection validation.

These reports ensure that each component of every subsystem works fine under extreme conditions like heat, wet conditions, extreme electrical stress, and other hazardous conditions (Figures 17.78–17.81).

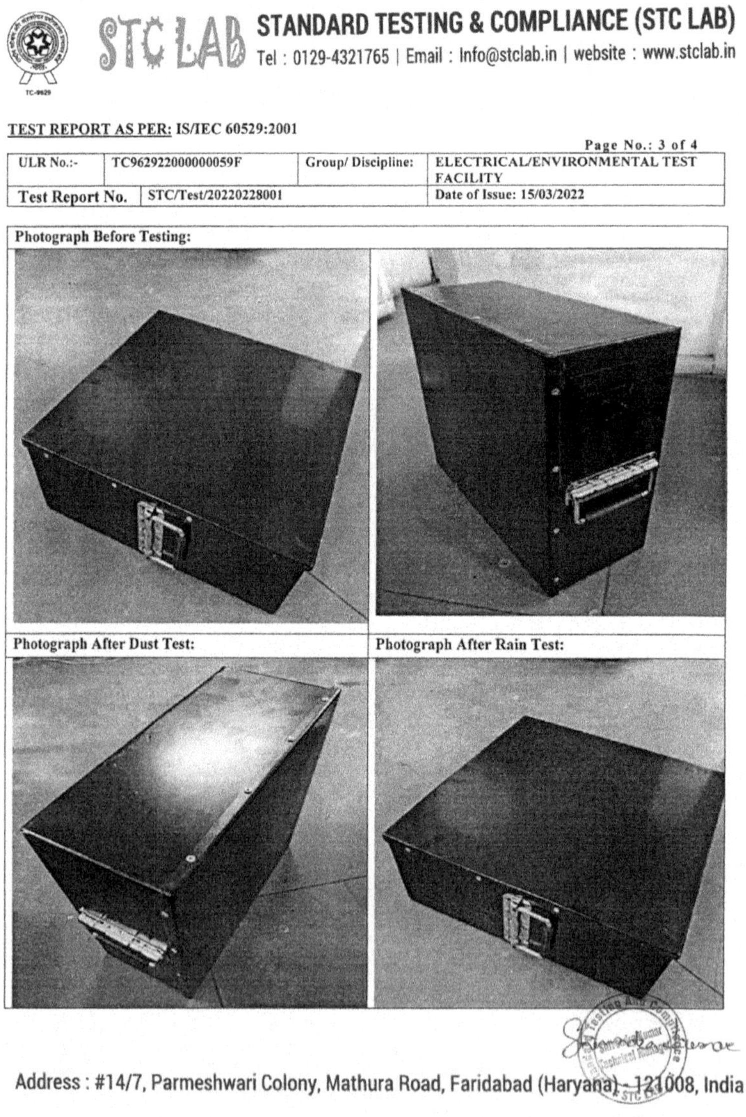
STC LAB STANDARD TESTING & COMPLIANCE (STC LAB)

Tel : 0129-4321765 | Email : Info@stclab.in | website : www.stclab.in

TC-9629

TEST REPORT AS PER: IS/IEC 60529:2001

Page No.: 3 of 4

ULR No.:-	TC962922000000059F	Group/ Discipline:	ELECTRICAL/ENVIRONMENTAL TEST FACILITY
Test Report No.	STC/Test/20220228001		Date of Issue: 15/03/2022

Photograph Before Testing:

Photograph After Dust Test:

Photograph After Rain Test:

Address : #14/7, Parmeshwari Colony, Mathura Road, Faridabad (Haryana) - 121008, India

FIGURE 17.78 Battery pack IP test results.

STC LAB STANDARD TESTING & COMPLIANCE (STC LAB)
Tel : 0129-4321765 | Email : Info@stclab.in | website : www.stclab.in

TC-9629

TEST REPORT AS PER: IS/IEC 60529:2001

Page No.: 3 of 3

ULR No.:-	TC962921000000020F	Group/ Discipline:	ELECTRICAL/ENVIRONMENTAL TEST FACILITY
Test Report No.	STC/Test/ 20211115001		Date of Issue: 17/11/2021

Photograph Before Testing

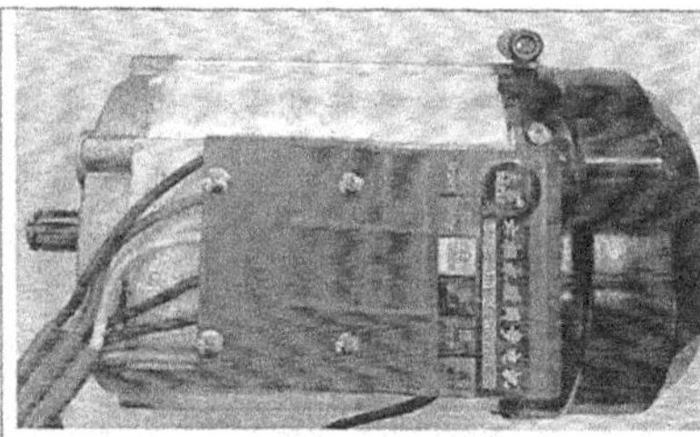

Photograph After Testing

PART D:- Nil

REMARKS:

Tested by	Approved by
(Bharti Rani/T.E)	(Chandra Prakash Chandan/Director)

***** END OF TEST REPORT *****

Address : #14/7, Parmeshwari Colony, Mathura Road, Faridabad (Haryana) - 121008, India

FIGURE 17.79 Motor test results.

STC LAB **STANDARD TESTING & COMPLIANCE (STC LAB)**

Tel : 0129-4321765 | Email : Info@stclab.in | website : www.stclab.in

TC-9829

TEST REPORT AS PER: IS/IEC 60529:2001

Page No.: 3 of 3

ULR No.:-	TC962921000000023F	**Group/ Discipline:**	**ELECTRICAL/ENVIRONMENTAL TEST FACILITY**
Test Report No.	STC/Test/ 20211118002		**Date of Issue:** 20/11/2021

Photograph Before Testing

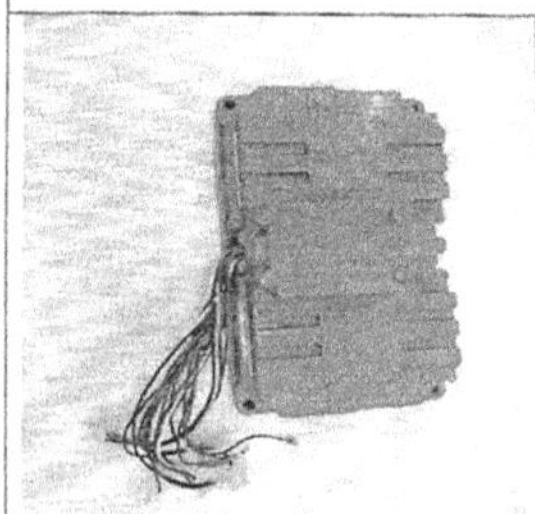

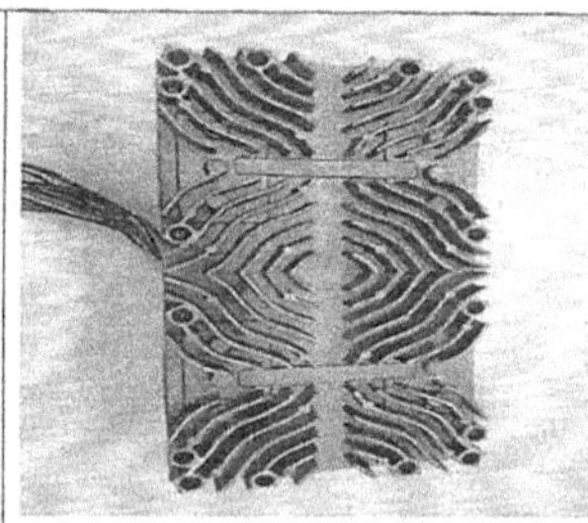

Photograph After Testing

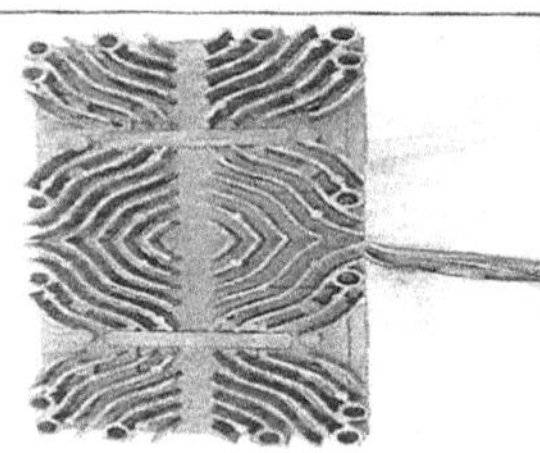

PART D:- Nil

REMARKS:

Tested by	**Approved by**
Bharti Rani	Chandan
(Bharti Rani/T.E)	**(Chandra Prakash Chandan/Director)**

******* END OF TEST REPORT *******

Address : #14/7, Parmeshwari Colony, Mathura Road, Faridabad (Haryana) - 121008, India

FIGURE 17.80 Motor controller test results.

STANDARD TESTING & COMPLIANCE (STC LAB)

Test Report AS Per:AIS048

Page No.: 7 of 7

ULR No.:-	--	Group/ Discipline:	ELECTRICAL /BATTERY
Test Report No.	STC/Test/ N20220228002		Date of Issue: 15/03/2022

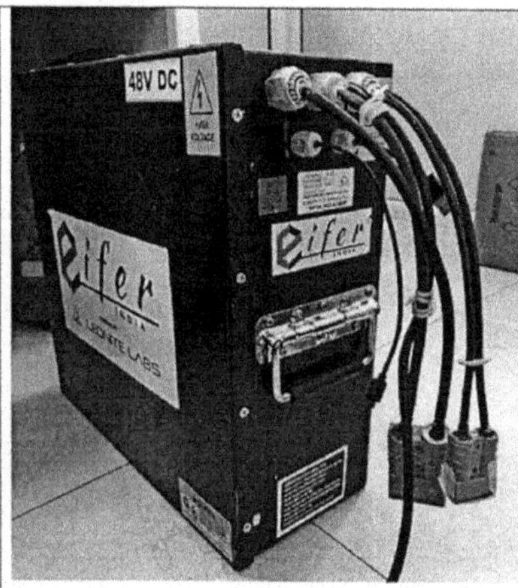

PART D:- Nil

REMARKS: We are Nabl accredited Laboratory.

Tested by	Approved by
(signature)	(stamp: Standard Testing And Compliance, Shaktiman Sangam QM / TM, STC LAB) (signature)
(Esha Tyagi/T.E.)	(Shaktiman Sangam/QM/TM)

***** END OF TEST REPORT *****

Address : #14/7, Parmeshwari Colony, Mathura Road, Faridabad (Haryana) - 121008, India

FIGURE 17.81 Battery test results.

17.27.4 Additional Mounting Components of the System

The motor is mounted to the chassis via a plate. For connecting the metal plate, an externally designed mount is used. The motor shaft is coupled to this motor mount, and the primary pulley of the CVT is connected at the other end, thus making a linear shaft arrangement (Figure 17.82).

The battery is mounted to the chassis using "L" bracket mounts. Battery stopper brackets are manufactured in-house (Figure 17.83).

The battery holding "L" brackets were in-house designed and manufactured (Figure 17.84).

The driver is safeguarded from the propeller shaft by implementing a propeller shaft casing made up of a thick aluminum sheet, which is blended into a hollow tunnel shape for protection (Figure 17.85).

A swapping cart is designed and fabricated to take the accumulator from one place to another. The dimensions, wheel size, and insulation properties were given according to rulebook C.9.2.2 (Figures 17.86 and 17.87).

FIGURE 17.82 Motor mount.

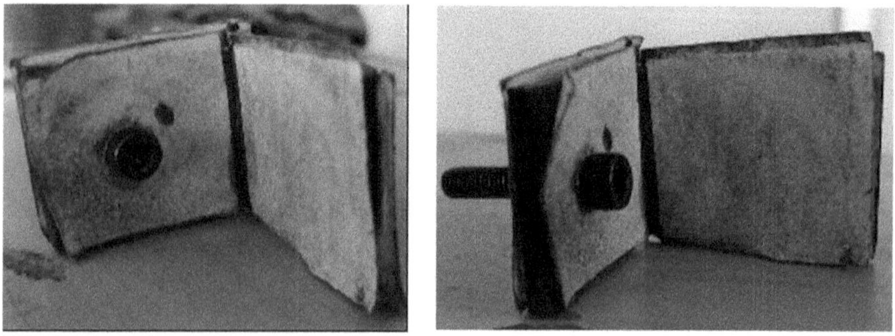

FIGURE 17.83 Battery stopper mounting brackets.

FIGURE 17.84 L-bracket mounting position and mounted L-brackets.

FIGURE 17.85 Propeller shaft casing guard.

17.27.5 Design Comparison of Previous Year Design and Current New Design

The comparison of the previous year's design with the new year's design enables the team to optimize the subsystem design flow. The sub-wings of the team identified the pros and cons and designed accordingly (Table 17.10).

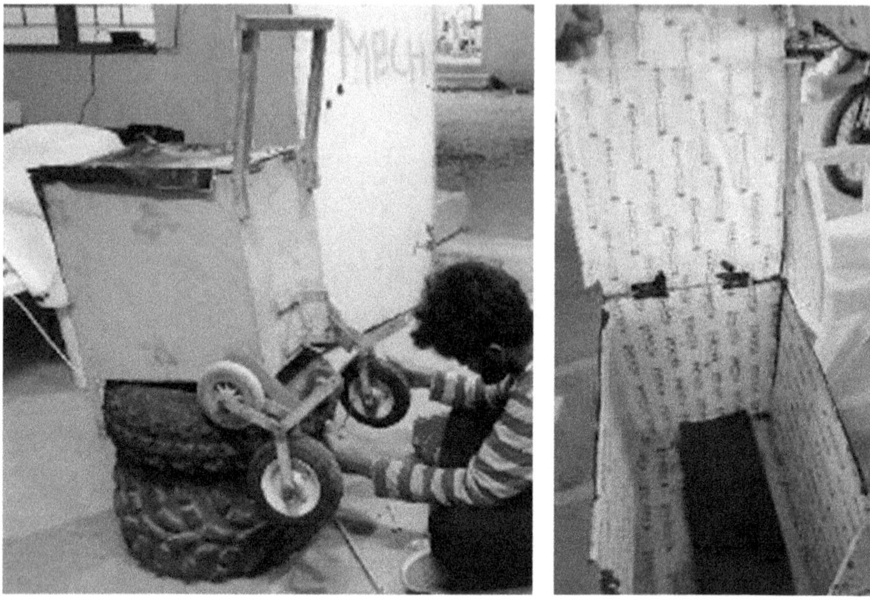

FIGURE 17.86 Hand cart fabrication and Insulation coating using Nomex "H" class material.

FIGURE 17.87 Vehicle with complete GLV and HV system activated.

TABLE 17.10
Design Comparison of Previous Year's and New Design

Previous Year's Design	New Design

Last year, we opted for 2 Wheel Drive (2WD)	In the new design, we adopted a 4 Wheel Drive (4WD)
Traction was reasonable	More traction is achieved due to the 4WD setup, which helps to drive the vehicle on slippery surfaces, with comparatively minimal tractive effort
Pin-type connector blocks used	Ring connector type pins and blocks used

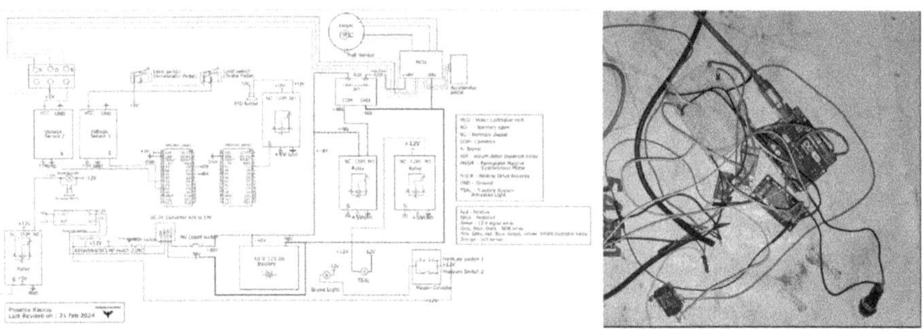

FIGURE 17.88 Schematic circuit diagram developed in EasyEDA software and the circuit implemented at small level.

17.27.6 Implementation of Compact PCB-Based Arduino Microcontroller Controlled Main Control Unit for New Design

For the upcoming year, buggies which are yet to be designed as per previous years' feedback, a new custom-designed Arduino microcontroller-controlled Printed Circuit Board (PCB)-based main controller is proposed.

A setup of the same is designed using a dot board and has been implemented and successfully tested in real-time driving and found to be working fine. In the future, it will be improved as a more compact PCB-based board with solid-state relays or Metal-Oxide-Semiconductor Field-Effect Transistors (MOSFETs), making the system very compact, weightless, and efficient (Figures 17.88 and 17.89).

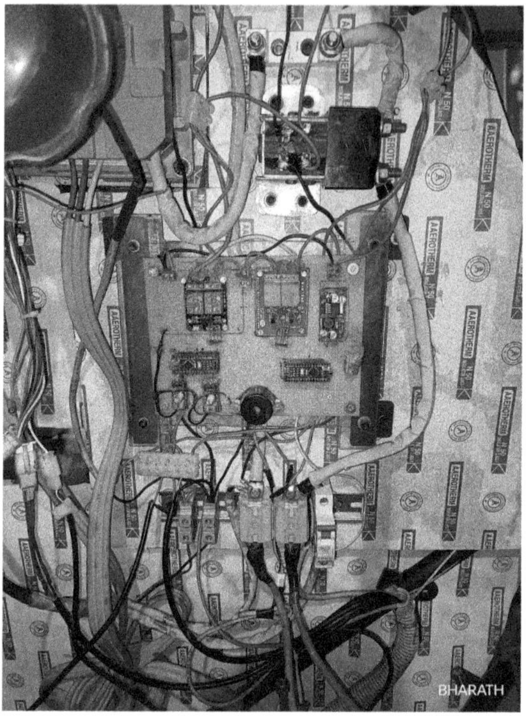

FIGURE 17.89 Circuit developed on dot board and fitted in the vehicle's electrical and electronics system.

17.28 SCALABILITY AND APPLICABILITY OF THE 4WD ELECTRIC POWERTRAIN TO OTHER ELECTRIC VEHICLES

The design and implementation of the 4WD electric powertrain for the E-ATV offer a robust framework that can be adapted to various types of electric vehicles (EVs) beyond all-terrain vehicles. The fundamental components—including the battery pack, motor, motor controller, battery management system (BMS), and power electronics—can be modified and optimized to meet the performance, efficiency, and operational requirements of different EV categories.

The scalability of an EV powertrain is primarily determined by its power architecture, drivetrain configuration, battery management strategies, and control algorithms. A well-designed powertrain should exhibit modular adaptability, allowing integration into vehicles with different operational requirements, ranging from urban electric cars and cargo vehicles to high-performance off-road EVs.

17.28.1 Key Factors Influencing Powertrain Scalability

The modularity and scalability of an electric powertrain are influenced by several key parameters. Different vehicle types require varying levels of power output and torque delivery, necessitating adaptable motor and transmission configurations.

So, identifying power and torque requirements for a given application is crucial. Battery architecture is also another important key factor. The voltage levels, energy capacity, and charging strategies must be modified to align with vehicle range and performance needs. Drivetrain configuration, such as transitioning between 4WD, 2WD (2-wheel drive), and AWD setups based on vehicle type and operational conditions like on-road/off-road/mix of both directly influence the Powertrain and associated systems' designs. Thermal management methods such as heat dissipation techniques, should be adapted based on the operating environment, battery chemistry, and cooling requirements. Software and control systems mainly ensure energy efficiency and, proper power delivery within the vehicle. Adaptive motor controllers and vehicle management systems allow efficient energy usage across different EV types. By addressing these factors, the 4WD E-ATV powertrain can be optimized for broader applications, making it a viable platform for various categories of electric vehicles.

17.28.2 Applicability to Different Electric Vehicle Segments

17.28.2.1 Off-Road and Heavy-Duty Electric Vehicles (Utility Terrain Vehicles, E-Tractors, and Military EVs)

The high torque output and rugged design of the E-ATV powertrain make it particularly suited for utility vehicles, electric tractors, and military-grade EVs that require high load-bearing capacity and superior traction on rough terrains.

17.28.3 Modifications Required for Heavy-Duty Applications

Battery pack scaling can be considered for modification of the given vehicle. The 48 V, 120 Ah battery used in the E-ATV may be replaced with higher-capacity packs (72–96 V, 300 Ah+) to support longer operational periods. High-torque motor integration is also a good choice for modification. Instead of a 5 kW PMSM motor, an increased power motor (10–20 kW, higher torque density) can be utilized for towing and heavy payload handling. Thermal management enhancements, such as liquid cooling systems for motors and phase-change cooling for battery packs, can help maintain efficiency in extreme conditions. Such modifications enable the powertrain to transition from a lightweight ATV system to a heavier-duty industrial or agricultural vehicle with minimal structural changes (Figure 17.90).

17.28.4 Urban and Passenger Electric Vehicles (Electric Cars, Three-Wheelers, and 2WD Configurations)

Urban EVs prioritize energy efficiency, lightweight components, and compact designs over high torque and rugged performance. The E-ATV powertrain can be downsized and optimized for small passenger vehicles, such as electric three-wheelers and micro-EVs.

FIGURE 17.90 The 4WD powertrain of the E-ATV and a typical EV, adaptable for off-road vehicles like quadbikes and ranger trucks, which currently use fossil fuel-based powertrains.

17.28.5 Key Adaptations for Urban Electric Vehicles

Transition from 4WD to 2WD, such as reducing the number of motors to a single-motor or dual-motor configuration, can be done to enhance energy efficiency. Integration of regenerative braking, such as implementing regenerative braking algorithms to improve range and battery efficiency.

Lightweight battery configurations, such as a shift from heavy-duty NMC batteries to LFP (Lithium Iron Phosphate) cells, can improve longevity and safety. Software modifications, such as implementing smart motor controllers, can optimize power delivery based on city driving conditions. By making these changes, the powertrain architecture can be effectively applied to compact electric passenger vehicles, particularly for last-mile transportation and urban mobility solutions.

17.28.6 Commercial Electric Vehicles (Delivery Vans, Cargo Transporters, and Fleet Vehicles)

Electric delivery vans and cargo EVs require a balance between torque delivery and energy efficiency, making the modular approach of the E-ATV powertrain highly beneficial. The ability to distribute torque across multiple wheels, as in a 4WD configuration, ensures better traction on variable terrains, making it ideal for commercial EV fleets operating in mixed urban and off-road conditions.

17.28.7 Optimizations for Commercial EVs

Implementing a modular, swappable battery system can be done to support extended operations. Integrating IoT (Internet of Things)-based fleet-level smart BMS offers an opportunity to optimize fleet-wide charging cycles and energy consumption. Adjusting gear ratios to optimized values helps balance speed and torque for cargo load distribution and efficient highway operation. Reducing complexity by replacing centralized powertrains with in-wheel hub motors can enhance cargo space utilization. By customizing the drivetrain configuration and energy management system, this powertrain architecture can be scaled to larger commercial applications, such as electric trucks, e-rickshaws, and logistics vehicles.

17.28.8 Future Research Directions on Powertrain Scalability

To further enhance the scalability of the E-ATV electric powertrain, future research should focus on several areas. The research areas like Development of a modular powertrain platform, such as creating a flexible powertrain architecture that allows customization based on specific vehicle categories or a Multi-mode drive systems such as implementing dynamic switching between AWD, 2WD, and energy-saving modes based on driving conditions have the potential to get improved by research. AI-driven power management can also be explored, such as utilizing machine learning algorithms to optimize torque delivery, battery consumption, and thermal regulation. Standardized battery interfaces, such as developing universal battery packs that can be interchanged across multiple EV models can also be looked at so that standardized battery usage can be implemented. Integration with renewable energy sources can also be explored, including solar-assisted charging and regenerative drive cycles for enhanced sustainability.

These advancements will contribute to improving EV performance, increasing adoption rates, and enhancing energy efficiency across multiple vehicle types.

Hence, by modifying key parameters such as battery capacity, drivetrain configuration, and thermal management strategies, the powertrain can be scaled to meet the operational demands of off-road vehicles, passenger EVs, and commercial electric fleets.

Future work should focus on developing modular and AI-driven powertrain solutions that facilitate flexible adaptation, ensuring widespread applicability in the evolving electric mobility sector. By doing so, the proposed powertrain can contribute to the advancement of sustainable and high-performance electric vehicle technologies.

17.29 ENVIRONMENTAL IMPACT OF MATERIAL SELECTION AND COMPONENT SOURCING IN ELECTRIC POWERTRAIN DEVELOPMENT

As the global automotive industry transitions toward sustainability, the environmental footprint of EV powertrains has become a critical area of focus. While electric powertrains offer zero tailpipe emissions, their lifecycle environmental impact—including material extraction, manufacturing, sourcing, and disposal—must be carefully assessed. The selection of battery materials, motor components, insulation, and structural materials significantly influences carbon emissions, resource depletion, and recyclability.

17.30 BATTERY PACK AND ITS ENVIRONMENTAL IMPACT

17.30.1 Material Extraction and Resource Scarcity

The battery pack used in this study consists of Nickel Manganese Cobalt (NMC) lithium-ion cells, which are widely adopted in high-performance EV applications. However, the sourcing of lithium (Li), nickel (Ni), and cobalt (Co) presents significant environmental and ethical concerns.

Lithium Extraction: Primary sources of lithium are lithium brine (Chile, Argentina) and hard rock mining (Australia, China). Environmental concerns related to lithium include high water consumption, groundwater depletion, soil degradation, and chemical contamination. Over 70% of the world's cobalt is sourced from the Democratic Republic of the Congo (DRC), where unregulated mining practices lead to ecological destruction and human rights violations. Significant carbon emissions are also associated with nickel refining due to its high-energy extraction process.

17.30.2 Sustainable Alternatives for Battery Materials

Given these environmental challenges, the industry is actively researching low-impact alternatives. Lithium iron phosphate (LFP) batteries have advantages such as no cobalt, improved cycle life, and lower environmental impact. But they do have some challenges such as lower energy density than NMC. Solid-state batteries are an emerging technology that eliminate toxic liquid electrolytes and offer higher energy density with reduced material dependence.

Battery recycling and second-life applications are also good solutions. This enables repurposing used EV batteries for stationary energy storage, extending lifespan and reducing mining demand. Technologies such as hydrometallurgical and direct cathode recycling enable the recovery of lithium, cobalt, and nickel with lower emissions, which can be used for another life cycle. This supports a circular economy.

17.31 ELECTRIC MOTOR AND RARE EARTH ELEMENT CONCERNS

17.31.1 Rare Earth Dependency in Permanent Magnet Motors

The permanent magnet synchronous motor (PMSM) used in the E-ATV contains rare earth elements (REEs), primarily neodymium (Nd) and dysprosium (Dy). These elements are crucial for high-efficiency magnetics but have significant environmental and geopolitical challenges. China supplies over 85% of global rare earth elements, leading to monopoly risks and trade instability. Rare earth mining generates radioactive waste and heavy metal contamination, severely impacting local ecosystems.

17.31.2 Alternative Motor Technologies for Environmental Sustainability

To mitigate the environmental impact of rare earth elements, research focuses on alternative motor designs. Induction motors (IMs) do not require rare earth materials but have higher electrical losses compared to PMSMs. Switched reluctance motors (SRMs) are high-efficiency and fully rare-earth-free, but require advanced control algorithms. Ferrite-based permanent magnet motors can replace neodymium with cheaper and more abundant ferrite magnets, reducing resource dependency. Synchronous reluctance motors (SynRMs) are also promising alternatives with no or very few rare-earth-based magnets. But they do offer good efficiency only at larger volumes, and thus their energy density is lower compared to rare-earth-based motors. More research is currently ongoing in this field, similar to the battery field. Permanent magnet-assisted Synchronous Reluctance Motors (PMaSynRMs) give good performance with a reduced amount of magnets.

17.32 STRUCTURAL AND INSULATION MATERIALS: IMPACT AND ALTERNATIVES

17.32.1 Chassis and Structural Components

The roll cage and battery mounting brackets in the E-ATV are constructed from AISI 1050 high-carbon steel due to its high strength-to-weight ratio. However, steel production is responsible for nearly 7%–9% of global CO_2 emissions.

17.32.2 Sustainable Alternatives

Aluminum alloys can reduce weight and improve recyclability but have higher energy refining costs. Advanced composites (e.g., carbon fiber, bio-composites) offer superior strength-to-weight benefits but remain costly and difficult to recycle. Recycled high-strength steel, such as using secondary steel, reduces emissions by up to 50% compared to virgin steel production.

17.32.3 Wire Insulation and Fireproofing Materials

The E-ATV uses Nomex "H" Class insulation to protect HV wiring and prevent thermal runaway. While Nomex is a high-performance material, its synthetic fiber production process generates significant carbon dioxide (CO_2) emissions.

17.32.4 Potential Eco-Friendly Substitutes

Bio-based polymers (e.g., polylactic acid (PLA), polyhydroxyalkanoates (PHAs)) can reduce dependency on petroleum-based plastics. Recycled polyethylene terephthalate (PET)-based insulation provides fire-resistant properties with a lower carbon footprint.

17.33 RESPONSIBLE COMPONENT SOURCING AND CIRCULAR ECONOMY CONSIDERATIONS

17.33.1 Ethical and Sustainable Sourcing Practices

Manufacturers must adopt transparent and sustainable sourcing practices to mitigate environmental and ethical concerns. They should adopt responsible mining certifications (e.g., IRMA, RMI) for lithium, cobalt, and nickel procurement. They should ensure supply chain traceability to prevent materials from conflict-prone regions. They should also encourage regional sourcing to minimize transportation emissions.

17.33.2 Implementing a Circular Economy Approach

To reduce waste and enhance material recovery, a circular economy strategy should be incorporated into powertrain manufacturing. Remanufacturing and refurbishing powertrain components should be considered. Extending the lifespan of motors, batteries, and control units through remanufacturing reduces electronic waste. Battery recycling and end-of-life management reduces carbon footprint by extracting the materials from already used ones. Companies like Redwood Materials and Li-Cycle are developing closed-loop lithium-ion recycling to reclaim up to 95% of lithium, cobalt, and nickel. Designing for disassembly is a good choice in reducing carbon emissions. Designing modular powertrain design enables easier component recycling and part replacement (Figure 17.91).

17.33.3 Life Cycle Assessment (LCA) of the E-ATV Powertrain

A comprehensive life cycle assessment (LCA) can quantify the carbon footprint, energy consumption, and ecological impact of the powertrain across four key phases. In raw material extraction stage, the carbon footprint of lithium, cobalt, and steel should be calculated. In the manufacturing and assembly stage, energy-intensive processes are used for battery production and motor assembly. Energy efficiency and emissions reduction should be compared to internal combustion alternatives. The carbon emissions should also be considered in the end-of-life disposal and recycling stage, including the environmental impact of battery disposal, motor dismantling, and component reuse.

Preliminary LCA studies indicate that the largest environmental impact of EVs occurs during raw material extraction and battery production, reinforcing the need for sustainable sourcing and recycling initiatives.

17.34 FUTURE CONSIDERATIONS

The transition to sustainable powertrain technologies requires a multi-faceted approach that addresses material selection, supply chain ethics, and end-of-life management. While the current 4WD electric powertrain design prioritizes performance and safety, future iterations should focus on all the above-discussed areas. By integrating these sustainability-driven optimizations, the environmental footprint of EV

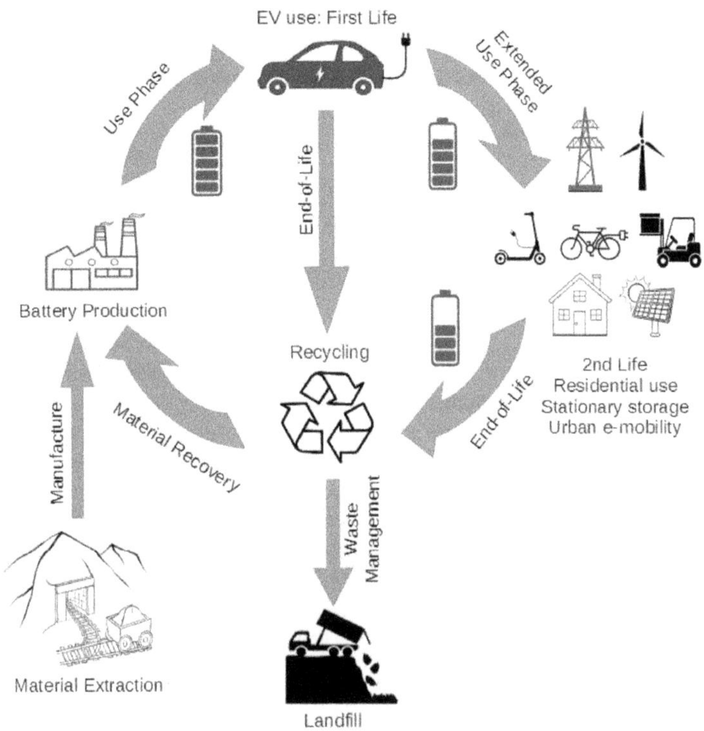

FIGURE 17.91 Depiction of circular economy as depicted in "Pathways to Circular Economy for Electric Vehicle Batteries" by Antony Jose et al. (2024).

powertrain systems can be significantly reduced, paving the way for a greener and more resource-efficient future in electric mobility.

17.35 REDUCE–REUSE–RECYCLE (RRR) ENVIRONMENTAL MANAGEMENT CONSIDERATION OF EBAJA SAEINDIA 2023 VEHICLE FOR GO-GREEN VERTICAL AWARD

The team also prepared a small report of the measures and methods it undertook for environmental management considerations. The teams are evaluated for their measures by the BAJA SAE India authority for the Go-Green Award. The team was shortlisted in the finals of the same (Table 17.11).

17.35.1 Preparation of 3R Sheet

Forthe2024event,theauthorityaskedtheteamstopreparea3R(Reduce-Reuse-Recycle) report. The report should contain embodied energy, CO_2,and water identification for each subsystem during both the primary production and secondary production processes. The team also identified the same. For some subsystems, the embodied elements were not calculated since they had already been used for the past 2 years

TABLE 17.11
Table Depicting the Measures Taken by the Team for the Given Carbon Reduction Method

Area	Proposed Target/Initiatives	Actual Action/Initiative Taken by Team
Environmental Vision Statement	Preparation of the Vision statement	To reduce the CO_2 emissions drastically and promote E-Mobility
Environmental Management Team	Team formation	Team Chairman-Secretary-Environmnetal Committee-Junior Members
Environmental Awareness	[INPUT] Identification of energy consumed in manufacturing for the components machined/fabricated in-house: Renewable Versus Environmentally harmful. [OUTPUT] Life cycle of the product (motor, battery, controller, DC-DC converter, roll cage frame with brackets, transmission, it's components, CVT, suspension elements, wheel rims, tires, steering components, brake system components/fluid, pedals, controls and dashboard components, cooling system, fire extinguisher, seat belt, seat, body panels, wiring harness, etc. For all the above components and assemblies materials used with approximate weights: steels, non-ferrous metals (aluminum/copper, brass and alloys), types of plastic, FRP, rubber, etc.	The team manufactured some components in-house, such as 3D printing the fixtures used for aligning the chassis for welding from waste plastic materials. The brake rotor, CVT casing, and knuckle hub were locally machined instead of being sourced from factories.
Reduce Emmissions	Calculation/plan for reduction of emission levels of CO_2 for products that are environmentally harmful materials and reducing emissions during the process of manufacturing.	Monitoring energy uage, reducing energy usage, reducing water usage, reviewing logistics, and switching to green mobility.
Reduce Waste	Reduction in waste material by the use of forged parts compared to machined parts.	
Reduce Primary Material and Secondary Material	Reduce environmentally harmful material/substances used in the product ATV.	Usage of biodegradable 3D printing filaments.
Effective Use of Resources	Over all water and electricity usage/energy consumption	Noting down electricity and water usage periodically from Garage's supply meters
Re-cycle Material	Parts made from recycled material or a list of parts of ATV that can be recycled	Aluminum sheets, old tires can be retreaded and used again.

(Continued)

TABLE 17.11 (*Continued*)
Table Depicting the Measures Taken by the Team for the Given Carbon Reduction Method

Area	Proposed Target/Initiatives	Actual Action/Initiative Taken by Team
Re-use Material	List of parts re-used, if any. Or plans to reuse next year's ATV fabrication	suspension strut, driveshaft, universal (UV) joints, rims, master cylinders
Control and Reduce Environmentally Harmful Substances Contained in the Product	Identification and preparation of list and control plan of environmentally harmful materials	Brake fluids, thinners, lubricants, damaged tires, rubber hoses, damaged lines, rubber beadings, aluminum scrap, electronic waste.

(2022 and 2023). The per-unit values of the elements for different materials and production processes were identified from the Annexure library given by the BAJA SAE India authority (Figures 17.92 and 17.93).

The embodied elements were identified using the following formulae (Figure 17.94).

$$\text{Embodied Energy of the material} = \text{Weight of the material} * \text{Quantity} \\ * \text{Embodied energy in Mega Joule} \\ (\text{MJ}) \text{ for unit material weight} (\text{kg})$$

$$\text{Embodied CO}_2 \text{ of the material} = \text{Weight of the material} * \text{Quantity} \\ * \text{Embodied CO}_2 (\text{kg}) \text{ for unit material weight} (\text{kg})$$

$$\text{Embodied Water of the material} = \text{Weight of the material} \\ * \text{Quantity} * \text{Embodied water} (\text{L}) \text{ for unit weight} (\text{kg})$$

$$\text{Embodied Water of the material} = \text{Weight of the material} \\ * \text{Quantity} * \text{Embodied water} (\text{L}) \text{ for unit weight} (\text{kg})$$

$$\text{Embodied Energy of process} \\ \text{done for the material} = \text{Weight of the material removed by that particular} \\ \text{machining process} * \text{Quantity} * \text{Embodied energy} \\ \text{in Mega Joule } (\text{MJ}) \text{ required for that particular machining} \\ \text{process to remove per unit weight} (\text{kg})$$

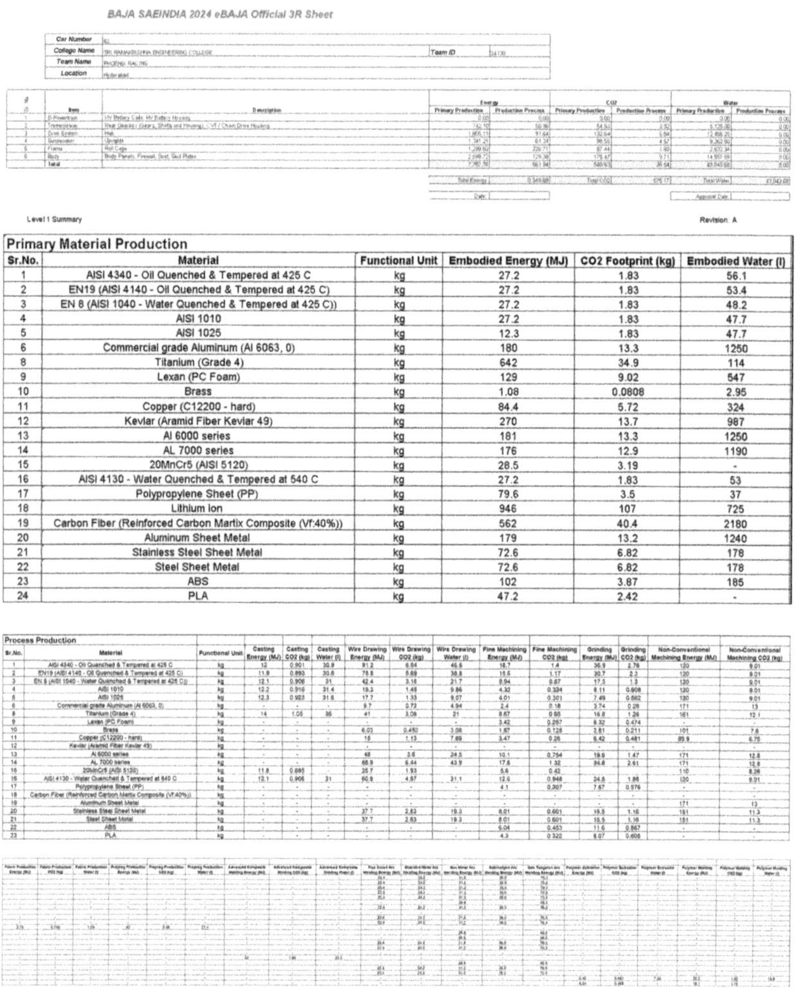

Primary Material Production					
Sr.No.	Material	Functional Unit	Embodied Energy (MJ)	CO2 Footprint (kg)	Embodied Water (l)
1	AISI 4340 - Oil Quenched & Tempered at 425 C	kg	27.2	1.83	56.1
2	EN19 (AISI 4140 - Oil Quenched & Tempered at 425 C)	kg	27.2	1.83	53.4
3	EN 8 (AISI 1040 - Water Quenched & Tempered at 425 C))	kg	27.2	1.83	48.2
4	AISI 1010	kg	27.2	1.83	47.7
5	AISI 1025	kg	12.3	1.83	47.7
6	Commercial grade Aluminum (Al 6063, 0)	kg	180	13.3	1250
8	Titanium (Grade 4)	kg	642	34.9	114
9	Lexan (PC Foam)	kg	129	9.02	547
10	Brass	kg	1.08	0.0808	2.95
11	Copper (C12200 - hard)	kg	84.4	5.72	324
12	Kevlar (Aramid Fiber Kevlar 49)	kg	270	13.7	987
13	Al 6000 series	kg	181	13.3	1250
14	AL 7000 series	kg	176	12.9	1190
15	20MnCr5 (AISI 5120)	kg	28.5	3.19	-
16	AISI 4130 - Water Quenched & Tempered at 540 C	kg	27.2	1.83	53
17	Polypropylene Sheet (PP)	kg	79.6	3.5	37
18	Lithium Ion	kg	946	107	725
19	Carbon Fiber (Reinforced Carbon Martix Composite (Vf:40%))	kg	562	40.4	2180
20	Aluminum Sheet Metal	kg	179	13.2	1240
21	Stainless Steel Sheet Metal	kg	72.6	6.82	178
22	Steel Sheet Metal	kg	72.6	6.82	178
23	ABS	kg	102	3.87	185
24	PLA	kg	47.2	2.42	-

Process Production																
Sr.No.	Material	Functional Unit	Casting Energy (MJ)	Casting CO2 (kg)	Casting Water (l)	Wire Drawing Energy (MJ)	Wire Drawing CO2 (kg)	Wire Drawing Water (l)	Fine Machining Energy (MJ)	Fine Machining CO2 (kg)	Grinding Energy (MJ)	Grinding CO2 (kg)	Non-Conventional Machining Energy (MJ)	Non-Conventional Machining CO2 (kg)
1	AISI 4340 - Oil Quenched & Tempered at 425 C	kg	12	0.901	30.9	91.2	6.84	46.5	18.7	1.4	36.9	2.76	120	9.01
2	EN19 (AISI 4140 - Oil Quenched & Tempered at 425 C)	kg	11.9	0.893	30.6	78.8	5.89	38.8	15.6	1.17	30.7	2.3	120	9.01
3	EN 8 (AISI 1040 - Water Quenched & Tempered at 425 C))	kg	12.1	0.906	31	42.4	3.18	21.7	8.94	0.67	17.3	1.3	120	9.01
4	AISI 1010	kg	12.2	0.916	31.4	19.3	1.48	9.86	4.32	0.324	8.11	0.608	120	9.01
5	AISI 1025	kg	12.3	0.923	31.6	17.7	1.33	9.07	4.01	0.301	7.49	0.562	120	9.01
6	Commercial grade Aluminum (Al 6063, 0)	kg	-	-	-	9.7	0.72	4.94	2.4	0.18	3.74	0.28	171	13
8	Titanium (Grade 4)	kg	14	1.05	36	41	3.08	21	8.67	0.65	16.8	1.26	161	12.1
9	Lexan (PC Foam)	kg	-	-	-	-	-	-	3.42	0.257	6.32	0.474	-	-
10	Brass	kg	-	-	-	6.03	0.452	3.08	1.67	0.125	2.81	0.211	101	7.6
11	Copper (C12200 - hard)	kg	-	-	-	15	1.13	7.69	3.47	0.26	6.42	0.481	89.9	6.75
12	Kevlar (Aramid Fiber Kevlar 49)	kg	-	-	-	-	-	-	-	-	-	-	-	-
13	Al 6000 series	kg	-	-	-	48	3.6	24.5	10.1	0.754	19.6	1.47	171	12.8
14	AL 7000 series	kg	-	-	-	68.9	6.44	43.9	17.6	1.32	34.8	2.61	171	12.8
15	20MnCr5 (AISI 5120)	kg	11.8	0.885	-	25.7	1.93		5.6	0.42			110	8.26
16	AISI 4130 - Water Quenched & Tempered at 540 C	kg	12.1	0.906	31	60.9	4.57	31.1	12.6	0.948	24.8	1.86	120	9.01
17	Polypropylene Sheet (PP)	kg	-	-	-	-	-	-	4.1	0.307	7.67	0.576	-	-
18	Carbon Fiber (Reinforced Carbon Martix Composite (Vf:40%))	kg	-	-	-	-	-	-	-	-	-	-	-	-
19	Aluminum Sheet Metal	kg	-	-	-	-	-	-	-	-	-	-	171	13
20	Stainless Steel Sheet Metal	kg	-	-	-	37.7	2.83	19.3	8.01	0.601	15.5	1.16	151	11.3
21	Steel Sheet Metal	kg	-	-	-	37.7	2.83	19.3	8.01	0.601	15.5	1.16	151	11.3
22	ABS	kg	-	-	-	-	-	-	6.04	0.453	11.6	0.867	-	-
23	PLA	kg	-	-	-	-	-	-	4.3	0.322	8.07	0.605	-	-

FIGURE 17.92 Total cumulative list of embodied elements for all subsystems, embodied elements released for each primary material production process and secondary process.

$$\text{Embodied } CO_2 \text{ of process done for the material} = \text{Weight of the material removed by that particular machining process} * \text{Quantity} * \text{Embodied } CO_2\ (\text{kg}) \text{ released by that particular machining process to remove per unit weight (kg)}$$

Recycling Material Production				
Sr.No.	Material	Functional Unit	Embodied Energy (MJ)	CO2 Footprint (kg)
1	AISI 4340 - Oil Quenched & Tempered at 425 C	kg	7.48	0.588
2	EN19 (AISI 4140 - Oil Quenched & Tempered at 425 C)	kg	7.48	0.588
3	EN 8 (AISI 1040 - Water Quenched & Tempered at 425 C))	kg	7.48	0.588
4	AISI 1010	kg	7.48	0.588
5	AISI 1025	kg	7.48	0.588
6	Commercial grade Aluminum (Al 6063, 0)	kg	31.5	2.47
8	Titanium (Grade 4)	kg	82.7	6.5
9	Lexan (PC Foam)	kg	-	-
10	Brass	kg	17	1.34
11	Copper (C12200 - hard)	kg	17.7	1.39
12	Kevlar (Aramid Fiber Kevlar 49)	kg	-	-
13	Al 6000 series	kg	31.6	2.48
14	AL 7000 series	kg	30.9	2.43
15	20MnCr5 (AISI 5120)	kg	7.76	0.61
16	AISI 4130 - Water Quenched & Tempered at 540 C	kg	7.48	0.588
17	Polypropylene Sheet (PP)	kg	27.1	1.19
18	Aluminum Sheet Metal	kg	29	2.28
19	Stainless Steel Sheet Metal	kg	15.8	1.24
20	Steel Sheet Metal	kg	15.8	1.24
23	ABS	kg	34.7	1.32
24	PLA	kg	16	0.823

Sr.No	Reused Components
1	Gearbox (Gears, Housing)
2	Drive shafts
3	Propellor shaft
4	Hv battery pack (HV Battery Cells, HV Battery Housing)
5	Transmission - CVTsetup
6	Body - Seat

FIGURE 17.93 Embodied elements for remaining production processes and recycling processes, calculation of embodied elements for unique parts of each subsystem.

$$\text{Embodied water of process done for the material} = \text{Weight of the material removed by that particular machining process} * \text{Quantity} * \text{Embodied water}(\text{L})\text{ required for that particular machining process to remove per unit weight}(\text{kg})$$

FIGURE 17.94 Old and new vehicles at event sites: E33 (2022, NATRAX, MP) won Fastest Vehicle award; E20 (2023, Chitkara University, Himachal Pradesh) and E62 (2024, BVRIT, Hyderabad) — E62 secured AIR 2 in Sled Pull event.

ABBREVIATIONS

2WD	Two-wheel drive
4WD	Four-wheel drive
AC	Alternating current
Ah	Ampere hour
AIR	Accumulator isolation relay
AIS	Automotive Industry Standards
ATV	All-terrain vehicle
BLC	Overhead lateral cross member
BLDC	Brushless direct current motor
BMS	Battery Management System
CAD	Computer aided design
CAE	Computer aided engineering

COM	Common
CVT	Continuous variable transmission
D0D	Depth of discharge
dB	decibel
DC	Direct current
DFMEA	Design failure mode and effect analysis
DVP	Design validation plan
FOS	Factor of safety
GLV	Grounded low voltage
HV	High voltage
IC	Integrated circuit
ID	Inner diameter
IEC	International Electrotechnical Commission
IM	Induction motors
IP	Ingress protection
kW	kilo watt
LCA	Life cycle assessment
LDB	Lateral diagonal bracing
LED	Light emitting diode
LFP	Lithium ferro phosphate
LMO	Lithium Manganese oxide
LTO	Lithium titanate
MCB	Miniature circuit breaker
NC	Normally connected
NCA	Lithium Nickel Cobalt Aluminum oxide
NMC	Nickel Manganese Cobalt oxide
NO	Normally open
OD	Outer diameter
OEM	Original equipment manufacturer
PET	Poly ethylene terephthalate
PFMEA	Process failure mode and effect analysis
PHA	Poly hydroxy alkanoates
PLA	Poly lactic acid
PMSM	Permanent magnet synchronous motor
RC	Ring connector
RPM	Revolution per minute
RRH	Rear roll hoop
RTDS	Ready to drive sound
SAE	Society of Automotive Engineers
SIM	Side impact member
SoC	State of charge
SoH	State of health
SynRM	Synchronous reluctance motors
TIG	Tungsten inert gas welding
TSAL	Tractive system activated light
TTE	Total tractive effort
WKT	We know that

REFERENCES

Antony Jose, S., Dworkin, L., Montano, S., Noack, W. C., Rusche, N., Williams, D., Menezes, P. L. (2024). "Pathways to Circular Economy for Electric Vehicle Batteries," *Recycling* 9(5), 76.

BAJA SAE India Committee. (2022). BAJA SAEINDIA - MATLAB & Simulink, https://in.mathworks.com/academia/student-competitions/baja-saeindia.html

Baja SAE India Rulebook 2023_Rev0_1660412649. https://bajasaeindia.org/upload/Resource/BAJA%20SAEINDIA%20RULEBOOK%202023_Rev0_1660412649.pdf

Chidambaram, K., Ashok, B., Dighe, H., Agarwal, S., Kapoor, D., Kasana, A., and Mathew, A. (2022). Design, Analysis, and Simulation of a Four-Wheel-Drive Transmission for an All-Terrain Vehicle, *SAE Technical Paper Series*. doi:10.4271/2022-01-5039.

Dick, W.M. (1995). "All-Wheel and Four-Wheel-Drive Vehicle Systems," *SAE Technical Paper* 952600, https://doi.org/10.4271/952600

Gupta, J. B. (2015). *Theory and Performance of Electrical Machines*, [[Tab]]15th edition, S.K. Kataria & Sons. ISBN 978-93-5014-277-6.

Kumar, Y. et.al. (2020). "Design and Development of Electrical Powertrain of ATV," *International Research Journal of Engineering and Technology (IRJET)* 7(4), 1181–1188.

Miri, I., Fotouhi, A., & Ewin, N. (2020). "Electric Vehicle Energy Consumption Modelling and Estimation—A Case Study," *International Journal of Energy Research* 45. doi:10.1002/er.5700.

Nagubandi, R. (2015). *Baja SAE UTEP 2014 Design Report - Team Jaabaz*, VIT University. doi:10.13140/RG.2.1.4307.8889.

Naunheimer, H. et al. (2011). *Automotive Transmissions - Fundamentals, Selection, Design and Application* 2nd Edition, Springer Berlin Heidelberg, p. 694. https://doi.org/10.1007/978-3-642-16214-5

Parmar, K., Desai, J., Patel, J., & Patel, N. (2020). *Information and Communication Technology for Intelligent Systems: Smart Innovation, Systems and Technologies*, Vol. 196, Springer: Singapore, pp. 481–491. https://doi.org/10.1007/978-981-15-7062-9_47

Professors of PSG College of Technology. (1968). *Design Data – Databook of Engineers*, Kalaikathir Achchagam: Coimbatore, India. ISBN 978-81-927355-0-4.

Sunil, S. R., Jain, R. K., & Saluja, H. L. (1975). *Electrical Safety, Fire Safety Engineering and Safety Management*, Khanna Publishers.

Tomar, P. et al. (June 2012). *Proceedings of the National Conference on Emerging Trends in Mechanical Engineering (ETME-2012)*. ITM University: Gurgaon, India, pp. 1–8.

Xu, X., Dong, P., Liu, Y., and Zhang, H. (2018). "Progress in Automotive Transmission Technology," *Automotive Innovation* 1, 187–210, https://doi.org/10.1007/s42154-018-0031-y

Index

For Product Safety Concerns and Information please contact our EU
representative GPSR@taylorandfrancis.com
Taylor & Francis Verlag GmbH, Kaufingerstraße 24, 80331 München, Germany

www.ingramcontent.com/pod-product-compliance
Ingram Content Group UK Ltd.
Pitfield, Milton Keynes, MK11 3LW, UK
UKHW022212190926
475702UK00003B/319
* 9 7 8 1 0 3 2 9 5 0 7 7 8 *